AF581795

Sustainable Construction Materials: From Paste to Concrete

Sustainable Construction Materials: From Paste to Concrete

Editor

Yeonung Jeong

MDPI • Basel • Beijing • Wuhan • Barcelona • Belgrade • Manchester • Tokyo • Cluj • Tianjin

Editor
Yeonung Jeong
Construction Technology
Research Center
Korea Conformity
Laboratories (KCL)
Seoul
Korea, South

Editorial Office
MDPI
St. Alban-Anlage 66
4052 Basel, Switzerland

This is a reprint of articles from the Special Issue published online in the open access journal *Materials* (ISSN 1996-1944) (available at: www.mdpi.com/journal/materials/special_issues/sustain_construct_mater).

For citation purposes, cite each article independently as indicated on the article page online and as indicated below:

LastName, A.A.; LastName, B.B.; LastName, C.C. Article Title. *Journal Name* **Year**, *Volume Number*, Page Range.

ISBN 978-3-0365-6861-4 (Hbk)
ISBN 978-3-0365-6860-7 (PDF)

Contents

About the Editor

Yeonung Jeong

Dr. Yeonung Jeong obtained a doctorate from the Ulsan National Institute of Science & Technology (UNIST) with a study on the mechanical performance and microstructure of alkaline earth-activated slag cements. He is working as a senior researcher at Korea Conformity Laboratories (KCL). Currently, he is mainly researching the development of construction materials using nanomaterials, the development of dredged soil treatment technology for contaminated sediments, the development of carbon-neutral construction materials, and the analysis technology of cement chemistry-based construction materials.

 materials

Article

Effect of Solution-to-Binder Ratio and Alkalinity on Setting and Early-Age Properties of Alkali-Activated Slag-Fly Ash Binders

Ali Naqi [1], Brice Delsaute [1], Markus Königsberger [1,2,*] and Stéphanie Staquet [1]

[1] BATir Department, Université Libre de Bruxelles, CP194/02, 50 Avenue F.D. Roosevelt, 1050 Brussels, Belgium
[2] Institute for Mechanics of Materials and Structures, TU Wien, Karlsplatz 13/202, 1040 Vienna, Austria
* Correspondence: markus.koenigsberger@tuwien.ac.at

Abstract: The growing use of blends of low- and high-calcium solid precursors in combination with different alkaline activators requires simple, efficient, and accurate experimental means to characterize their behavior, particularly during the liquid-to-solid transition (setting) at early material ages. This research investigates slag-fly ash systems mixed at different solution-to-binder (s/b) ratios with sodium silicate/sodium hydroxide-based activator solutions of varying concentrations. Therefore, continuous non-destructive tests—namely ultrasonic pulse velocity (UPV) measurements and isothermal calorimetry tests—are combined with classical slump flow, Vicat, and uniaxial compressive strength tests. The experimental results highlight that high alkali and silica contents and a low s/b ratio benefit the early-age hydration, lead to a faster setting, and improve the early-age strength. The loss of workability, determined from the time when the slump flow becomes negligible, correlates well with ultrasonic P-wave velocity evolutions. This is, however, not the case for Vicat or calorimetry tests.

Keywords: alkali-activated materials; setting time; ultrasonic measurements; calorimetry; strength

Citation: Naqi, A.; Delsaute, B.; Königsberger, M.; Staquet, S. Effect of Solution-to-Binder Ratio and Alkalinity on Setting and Early-Age Properties of Alkali-Activated Slag-Fly Ash Binders. *Materials* **2023**, *16*, 373. https://doi.org/10.3390/ma16010373

Academic Editor: Yeonung Jeong

Received: 28 November 2022
Revised: 24 December 2022
Accepted: 26 December 2022
Published: 30 December 2022

1. Introduction

As an essential construction material, concrete and its main constituent, cement, were widely used in the construction industry for over a century. With rapid infrastructural growth, global cement production has now reached over 4 billion tonnes per annum [1] contributing to 8% of the planet's total CO_2 emissions [2]. Alkali-activated materials (AAMs) are promising alternatives to traditional ordinary Portland cement (OPC) systems, as they not only lower CO_2 emissions but also foster the development of a circular economy by utilizing waste, such as magnesia iron slags [3] or tungsten mine waste [4].

Blast furnace slag and fly ash are the most commonly used solid precursors. They are activated with aqueous hydroxides and/or silicate solutions so that more and more reaction products (hydrates) form and lead to strong and durable alkali-activated materials. Alkaline activation of calcium-rich slag (AAS) forms calcium-aluminosilicate hydrate (C-A-S-H) gel [5,6], which provides high early-age strength [7–9] but often leads to non-workable and quick-setting mixes [10]. As for the activation of low-calcium fly ash (AAF), the main hydration product is sodium-aluminosilicate hydrate (N-A-S-H) gel [11], which leads to the desired workability. However, these mixes suffer from low early-age strength [12–14], which may be overcome by curing at elevated temperatures [15]. Blending slag and fly ash, together with the choice of suitable activators, overcomes the aforementioned drawbacks and provides somewhat contradicting requirements for modern AAMs: workable mixes and high early-age strengths [16–19].

Workability is typically assessed by the characterization of the setting behavior of the material. Setting, in turn, refers to the solidification process of a previously liquid material due to a gradual stiffening [20]. The beginning of solidification of the fresh paste, often termed as the "initial set", refers to the time when the paste loses its plasticity or becomes unworkable [21]. The "final set" refers to the time when the solidification is completed or when measurable mechanical properties start to develop [21,22]. Standardized test methods for the determination of an initial and final set of cement-based materials are based on thresholds for penetration resistance using a standardized test apparatus—see EN 196-3 (for paste) and ASTM C403 (for mortar). These two somewhat arbitrary penetration resistance thresholds do not sufficiently describe the continuous fluid-to-solid transition, particularly not for AAMs, where the setting behavior is often unexpected.

To better understand and monitor the setting process, continuous and, thus, non-destructive tests (NDT) have been employed, among which the ultrasonic pulse velocity (UPV) experiments are widely used [23–30]. Several suggestions have been made to analyze the measured wave velocities [28,30–34]—see Table 1. Chotard [35] and Smith [31] highlighted two characteristic transition points in the temporal evolution of the P-wave velocity of cement-paste samples and linked them to the setting process. The first transition point (initial set) refers to the time when the P-wave velocity increases significantly and the second transition point signifies a slow increase in velocity on the UPV curve. Trtnik et al. [36] defined the initial set by the first inflection (maximum of the first derivative of the P-wave velocity) for cement-paste mixes. By analogy, Reinhardt et al. [28] successfully correlated the first inflection point of the P-wave velocity to the initiation of the setting and suggested a velocity threshold of 1500 m/s for the final set. De Belie et al. [30] investigated the energy changes of the ultrasound wave on mortar samples and correlated the local maximum of the temporal evolution of the ultrasound energy to the end of workability. Krüger et.al [37] suggested that a dynamic shear modulus of about 0.1 GPa represents a suitable indicator for the initial setting.

For alkali-activated materials, very few studies are available [38–44], despite the complex setting behavior and the large variety of material systems. Another frequently used continuous NDT technique is calorimetry testing, i.e., the measurement of the heat emitted during the exothermal chemical reaction of binder and solution. For OPC-based materials, setting occurs in the acceleration stage [27,45–47], during which the heat flow increases continuously with time. For AAMs, e.g., for alkali silicate-activated binders, however, setting occurs well before the acceleration period [48]. Herb [49] proposed a bilinear approximation approach, considering UPV as a function of the calorimetry-derived reaction degree to indicate the final setting, while Schindler's [50] approach included additional consideration of w/c ratios to predict the initial and final setting times. While most of the work on setting identification and mechanical evolution is reported for OPC-based materials, studies on AAMs are very limited: especially considering the influence of precursor nature and activator type on the setting process. A list of criteria, defined in the literature, for initial and final setting determination using different references and experimental methods are summarized in Table 1.

In this study, the efficacy of both ultrasonic and calorimetric tests to decipher the setting of alkali-activated slag fly-ash binders was investigated. These continuous NDT techniques were accompanied with classical penetration-resistance (Vicat) tests and slump flow measurements. To monitor the development of mechanical properties of the fragile samples at early ages, close to the setting, uniaxial compressive strength tests were performed with utmost care so as to not damage the samples before testing. This way, characteristic features (such as inflection points) found in the temporal evolution of the ultrasound properties, as well as the heat release, can be directly correlated to changes in the load-bearing capacity of the material. Five different slag-fly ash mixes are compared. The influence of precursors, activator concentration, alkali and silica content, and solution-to-binder ratio (s/b) was thoroughly studied.

Table 1. Initial and final setting estimation by different evaluation techniques for OPC systems and AAMs.

Method	Criterion	Mix Scale	Initial Set	Final Set
UPV	Velocity threshold (m/s)	Paste Mortar	1500 [36,51], 1440–1550 [52] * 800–980 [29]	1750–1850 [51], 1650–1725 [52] * 1200–1400 [29], 1500 [28]
		Concrete	1000–1500 [53], 2300–2700 [49], 1100–2000 [29]	3000 [53], 2790–3180 [49], 2000–3000 [29]
UPV	First inflection point		[28,36,54–56] *	-
UPV	Intersection of tangent lines		[31,35,42,57] *	
UPV & Calorimetry	Minimum reaction degree required for setting to occur (%)	Concrete	Schindler [50] prediction = 7.5%	Schindler [50] = 13% Herb [49] bilinear approx. = 8.5%
		Paste	Shiva [52] * approximation = 2.1–3.2%	Shiva [52] * = 3.5–4.5%

* Indicate AAMs systems.

2. Raw Materials and Experimental Methods

2.1. Raw Materials

Ground granulated blast furnace slag (GGBFS) and class F fly ash (FA) were the main precursors used in this study. The oxide compositions of all the precursor are summarized in Table 2. The Blaine fineness of GGBFS and FA were 516 m^2/kg and 921 m^2/kg, respectively. The densities of GGBFS and FA were 2.92 and 2.32 (g/cm^3), respectively. Sodium hydroxide (NaOH) with 97% purity, in powdered form, and liquid sodium silicate (Na_2SiO_3) solution with 28.50% SiO_2, 18% Na_2O, and 53.50% H_2O, by wt., were used as alkaline activators.

Table 2. Oxide compositions of precursors (wt.%) [58].

Oxide Composition	Slag [%]	Fly Ash [%]
CaO	40.80	4.33
SiO_2	33.30	56.70
Al_2O_3	12.30	23.50
MgO	7.84	1.43
SO_3	2.30	1.16
TiO_2	1.29	1.23
K_2O	0.67	2.65
Na_2O	0.44	0.91
Fe_2O_3	0.39	5.92
MnO	0.36	-
BaO	0.11	0.21
SrO	-	0.15
P_2O_5	-	1.49

2.2. Mix Design and Specimen Preparation

The slag-fly ash ratio was kept constant at 50/50 by weight for all the mixes. Alkaline activators (mixture of NaOH solution and liquid sodium silicate) were used to prepare all mixes. The concentration of NaOH solution was 8 M and the modulus of sodium silicate (Ms) value (defined as the mass ratio SiO_2/Na_2O) was adjusted to 1.44 by adding NaOH solution. The NaOH solution was prepared a day ahead of casting to allow the heat stemming from the exothermic reaction to dissipate (cooled down to room temperature). The solution (combined activator + additional water) to binder (slag + fly ash) ratio varied from s/b = 0.47 to s/b = 0.70, while water to binder ratio ranged from w/b = 0.36 to w/b = 0.53, as shown in Table 3. The mixtures were designated with initials from the precursors, followed by the s/b ratio (e.g., first labelled mix, SF—050SB, where the S symbol denotes slag, F for fly ash followed by 0.50 s/b). In the case of the last two mixes, SF—047SB—LA, and SF—070SB—HA, the additional labels LA and HA indicate the lower and higher alkali content, respectively.

Table 3. Mix design for alkali-activated slag-fly ash binders (100 g of binder).

Parameters	SF—050SB	SF—055SB	SF—064SB	SF—047SB—LA	SF—070SB—HA
Slag [g]	50	50	50	50	50
Fly ash [g]	50	50	50	50	50
NaOH (8 M aq) [g]	2.49	2.49	2.49	2.13	3.11
Na_2SiO_3 (aq) [g]	26.53	26.53	26.53	22.91	33.79
Water [g]	20.85	25.85	34.85	22.07	32.47
s/b [-]	0.50	0.55	0.64	0.47	0.70
w/b [-]	0.37	0.42	0.51	0.36	0.53
Na_2O content [%]	5.26	5.26	5.26	4.54	6.69
SiO_2 content [%]	7.56	7.56	7.56	6.53	9.63
Ms value [-]	1.44	1.44	1.44	1.44	1.44

Mixing was carried out in a Hobart mixer, carefully sticking to the following procedure: The activating solution was poured into the bowl, followed by adding all the precursors. Mixing was carried out for one minute at low-speed (140 ± 5 rpm) and switched to high-speed (185 ± 10 rpm) for another minute. Thereafter, the mix rested for one and a half minutes and, at the same time, any material attached to the mixing paddle, sides, or at the bottom of the bowl was scrapped and added back to the mix. The mixing process ended with one minute of high-speed mixing. The mixture was poured into the molds, as required for the specific test, wrapped in plastic sheets to avoid any evaporation, and placed into a climatic chamber at 20 ± 1 °C and 70% RH until testing.

2.3. Experimetnal Methods

2.3.1. Slump Flow

The workability loss of fresh slag-fly ash binders was investigated using a mini-slump flow test [59,60]. Immediately after mixing, the fresh paste was poured into a conical mold with a top opening with an inner diameter of 70 ± 0.5 mm), a height of 50 ± 0.5 mm, and a bottom opening with an inner diameter of 100 ± 0.5 mm, as described in ASTM C230 [61]. The mold was placed in the center of a clean and lubricated plate, filled with fresh paste, and, subsequently, lifted to allow the paste to spread. The average of four spread diameter readings was recorded as soon as the flow stopped. The procedure was repeated at intervals of 10 to 15 min until virtually no spreading was further observed. The test was conducted once for each composition.

2.3.2. Vicat Test

Setting-time characterization was carried out using a classical Vicat needle, in accordance with EN 196–3 [62]. Immediately after mixing, the fresh paste was poured into a lightly oiled conical rubber mold, with a top inner diameter of 70 ± 0.5 mm, a height of 40 ± 0.2 mm, and a bottom inner diameter of 80 ± 0.5 mm. and placed on a base plate. The paste-filled molds were immersed in water controlled at 20 ± 1 °C. The initial setting time was recorded as the time elapsed between the starting time (when the precursors first came in contact with the alkaline activator) and the time when the distance between the penetrating needle and base plate was 6 ± 3 mm. The final setting time was recorded as the time measured from the starting time to the time when the needle was only able to penetrate 0.5 mm into the paste. Vicat tests were performed once for each composition.

2.3.3. Isothermal Calorimetry

The heat-flow evolution related to the exothermal chemical reaction was continuously monitored using an eight-channel TAM Air isothermal calorimeter. Heat-flow q was measured and translated to cumulative heat Q by integration, reading as:

$$Q\,(t) = \int_0^t q(t)\cdot dt \tag{1}$$

After mixing, 2 samples of about 10 g were collected and directly inserted into the isothermal calorimeter. The test was started at an age of about 15 min for each composition. The heat-flow data were recorded up to ages of 96 h at a fixed container temperature of 20 ± 0.01 °C. Later calorimetry measurements (more than 96 h) were not sufficiently precise, due to decreasing heat flow values. The test was performed once for each composition.

2.3.4. Ultrasonic Measurements

The measurement of P-wave transmission velocity was carried out using a FreshCON system [28] equipped for monitoring P-wave transmission on cementitious materials, as shown in Figure 1. The system consisted of two polymethacrylate (PMMA) walls with semi-embedded sensors on both external sides of the walls, tied together by four screws and separated by a U-shape rubber foam. An ultrathin polyimide film was glued onto the sensors as protection. The mold thickness for the paste mixtures was 2.5 cm, the width was 10 cm, and the height was 11 cm. Every minute, a high-voltage pulse generator produced a 5 µs pulse that was sent through the mix-filled container by a cylindrical broadband piezo-electric transducer (with 0.5 MHz center frequency). The signals produced by the sensor situated on the other side of the sample (receiver) were sent to the computer through a DAQ (data acquisition) card. The ultrasonic measurements started immediately after casting and were performed continuously until the material aged 72 h. The containers were plastic-wrapped and kept in a climatic chamber at 20 ± 1 °C throughout the entire testing. The sought P-wave velocity V_P was computed as:

$$V_P = \frac{D}{t-d} \tag{2}$$

where D is the distance between the sensors, t is the picked time at signal onset and d is the delay time defined by the container. Testing was carried out once for each mix.

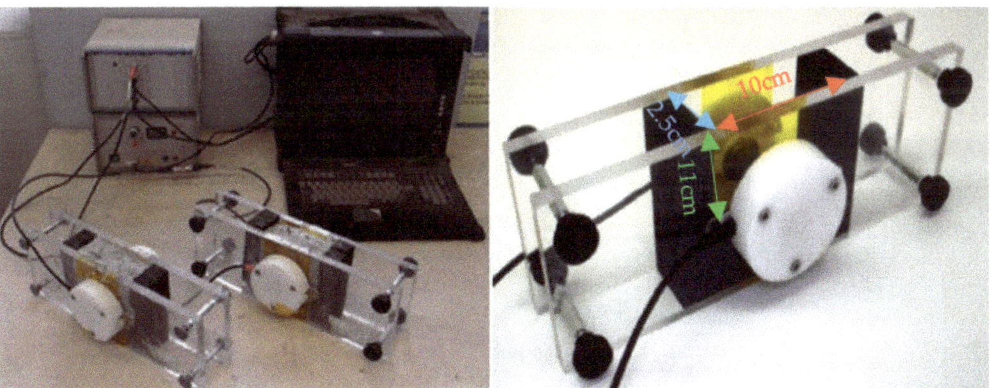

Figure 1. FreshCON system [63] for measuring the ultrasonic pulse velocity.

2.3.5. Compressive Strength

The compressive strength of slag-fly ash paste samples was tested on 50 mm large cubes. The samples were demolded after 24 h, wrapped in a plastic sheet, and put back into the climatic chamber until testing. The compressive strength was tested at the desired ages (every 7 h from the time of casting until 70 h and 168 h). Two samples were tested at each age.

3. Individual Test Results

3.1. Slump Flow

Figure 2 shows the slump flow evolution from various slag fly-ash binders. The first recorded flow diameters after 10–15 min amount to 28–45 cm, whereby the slump flow, as expected, was higher for mixes with high s/b ratios. The flow diameters decreased rapidly for all mixes, and the flow was lost (the flow diameter was equal to the diameter of the cylinder, amounting to 10 cm) after 45–60 min. Compositions with low s/b ratios reached the 10 cm threshold earlier than compositions with a high s/b ratio, except for the composition SF–070SB–HA. Its flow decreased rapidly, due to the high Na_2O and SiO_2 content (see Table 3). This observation was supported by previous experimental findings [59,64], where high Na_2O dosage was found to promote the dissolution of slag at a very early age, in addition to increased SiO_4^{4-} ion concentration that assisted in the rapid formation of a solid percolation path. Moreover, the time at which no virtual paste spreading was observed in the slump flow test was taken as a criterion for the initial setting time (t_{sf}) for all the studied mixes. In other words, the (initial) setting time, obtained from the slump flow tests, was considered as the time when the flow diameter was equal to 10 cm, which was the diameter of the bottom opening of the cone.

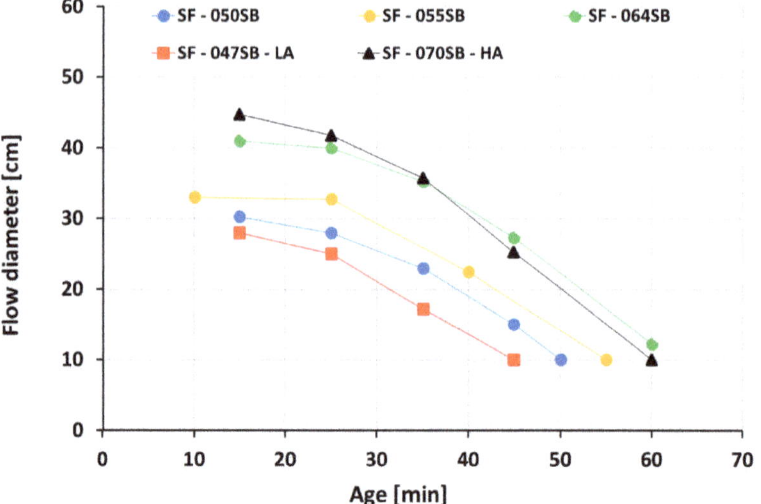

Figure 2. Slump flow evolution as a function of age.

3.2. Vicat Test

Figure 3 shows the influence of varying s/b ratios and alkali contents on setting times of slag fly-ash binders, determined by the Vicat needle test. The initial setting times (t_i) ranged from 45 to 100 min while the final setting times (t_f) recorded were between 80 and 175 min. By analogy to the slump flow, for compositions with the same alkali content (SF—050SB, 055SB, and 064SB), the setting times increased with an increasing s/b ratio. Lower solution quantities resulted in higher initial solid volume fractions, enabled a faster formation of a continuous network of reaction products, and, thus, resulted in a faster setting. However, for the mix with the lowest solution–binder ratio (SF—047SB—LA), initial and final setting times were comparable to SF—050SB, due to low alkali and silica content in the mix, which decelerated the very early age reaction. In the case of SF—070SB—HA mix, final setting times were comparable to SF—064SB, due to the presence of higher silica content in the activating solution that formed additional early-reaction products [65]. The initial setting time for SF—070SB—HA, however, was still higher than that of SF—064SB, revealing that this reaction boost due to additional alkalis did not compensate for the initial setting delay due to the higher s/b ratio.

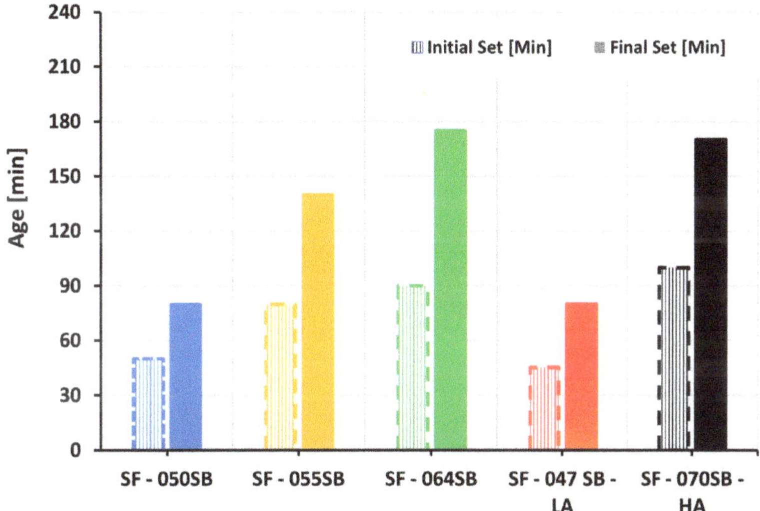

Figure 3. Initial and final setting times obtained from Vicat tests.

3.3. Isotheral Calorimetry

The evolutions of the calorimetry-determined heat flow q (t) and the cumulative heat Q (t) are depicted in Figure 4. Notably, due to external mixing, only a part of the first peak was recorded. This initial peak was related to the rapid dissolution of the slag as soon as it came in contact with the alkaline activators [65]. The peak was followed, for all mixes, by a rapid decrease in the heat flow until a minimum was reached between 12 and 24 h. This, in turn, was followed by a second, very pronounced heat-flow peak, which was known to be a result of the polymerization of dissolved silicate and aluminate units [52,66]. There was a very noticeable shift of this main reaction peak, both in terms of age of occurrence and amplitude. For the mixes with the same alkali content, the blue lines in Figure 4a may be compared. The peak occurred earlier and was higher but less wide the lower the s/b ratio was. Interestingly, however, the three mixes exhibited virtually the same amount of cumulative heat of 120 J/g at the end of the monitoring at an age of 96 h. This indicated that the reaction for the mix with the lowest s/b ratio was already limited by the availability of the solution, allowing the others to catch up. The mix with the highest s/b ratio (SF—070SB—HA) showed a shorter dormant period than the mix SF—064SB, due to high alkali and silica contents that facilitated the formation of reaction products and manifested in higher cumulative heats (Figure 4b) at 96 h, compared to the other mixes. The main heat-flow peak for the low-alkali mix (SF—047SB—LA) occurred later than that of the mix SF—050SB, despite the lower s/b ratio, demonstrating that a shortage of alkali decelerated the hydration.

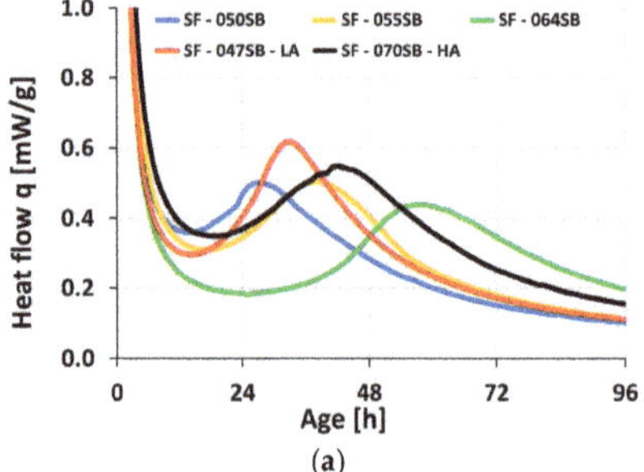

(**a**)

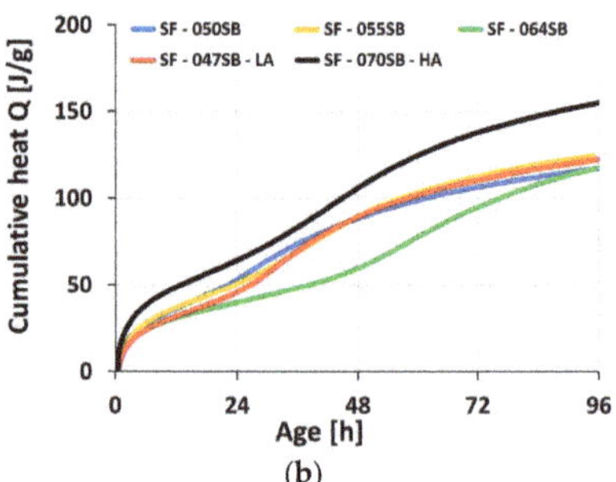

(**b**)

Figure 4. Isothermal calorimetry as a function of age: (**a**) heat flow; (**b**) cumulative heat release.

Next, we aimed to identify a setting time based on the heat-flow curves. However, all characteristic points (the start of the second acceleration, as well as the second peak) occurred well after the setting of the material. In more detail, an early-age minimum heat flow (equal to the start of the second acceleration) was reached at 12 h or even later, while the slump flow was already lost after roughly one hour (see Figure 2), and Vicat tests resulted in setting times of 45 minutes to two hours (Figure 3). In conclusion, isothermal calorimetry did not allow for extracting quantitative information on the setting.

As for the evaluation and discussion of the results, we defined a degree of reaction r as the ratio between the cumulated heat release Q(t) at any age and the ultimate heat release Q_∞ at an infinite time [67,68]:

$$r\,(t) = \frac{Q(t)}{Q_\infty} \tag{3}$$

To estimate the ultimate heat release Q_∞, the heat flow was extrapolated. Therefore, we followed [69] and plotted the cumulative heat Q as a function of the inverse square root of time and considered that Q linearly increased with decreasing the inverse square root of time—see Figure 5a for the extrapolation procedure for the mix SF—047SB—LA. This yielded the sought ultimate heat, Q_∞, as the intercept with the vertical axis. This allowed us to plot cumulative heat (from the actual measurement and extrapolation) and the respective degree of reaction up to later ages—see Figure 5b for the evolution of mix SF—047SB—LA. Comparing the resulting ultimate heat of all five mixes (Table 4) shows that the mixes with increasing s/b ratio reached higher ultimate heat at an infinite time. This was due to higher solution availability in the system, resulting in a higher reaction degree at an infinite time. Mixes with high s/b ratios, SF—070SB—HA, achieved higher ultimate cumulative heats among all studied mixes at an infinite time, due to the higher availability of the activating solution with high alkali and silica contents that facilitated higher product formation over time.

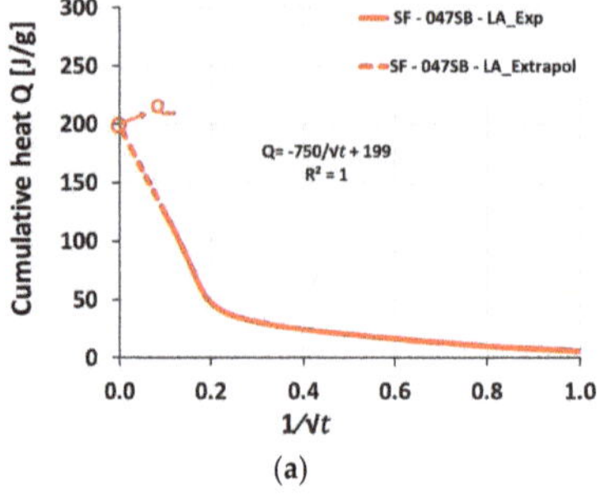

(a)

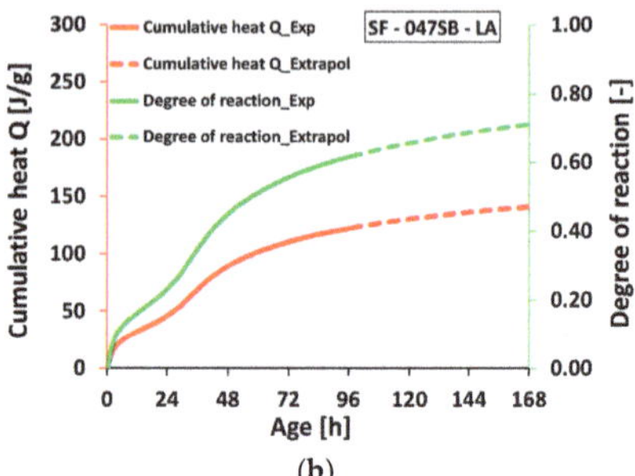

(b)

Figure 5. Extrapolation procedure for determination of the ultimate heat after [68] for mix SF—047SB—LA: (**a**) cumulative heat as a function of $1/\sqrt{t}$; (**b**) evolution of cumulative heat, as well as the respective degree of reaction, including the actual measurements.

Table 4. Estimated ultimate heat release Q_∞ through extrapolation of calorimetry results.

Ultimate Heat Release	SF—050SB	SF—055SB	SF—064SB	SF—047SB—LA	SF—070SB—HA
Q_∞ [J/g]	187.6	204.7	258.8	199.2	262.2

3.4. Ultrasonic Measurements

Figure 6 depicts the ultrasonic pulse velocity (UPV) evolution of slag-fly-ash binders for 3 days (72 h). To highlight the very-early-age velocity variations on the UPV curves, a logarithmic scale was used, as shown in Figure 7a. The initial P-wave velocity was around 190–250 m/s for all the studied mix compositions, which was lower than the velocity in the air (340 m/s) and alkaline solution (1450 m/s) [52]. This low UPV velocity was due to the air bubbles entrapped during solution and sample preparation [70]. Wetting and dissolution (stage 01) of slag and fly ash occurred in this stage, resulting in aluminosilicate monomers in solution [71]. All the slag-fly ash binder mixes followed a constant velocity until the end of the dissolution period, followed by a steep increase in P-wave velocity (stage 02—acceleration or condensation period). This sharp increase in velocity was attributed to the continuous condensation reaction and enhanced connected solid volume fraction. Among the three mixes, SF—050SB, SF—055SB, and SF—064SB, with the only difference being the s/b ratio, SF—050SB showed an earlier increase in velocity on a UPV curve, indicating a faster setting time. This implied that a lower solution quantity facilitated a quicker solid percolation path, allowing the formation of a continuous chain of hydration products and resulting in a denser microstructure to better propagate the P-wave.

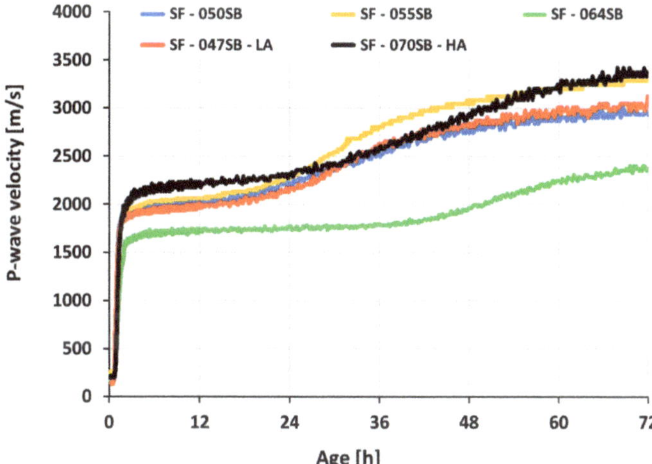

Figure 6. UPV evolution as a function of age.

For the SF—047—LA mix, even though s/b was the lowest among all of the mixes, corresponding transitions on the UPV curve were comparable to the SF—050SB mix. One of the possible reasons for this effect may be the reduced alkali content that slows down the alkali activation in the mix [72]. In contrast, the sample with the highest solution-to-binder ratio and high alkali content, SF—070SB—HA, demonstrated a faster transition of P-wave velocity in the acceleration period on the UPV curve, compared to SF—064SB. This was mainly due to increased silica and alkali content in the mix that assisted in improving reaction products at an early age. Moreover, a higher Na_2O dosage increased the alkalinity of the mix, promoting a higher dissolution of slag [73]. An induction period (stage 03) was noticed for all mixes as the P-wave velocity increased slowly followed by a secondary acceleration period (stage 04), due to the gradually increasing total solid volume fraction.

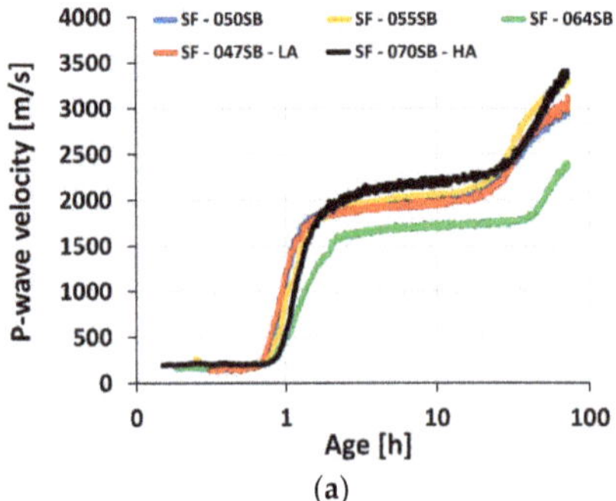

(a)

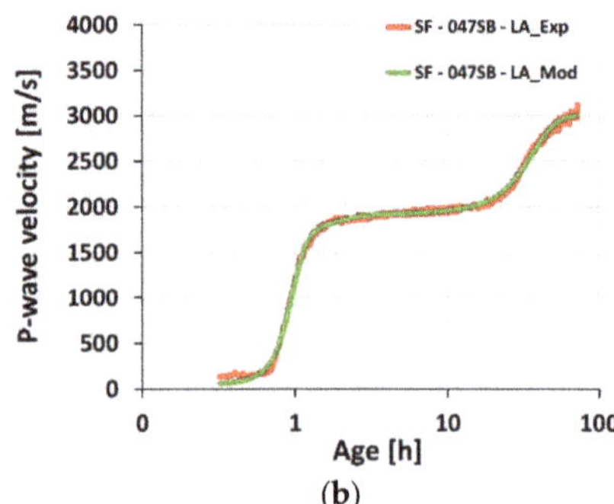

(b)

Figure 7. UPV evolution as a function of age: (**a**) all studied mixes in logarithmic scale; (**b**) reference modeled UPV curve for mix SF—047SB—LA.

For all five mixes, the UPV evolved in a characteristic double S-shaped fashion, motivating a mathematical treatment to analyze the measurements and compare the different mixes. The measured UPV evolution was fitted by means of superimposing three logistic functions, reading, as in [29]:

$$V(t) = \sum_{i=1}^{3}\left(\frac{V_i}{1 + e^{(t-t_i)/g_i}}\right) + c \tag{4}$$

with gradients g_i, inflection points t_i, plateaus or asymptotic values V_i, and an additive constant c. Table 5 shows the results in root mean square error (RMSE) for all studied mixes. Detailed information on the fitting procedure can be found in [74]. The quality of the fit was excellent; RMSE was lower than 30 m/s, which was demonstrated to be exemplary for mix SF—047SB—LA (see Figure 7b).

Table 5. RMSE of the modeled curves of UPV evolution (P-wave) for all mix compositions.

	SF—050SB	SF—055SB	SF—064SB	SF—047SB—LA	SF—070SB—HA
RMSE [m/s]	19.83	27.85	18.88	25.46	26.33

The (modeled) UPV curve allows for identifying characteristic stages, as described next. Therefore, three straight lines were introduced (see Figure 8). The first one was fitted to the very first UPV measurements. The second line was a tangent through the first inflection point (IP_1) at t_1. The third straight line, in turn, was a tangent through the second inflection point (IP_2) at t_2. The intersection of the first and second regression lines marked the transition of the dissolution period (stage 01) to the acceleration or condensation period (stage 02) and the transition point was a candidate for the ultrasound-derived initial setting time [35,57], labeled $t_{Vp(int\text{-}1)}$. Another candidate for the beginning of setting or initial setting was the time at the first inflection point itself [75,76], labeled $t_{Vp'(max)}$. The intersection of the second and third straight line marked the transition to the dormant or induction period (stage 03), and was considered a candidate for the final setting time [35,57] and labeled $t_{Vp(int\text{-}2)}$. Alternatively, the final setting was assigned to the time $t_{0.2^{*}Vp'(max)}$ when the derivative of the P-wave velocity dropped down to 20% of its maximum [34]. This was followed by the so-called secondary acceleration period (stage 04), which ended at the third inflection point (IP_3). Some studies suggested an additional stage (stage 05—stable stage), where the P-wave velocity gradually increases until it reaches an asymptotic value [34,55]. In this particular study, this stage was not found within the 72 h-long measurements. Derivative method and intersection method are two terms that are used hereafter for derivative-based and intersection-based identification of characteristic stages, linked to the setting time, on a UPV curve. Initial and final setting times obtained through the different criteria are listed in Table 6.

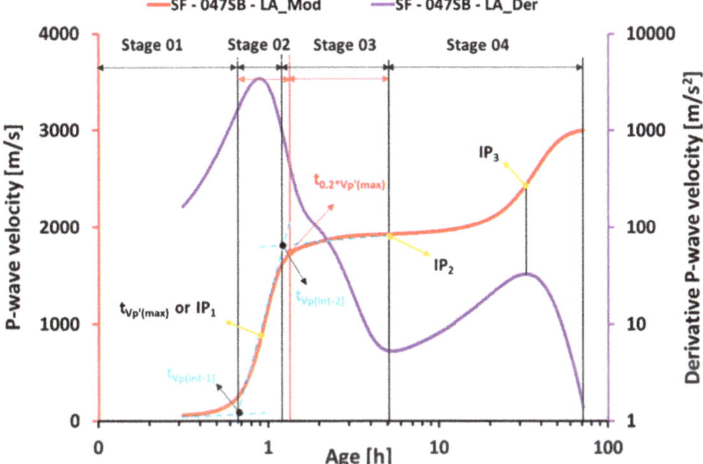

Figure 8. Different stages, intersections, and inflection points on the UPV curve as a function of age.

Table 6. Initial and final setting times obtained from different techniques.

Mix ID	Initial Setting Time [min]				Final Setting Time [min]		
	Slump Flow $[t_{sf}]$	Vicat Test $[t_i]$	FreshCON $[t_{Vp'(max)}]$	FreshCON $[t_{Vp(int\text{-}1)}]$	Vicat Test $[t_f]$	FreshCON $[t_{0.2{*}Vp'(max)}]$	FreshCON $[t_{Vp(int\text{-}2)}]$
SF—050SB	50	50	58	45	80	77	73
SF—055SB	55	80	62	46	140	88	85
SF—064SB	60	90	75	48	175	118	110
SF—047SB—LA	45	45	55	41	80	78	73
SF—070SB—HA	60	100	71	54	170	95	92

As for a final evaluation of the UPV evolutions, we aimed at discussing the eligibility of the frequently adopted velocity thresholds [28,29,36,49,51–53] (see also Table 1) to quantify the initial and final settings. Figure 9 depicts and Table 7 lists the P-wave velocities at the setting times obtained from slump flow, Vicat test, and the characteristic points from the UPV tests, respectively, for all the slag-fly ash binder mixes. Interestingly, the initial setting times from the Vicat and slump flow tests corresponded to a wide range of P-wave velocities within the interval from 411 to 1847 m/s. This showed that intrinsic velocity thresholds that correspond to the initial setting could not be defined for our different mixes. By analogy, the P-wave velocities that corresponded to the final setting from Vicat tests were not at all constant for all mixes, making the definition of a threshold impossible. Even the measured P-wave velocities that corresponded to the UPV-determined characteristic points were not constant for the different mixes, particularly not for the mixes with different alkalinities.

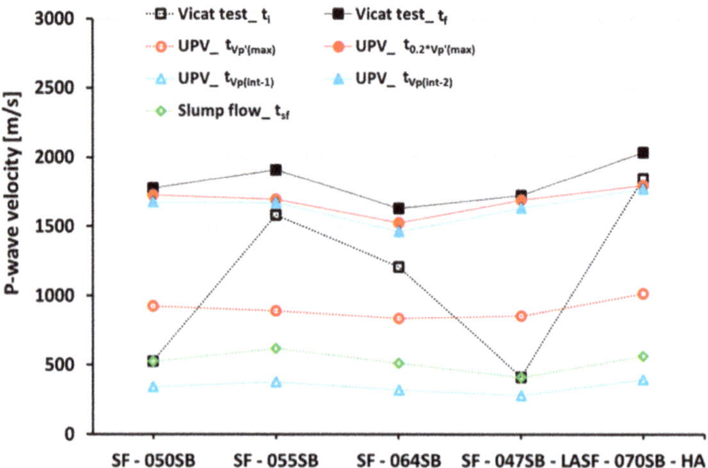

Figure 9. P-wave velocities that correspond to the setting times determined by slump flow tests, Vicat tests, and UPV tests, plotted for all five mixes. Hollow markers with dashed lines depict the initial setting, while solid markers with solid lines show the final setting.

Table 7. Summary P-wave velocity that corresponds to the setting times determined by slump flow tests, Vicat tests, and UPV tests, plotted for all five mixes [m/s].

Mix ID	UPV Values at Initial Set [m/s]				UPV Values at Final Set [m/s]		
	Slump Flow [t_{sf}]	Vicat Test [t_i]	FreshCON [$t_{Vp'(max)}$]	FreshCON [$t_{Vp(int-1)}$]	Vicat Test [t_f]	FreshCON [$t_{0.2*Vp'(max)}$]	FreshCON [$t_{Vp(int-2)}$]
SF—050SB	525	525	928	344	1775	1730	1678
SF—055SB	620	1581	892	380	1910	1696	1675
SF—064SB	516	1209	840	322	1631	1526	1466
SF—047SB—LA	411	411	857	282	1725	1694	1635
SF—070SB—HA	565	1847	1020	397	2037	1798	1771

3.5. Compressive Strength Evolution

Figure 10 shows the compressive strength development of slag-fly ash binders. The samples were tested every 7 h from the time of casting, up to 70 h, and additionally at 168 h (7 days). The largest compressive strength was observed, at any age, for the mix SF—050SB. At 7 days, the mix exhibited a strength of roughly 67 MPa, thus clearly outperforming the other mixes. As expected, the mixes with a larger s/b ratio exhibited a lower strength. Interestingly, the strength of the SF—064SB mix remained below 5 MPa, even after an age of 48 h. This demonstrated that the higher water content in the mix implied a weak paste at an early age. On the contrary, a lower w/b ratio resulted in a higher solid volume fraction, and the formation of more reaction products that link together to produce a dense microstructure, which, in turn, resulted in a faster increase of the compressive strength over time. In the case of SF—047SB—LA, even though the s/b ratio was the lowest among all the mixes, the strength results were lower than those of SF—050SB and comparable to those of SF—055SB. This observation can be explained by the smaller alkali content in the mix that resulted in decelerated slag dissolution at a very early age and, thus, fewer reaction products (at the same age) compared to those of SF—050SB. The mix with the highest s/b ratio, SF—070SB—HA, showed comparable strengths to SF—055SB at an early age, due to the presence of high alkali content. Higher silica content in the activating solution promoted Si-rich C-A-S-H precipitation [48,77] and, hence, an increase in compressive strength.

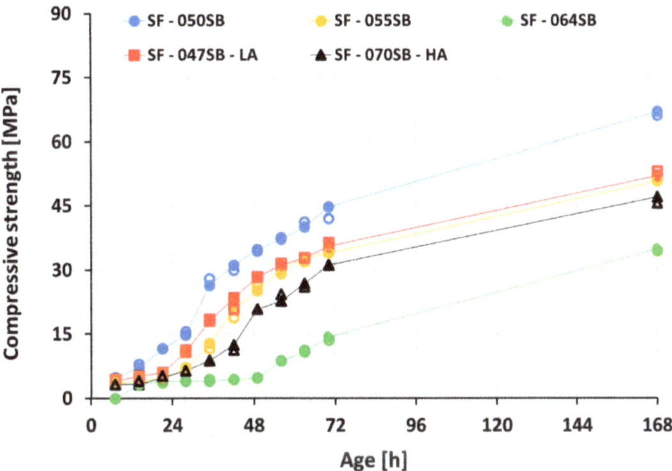

Figure 10. Compressive strength development as a function of age. All strength tests were performed on two samples—compare the hollow and solid markers for the individual results.

Next, mechanical percolation thresholds (MPT) were defined. Therefore, the measured strength evolutions were studied as a function of the calorimetry-based evolution of the degree of reaction (see Figure 11). To quantify the MPTs, two different functions were fitted to the depicted strength (f_c) vs. degree of reaction (α) evolutions, as was done in earlier works [69] and [78]:

- A linear function of the form

$$f_c(\alpha) = (\alpha - \alpha_{lf})\cdot k \tag{5}$$

with two fitting parameters α_{lf} denoting the degree of reaction at the mechanical percolation threshold and k as slope (see Table 8 for numerical values). The resulting straight-line fits represented the observed strength evolution fairly well (Figure 11a), as quantified by average RMSE, amounting to 2.31 MPa.

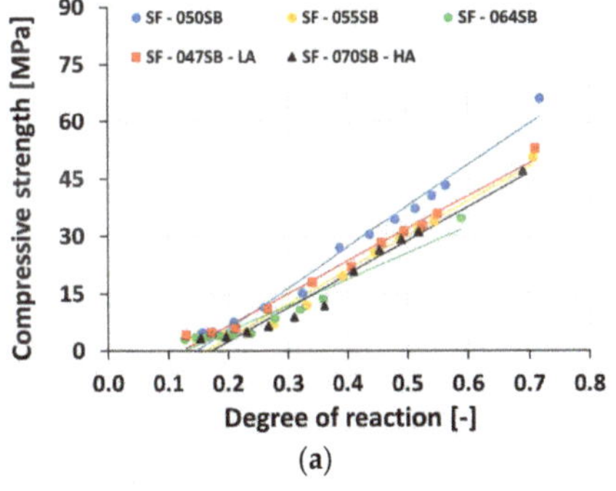

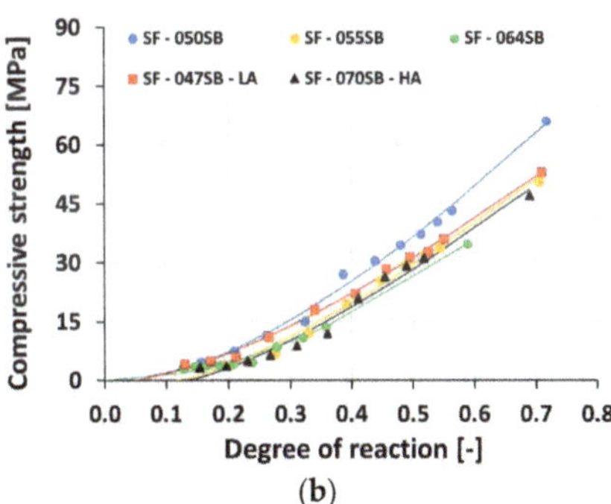

Figure 11. Compressive strength evolution as a function of the degrees of reaction: (**a**) linear fit; (**b**) power-law fit.

Table 8. Parameters of the fitting function, Equations (5) and (6), for the compressive strength evolution as a function of the degree of reaction.

Mix ID	Linear Function, MPT_{lf}				Power Function, MPT_{pf}				
	Slope k	RMSE	Reaction Degrees α_{lf}	Age	Compressive Strength $f_{c,\infty}$	Power Law Exponent β	RMSE	Reaction Degrees α_{pf}	Age
	[MPa]	[MPa]	[-]	[min]	[MPa]	[-]	[MPa]	[-]	[min]
SF—050SB	64.89	2.53	0.147	370	70.31	1.46	1.48	0.052	70
SF—055SB	52.81	2.28	0.161	530	56.94	1.30	1.51	0.118	260
SF—064SB	42.90	2.32	0.126	820	54.72	1.86	0.85	0.000	00
SF—047SB—LA	53.46	1.81	0.122	360	57.59	1.38	0.93	0.050	90
SF—070SB—HA	50.97	2.62	0.172	570	55.50	1.30	1.20	0.132	280

- Alternatively, a power function of the form

$$f_c(\alpha) = f_{c,\infty}\left(\frac{\alpha - \alpha_{pf}}{\alpha_\infty}\right)^\beta \tag{6}$$

was fitted to the experimental data, where $f_{c,\infty}$ is the compressive strength for a reaction degree of $\alpha_\infty = 0.75$, α_{pf} denotes the degree of reaction at the mechanical percolation threshold, and β is the power law exponent. Numerical values of the three fitting parameters, $f_{c,\infty}$, α_{pf} , and β, along with RMSE, are listed in Table 8. It is also important to note that RMSE was generally smaller with the power law fit, compared to the linear fit for all studied mixes; the average RMSE amounted to 1.35 MPa. This indicated the quality of the fit, using this model, to link the strength evolution to the degree of reaction of all the studied mix compositions.

Figure 11a showed a linear fit between compressive strength evolution and corresponding reaction degrees of all slag-fly ash binders. It was evident that reaction degrees were overestimated in this case for mechanical percolation (MPT_{lf}) to occur, as the compressive strength values showed a significant development already. On the other hand, for a power-law fit; to determine the mechanical percolation threshold (MPT_{pf}), Figure 11b, compressive strength evolutions are in good agreement with the corresponding degrees of reaction of all studied mixes except SF—064SB. The power-law fit for the SF—064SB mix showed a mechanical percolation (MPT_{pf}) to occur at the reaction degree of $\alpha_{pf} = 0$. This value is unrealistic as it implies already a strength gain when the first ions dissolve and generate heat. However, non-zero strength requires the precipitation of enough reaction products. The degrees of reaction at the mechanical percolation threshold, criteria obtained by a linear function and power function, are listed in Table 8. The power-law fit results are consistent with other experimental studies [79,80] which at the final setting and compressive strengths amounting to roughly 0.50 MPa [81]. These low degrees of reaction at the final set can be attributed to the very early age slag dissolution and rapid reaction product formation (primary C-S-H formation by dominant silicate ions) around the unreacted slag-fly ash particles that hinder the continuation of reaction (pre-dormant period on heat flow curves) until the critical concentration of dissolved units was achieved to produce further reaction products (post-dormant period) [68].

In contrast, previous studies on OPC-based concrete with s/b = 0.5, identified reaction degrees of 0.075–0.13 for initial and final setting, respectively [50,82]. These significant differences can be attributed to two phenomena: firstly, to different scales of testing (paste, mortar, or concrete mixtures), and secondly, to different ultimate heats Q_∞. In our current study, the value was approximated through extrapolation of the experimental heat flow data at an infinite time, while those previously mentioned studies opted for the maximum cumulative heat measured at 7 days to calculate the reaction degrees.

Furthermore, all samples achieved a considerable compressive strength before the dormant period. We conclude that setting occurred in the pre-dormant stage. Interestingly, these results are in contrast to findings of Bernal et.al [66] and Yang et.al [72] who determined the final setting in the acceleration stage (similar to conventional Portland cement [20]). The former study used slag-metakaolin blends (secondary peak on heat flow curves obtained within 2 h) while the latter utilized slag-fly ash along with the varying quantities of fly ash microspheres (acceleration period achieved within 1–2 h) in their study.

4. Discussion

Herein, the setting times, as identified by means of the different tests, are compared in detail (see Figure 12 for a schematic overview of the test results). In the case of the first three mixtures, samples with the same alkali content but varying s/b ratio, both the initial and final setting times increased with the increasing s/b ratio (in all the studied criteria) due to the increasing distance between the precursor particles in the fluid mix. The low alkali mix SF—047SB—LA consistently showed the earliest setting out of all five, given that it exhibited the lowest s/b ratio. The setting of the high alkali mix SF—070SB—HA was very close to the setting of the SF—064SB mix; depending on the test technique, it sometimes may even have occurred earlier, despite the larger s/b ratio. This underlines that the high content of alkali accelerated the formation of reaction products, which was particularly visible when comparing the setting durations, i.e., the difference between the final and initial setting times. This difference was much smaller for the SF—070SB—HA, compared with that of SF—064SB.

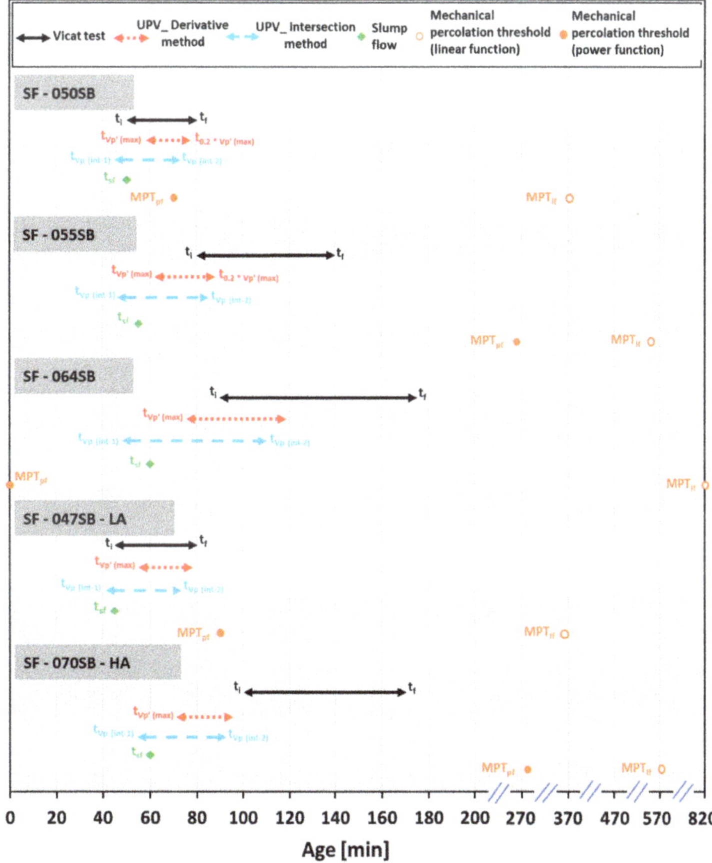

Figure 12. Timeline for Initial and final setting time values as obtained through various techniques.

Next, the setting times for the different techniques were compared. For the two mixes with SF—047SB—LA and SF—050SB, the initial and final setting times determined from different criteria were in excellent agreement. The initial setting times from the UPV and Vicat tests were particularly close to the slump loss, which indicated that either test yields trustworthy setting results for low s/b mixes.

However, for mixes with a higher s/b ratio, SF—055SB, SF—064SB and SF—070SB—HA, the different techniques yielded rather different results. The Vicat tests indicated that initial setting occurred much later, while the slump flow and UPV techniques, particularly the intersection method results, were relatively close. This could be attributed to the fact that the loss of the flowability in a slump test and the first intersection point of the ultrasonic pulse velocity test were both sensitive to the first solid percolation path, which occurred at the beginning of the acceleration period (Stage 02 in Figure 8) [83]. Even though the precursor grains connected, the connectivity or the cohesion was not enough to indicate the beginning of a mechanical set [84,85]. Thus, we concluded that setting determination in high s/b systems is more accurate with slump flow and UPV tests, while the sensitivity of Vicat tests is not sufficiently high. The non-destructive UPV tests were particularly interesting, because they were sensitive to changes in the microstructure of the hardening mix. For example, in the case of SF—070SB—HA, the presence of high alkali and silica content in the activating solution promoted Si-rich reaction products at a very early age, and this microstructure change was tracked by the earlier occurrence of the first inflection point on the UPV curve (see Table 6 and Figure 12).

The mechanical percolation threshold criterion obtained through linear as well as power/law fitting overestimated the time of occurrence of a mechanical set, compared to the time obtained via other criteria. This showed that any extrapolation attempts from destructive strength testing at early ages fails to provide accurate information on setting.

5. Conclusions and Outlook

This study focused on investigating the capability of five different destructive and non-destructive measurements on the determination of the setting behavior of alkali-activated slag-fly ash binders, prepared with various amounts of alkali, silica, and s/b ratio. The main conclusions drawn from this study are as follows:

1. The ultrasonic pulse velocity (UPV) measurements for all slag-fly ash binders revealed four characteristic stages of velocity evolutions, clearly sensitive to the alkali and silica dosages as well as to the s/b ratio.
2. The UPV-derived initial setting times showed good agreement with the slump flow measurements for all the studied mixes. Interestingly, the classical Vicat tests were accurate for mixes with low s/b ratios of 0.5 or smaller; for increasing s/b ratios, Vicat initial setting times occurred significantly later than UPV-derived initial setting times, and then the slump loss.
3. P-wave velocities at the identified (initial and final) setting times are very different from one mix to another. Thus, P-wave velocity thresholds cannot be used for setting quantification of the studied AASF mixes.
4. Heat flow evolutions obtained from isothermal calorimetry could not be used to determine setting times, as the characteristic minimum and maximum heat flow occurred at later ages (dormant period, acceleration stage, and appearance of main hydration peak).
5. For mixes with the same alkali and silica content, the compressive strength decreased with an increasing solution-to-binder (s/b) ratio. This trend was expected, as a lower s/b ratio facilitates the closer packing of precursor particles, fewer pores, and, thus, a stronger microstructure. The compressive strength increased progressively with increasing calorimetry-derived reaction degrees. The correlation could be fitted reasonably well using a power-law function, but extrapolation attempts to predict the setting were not accurate.

As a perspective for characterizing the setting and hardening behavior of the mixes, a Temperature Stress Testing Machine (TSTM) can be used. This allows for the continuous monitoring of the evolution of elastic stiffness since the earliest age [26,86]. This way, the onset of mechanical percolation can be determined and linked to the final setting time determined herein, based on indirect techniques.

6. Limitation of the Study

This study is limited to slag-fly ash paste activated by sodium hydroxide and sodium silicate at a curing temperature of 20 °C.

Author Contributions: Conceptualization, A.N., B.D., M.K. and S.S.; methodology, A.N. and M.K.; formal analysis, A.N., B.D., M.K. and S.S.; investigation, A.N.; resources, S.S.; writing—original draft preparation, A.N.; writing—review and editing, A.N., B.D., M.K. and S.S.; visualization, A.N.; supervision, B.D., M.K. and S.S.; project administration, S.S.; funding acquisition, S.S. All authors have read and agreed to the published version of the manuscript.

Funding: This paper is the result of research carried out within the framework of the FNRS-FWO-EOS project 30439691, "INTERdisciplinary multiscale Assessment of a new generation of Concrete with alkali-activated maTerials" (INTERACT). The financial support by FNRS-FWO-EOS is gratefully acknowledged.

Institutional Review Board Statement: Not applicable.

Informed Consent Statement: Not applicable.

Data Availability Statement: Not applicable.

Conflicts of Interest: The authors declare no conflict of interest.

References

1. United States Geological Survey. *Mineral Commodity Summaries 2020*; United States Geological Survey: Reston, VA, USA, 2020; ISBN 9781411343627.
2. Miller, S.A.; Horvath, A.; Monteiro, P.J.M. Readily Implementable Techniques Can Cut Annual CO_2 Emissions from the Production of Concrete by over 20%. *Environ. Res. Lett.* **2016**, *11*, 074029. [CrossRef]
3. Zosin, A.P.; Priimak, T.I.; Avsaragov, K.B. Geopolymer Materials Based on Magnesia-Iron Slags for Normalization and Storage of Radioactive Wastes. *At. Energy* **1998**, *85*, 510–514. [CrossRef]
4. Pacheco-Torgal, F.; Castro-Gomes, J.; Jalali, S. Investigations about the Effect of Aggregates on Strength and Microstructure of Geopolymeric Mine Waste Mud Binders. *Cem. Concr. Res.* **2007**, *37*, 933–941. [CrossRef]
5. Fernández-Jiménez, A.; Palomo, J.G.; Puertas, F. Alkali-Activated Slag Mortars: Mechanical Strength Behaviour. *Cem. Concr. Res.* **1999**, *29*, 1313–1321. [CrossRef]
6. Kumar, S.; Kumar, R.; Mehrotra, S.P. Influence of Granulated Blast Furnace Slag on the Reaction, Structure and Properties of Fly Ash Based Geopolymer. *J. Mater. Sci.* **2010**, *45*, 607–615. [CrossRef]
7. Li, Z.; Alfredo Flores Beltran, I.; Chen, Y.; Šavija, B.; Ye, G. Early-Age Properties of Alkali-Activated Slag and Glass Wool Paste. *Constr. Build. Mater.* **2021**, *291*, 123326. [CrossRef]
8. Bakharev, T.; Sanjayan, J.G.; Cheng, Y.B. Alkali Activation of Australian Slag Cements. *Cem. Concr. Res.* **1999**, *29*, 113–120. [CrossRef]
9. Chen, W.; Xie, Y.; Li, B.; Li, B.; Wang, J.; Thom, N. Role of Aggregate and Fibre in Strength and Drying Shrinkage of Alkali-Activated Slag Mortar. *Constr. Build. Mater.* **2021**, *299*, 7–9. [CrossRef]
10. Palacios, M.; Banfill, P.F.G.; Puertas, F. Rheology and Setting of Alkali-Activated Slag Pastes and Mortars: Effect If Organic Admixture. *ACI Mater. J.* **2008**, *105*, 140–148. [CrossRef]
11. Palomo, A.; Alonso, S.; Fernández-Jiménez, A.; Sobrados, I.; Sanz, J. Alkaline activation of fly ashes: NMR study of the reaction products. *J. Am. Ceram. Soc.* **2004**, *87*, 1141–1145.
12. Ruengsillapanun, K.; Udtaranakron, T.; Pulngern, T.; Tangchirapat, W.; Jaturapitakkul, C. Mechanical Properties, Shrinkage, and Heat Evolution of Alkali Activated Fly Ash Concrete. *Constr. Build. Mater.* **2021**, *299*, 123954. [CrossRef]
13. Somna, K.; Jaturapitakkul, C.; Kajitvichyanukul, P.; Chindaprasirt, P. NaOH-Activated Ground Fly Ash Geopolymer Cured at Ambient Temperature. *Fuel* **2011**, *90*, 2118–2124. [CrossRef]
14. Das, B.B.; Narayanan, N. *Sustainable Construction and Building Materials: Select Proceedings of ICSCBM 2018*; Springer: Berlin/Heidelberg, Germany, 2019; Volume 25, ISBN 978-981-13-3317-0.
15. Puertas, F.; Martínez-Ramírez, S.; Alonso, S.; Vázquez, T. Alkali-Activated Fly Ash/Slag Cements. Strength Behaviour and Hydration Products. *Cem. Concr. Res.* **2000**, *30*, 1625–1632. [CrossRef]

16. Lee, N.K.; Lee, H.K. Setting and Mechanical Properties of Alkali-Activated Fly Ash/Slag Concrete Manufactured at Room Temperature. *Constr. Build. Mater.* **2013**, *47*, 1201–1209. [CrossRef]
17. Fang, G.; Bahrami, H.; Zhang, M. Mechanisms of Autogenous Shrinkage of Alkali-Activated Fly Ash-Slag Pastes Cured at Ambient Temperature within 24 h. *Constr. Build. Mater.* **2018**, *171*, 377–387. [CrossRef]
18. Jang, J.G.; Lee, N.K.; Lee, H.K. Fresh and Hardened Properties of Alkali-Activated Fly Ash/Slag Pastes with Superplasticizers. *Constr. Build. Mater.* **2014**, *50*, 169–176. [CrossRef]
19. Sarıdemir, M.; Çelikten, S. Investigation of Fire and Chemical Effects on the Properties of Alkali-Activated Lightweight Concretes Produced with Basaltic Pumice Aggregate. *Constr. Build. Mater.* **2020**, *260*, 119969. [CrossRef]
20. Hewlett, P. *Lea's Chemistry of Cement and Concrete*, 4th ed.; Butterworth-Heinemann: Oxford, UK, 2003.
21. Mehta, P.K.; Monteiro, P.J.M. *Concrete: Microstructure, Properties, and Materials*, 3rd ed.; McGraw-Hill Education: New York, NY, USA, 2014.
22. Pinto, R.C.A.; Schindler, A.K. Unified Modeling of Setting and Strength Development. *Cem. Concr. Res.* **2010**, *40*, 58–65. [CrossRef]
23. Struble, L.; Kim, T.Y.; Zhang, H. Setting of Cement and Concrete. *Cem. Concr. Aggregates* **2001**, *23*, 88–93.
24. Fernandez-Jimenez, A.; Puertas, F. Setting of Alkali-Activated Slag Cement. Influence of Activator Nature. *Adv. Cem. Res.* **2001**, *13*, 115–121. [CrossRef]
25. Darquennes, A.; Staquet, S.; Espion, B. Determination of Time-Zero and Its Effect on Autogenous Deformation Evolution. *Eur. J. Environ. Civ. Eng.* **2011**, *15*, 1017–1029. [CrossRef]
26. Boulay, C.; Staquet, S.; Delsaute, B.; Carette, J.; Crespini, M.; Yazoghli-Marzouk, O.; Merliot, É.; Ramanich, S. How to Monitor the Modulus of Elasticity of Concrete, Automatically since the Earliest Age? *Mater. Struct. Constr.* **2014**, *47*, 141–155. [CrossRef]
27. Serdar, M.; Gabrijel, I.; Schlicke, D.; Staquet, S. *Advanced Techniques for Testing of Cement-Based Materials*; Springer: Berlin/Heidelberg, Germany, 2020; ISBN 978-3-030-39737-1.
28. Reinhardt, H.W.; Grosse, C.U. Continuous Monitoring of Setting and Hardening of Mortar and Concrete. *Constr. Build. Mater.* **2004**, *18*, 145–154. [CrossRef]
29. Lee, H.K.; Lee, K.M.; Kim, Y.H.; Yim, H.; Bae, D.B. Ultrasonic In-Situ Monitoring of Setting Process of High-Performance Concrete. *Cem. Concr. Res.* **2004**, *34*, 631–640. [CrossRef]
30. De Belie, N.; Grosse, C.U.; Kurz, J.; Reinhardt, H.W. Ultrasound Monitoring of the Influence of Different Accelerating Admixtures and Cement Types for Shotcrete on Setting and Hardening Behaviour. *Cem. Concr. Res.* **2005**, *35*, 2087–2094. [CrossRef]
31. Smith, A.; Chotard, T.; Gimet-Breart, N.; Fargeot, D. Correlation between Hydration Mechanism and Ultrasonic Measurements in an Aluminous Cement: Effect of Setting Time and Temperature on the Early Hydration. *J. Eur. Ceram. Soc.* **2002**, *22*, 1947–1958. [CrossRef]
32. Lawson, J.L. On the Determination of the Elastic Properties of Geopolymeric Materials Using Non-Destructive Ultrasonic Techniques. *Mech. Eng.* **2008**, *121*, 1457480.
33. D'Angelo, R.; Plona, T.J.; Schwartz, L.M.; Coveney, P. Ultrasonic Measurements on Hydrating Cement Slurries. *Adv. Cem. Based Mater.* **1995**, *2*, 8–14. [CrossRef]
34. Carette, J.; Staquet, S. Monitoring the Setting Process of Mortars by Ultrasonic P and S-Wave Transmission Velocity Measurement. *Constr. Build. Mater.* **2015**, *94*, 196–208. [CrossRef]
35. Chotard, T.; Gimet-Breart, N.; Smith, A.; Fargeot, D.; Bonnet, J.P.; Gault, C. Application of Ultrasonic Testing to Describe the Hydration of Calcium Aluminate Cement at the Early Age. *Cem. Concr. Res.* **2001**, *31*, 405–412. [CrossRef]
36. Trtnik, G.; Turk, G.; Kavčič, F.; Bosiljkov, V.B. Possibilities of Using the Ultrasonic Wave Transmission Method to Estimate Initial Setting Time of Cement Paste. *Cem. Concr. Res.* **2008**, *38*, 1336–1342. [CrossRef]
37. Krüger, M.; Bregar, R.; David, G.A.; Juhart, J. Non-Destructive Evaluation of Eco-Friendly Cementituous Materials by Ultrasound. In Proceedings of the International RILEM Conference: Materials, Systems and Structures in Civil Engineering, Segment on Service Life of Cement-Based Materials and Structures, Lyngby, Denmark, 22–24 August 2016; pp. 503–512.
38. Buchwald, A.; Tatarin, R.; Stephan, D. Reaction Progress of Alkaline-Activated Metakaolin-Ground Granulated Blast Furnace Slag Blends. *J. Mater. Sci.* **2009**, *44*, 5609–5617. [CrossRef]
39. Kar, A.; Halabe, U.B.; Ray, I.; Unnikrishnan, A. Nondestructive Characterizations of Alkali Activated Fly Ash and/or Slag Concrete. *Eur. Sci. J.* **2013**, *9*, 1857–7881.
40. Ren, W.; Xu, J.; Bai, E. Strength and Ultrasonic Characteristics of Alkali-Activated Fly Ash-Slag Geopolymer Concrete after Exposure to Elevated Temperatures. *J. Mater. Civ. Eng.* **2016**, *28*, 04015124. [CrossRef]
41. Panchmatia, P.; Zhou, N.S.; Juenger, M.; van Oort, E. Monitoring the Strength Development of Alkali-Activated Materials Using an Ultrasonic Cement Analyzer. *J. Pet. Sci. Eng.* **2019**, *180*, 538–544. [CrossRef]
42. Ryu, G.S.; Choi, S.; Koh, K.T.; Ahn, G.H.; Kim, H.Y.; You, Y.J. A Study on Initial Setting and Modulus of Elasticity of AAM Mortar Mixed with CSA Expansive Additive Using Ultrasonic Pulse Velocity. *Materials* **2020**, *13*, 4432. [CrossRef]
43. Tekle, B.H.U.; Hertwig, L.; Holschemacher, K. Setting Time and Strength Monitoring of Alkali-Activated Cement Mixtures by Ultrasonic Testing. *Materials* **2021**, *14*, 1889. [CrossRef]
44. DENER, M. Effect of Alkali Modulus on the Compressive Strength and Ultrasonic Pulse Velocity of Alkali-Activated BFS/FS Cement. *Türk Doğa ve Fen Derg.* **2022**, *11*, 63–68. [CrossRef]
45. Nelson, E.B. *Well Cementing*, 2nd ed.; Schlumberger: Houston, TX, USA, 1990.

46. Bentz, D.P.; Coveney, P.V.; Garboczi, E.J.; Kleyn, M.F.; Stutzman, P.E. Cellular Automaton Simulations of Cement Hydration and Microstructure Development. *Model. Simul. Mater. Sci. Eng.* **1994**, *2*, 783–808. [CrossRef]
47. Ye, G. Experimental Study and Numerical Simulation of the Development of the Microstructure and Permeability of Cementitious Materials. Ph.D. Thesis, Delft University of Technology, Delft, The Netherlands, 2003; ISBN 9040724350.
48. Li, Z.; Wyrzykowski, M.; Dong, H.; Granja, J.; Azenha, M.; Lura, P.; Ye, G. Internal Curing by Superabsorbent Polymers in Alkali-Activated Slag. *Cem. Concr. Res.* **2020**, *135*, 106123. [CrossRef]
49. Herb, A.T. Indirekte Beobachtung des Erstarrens und Erhärtens von Zementleim, Mörtel und Beton Mittels Schallwellenausbreitung. Ph.D. Thesis, University of Stuttgart, Stuttgart, Germany, 2003.
50. Schindler, A.K. *Concrete Hydration, Temperature Development, and Setting at Early-Ages*; University of Texas: Austin, TX, USA, 2002.
51. Kamada, T.; Uchida, S.; Rokugo, K. Nondestructive Evaluation of Setting and Hardening of Cement Paste Based on Ultrasonic Propagation Characteristics. *J. Adv. Concr. Technol.* **2005**, *3*, 343–353. [CrossRef]
52. Uppalapati, S. Early-Age Structural Development and Autogenous Shrinkage of Alkali-Activated Slag/Fly Ash Cements. Ph.D. Thesis, KU Leuven, Leuven, Belgium, 2020.
53. Van der Winden, N.G. Ultrasonic measurement for setting control of concrete. In *Testing during Concrete Construction*; CRC Press: Boca Raton, FL, USA, 1991; pp. 122–137.
54. Voigt, T.; Grosse, C.U.; Sun, Z.; Shah, S.P.; Reinhardt, H.W. Comparison of Ultrasonic Wave Transmission and Reflection Measurements with P- and S-Waves on Early Age Mortar and Concrete. *Mater. Struct. Constr.* **2005**, *38*, 729–738. [CrossRef]
55. Robeyst, N.; Gruyaert, E.; Grosse, C.U.; De Belie, N. Monitoring the Setting of Concrete Containing Blast-Furnace Slag by Measuring the Ultrasonic p-Wave Velocity. *Cem. Concr. Res.* **2008**, *38*, 1169–1176. [CrossRef]
56. Dai, X.; Aydin, S.; Yardimci, M.Y.; Lesage, K.; De Schutter, G. Early Age Reaction, Rheological Properties and Pore Solution Chemistry of NaOH-Activated Slag Mixtures. *Cem. Concr. Compos.* **2022**, *133*, 104715. [CrossRef]
57. Garnier, V.; Corneloup, G.; Sprauel, J.M.; Perfumo, J.C. Setting Time Study of Roller Compacted Concrete by Spectral Analysis of Transmitted Ultrasonic Signals. *NDT E Int.* **1995**, *28*, 15–22. [CrossRef]
58. Dai, X.; Aydın, S.; Yardımcı, M.Y.; Lesage, K.; De Schutter, G. Effects of Activator Properties and GGBFS/FA Ratio on the Structural Build-up and Rheology of AAC. *Cem. Concr. Res.* **2020**, *138*, 106253. [CrossRef]
59. Nedeljković, M.; Li, Z.; Ye, G. Setting, Strength, and Autogenous Shrinkage of Alkali-Activated Fly Ash and Slag Pastes: Effect of Slag Content. *Materials* **2018**, *11*, 2121. [CrossRef]
60. Tan, Z.; Bernal, S.A.; Provis, J.L. Reproducible Mini-Slump Test Procedure for Measuring the Yield Stress of Cementitious Pastes. *Mater. Struct. Constr.* **2017**, *50*, 235. [CrossRef]
61. *ASTM C230*; Standard Specification for Flow Table for Use in Tests of Hydraulic Cement. ASTM International: West Conshohocken, PA, USA, 2010; pp. 4–9. [CrossRef]
62. *BS EN 196*; Part 3, 1995 Methods of Testing Cement. Part 3: Determination of Setting Time and Soundness. Mark Wright, Bechtel Ltd.: London, UK, 1995; p. 10.
63. Carette, J.; Staquet, S. Monitoring the Setting Process of Eco-Binders by Ultrasonic P-Wave and S-Wave Transmission Velocity Measurement: Mortar vs Concrete. *Constr. Build. Mater.* **2016**, *110*, 32–41. [CrossRef]
64. Adam, A. Strength and Durability Properties of Alkali Activated Slag and Fly Ash-Based Geopolymer Concrete. Ph.D. Thesis, RMIT University, Melbourne, Australia, 2009.
65. Reddy, K.C.; Subramaniam, K.V.L. Investigation on the Roles of Solution-Based Alkali and Silica in Activated Low-Calcium Fly Ash and Slag Blends. *Cem. Concr. Compos.* **2021**, *123*, 104175. [CrossRef]
66. Bernal, S.A.; Provis, J.L.; Rose, V.; Mejía De Gutierrez, R. Evolution of Binder Structure in Sodium Silicate-Activated Slag-Metakaolin Blends. *Cem. Concr. Compos.* **2011**, *33*, 46–54. [CrossRef]
67. Carette, J.; Staquet, S. Monitoring and Modelling the Early Age and Hardening Behaviour of Eco-Concrete through Continuous Non-Destructive Measurements: Part I. Hydration and Apparent Activation Energy. *Cem. Concr. Compos.* **2016**, *73*, 10–18. [CrossRef]
68. Uppalapati, S.; Vandewalle, L.; Cizer, Ö. Monitoring the Setting Process of Alkali-Activated Slag-Fly Ash Cements with Ultrasonic P-Wave Velocity. *Constr. Build. Mater.* **2021**, *271*, 121592. [CrossRef]
69. Gargouri, A.; Daoud, A.; Loulizi, A.; Kallel, A. Laboratory Investigation of Self-Consolidating Waste Tire Rubberized Concrete. *ACI Mater. J.* **2016**, *113*, 661–668. [CrossRef]
70. Povey, M. *Ultrasonic Techniques for Fluids Characterization*; Elsevier: Amsterdam, The Netherlands, 1997.
71. Shi, Z.; Shi, C.; Wan, S.; Li, N.; Zhang, Z. Effect of Alkali Dosage and Silicate Modulus on Carbonation of Alkali-Activated Slag Mortars. *Cem. Concr. Res.* **2018**, *113*, 55–64. [CrossRef]
72. Yang, T.; Zhu, H.; Zhang, Z.; Gao, X.; Zhang, C.; Wu, Q. Effect of Fly Ash Microsphere on the Rheology and Microstructure of Alkali-Activated Fly Ash/Slag Pastes. *Cem. Concr. Res.* **2018**, *109*, 198–207. [CrossRef]
73. Sarıdemir, M.; Çelikten, S. Effects of Ms Modulus, Na Concentration and Fly Ash Content on Properties of Vapour-Cured Geopolymer Mortars Exposed to High Temperatures. *Constr. Build. Mater.* **2023**, *363*, 129868. [CrossRef]
74. Cavallini, F. Fitting a Logistic Curve to Data. *Coll. Math. J.* **1993**, *24*, 247–253. [CrossRef]
75. Trtnik, G.; Kavčič, F.; Turk, G. Prediction of Concrete Strength Using Ultrasonic Pulse Velocity and Artificial Neural Networks. *Ultrasonics* **2009**, *49*, 53–60. [CrossRef]

76. Zhang, S.; Zhang, Y.; Li, Z. Ultrasonic Monitoring of Setting and Hardening of Slag Blended Cement under Different Curing Temperatures by Using Embedded Piezoelectric Transducers. *Constr. Build. Mater.* **2018**, *159*, 553–560. [CrossRef]
77. Uppalapati, S.; Vandewalle, L.; Cizer, Ö. Autogenous Shrinkage of Slag-Fly Ash Blends Activated with Hybrid Sodium Silicate and Sodium Sulfate at Different Curing Temperatures. *Constr. Build. Mater.* **2020**, *265*, 121276. [CrossRef]
78. De Schutter, G.; Taerwe, L. Degree of Hydration-Based Description of Mechanical Properties of Early Age Concrete. *Mater. Struct. Constr.* **1996**, *29*, 335–344. [CrossRef]
79. Zhu, J.; Cao, J.N.; Bate, B.; Khayat, K.H. Determination of Mortar Setting Times Using Shear Wave Velocity Evolution Curves Measured by the Bender Element Technique. *Cem. Concr. Res.* **2018**, *106*, 1–11. [CrossRef]
80. Carette, J.; Staquet, S. Monitoring and Modelling the Early Age and Hardening Behaviour of Eco-Concrete through Continuous Non-Destructive Measurements: Part II. Mechanical Behaviour. *Cem. Concr. Compos.* **2016**, *73*, 1–9. [CrossRef]
81. Latif Al-Mufti, R.; Fried, A.N. The Early Age Non-Destructive Testing of Concrete Made with Recycled Concrete Aggregate. *Constr. Build. Mater.* **2012**, *37*, 379–386. [CrossRef]
82. Robeyst, N. Monitoring Setting and Microstructure Development in Fresh Concrete with the Ultrasonic Through-Transmission Method Opvolging van Binding En Microstructuurontwikkeling in Vers Beton Met de Ultrasone-Transmissiemethode. Ph.D. Thesis, Ghent University, Ghent, Belgium, 2010.
83. Lu, Y.; Ma, H.; Li, Z. Ultrasonic Monitoring of the Early-Age Hydration of Mineral Admixtures Incorporated Concrete Using Cement-Based Piezoelectric Composite Sensors. *J. Intell. Mater. Syst. Struct.* **2015**, *26*, 280–291. [CrossRef]
84. Bentz, D.P.; Garboczi, E.J. Percolation of Phases in a Three-Dimensional Cement Paste Microstructural Model. *Cem. Concr. Res.* **1991**, *21*, 325–344. [CrossRef]
85. Torrenti, J.M.; Benboudjema, F. Mechanical Threshold of Cementitious Materials at Early Age. *Mater. Struct. Constr.* **2005**, *38*, 299–304. [CrossRef]
86. Delsaute, B. Influence of Cyclic Movement on the Hardening Process of Grout: Case of Offshore Wind Turbine Installation. In Proceedings of the 10th International Conference on Fracture Mechanics of Concrete and Concrete Structures, Bayonne, France, 24–26 June 2019.

Article

Design of Performance-Based Concrete Using Sand Reclaimed from Construction and Demolition Waste–Comparative Study of Czechia and India

Tereza Pavlů [1,*,†], Namratha V. Khanapur [2,†], Kristina Fořtová [1], Diana Mariaková [1], Bhavna Tripathi [2,*], Tarush Chandra [3] and Petr Hájek [1]

1 University Centre for Energy Efficient Building, Czech Technical University in Prague, 27343 Bustehrad, Czech Republic
2 School of Civil and Chemical Engineering, Manipal University Jaipur, Jaipur 303007, Rajasthan, India
3 Department of Architecture & Planning, Malaviya National Institute of Technology Jaipur, Jaipur 302017, Rajasthan, India
* Correspondence: tereza.pavlu@cvut.cz (T.P.); bhavna.tripathi@jaipur.manipal.edu (B.T.)
† These authors contributed equally to this work.

Abstract: The main goal and novelty of this study is to show the transferability of practices and experiences with the use of reclaimed sand worldwide in the case in two different regions, the Czech Republic and India, which is necessary for both regions due to the sand availability (Czech Republic) and illegal sand mining including criminal offences (India). Due to the deteriorating environmental impacts associated with sand mining, finding substitution possibilities for natural sand is becoming more important worldwide. It is realized that the reuse of construction demolition waste concrete is inevitable in the pursuit of circular concrete and cleaner production, envisioned by the United Nations (UN) as the attainment of ensuring sustainable consumption and production patterns (Sustainable Development Goal 12-SDG 12) with an inclusive approach of partnerships to achieve the goal (Sustainable Development Goal 17-SDG 17) for the validation of results. The basic material properties of reclaimed sand were examined, and its impact on the physical, mechanical, and durability properties of concrete with complete replacement of sand was evaluated. Generally, a slight decline in properties of concrete with fine recycled aggregate was found. No significant decrease was found from usage possibility in the point of view of its utilization in specific structures and conditions. The research shows the slight differences of results between the Czech and Indian investigations, which are not essential for the transferability of the results.

Keywords: fine recycled aggregate; construction and demolition waste; recycled concrete aggregate; recycled masonry aggregate; recycled aggregate concrete

Citation: Pavlů, T.; Khanapur, N.V.; Fořtová, K.; Mariaková, D.; Tripathi, B.; Chandra, T.; Hájek, P. Design of Performance-Based Concrete Using Sand Reclaimed from Construction and Demolition Waste–Comparative Study of Czechia and India. *Materials* **2022**, *15*, 7873. https://doi.org/10.3390/ma15227873

Academic Editor: Yeonung Jeong

Received: 4 October 2022
Accepted: 6 November 2022
Published: 8 November 2022

Publisher's Note: MDPI stays neutral with regard to jurisdictional claims in published maps and institutional affiliations.

1. Introduction

Concrete is the largest consumer of natural resources in the construction sector, which is completely dependent on primary resources, where natural aggregates such as crushed rock, river sand, gravel, cement, and water are most used. Excessive extraction of natural sand (fNA) leads to serious ecological and economic problems, for example, changing the water direction, coastal erosion, building dead-end diversions and holes [1] and, furthermore, illegal sand mining including criminal offences [2]. Due to the concrete demand, the impact of extracting fine natural aggregate (0–4 mm) from rivers and seas becomes essential for finding alternative sources for future concrete production. Besides the ecological and regulatory issues, the production of natural aggregates is related to energy consumption and emission. By the replacement of natural aggregate in concrete mixture with recycled aggregate, the amount of CO_2-eq. emissions related to the aggregate production decreased from 33 kg to 12 kg [3]. The approach of reusing or recycling the structures that are demolished or renovated is the key factor to achieve circular economy and to attain the targets set for sustainable development goal (SDG) 12 [4–8].

As mentioned before, the substitution of natural sand in concrete production has many advantages. However, it is necessary to mention that the wider use of fine recycled aggregate (fRA) also brings challenges. First, there is high variability of the physical and chemical properties of fRA due to the presence of soil, dust, clay, glass, wood, plastics, etc. Assuming that the pure waste material, concrete or masonry, without unwanted impurities is used for preparing fRA as an aggregate for concrete, the decline of concrete properties becomes acceptable. Even so, if the impurities are removed, there is still a wide range of properties, mostly due to the composition of the parent material, which are dependent on local materials, construction, demolition, and recycling processes' habits. The fRA particle shape, gradation, specific surface area, and chemical composition of the parent material are the important aspects that lead to inconsistencies in the mechanical properties and durability of mortars and concrete prepared using fRA.

The fRA (<4 mm) can be divided into three types: (1) fine recycled concrete aggregate (fRCA), which originates from waste concrete; (2) fine recycled masonry aggregate (fRMA), which originates from masonry; and (3) fine mixed recycled aggregate (fMRA), which originates from the construction and demolition waste (CDW).

fRCA is produced by crushing waste concrete from CDW [9–12] and the pre-cast industry [13]. It is made up of natural aggregate particles and old cement paste, mostly attached to the aggregate surface. The decline in properties relates to the old cement mortar. This leads to higher porosity and consequently to higher water absorption and weak interfacial transition zones (ITZ). Furthermore, a higher percentage of fine fraction in fRCA is generally observed with a higher specific surface area [14]. The use of fRCA is usually limited to low-grade applications, such as a filling material for soil stabilization, geosynthetic structures, and road constructions [1]. Furthermore, the fRAC as a substituent for natural sand in cementitious renderings and masonry mortars was verified [15–22]. To allow for more sophisticated applications, there are possible ways to improve the quality of fRCA by removing the old cement mortar by adopting a multistage mechanical process, thermal or chemical treatment, separation using microwaves, or a combination of these processes [23]. On the one hand, these processes could improve the quality of fRCA and provide opportunities for further use of cement paste. On the other hand, it would be more economically and environmentally demanding. In conclusion, optimization of processes and usage is necessary for a meaningful solution to the utilization of fRCA, especially considering the possibilities of practical use in the concrete industry.

fRMA originates from waste masonry and the main constituents are red clay bricks, ceramic, mortar, plaster, and, very often, also waste concrete with aggregate particles and cement mortar. Similarly, with fRCA, the fRMA is more porous compared to natural sand and has a higher water absorption. Generally, the porosity and water absorption of fRMA are higher compared to those of fRCA [24–27].

For these mentioned reasons, the applications of various fRAs in the concrete industry, due to the unknown origin and properties of parent material, upscaling, and lack of guidelines for testing essential properties such as water absorption influencing the workability and effective water-to-cement ratio, make the practical application extremely challenging.

In contrast to the utilization of coarse fraction for recycled concrete aggregate (RCA) as a substitution of aggregate in concrete, where the investigation has been clearly concluded with the description of all negative effects on the properties of fresh and hardened concrete and durability, such as high porosity and consequently high water absorption and weak interfacial transition zones [1,14,28–30], the utilization of fRA has demonstrated inconsistent results. It is reported in previous studies that the use of recycled aggregate (RA) negatively influences the workability of concrete due to its higher water absorption, which consequently leads to a negative effect on mechanical properties, mostly the compressive strength as the key material property of concrete. In the case of fRA, the utilization is more complicated by the fact that the method of measurement of water absorption has not been clearly developed, where the differences between various evaluation methods are huge, and the absorbability of fRA during concrete manufacturing is also not known. Due to these facts, the use of fRA in concrete is quite challenging. Consequently, the standards around the world respond by essentially not allowing the use of fRA (<4 mm), contrary to a coarse fraction, as a possible substitution of natural aggregate in concrete.

The main goal of this study is to evaluate the possibilities of using fRCA as a substitute for natural sand in concrete with an inclusive approach of partnerships between two different regions by following the same research approach. Furthermore, the transferability of practices and experiences is verified. The results, as mentioned in previous studies, of the influence of the recycling technology and properties of the parent concrete on fRA are essential for its future use. For these reasons, the basic material properties of fRA and fine recycled aggregate concrete (fRAC) were examined and compared to find differences in this investigation.

2. Background

The use of fRCA [22,31–41] and fRMA [24–27] as a substitute for natural sand in concrete has been studied by few research groups worldwide.

Nedeljkovic et al. [1] reviewed 171 studies, in which fRCA is proposed to be used for low-grade applications such as cementitious render, masonry mortars, road constructions, soil stabilization, and as a filling material for geosynthetic-reinforced structures, except structural concrete. Studies reveal that the properties of RCA concrete depend on the type and size of the natural aggregate in the parent concrete, the strength of the parent concrete, and the method and number of crushing steps [35,42–45]. The main complications associated with fRCA are the presence of mortar, a highly angular and irregular shape, and a porous and rough particle structure [31,46,47]. Few studies recommend restricting the replacement of fRCA with NA by up to 30% [31,33,38,46,48,49]. However, to address the problem of CDW, procedures must be developed and researched to administer the full replacement of NA with fRCA. In this regard, few attempts have been made to improve the properties of fRCA concrete (fRCAC) by adopting different approaches to proportion concrete [50,51], control the particle gradation [44], and adjust the method of adding water and the saturation level of fRCA [37,52]. The improvement of the properties of superplasticizers has been also verified [34,53,54].

Evangelista and de Brito [31,37] studied the mechanical and durability properties of concrete with different replacement ratios of natural sand by fRCA. During the mixing procedure, the additional water, which was estimated to be absorbed, was added to the mixture and the effective water-to-cement ratio was estimated. The mechanical behavior was studied for replacement ratios of 10%, 20%, 30%, 50%, and 100% [31]. A slight increase in compressive strength was observed, with a maximum replacement ratio 100%. The mechanical properties (tensile strength and modulus of elasticity) were observed to slightly decrease with increasing replacement ratio with a maximum decrease of 30% of tensile strength and 18% of modulus of elasticity for a 100% replacement rate. The decline of properties for replacement ratios of up to 30% was decided as acceptable. In the case of durability [37], the water absorption by immersion and capillary sorption was observed to increase with increasing replacement ratios. The water absorption by immersion increased to a maximum of 46% and the capillarity sorption increased by 70% for concrete made with 100% fRCA compared to control concrete. On the contrary, it has been found that the depth of carbonation increased by about 110% for the control concrete. In the case of mechanical and durability properties, the authors suggested that 30% is the best replacement ratio for the quality of fRCAC.

Geng et al. [41] investigated the effect of fRCA on the carbonation resistance of fRCAC. Three phenomena were evaluated: (1) the influence of the replacement ratio, which was 20%, 40%, 60%, and 80%; (2) the influence of the different types of fRCA with different fineness moduli; and (3) mixtures with similar workability and different replacement ratios. In this study, the carbonation depth of fRCAC with a 20% replacement ratio was similar to that of the control concrete. As the replacement ratio increased, the increase in the depth of carbonation became evident with the depth of carbonation of concrete with the 80% replacement ratio being 10 times higher than the control mixture. In this study, the influence of the finest particles was also tested. It was found that the fineness modulus also influences the depth of carbonation as a result of the old cement paste contained in the fRCA. Furthermore, the influence of water content was verified, where the workability was improved by adding more water to the concrete mixture; however, the carbonation resistance was reduced drastically. As reported in previous studies [33,55], the higher water-to-cement ratio is expected to increase the porosity of concrete resulting in an increase in the depth of carbonation. In addition, in this study, the workability and compressive strength were also tested. The decreasing workability was attributed to an increasing replacement ratio with a constant water-to-cement ratio; consequently, the decline of compressive strength was more than 50% with an 80% replacement rate. Furthermore, a higher reduction in compressive strength was witnessed by optimizing workability by adding water.

Fan et al. [56] studied the fRCA influence of the crushing method on the properties of fRCAC. Two crushing approaches were compared: (1) one-stage crushing and (2) multistage crushing. It was found that the multistage crushing produces better quality fRCA with a higher density, lower water absorption, lower fineness modulus, and, consequently, a lower impact on the fRCAC properties. The maximum decrease in compressive strength of fRCAC was not more than 20% with the complete replacement of natural sand.

Khatib [36] studied the effect of substitution of natural sand with crushed concrete (fRCA) and crushed bricks (fRMA) for replacement ratios of 0%, 25%, 50%, 75%, and 100%. The amount of cement in the concrete mixture was reduced in proportion to its presence in fRCA. A higher decline of compressive strength (maximum 35%) was witnessed for fRCAC with higher fRCA than that for fRMA. However, in this case, the reduction of compressive strength may be attributed to the lower amount of cement in mixtures.

3. Recycling of CDW in the Czech Republic and India

In the case of the Czech Republic, the use of recycled aggregate from construction and demolition waste became increasingly desirable over the last few years. The primary reason is the decreasing amount of available natural resources, which is mostly caused by mining closure and restrictions on the opening of new or expansion of existing quarries, which in turn has escalated the price of natural aggregates. Secondly, owing to the interest in promoting circular practices, demolition and construction companies are increasingly approaching the sorting of individual waste, such as waste concrete and masonry, for on-site use, especially for landscaping. This approach, although not ideal, is satisfactory in many respects, especially if landscaping is necessary on-site. For this reason, the amount of mineral CDW (concrete, bricks, ceramics, etc.) reported as received in a landfill or at a recycling center is relatively small, approximately 4.6 million tonnes per year (2020), which is approximately 450 kg per person, and year-on-year has a declining trend. This means that the amount of CDW reported in landfills and recycling centers is decreasing. On the contrary, the extraction of primary raw materials for the construction industry is still growing and is around 72 million tonnes. For comparison, the extraction of primary raw materials in 2015 was 6400 million kg per person per year, which rose to 6700 million kg per person per year by 2020. As is evident from the statistics compiled annually by the Czech Statistical Office [57], even if there is complete utilization of recycled aggregates from waste, the need for construction aggregates cannot be fulfilled. The Czech Republic covers 4% of the aggregate requirement from concrete and masonry waste; however, if unsorted waste (preferably assuming it gets sorted) is used, it can cover around 7% of the requirement. From the point of view of the requirements of the Czech standard, which corresponds with the EN standard, it is possible to partially replace the coarse fraction in concrete with a coarse fraction of RCA, containing more than 90% of the waste concrete and natural aggregate. The maximum replacement rate is 30% for selected classes of concrete mostly without any environmental burdens. This corresponds to the published results from the investigations carried out worldwide that the replacement of coarse fractions up to 30% does not significantly influence the properties of RAC [52]. However, owing to the problematic quality assurance and water absorption determination, the use of a fine fraction of RCA or RMA, in general, is not allowed by a standard. For all these reasons, it is becoming more and more important to optimize demolition and recycling technology to obtain as many quality materials as possible, which will stop landfilling because natural raw materials are not worth landfilling.

In India, the management and reuse of CDW is a prime concern. A study conducted by the Building Material and Technology Promotion Council [58], New Delhi in the year 2018 indicates that the quantity of CDW in India varies from 1215 to 25–30 million tonnes per year (BMTPC, 2018). The report mentions that the estimated quantity of CDW from new construction is approximately 40–60 kg/m^2 of the built-up area and that from the demolition of constructed structures is around 300–500 kg/m^2 of the built-up area. In order to tackle the problem of CDW, recycling plants are set up in a few cities in India. There are four operational recycling plants in India [58]; the first operational large-scale CDW recycling facility was set up in Burari, New Delhi in 2009, followed by another plant in East Kidwai Nagar, New Delhi, and one in Ahmedabad, Gujarat. Most of the other cities have not set up CDW recycling facilities despite having CDW management rules issued by the Ministry of Environment, Forest and Climate Change published by the Central Pollution and Control Board [59]. As per the guidelines, all construction projects and facilities generating more than 20 tons of CDW in a day or 300 tons in a month are identified as bulk generators of CDW and are required to implement a waste management plan from the point of view of the requirements of the Indian standards with the latest revision of guidelines in 2016. The use of RCA as coarse fraction has been permitted up to 50%, 25%, and 100%, respectively, for plain cement concrete, reinforced concrete, and lean concrete with compressive strength less than 15 MPa [60]. Similarly, the use of fRCA is permitted up to 25%, 20%, and 100%, respectively, for plain cement concrete, reinforced concrete with compressive strength less than 25 MPa, and lean concrete with compressive strength less than 15 MPa. On the contrary, the use of RA is not permitted either as coarse or as fine aggregate to produce plain cement concrete and reinforced concrete. RA is allowed for use as coarse aggregate in lean concrete (<15 MPa compressive strength) only.

4. Materials and Methods

In total, 13 concrete mixtures (7 from the Czech part and 6 from the Indian part) were manufactured and examined to verify the possible replacement of natural sand by fRA. Two concrete strength classes were chosen for comparison: the concrete class with compressive strength of 20 MPa for plain concrete and the concrete class with compressive strength of 30 MPa for structural (reinforced) concrete. All concrete mixtures contained coarse natural aggregate and recycled fine aggregate. Natural river sand and crushed stone sand (India) in these mixtures were replaced by fine recycled aggregate (fRA) originating from the CDW from waste concrete (fRCA) by both the India and Czech teams and waste masonry (fRMA) by the Czech team. The basic physical properties (density, water absorption), mechanical properties (compressive strength, flexural strength, and modulus of elasticity), and durability (freeze–thaw resistance and carbonation resistance) of concrete were verified and compared.

4.1. Fine Recycled Aggregate

As described above, the measurement method of fRA's density and water absorption has not yet been established, leading to the high differences between published results. Previous studies have reported that the dry density of fRMA ranges from 2000 to 2500 kg/m^3 and water absorption (WA) ranges from 12% to 15% [24,36,61,62]. The dry densities of fRCA range between 1630 and 2560 kg/m^3 and WA varies between 2.38%

and 19.3% [14,33,34,45,47,63–67]. For comparison, the presented values of the densities of natural sand range between 2530 and 2720 kg/m^3 and WA for the natural fine aggregate ranged between 0.15 and 4.1%. In conclusion, fRMA and fRCA have lower densities than natural sand [1]. The evaluation methodology for the determination of WA of fRA has not been established, which differs from coarse RA, where the methodology for the evaluation of the properties is clearly defined. This leads to the unclear and non-comparable results presented in available literature, where the measured values differ up to 60% when tested by different operators and methods. Furthermore, it is known that fRA does not absorb water to its full capacity during mixing. As has been published in previous studies, it has been estimated that, during mixing, the amount of water absorbed ranges between 49% and 89% [14]. For this reason, the determination of the effective water-to-cement ratio, which influences the workability and, consequently, the mechanical properties of the RAC, becomes complicated.

This study presents the possibility of replacing the whole fine fraction of natural sand (fNA) with fine recycled aggregate (fRA). The Czech team used one type of coarse NA (NA1 fractions 4–8 mm and 8–16 mm) and natural mined sand (fNA1 0–4 mm), two types of fRCA, and one type of fRMA (fractions 0–4 mm), as shown in Figure 1. A type of fRCA1 and fRMA were prepared by a Czech recycling company; the origin of these aggregates was building structures and the aggregate was washed during the recycling process. The other type of fRCA2 was prepared in the laboratory by crushing waste concrete originating from floor structures. For comparison, the India team used one type of coarse NA (NA2 fractions 4.75–10 mm and 10–20 mm), two types of fNA, natural river sand (fNA2) and crushed stone sand (CSS), and one type of fRA (fractions 0–4.75 mm), as shown in Figure 1. fRCA3 was obtained by crushing waste concrete obtained from precast plant set-up for a bridge construction project in Jaipur, India. The waste concrete was crushed in the laboratory and was not washed during the preparation of fRCA3. In this research, the fine fraction of RA (0–4 mm) was used and the coarse fractions were not replaced and remained NA for all mixtures. The main component of fRCA was waste concrete, containing particles of natural aggregate and old cement mortar, and fRMA mainly contained waste masonry (red brick, aerated concrete, and plaster). All tested properties of fRA differed from fNA, especially WA, which was higher and ranged from 3.6 to 8.9% for f RA, while the value for fNA was 1.0% and CSS was 2.8%. This evaluation shows a slightly lower WA of fRMA compared to the results of previous studies [14,33,34,45,47,63–67], which was probably caused by an inconsistent method of measuring fRA WA. The results of fRA WA confirm the conclusions of many previous studies, such as the influence of WA by parent concrete and recycling technology [14]. The lower WA was measured for fRCA1 that originated from normal-strength concrete and was washed during the recycling procedure, so it is assumed that the content of cement mortar was low. In contrast, the higher WA that was measured for fRCA3 may be firstly attributed to the properties of parent concrete, which was high-strength and contained a high amount of cement mortar, and secondly to not washing fRCA3 during preparation. Furthermore, different WA measurement evaluation procedures were used.

The properties of NA and RA fractions used in this study by both the working teams are given in Table 1 and the valid Czech European standard and Indian standards used for testing the properties are mentioned in Table 2. Figures 2 and 3 present the particle size distributions of all types of NA and RA fractions with limits defined in the European (EN 12620) and Indian standards (BIS:383, 2016) used for preparing concrete mixtures, respectively. The oven-dried particle density of fRA ranges from 2052 kg/m^3 to 2430 kg/m^3, where the decline of density in comparison with NA is up to 20%, corresponding with the results of previous studies [14,33,34,45,47,63–67]. Furthermore, fRMA, fRCA1, and fRCA2 do not meet the requirements of the Standard [68] due to difference in granulometry and the presence of fine particles compared to NA (see Table 1 and Figure 2).

Figure 1. NA and RA used in concrete mixtures.

Table 1. The basic physical properties of each fraction of aggregates used for concrete mixtures.

RA Types	Grading (mm)	Finest Particles Content f (%)	Oven-Dried Particle Density ρ_{RD} (kg/m^3)	σ	Water Absorption Capacity WA_{24} (%)	σ	Saturation Level (%)
Natural aggregate (NA1)	0–4	0.3	2570	81	1.0	0.0	0.0
	4–8	0.3	2530	12	1.7	0.3	0.0
	8–16	0.4	2540	12	1.9	0.2	0.0
Natural aggregate (NA2)	0–4.75	0.0	2581	23	0.81	0.00	0.0
	4.75–10	0.0	2670	11	0.45	0.01	0.0
	10–20	0.0	2690	06	0.45	0.05	0.0
Crushed stone sand (CSS)	0–4.75	0.0	2596	83	2.78	0.18	0.0
Fine recycled masonry aggregate (fRMA)	0–4	1.0	2320	130	6.6	0.8	4.7
Fine recycled concrete aggregate (fRCA1)	0–4	0.6	2430	60	3.6	0.8	1.6
Fine recycled concrete aggregate (fRCA2)	0–4	2.0	2220	80	6.9	0.5	2.5
Fine recycled concrete aggregate (fRCA3)	0–4.75	0.0	2052	12	8.90	0.15	0.0

Table 2. The overview of test methods for aggregates.

Tests/Standards	The Czech Team	The Indian Team
Specific gravity/dry density	EN 1097-6	BIS 2386-3 (1963)
Water absorption of aggregates	EN 1097-6	BIS 2386-3 (1963)
Particle size distribution	EN 933-1	BIS 2386-1 (1963)

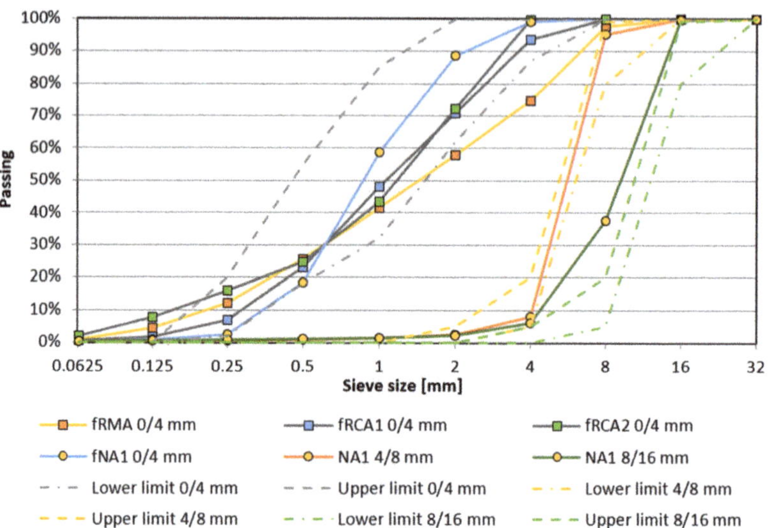

Figure 2. Particle size distribution of aggregates used by the Czech team.

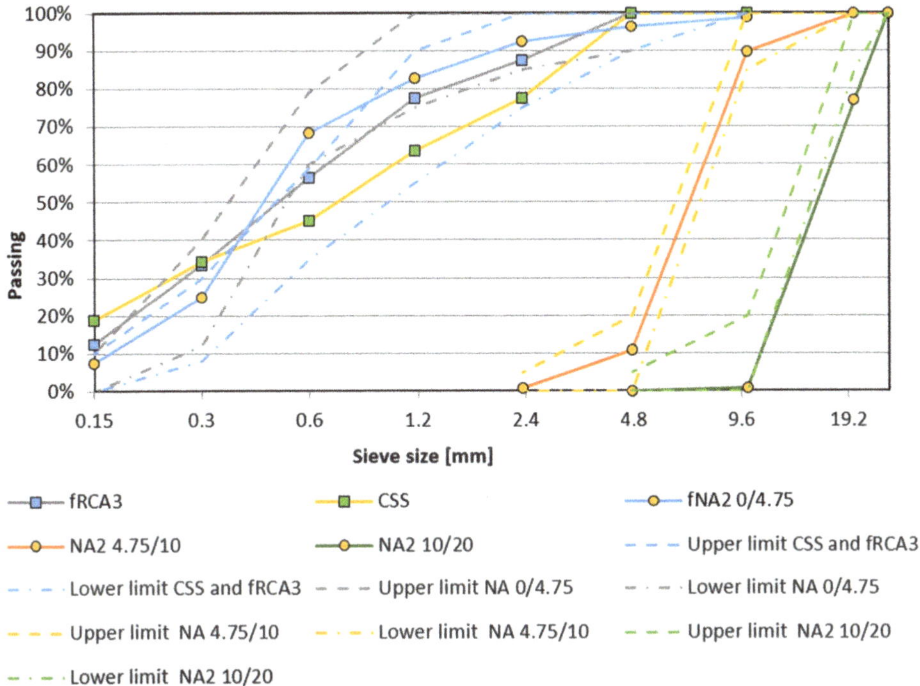

Figure 3. Particle size distribution of aggregates used by the Indian team.

4.2. Recycled Aggregate Concrete Mixtures

The 13 concrete mixtures were manufactured for laboratory verification of the properties. The cement CEM I 42.5 R content was 260 kg/m^3 for the mixtures labelled I and III, 300 kg/m^3 for II, and 320 kg/m^3 for IV. The mixture proportions are given in Table 3. Six control mixtures of conventional concrete (NAC IA, NAC IIA, NAC IB, NAC IIB, CSSC IB, and CSSC IIB), three mixtures of strength-class-corresponding compressive strengths of 20 MPa and three mixtures of 30 MPa, respectively, with only NA up to a particle size of 16 mm were produced. In these mixtures, three types of fNA were used: (1) mined sand by the Czech team; (2) river sand; and (3) crushed stone sand by the Indian team. For comparison, in the further 7 mixtures for both concrete classes, the fNA was fully replaced by the different types of fRA: (1) fRCA1, fRCA2, and fRMA by the Czech team and (2) fRCA3 the by Indian team.

The Bolomey particle size distribution curve was used for optimizing the skeleton of the concrete mixtures. The mixing procedure used by the Czech team was similar to that reported by Evangelista and de Brito [65]. Here, a two-stage mixing technique was used, wherein the first stage comprised of 10-minute-long mixing of fRAs with water estimated to be absorbed and 2/3 of the mixing water; subsequently, in the second stage, the remaining constituents were added [1]. The fRAs were mixed with part of the water (water estimated to be absorbed) for 10 min, and after this stage, the remaining constituents and the mixing water were added to the concrete mixture. The water estimated to be absorbed was calculated as a difference between WA of fRA and current levels of aggregate saturation before mixing. The effective water-to-cement ratio for mixtures in the Czech part of the study was estimated as 0.65 for compressive strengths of 20 MPa and 0.55 for 30 MPa reversal. In the case of the Indian part, the effective water-to-cement ratio was estimated as 0.50 and 0.45 for compressive strengths of 20 MPa and 30 MPa, respectively, and a superplasticizer was added to maintain uniform workability of all the concrete mixtures.

Table 3. Concrete mix proportion, per cubic meter.

Concrete Mixture	Cement	Water Mixing + Additional	W/C Ratio	SP	Natural Aggregate		Recycled Aggregate
					Fine	Coarse	Fine
	(kg/m³)	(kg/m³)	(-)	(kg/m³)	(kg/m³)	(kg/m³)	(kg/m³)
NAC IA	260	169 + 0	0.65	-	709	1130	0
fRMAC IA	260	169 + 18	0.72	-	0	766	971
fRCAC1 IA	260	169 + 17	0.71	-	0	949	843
fRCAC2 IA	260	169 + 34	0.78	-	0	946	773
NAC IB	260	130 + 12	0.55	2.3	813	1266	0
CSSC IB	260	130 + 28	0.61	3.1	813	1266	0
fRCAC3 IB	260	130 + 78	0.80	0.3	0	1266	813
NAC IIA	300	165 + 0	0.55	-	671	1167	0
fRMAC IIA	300	165 + 17	0.61	-	0	822	920
fRCAC1 IIA	300	165 +16	0.60	-	0	994	800
NAC IIB	320	144 + 12	0.49	1.6	779	1213	0
CSSC IIB	320	144 + 27	0.53	3.2	779	1213	0
fRCAC3 IIB	320	144 + 75	0.68	0.3	0	1213	779

4.3. Evaluation Methodology

The physical, mechanical, and durability properties were examined by both teams. The dimensions of specimens and testing standards used in the experimental work are shown in Table 4. Testing procedures were designed to be as similar as possible to regional habits. However, the test procedures and their differences are described below. At the age of 28 days, physical and mechanical properties were verified according to Czech and Indian standards. Furthermore, durability (freeze–thaw resistance and accelerated ageing due to CO_2) and long-term strength development (at the ages of 90 and 180 days) were tested.

Table 4. The overview of test methods for concrete samples.

Tests	Curing Period	The Czech Team		The Indian Team	
		Standards	Specimen Size	Standards	Specimen Size
	(Days)		(mm)	-	(mm)
Compressive strength	7, 28, 90, 180	EN 12390-3 (2003)	150 × 150 × 150	BIS 516 (1956)	150 × 150 × 150
Flexural strength	28	EN 12390-5 (2009)	100 × 100 × 400	BIS 516 (1956)	100 × 100 × 500
Static modulus of elasticity	28	EN 12390-13 (2014)	100 × 100 × 400	BIS 516 (1956)	150 diameter and 300 length
Dynamic modulus of elasticity	28	EN 12504-4 (2005)	100 × 100 × 400	BIS 13311-1 (1992)	150 diameter and 300 length
Carbonation	28	Inspired by ČSN EN 12390-12	100 × 100 × 200	RILEM CPC-18 (1988)	100 × 100 × 100
Freeze–thaw resistance	28	ČSN 73 1322 (1969)	100 × 100 × 400	Inspired by C666/C666M– 15 (2015)	100 × 100 × 500
Water absorption by immersion	28	Usual procedure of examination	100 × 100 × 100	Usual procedure of examination	100 × 100 × 100
Sorptivity	28	Inspired by ASTM C1585-20	100 × 100 × approx. 200	Inspired by ASTM C1585-20	100 × 100 × 100

The water absorption capacity by immersion, which describes the transport behavior of the material, was obtained on 100 × 100 × 100 mm cube specimens. The specimens were immersed in a water chamber until constant weight and thereafter dried in an oven at 105 ± 2 °C until constant weight. The sorptivity of the concrete specimens of size 100 × 100 × 100 mm with time was determined by conditioning the samples at 105 °C in an oven until their weight stabilized. The specimens were placed on a support device by exposing one of the surfaces to water. The change in mass of specimens was noted at intervals of 0, 1, 10, 30, 60, 120, 240, 1440, 2160, and 4320 min. The slope of the line obtained by plotting absorption against the square root of time gives the sorptivity of the concrete as per ASTM C1585-20.

Prismatic samples of 100 × 100 × 400 mm and 100 × 100 × 500 mm, respectively, cured in water for 28 days were used for assessing the freeze–thaw resistance. The frost resistance was measured according to ČSN 73 1322 (1969) and ASTM C666/C666M-15 (2015) for the Czech team and Indian teams, respectively. Both the teams followed testing by cyclic freezing and thawing at temperatures ranging from −15 °C to +20 °C, where one cycle takes 4 h of freezing and 2 h of thawing. The Czech specimens were subjected to rapid freezing in air and thawing in water and the Indian specimens were subjected to rapid freezing and thawing in water as per ASTM C666/C666M-15 (2015) procedure A. The freeze–thaw resistance was observed for a total of 100 cycles. The measuring of the dynamic modulus of elasticity was performed by the Czech team according to EN 12504-4 (2005) after each phase and by the Indian team according to BIS 13311-1 (1992) after the completion of a total of 100 cycles. The flexural strength was tested by both teams after 100 cycles according to EN 12390-5 (2009) and BIS 516 (1956), respectively, by the Czech and Indian teams.

Prismatic samples of 100 × 100 × 400 mm were stored in a constant laboratory environment for measuring the carbonation resistance. Half of the samples (100 × 100 × 200 mm) were placed for 28 days in laboratory equipment with CO_2 atmosphere CO2CELL (MMM group) with a concentration of CO_2 3.0 ± 0.2%. This test has been prepossessed by the standard ČSN EN 12390-12. Nevertheless, the testing process has been slightly modified and does not fully comply with the standard regulation. The pH drop in the concrete due to CO_2 was evaluated using the phenolphthalein indication method. In contrast, the accelerated CO_2 ageing test by the Indian team was performed on samples of different dimensions and in a slightly different way. The depth of carbonation in the concrete cube of size 100 × 100 × 100 mm was tested as per RILEM CPC-18. The samples were cut into four pieces of size 50 × 50 × 100 mm and air-dried for 14 days. The longitudinal sides were coated with epoxy paint and kept in a carbonation chamber at a condition of 3 ± 0.1% CO_2, 25 ± 2 °C, and 60 ± 5% relative humidity for 28 days. The depth of carbonation was determined by spraying the phenolphthalein indicator on the split surface of the sample.

5. Results and Discussion

In this section, the results of physical, mechanical, and durability properties examined by both the research groups are presented and compared.

5.1. Physical Properties

As reported in previous studies [69,70], the durability of the concrete is fundamentally influenced by its porosity and water absorption. Therefore, immersion-based water absorption and capillary water absorption were evaluated to determine their impact on durability properties. The porosity of concrete and, consequently, the proportion of water with fRCA increased with the increasing replacement level of fRCA [41,71]. The dry density and water absorption of different mixtures are shown in Table 5 and Figure 4. The slight decline in density of RAC in comparison with NAC can be observed for samples in both regions, with the highest decline of about 10%. In general, the water absorption of fRA concrete by immersion was found to be higher than the control mixtures, which corresponds to previous studies [33,37]. As was concluded in many previous studies, water absorption of fRCA and fRMA decreases due to the presence of old mortar. Slight differences between the mixtures developed by the Czech Republic and India were observed. The maximum increase of water absorption by immersion for mixture FRCAC3 IB, manufactured by the Indian team, was more than twice the control concrete. In the case of the Czech team, the maximum increase of 85% was noticed in the fRMAC IA due to the high porous materials, such as red clay bricks, aerated concrete, and mortar, contained in the fRMA. The water absorption by immersion of the fCRAC mixtures increased between 30% and 50%, which is slightly higher than the values presented in previous studies, where water absorption by immersion has been reported to increase from 15% [71,72] to 46% [37] for concrete with the complete substitution of natural sand by fRCA.

Table 5. Density and water absorption of concrete mixtures.

Recycled Concretes	Dry Density		Water Absorption by Immersion		Water Absorption by Capillarity	
Designation	(kg/m^3)	σ	(%)	σ	(kg/m^2)	Σ
NAC IA	2301	18	5.89	0.35	2.31	0.30
fRMAC IA	2181	14	10.89	0.38	3.34	0.75
fRCAC1 IA	2276	11	7.68	0.25	1.98	0.22
fRCAC2 IA	2250	6	8.34	0.78	2.17	0.29
NAC IB	2373	5	6.30	0.66	3.83	0.05
CSSC IB	2391	74	8.80	2.88	4.53	0.25
fRCAC3 IB	2332	26	14.60	0.57	3.00	0.62
NAC IIA	2324	13	5.03	1.11	1.17	0.14
fRMAC IIA	2191	12	10.30	0.32	0.45	0.08
fRCAC1 IIA	2278	5	7.70	0.06	0.76	0.44
NAC IIB	2380	14	6.90	0.59	3.33	0.45
CSSC IIB	2193	11	8.30	0.87	1.93	0.48
fRCAC3 IIB	2188	21	13.20	0.31	1.70	0.36

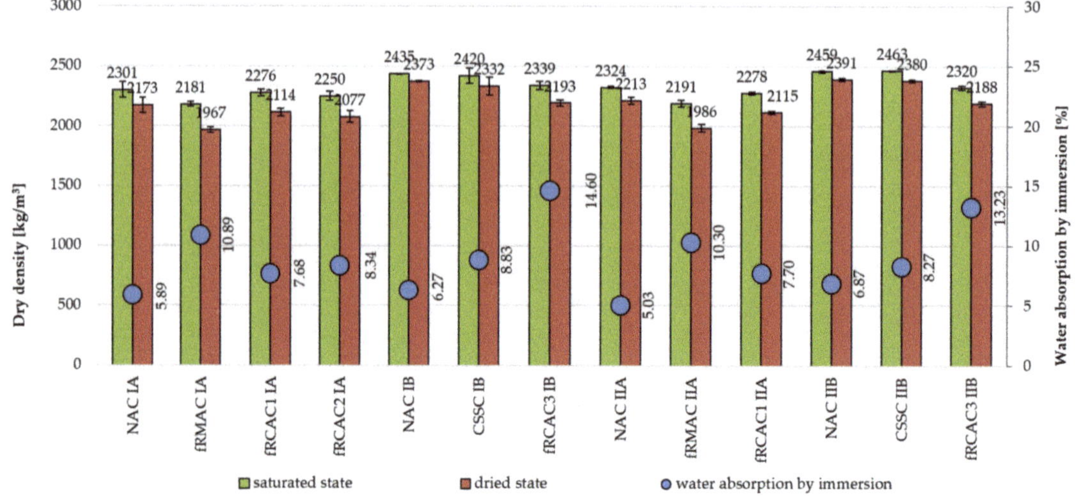

Figure 4. Physical properties of concrete mixtures—saturated surface dry density, oven dry density. and water absorption by immersion.

Although previous studies have reported a significant increase in capillary absorption from 46% to 95% for 100% fRCAC [53], this was not confirmed in this study. The capillary water absorption measured by both teams was lower than that of the control concretes. The only increase was found for fRMAC IA, which was 44%. On the contrary, the lower decrease was measured for fRMAC IIA with a decline of more than 60%. This decrease may be due to the filling of pores present in the concrete by the products of hydration of unhydrated cement present in the fRCA and, moreover, the water contained soaked in the concrete after curing due to the high WA of the fRA.

5.2. Mechanical Properties

5.2.1. Compressive Strength

The results of compressive strength, which is the key mechanical property of concrete, are presented in Tables 6 and 7, and Figures 5–7. It was observed that the compressive strength of fRCAC differs from previous studies. A study reported higher than control compressive strength [31], and few others reported a similar or lower [31,34,73–75] than control concrete. Generally, the sensitivity of compressive strength to the high replacement level of fRCA (100%) has been found, regardless of the strength class of concrete, mostly due to its inaccurately measured water absorption and unknown rate of water during the mixing procedure. For this reason, additional water is used to compensate for these two factors, which leads to the unknown effective water-to-cement ratio, which is only estimated in the case of fRAC. Despite these factors, the compressive strength could be positively affected by the filler effect of fRA, where the finest particles fill the pores and make the structure of the concrete denser, decrease internal tension, and early stress propagation. Moreover, the positive influence on mechanical properties could have an additional internal cure caused by the water absorbed in the aggregate. Furthermore, the fRA particles could have a better interlock between particles due to the rough surface and angular shape [1].

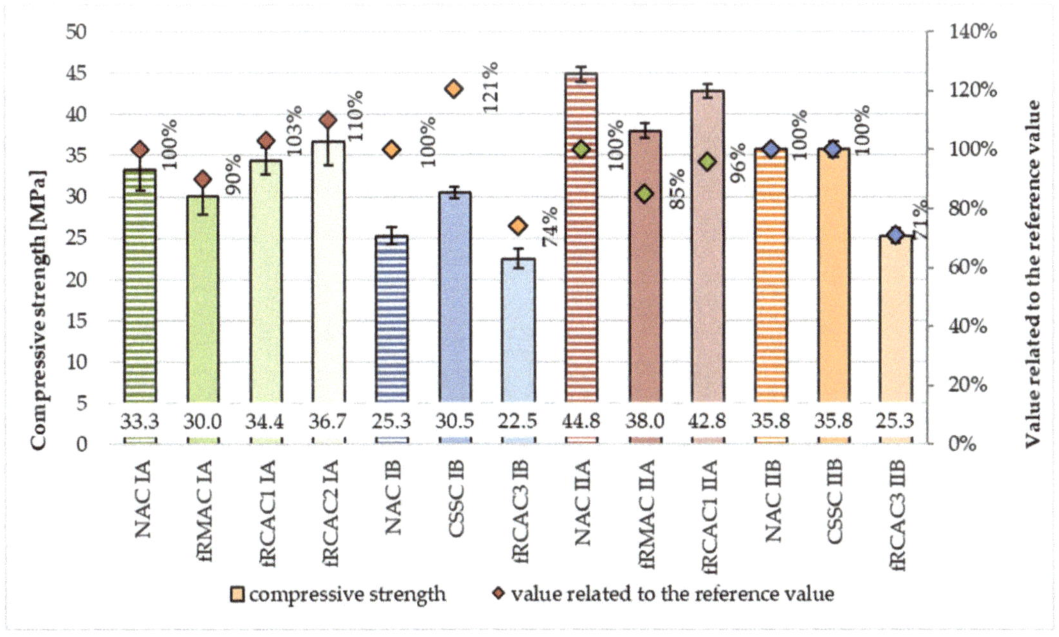

Figure 5. Comparison of compressive strength at 28 days of concrete containing fNA, fRCA, and fRMA with respect to control mixtures.

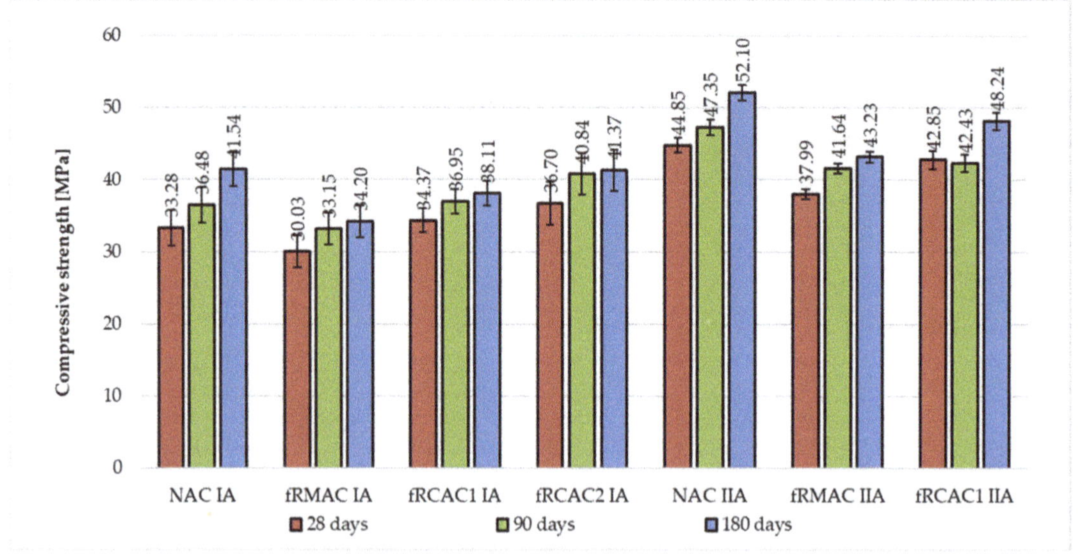

Figure 6. Comparison of long-term development of compressive strength of concrete mixtures prepared by the Czech research group.

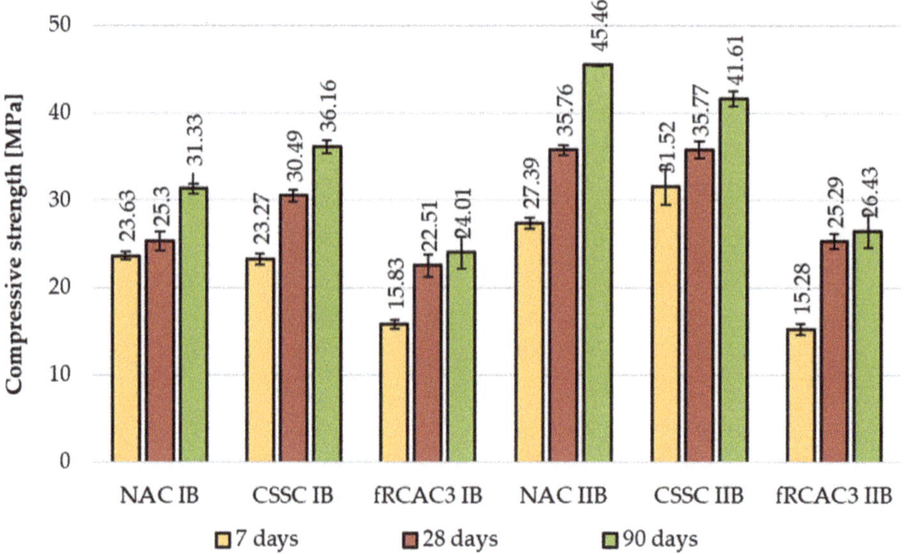

Figure 7. Comparison of long-term development of compressive strength of concrete mixtures prepared by the Indian research group.

Table 6. Average values and standard deviation of results of mechanical properties of concrete at age of 28 days.

Recycled Concrete Mixture	Compressive Strength		Flexural Strength		Static Modulus of Elasticity		Dynamic Modulus of Elasticity	
Designation	**(MPa)**	**σ**	**(MPa)**	**σ**	**(GPa)**	**σ**	**(GPa)**	**σ**
NAC IA	33.2	2.5	6.2	0.2	36.7 (1)	1.4	38.2 (1)	1.8
fRMAC IA	30.0	2.2	5.5	0.4	22.4 (1)	1.0	27.3 (1)	1.4
fRCAC1 IA	34.4	1.7	5.8	0.3	29.6 (1)	0.4	34.5 (1)	0.7
fRCAC2 IA	36.7	2.9	5.7	0.1	31.8 (1)	1.2	35.4 (1)	1.7
NAC IB	25.3	1.1	3.9	0.5	30.7 (2)	0.0	41.2 (2)	0.0
CSSC IB	30.5	0.7	4.3	0.7	24.4 (2)	0.0	26.6 (2)	0.0
FRCAC3 IB	22.5	1.2	6.5	0.6	25.3 (2)	0.0	26.4 (2)	0.0
NAC IIA	44.9	0.9	7.6	0.9	35.9 (1)	0.5	38.2 (1)	0.8
fRMAC IIA	38.0	0.9	6.8	0.6	25.3 (1)	0.2	30.0 (1)	0.9
fRCAC1 IIA	42.9	0.8	6.5	0.4	31.4 (1)	1.0	35.7 (1)	0.6
NAC IIB	35.8	0.6	4.3	0.1	33.2 (2)	0.0	44.5 (2)	0.0
CSSC IIB	35.8	0.9	5.3	1.1	29.3 (2)	0.0	29.7 (2)	0.0
FRCAC3 IIB	25.3	0.9	5.5	0.4	25.6 (2)	0.0	32.6 (2)	0.0

(1) Examined on prismatic specimen 100 × 100 × 400 mm^3.· (2) Examined on cylindric specimen of 150 mm diameter and 300 mm length

Table 7. Average values of results of long-term compressive strength development, including standard deviation.

Age of Samples	7 Days		28 Days		90 Days		180 Days	
Designation	**(MPa)**	**σ**	**(MPa)**	**σ**	**(MPa)**	**σ**	**(MPa)**	**Σ**
NAC IA	-	-	33.2	2.5	36.5	3.0	41.5	1.5
fRMAC IA	-	-	30.0	2.2	33.2	1.5	34.2	2.0
fRCAC1 IA	-	-	34.4	1.7	37.0	3.0	38.1	1.5
fRCAC2 IA	-	-	36.7	2.9	40.8	1.1	41.8	2.7
NAC IB	23.6	0.5	25.3	1.1	31.3	0.6	-	-
CSSC IB	23.3	0.7	30.5	0.7	36.2	0.8	-	-
fRCAC3 IB	15.8	0.5	22.5	1.2	24.0	1.9	-	-
NAC IIA	-	-	44.9	0.9	47.4	3.3	52.1	0.7
fRMAC IIA	-	-	38.0	0.9	40.6	1.8	43.2	1.1
fRCAC1 IIA	-	-	42.9	0.8	42.4	4.1	48.2	0.5
NAC IIB	27.4	0.6	35.8	0.6	45.5	0.1	-	-
CSSC IIB	31.5	2.0	35.8	0.9	41.6	0.9	-	-
fRCAC3 IIB	15.3	0.6	25.3	0.9	26.4	1.9	-	-

Figure 5 shows the compressive strength of concrete mixtures prepared by both the Czech and Indian teams. Similar to the previous findings, heterogeneous results for compressive strength were observed in this study. In the case of the Czech team, the compressive strength increased (equal to 10%) for concretes containing fRCA1 and 2 compared with control mixtures for lower concrete strength classes. On the contrary, the compressive strength of the mixture with fRCA1 in the higher strength class slightly decreased (4%). Furthermore, the compressive strengths of both fRCA3-containing mixtures were found to decrease in comparison with both control mixes, and, furthermore, the decline was greater compared to the fRCA1 and fRCA2 mixtures. This may be attributed to the difference in source and properties of fRCA3, as it was prepared in India from parent concrete consisting of crushed sandstone.

At 28 days, the strength of concrete fRCAC3 IB was observed to decrease with respect to the two controls, by 11% with respect to NAC IB and 26% with respect to CSSC IB. The strength of fRCAC3 IIB concrete was found to reduce by 29% in comparison with control NAC IIB and 29% with respect to CSSC IIB. The compressive strength of both concrete strength classes with fRMA slightly decreased (10% and 15%, respectively) compared to the control mix. Furthermore, the development of the compressive strength over time shows a higher rising of fRAC than control concrete.

The decrease in strength and differences between each mixture are probably caused by the presence of an undefined amount of adhered mortar and the amount of additional water to compensate for the higher water absorption and the ability of fRA to soak water during mixing. As previously written, the amount of cement mortar is influenced by the parent concrete and the recycling procedure [14]. In this case, it is assumed that in fRCA1 and fRMA the content of fines was reduced by washing. In contrast, fRCA3 originated from high-strength concrete, so the high amount of cement paste is assumed in the parent concrete and, consequently, the high fine content. In this case, the study confirms previous studies in which the negative effect due to lack of knowledge about fine particles and their influence on the effective water-to-cement ratio was described many times [1]. The maximum replacement rate in the case of compressive strength was stated as 30% [31,33,38,46,48,49].

The development of compressive strength was measured at 28, 90, and 180 days by the Czech team and 7, 28, and 90 days by the Indian team, respectively. The results show a lower increase for fRA mixtures in comparison with the references. The increase of the control concrete was maximally 25% at 180 days; however, in the case of fRA, the maximal increase was 16% (see Figures 6 and 7).

5.2.2. Flexural Strength

Past studies have reported a decrease in the flexural strength of fRCA concrete/mortar with the increase in the fRCA content [19,76]. The maximum reduction in tensile strength was 33% for concrete with a full replacement ratio. As it has been reported, the tensile strength declined with the increasing replacement ratio of natural sand with fRCA and with the water-to-cement ratio increase [77]. In contrast, the flexural strength of mortar at 28 days was found to be higher than the control by 13.7% [44]. However, the maximum replacement ratio in this case was 20% [78].

The comparison of the flexural strength of concrete mixtures at 28 days is presented in Figure 8. The flexural strength was observed to decrease for all concrete mixtures prepared by the Czech research group. The decrease in properties of the fRAC I mixtures ranged from 6% to 11% and the reduction for fRAC II mixtures was between 11% and 15%. Unexpectedly, in the case of flexural strength, the fRMAC achieved lower declines than both fRCACs. On the contrary, the flexural strength of fRAC mixtures was prepared by the Indian research group. The flexural strength of fRCAC3 IB was observed to increase by 67% and 51% compared with NAC IB and CSSC IB, respectively. The strength of fRCAC3 IIB was observed to increase by 28% and 4% compared with NAC IIB and CSSC IIB, respectively. The higher strength may be attributed to the better interlocking of the fRCA with the paste because of the presence of the uneven surfaces of fRCA.

5.2.3. Modulus of Elasticity

The static modulus of elasticity is the key characteristic for assessing the behavior of reinforced concrete structural elements. Past results of concrete containing coarse RA have reported that, among all the concrete properties, the modulus of elasticity was degraded the highest. The reductions in the static moduli of concretes where natural sand was replaced by fRA range between 9.5% and 17% [34,79]. Wang et al. [80] described that concrete with coarse NA and 100% fRCA showed a reduction in elastic modulus of 5.6–13.5%. Furthermore, a significant decline in modulus of elasticity was reported for low substitution levels (<30%) [81,82].

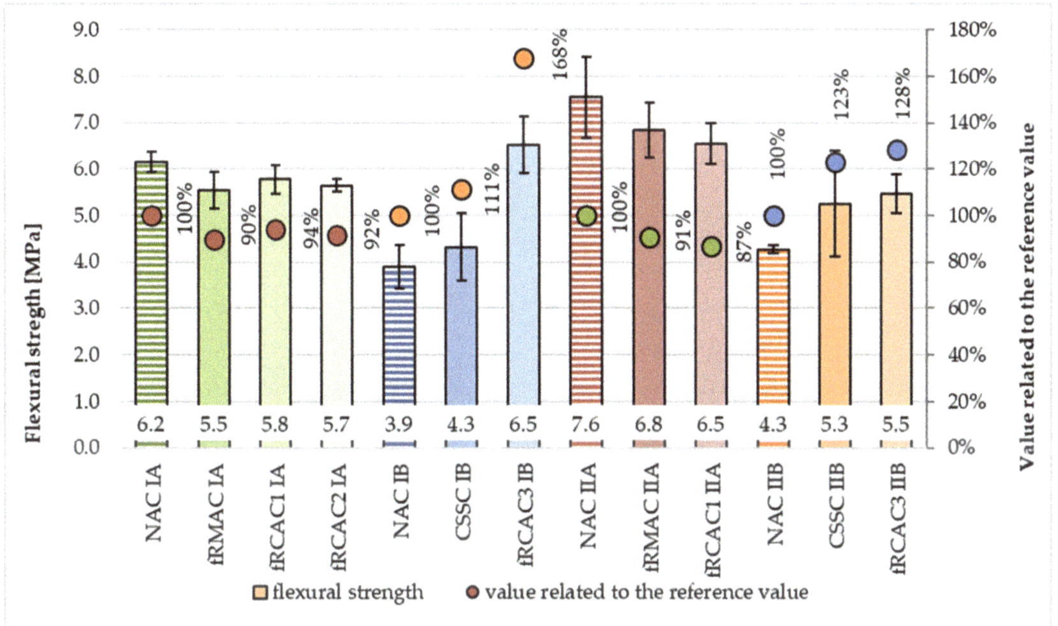

Figure 8. Comparison of flexural strength at age 28 days of concrete containing fNA, fRCA, and fRMA with respect to control mixtures.

The static and dynamic elastic moduli at 28 days of concrete prepared by the Czech and Indian teams, respectively is shown on Figures 9 and 10, respectively. The results show a similar phenomenon for concretes prepared in both regions corresponding with the previous studies. The results confirmed the decrease in static modulus of elasticity, which ranges from 13% to 39% with respect to fRAC I and up to 13% and 30% with respect to fRAC II. The dynamic modulus of elasticity decline was slightly lower and varied between 10% and 28% for fRAC I and between 6% and 21% for fRAC I. Similarly, lower static and dynamic elastic modulus for fRMA concrete for both classes of concrete were observed. The static modulus of elasticity of fRCAC3 IB was observed to decrease by 18% with respect to NAC IB and was similar to CSSC IB. The static modulus of fRCAC3 IIB decreased by 22% and 12% with respect to NAC IIB and CSSC IIB, respectively. The dynamic modulus of elasticity of fRCAC3 IB decreased by 36% with respect to NAC IB and was similar to CSSC IB. The dynamic modulus of elasticity of fRCAC3 IIB was found to decrease by 23% in comparison with NAC IIB and by 10% in comparison with CSSC IIB. The decrease may be attributed to the loss of mortar stiffness due to the presence of adhered mortar in fRCA. Similar findings were observed by past researchers [83].

5.3. Durability Properties

The most important factor that affects durability is concrete permeability, which is studied by the oxygen and water permeability test, water absorption by immersion, and capillarity [37,53,71,84–88]. In this study, water absorption by immersion, capillarity, freeze–thaw resistance, and carbonation of concrete containing fRA were verified.

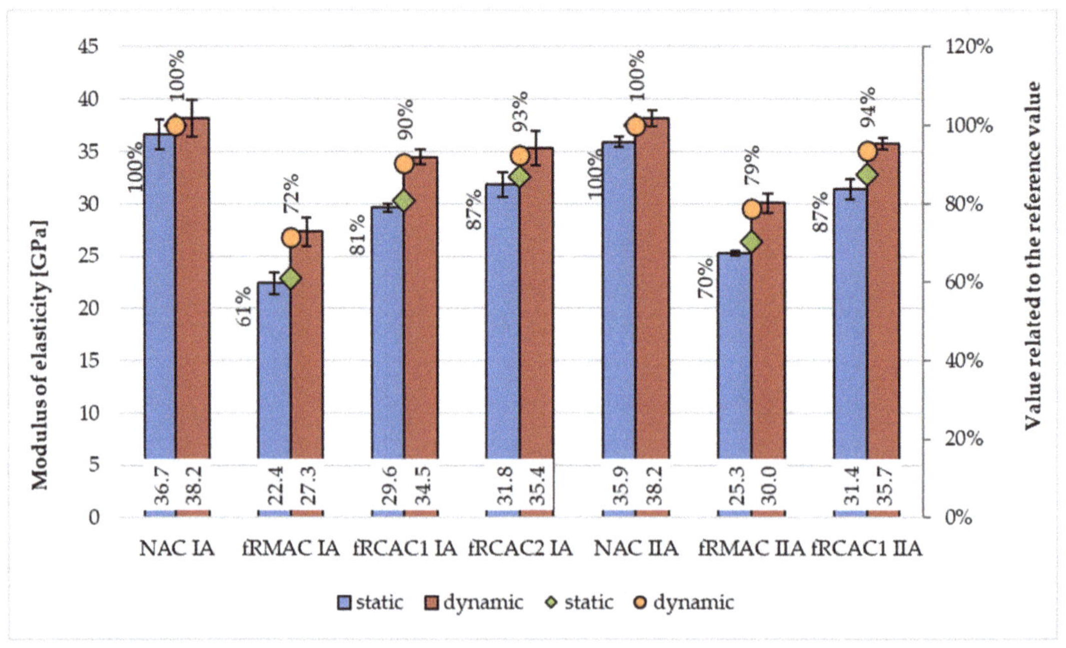

Figure 9. Comparison of static and dynamic moduli of elasticity at age 28 days for concrete mixtures prepared by the Czech team.

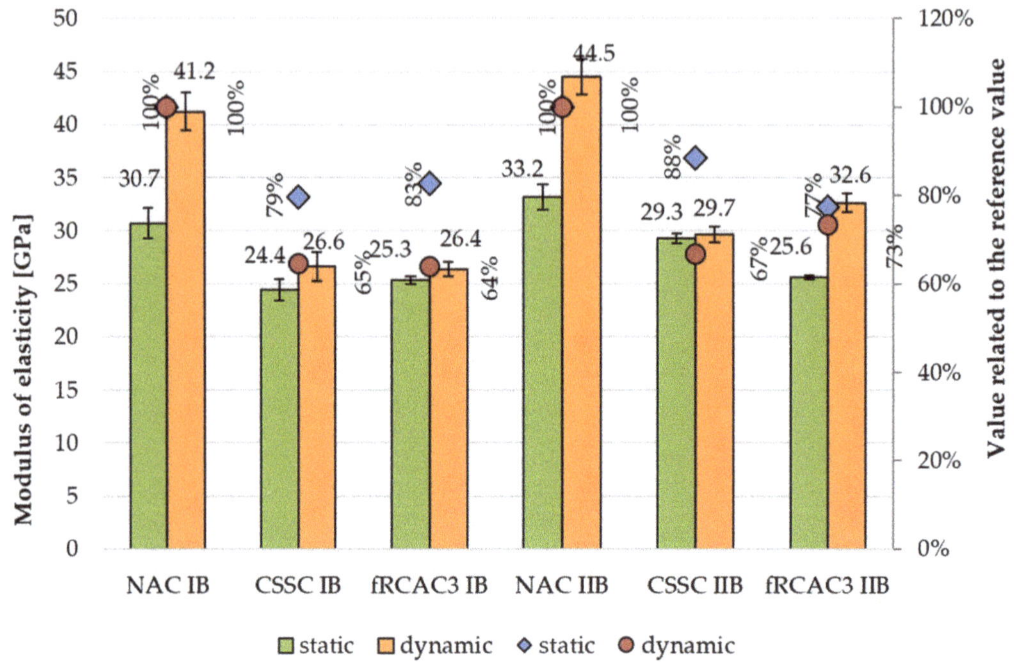

Figure 10. Comparison of static and dynamic moduli of elasticity at age 28 days for concrete mixtures prepared by the Indian team.

The frost resistance coefficient is according to the Czech Standard established from the flexural strength and dynamic elastic modulus before and after freezing and thawing cycles, respectively.

5.3.1. Freeze–Thaw Resistance

In the case of freeze–thaw resistance, the positive effect of fRA in the mixture has been found. This phenomenon is caused by the higher porosity of the fRCA, which can provide better hydraulic pressure dissipation. The negative influence of the affecting freezing and thawing could be observed on the fRAC surface, due to the less-resistant mortar, however, without loss of mechanical properties [76,78,89]. This investigation achieved the same results; the freeze–thaw resistance of all examined fRA concretes was similar to or slightly better than control concretes as depicted by the flexural strength, which was measured before and after freezing and thawing (see Table 8). In the case of the dynamic modulus of elasticity (Table 9), a slight decline of the frost resistance coefficient was observed with a maximum 13% reduction of the frost resistance coefficient with respect to control concretes. However, all mixtures conformed to the requirements defined in the Czech national standard, where the frost resistance coefficient must not decrease by more than 25% (see Figure 11). Additionally, the weight and dimensions of the fRCA specimens subjected to 100 freeze–thaw cycles were not significantly altered. The test procedure implemented by the Indian team was slightly different from the Czech team, however, similar results were achieved. The flexural strength of fRCAC3 IB after completion of 100 cycles was comparable with the controls. However, a decrease of 16% and 30% in the flexural strength of fRCA3 II was observed with respect to NAC IIB and CSSC IIB, respectively. The dynamic modulus of elasticity of fRCA concrete after 100 cycles was comparable with controls. These results are in accord with the previous findings [76].

Table 8. The comparison of the flexural strength before and after freezing–thawing, the frost resistance coefficient, and carbonation depth of concrete mixtures.

Recycled Concrete Mixture	Flexural Strength + Standard Deviation				Frost Resistance Coefficient	Freeze–Thaw Resistance	The Carbonation Depth
Designation	0 Cycles		100 Cycles		(-)	Cycles	(mm)
NAC IA	6.15	±0.22	6.87	±0.20	1.12	100	2.78 [(1)]
fRMAC IA	5.53	±0.39	5.85	±0.40	1.06	100	7.10 [(1)]
fRCAC1 IA	5.78	±0.30	6.57	±0.26	1.14	100	4.51 [(1)]
fRCAC2 IA	5.65	±0.14	6.22	±0.27	1.10	100	1.68 [(1)]
NAC IB	3.89	±0.46	3.44	±0.36	0.88	100	11.00 [(2)]
CSSC IB	4.32	±0.72	3.35	±0.09	0.78	100	11.83 [(2)]
fRCA3 IB	6.52	±0.61	3.15	±0.14	0.48	-	14.50 [(2)]
NAC IIA	7.55	±0.87	7.80	±0.12	1.03	100	0.77 [(1)]
fRMAC IIA	6.84	±0.60	6.78	±0.00	0.99	100	1.71 [(1)]
fRCAC1 IIA	6.54	±0.44	6.73	±0.10	1.03	100	0.57 [(1)]
NAC IIB	4.27	±0.08	4.00	±0.15	0.94	100	5.17 [(2)]
CSSC IIB	5.25	±1.14	4.78	±0.64	0.91	100	5.50 [(2)]
fRCA3 IIB	5.47	±0.41	3.36	±0.32	0.61	-	11.70 [(2)]

[(1)] Examined on prismatic specimen 100 × 100 × 200 mm. [(2)] Examined on specimen size 50 × 50 × 100 mm, longitudinal sides coated with epoxy paint.

Table 9. Dynamic modulus of elasticity after freezing and thawing.

Recycled Concretes	Dynamic Modulus of Elasticity (GPa) + Frost Resistance Coefficient (-)									Freeze–Thaw Resistance
Designation	0 Cycles	25 Cycles		50 Cycles		75 Cycles		100 Cycles		Cycles
NAC IA	37.6	36.4	0.97	36.5	0.97	35.8	0.95	36.9	0.98	100
fRMAC IA	19.7	17.0	0.86	18.4	0.94	17.6	0.89	19.3	0.98	100
fRCAC1 IA	35.1	32.3	0.92	30.9	0.88	32.4	0.92	31.3	0.89	100
fRCAC2 IA	37.3	34.3	0.92	33.3	0.89	33.2	0.89	33.1	0.89	100
NAC IB	29.31	-	-	-	-	-	-	25.94	0.89	100
CSSC IB	27.88	-	-	-	-	-	-	27.79	1.00	100
fRCA3 IB	23.89	-	-	-	-	-	-	30.78	1.29	100
NAC IIA	40.4	37.0	0.92	36.0	0.89	37.4	0.93	35.1	0.87	100
fRMAC IIA	31.6	28.4	0.90	25.9	0.82	29.8	0.94	28.0	0.88	100
fRCAC1 IIA	35.2	29.6	0.84	34.5	0.98	33.4	0.95	31.1	0.88	100
NAC IIB	25.37	-	-	-	-	-	-	21.17	0.83	100
CSSC IIB	18.91	-	-	-	-	-	-	24.42	1.29	100
fRCA3 IIB	24.84	-	-	-	-	-	-	22.59	0.91	100

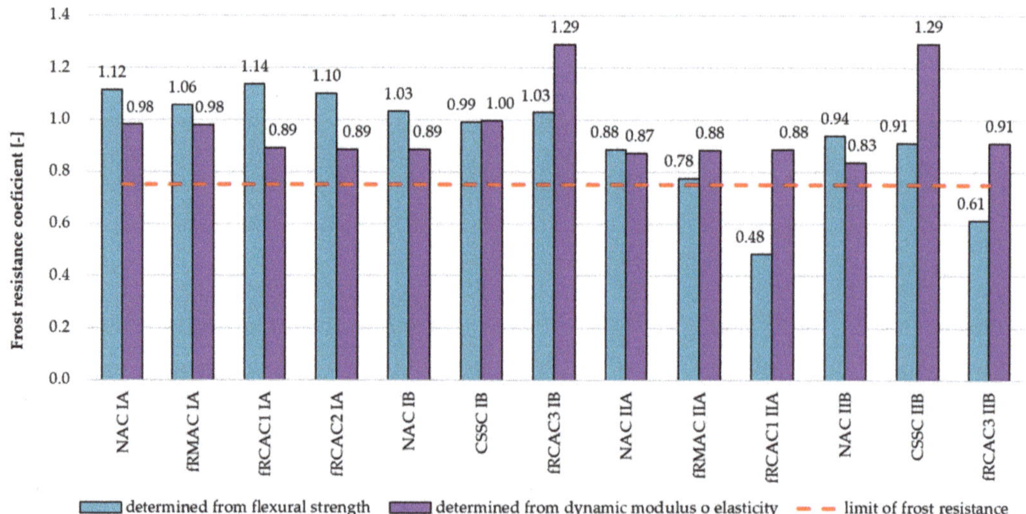

Figure 11. Comparison of frost resistant coefficient calculated from flexural strength and dynamic modulus of elasticity of concrete containing fNA, fRCA, and fRMA measured before and after 100 freeze–thaw cycles.

5.3.2. Carbonation Resistance

Carbonation depth determines the quality of concrete's protective cover over steel reinforcement bars. A poor carbonation resistance is bound to affect the service life of reinforced concrete structural elements [90]. In previous studies, the importance of using a reasonable amount of water was mentioned as essential for the carbonation resistance to carbonation of fRA concretes, especially when the amount of RFA exceeds 40%. It was reported that the higher amount of water unexpectedly did not improve the porosity of concrete but, on the contrary, worsened the carbonation resistance of fRAC. The optimal effective water-to-cement ratio was found to be essential for suitable resistance to carbonation of concrete [33,55]. This work confirmed the same observation. The depth of carbonation of concrete mixtures containing fNA, fRCA, and fRMA observed by both the teams is given in Table 8 and a comparison is presented in Figures 12 and 13. The mixtures with a lower estimated effective water-to-cement ratio were seen to achieve better carbonation resistance.

Moreover, the mixtures with a higher amount of cement seem to be more resistant to carbonation. The mixtures containing fRMA show a deeper penetration of CO_2 into the concrete, probably caused by the high porosity fRMA. The increase in carbonation depth was 155% for fRMAC IA and 123% for fRMAC IIA. In the case of the Indian concrete mixtures, fRCAC3 IB was observed to have a higher depth of carbonation by 32% and 23% compared with NAC IB and CSSC IB, respectively. A significant increase in carbonation depth was observed in fRCAC3 IIB with respect to NAC IIB and CSSC IIB by 126% and 112%, respectively. The increase in carbonation depth may be attributed to the presence of high mortar content and more pores present in the fRCA concrete. Similar observations were found by [37,53,85]. On the contrary, fRCAC mixtures prepared by the Czech team achieved more favorable results in carbonation resistance, where increase in carbonation depth was observed in one mixture (fRCAC1 IA) only. However, from the point of view of previous studies carried out by the same research group [91,92], the negative influence of fRA is significantly lower than the impact of coarse RA.

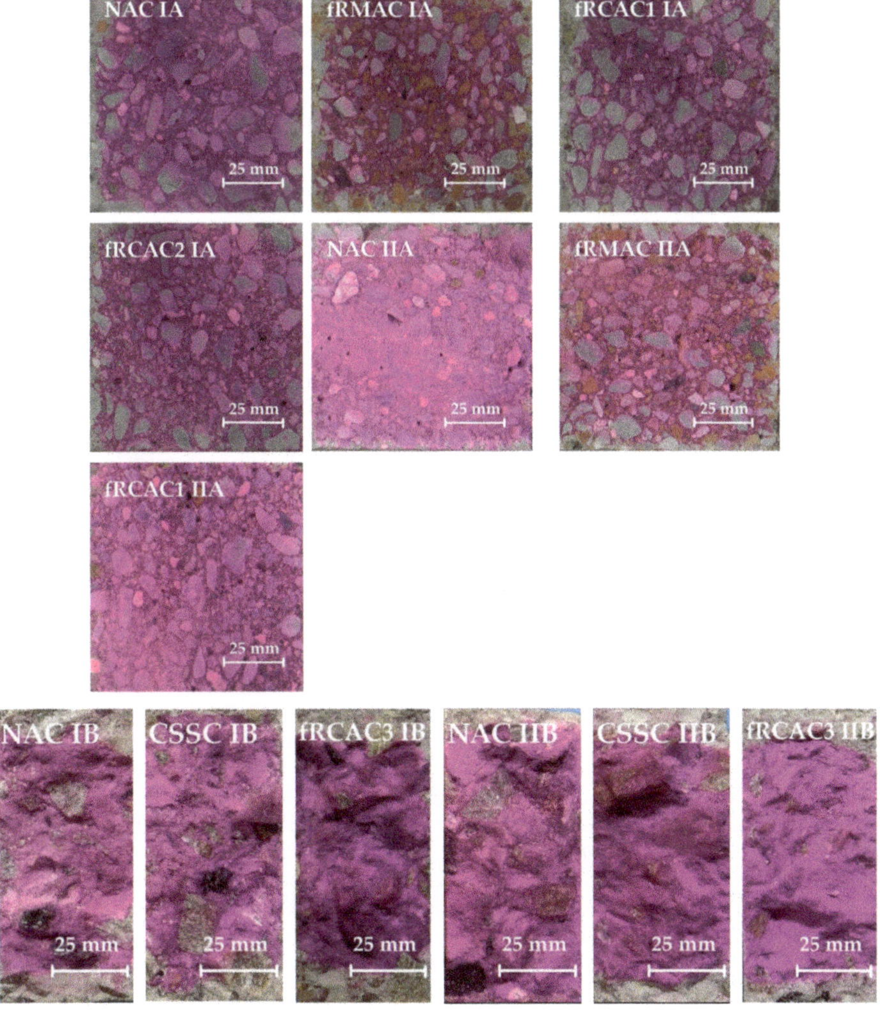

Figure 12. Carbonation depth of NAC and RAC.

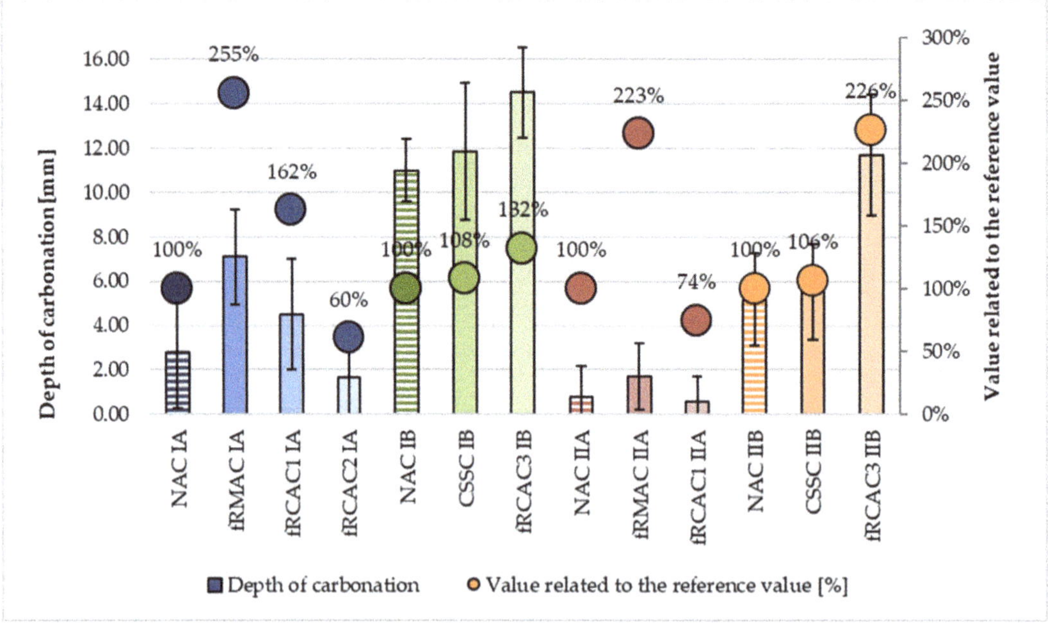

Figure 13. Comparison of carbonation depths of concrete containing fNA, fRCA, and fRMA with control mixtures.

6. Conclusions

In this study, the possibility of full replacement of natural sand in concrete by fine recycled aggregate was experimentally verified and discussed. This study was designed and implemented in the Czech Republic and India with the same research approach. The properties of fRA were examined by the validation of the physical and mechanical properties and durability of concrete containing fRA with a comparison between two different regions. The following conclusions are drawn.

- Differences were found in the properties of reclaimed sand originating from construction and demolition waste in two regions, which are probably caused by the parent concrete and differences in the recycling process. Moreover, the properties of concrete containing fRA slightly vary between the compared two regions, however, in both cases, the decline of properties in comparison with ordinary concrete is not limiting for finding satisfactory utilization.
- The density of fRA and, consequently, of fRAC decreases slightly compared with the natural sand and control mix, respectively. The water absorption of fRA and, consequently, of fRAC increases significantly compared to the natural sand control and control mix, respectively. On the contrary, the capillary water absorption decreases.
- The compressive strength mostly shows a slight decline; however, considering that the natural sand was fully replaced. For this reason, for future use of this material, the decline of compressive strength may be considered before a mixture design. The effect of the decline of the flexural strength is similar to compressive strength.
- In the case of modulus of elasticity, the highest decline in properties was found, which corresponds to previous studies, showing that the elastic modulus is the most decreased mechanical property of concrete with substitution by recycled aggregate in general. The static modulus of elasticity is the key characteristic of the material for the behavior of reinforced concrete structural elements because of its lower deflection of beams and slabs. For this reason, it is recommended to verify the strength and other

technical requirements prior to using fRA as a full replacement of natural sand in reinforced concrete structures.

- In contrast, durability properties were not worsened significantly with fRA. The freeze–thaw resistance was completely satisfactory, and, furthermore, the carbonation resistance, although slightly affected, was not essential in terms of structural use since the concrete cover is usually 2 mm or higher. However, due to the significant decline of the modulus of elasticity, the use of fRAC as a complete replacement of natural sand for reinforcement concrete structures may need attention.

The novelty of this study was the comparison of the properties of fRA and fRAC in different regions in accord with SDG 17, which encourages partnerships to achieve the goal. The main objective of this study was to verify the transferability of practices and experiences of the substitution of sand in concrete by comparison of the results in two different regions with the same research approach. As mentioned in previous studies, the recycling technology and properties of parent concrete influence the properties of fRA, therefore, the material properties of fRA and fRAC were examined and compared in this investigation. Although minor differences in material properties were found between the two regions, the differences were not essential for future utilization, which will be more suitable for the complete replacement of sand in plain concrete structural elements, such as foundation structures, cement, concrete screed, etc., due to the local standards, the availability of the material, and the results of this investigation. Overall, this work represents the efforts in the direction of attaining a circular economy and thereby addressing the SDG 12.

Author Contributions: Conceptualization, T.P. and B.T.; methodology, T.P.; validation, D.M., N.V.K. and K.F.; formal analysis, T.P. and B.T.; investigation, T.P., K.F. and N.V.K.; resources, T.P., N.V.K. and B.T.; data curation, K.F.; writing—original draft preparation, T.P., K.F., N.V.K. and B.T.; writing—review and editing, T.P.; visualization, K.F. and T.P.; supervision, P.H., B.T. and T.C.; project administration, P.H., B.T. and T.C.; funding acquisition, P.H. and B.T. All authors have read and agreed to the published version of the manuscript.

Funding: This work is a part of project entitled "Design of performance-based concrete using sand reclaimed from construction and demolition wastes", LTAIN19205, supported by the Ministry of Education, Youth and Sports of the Czech Republic through the program INTER-EXCELLENCE and "Design of Performance Based Concrete Using Sand Reclaimed from Construction and Demolition Waste" sanctioned by The Department of Science and Technology, New Delhi under India–Czech Republic international cooperation (Sanction number–DST/INT/Czech/P-13/2019) and the authors gratefully acknowledge the financial support given to accomplish this study.

Institutional Review Board Statement: Not applicable.

Informed Consent Statement: Not applicable.

Data Availability Statement: Not applicable.

Conflicts of Interest: The authors declare no conflict of interest.

References

1. Nedeljković, M.; Visser, J.; Šavija, B.; Valcke, S.; Schlangen, E. Use of Fine Recycled Concrete Aggregates in Concrete: A Critical Review. *J. Build. Eng.* **2021**, *38*, 102196. [CrossRef]
2. Welcome to Environment. Available online: http://environmentclearance.nic.in/ (accessed on 22 June 2022).
3. Hossain, M.U.; Poon, C.S.; Lo, I.M.C.; Cheng, J.C.P. Comparative Environmental Evaluation of Aggregate Production from Recycled Waste Materials and Virgin Sources by LCA. *Resour. Conserv. Recycl.* **2016**, *109*, 67–77. [CrossRef]
4. THE 17 GOALS | Sustainable Development. Available online: https://sdgs.un.org/goals (accessed on 26 June 2022).
5. Tošić, N.; Marinković, S.; Dašić, T.; Stanić, M. Multicriteria Optimization of Natural and Recycled Aggregate Concrete for Structural Use. *J. Clean. Prod.* **2015**, *87*, 766–776. [CrossRef]
6. Santos, S.; da Silva, P.R.; de Brito, J. Self-Compacting Concrete with Recycled Aggregates—A Literature Review. *J. Build. Eng.* **2019**, *22*, 349–371. [CrossRef]

7. De Brito, J.; Agrela, F.; Silva, R.V. 1—Construction and Demolition Waste. In *New Trends in Eco-Efficient and Recycled Concrete*; de Brito, J., Agrela, F., Eds.; Woodhead Publishing Series in Civil and Structural Engineering; Woodhead Publishing: Cambridge, UK, 2019; pp. 1–22, ISBN 978-0-08-102480-5.
8. Chen, W.; Jin, R.; Xu, Y.; Wanatowski, D.; Li, B.; Yan, L.; Pan, Z.; Yang, Y. Adopting Recycled Aggregates as Sustainable Construction Materials: A Review of the Scientific Literature. *Constr. Build. Mater.* **2019**, *218*, 483–496. [CrossRef]
9. Pedro, D.; de Brito, J.; Evangelista, L. Influence of the Use of Recycled Concrete Aggregates from Different Sources on Structural Concrete. *Constr. Build. Mater.* **2014**, *71*, 141–151. [CrossRef]
10. Ossa, A.; García, J.L.; Botero, E. Use of Recycled Construction and Demolition Waste (CDW) Aggregates: A Sustainable Alternative for the Pavement Construction Industry. *J. Clean. Prod.* **2016**, *135*, 379–386. [CrossRef]
11. Rao, A.; Jha, K.N.; Misra, S. Use of Aggregates from Recycled Construction and Demolition Waste in Concrete. *Resour. Conserv. Recycl.* **2007**, *50*, 71–81. [CrossRef]
12. Behera, M.; Bhattacharyya, S.K.; Minocha, A.K.; Deoliya, R.; Maiti, S. Recycled Aggregate from C&D Waste & Its Use in Concrete—A Breakthrough towards Sustainability in Construction Sector: A Review. *Constr. Build. Mater.* **2014**, *68*, 501–516. [CrossRef]
13. Thomas, C.; de Brito, J.; Gil, V.; Sainz-Aja, J.A.; Cimentada, A. Multiple Recycled Aggregate Properties Analysed by X-ray Microtomography. *Constr. Build. Mater.* **2018**, *166*, 171–180. [CrossRef]
14. Sosa, M.E.; Villagrán Zaccardi, Y.A.; Zega, C.J. A Critical Review of the Resulting Effective Water-to-Cement Ratio of Fine Recycled Aggregate Concrete. *Constr. Build. Mater.* **2021**, *313*, 125536. [CrossRef]
15. Silva, R.V.; de Brito, J.; Dhir, R.K. Performance of Cementitious Renderings and Masonry Mortars Containing Recycled Aggregates from Construction and Demolition Wastes. *Constr. Build. Mater.* **2016**, *105*, 400–415. [CrossRef]
16. Nováková, I.; Buyle, B.-A. Sand Replacement by Fine Recycled Concrete Aggregates as an Approach for Sustainable Cementitious Materials. In Proceedings of the International Conference of Sustainable Production and Use of Cement and Concrete, Villa Clara, Cuba, 23–30 June 2019; Martirena-Hernandez, J.F., Alujas-Díaz, A., Amador-Hernandez, M., Eds.; Springer: Cham, Switzerland, 2020; pp. 425–431.
17. Oliveira, R.; de Brito, J.; Veiga, R. Incorporation of Fine Glass Aggregates in Renderings. *Constr. Build. Mater.* **2013**, *44*, 329–341. [CrossRef]
18. Neno, C.; de Brito, J.; Veiga, R. Using Fine Recycled Concrete Aggregate for Mortar Production. *Mater. Res.* **2014**, *17*, 168–177. [CrossRef]
19. Kou, S.-C.; Poon, C.-S. Effects of Different Kinds of Recycled Fine Aggregate on Properties of Rendering Mortar. *J. Sustain. Cem.-Based Mater.* **2013**, *2*, 43–57. [CrossRef]
20. Ledesma, E.F.; Jiménez, J.R.; Ayuso, J.; Fernández, J.M.; de Brito, J. Maximum Feasible Use of Recycled Sand from Construction and Demolition Waste for Eco-Mortar Production—Part-I: Ceramic Masonry Waste. *J. Clean. Prod.* **2015**, *87*, 692–706. [CrossRef]
21. Vegas, I.; Azkarate, I.; Juarrero, A.; Frías, M. Design and Performance of Masonry Mortars Made with Recycled Concrete Aggregates. *Mater. Constr.* **2009**, *59*, 5–18. [CrossRef]
22. Dapena, E.; Alaejos, P.; Lobet, A.; Pérez, D. Effect of Recycled Sand Content on Characteristics of Mortars and Concretes. *J. Mater. Civ. Eng.* **2011**, *23*, 414–422. [CrossRef]
23. Tam, V.W.Y.; Soomro, M.; Evangelista, A.C.J. Quality Improvement of Recycled Concrete Aggregate by Removal of Residual Mortar: A Comprehensive Review of Approaches Adopted. *Constr. Build. Mater.* **2021**, *288*, 123066. [CrossRef]
24. Debieb, F.; Kenai, S. The Use of Coarse and Fine Crushed Bricks as Aggregate in Concrete. *Constr. Build. Mater.* **2008**, *22*, 886–893. [CrossRef]
25. Ge, Z.; Feng, Y.; Zhang, H.; Xiao, J.; Sun, R.; Liu, X. Use of Recycled Fine Clay Brick Aggregate as Internal Curing Agent for Low Water to Cement Ratio Mortar. *Constr. Build. Mater.* **2020**, *264*, 120280. [CrossRef]
26. Dang, J.; Zhao, J.; Pang, S.D.; Zhao, S. Durability and Microstructural Properties of Concrete with Recycled Brick as Fine Aggregates. *Constr. Build. Mater.* **2020**, *262*, 120032. [CrossRef]
27. Vieira, T.; Alves, A.; de Brito, J.; Correia, J.R.; Silva, R.V. Durability-Related Performance of Concrete Containing Fine Recycled Aggregates from Crushed Bricks and Sanitary Ware. *Mater. Des.* **2016**, *90*, 767–776. [CrossRef]
28. Arredondo-Rea, S.P.; Corral-Higuera, R.; Gómez-Soberón, J.M.; Gámez-García, D.C.; Bernal-Camacho, J.M.; Rosas-Casarez, C.A.; Ungsson-Nieblas, M.J. Durability Parameters of Reinforced Recycled Aggregate Concrete: Case Study. *Appl. Sci.* **2019**, *9*, 617. [CrossRef]
29. Poon, C.S.; Shui, Z.H.; Lam, L. Effect of Microstructure of ITZ on Compressive Strength of Concrete Prepared with Recycled Aggregates. *Constr. Build. Mater.* **2004**, *18*, 461–468. [CrossRef]
30. Sáez del Bosque, I.F.; Zhu, W.; Howind, T.; Matías, A.; Sánchez de Rojas, M.I.; Medina, C. Properties of Interfacial Transition Zones (ITZs) in Concrete Containing Recycled Mixed Aggregate. *Cem. Concr. Compos.* **2017**, *81*, 25–34. [CrossRef]
31. Evangelista, L.; de Brito, J. Mechanical Behaviour of Concrete Made with Fine Recycled Concrete Aggregates. *Cem. Concr. Compos.* **2007**, *29*, 397–401. [CrossRef]
32. Kou, S.C.; Poon, C.S. Properties of Self-Compacting Concrete Prepared with Recycled Glass Aggregate. *Cem. Concr. Compos.* **2009**, *31*, 107–113. [CrossRef]
33. Zega, C.J.; Di Maio, Á.A. Use of Recycled Fine Aggregate in Concretes with Durable Requirements. *Waste Manag.* **2011**, *31*, 2336–2340. [CrossRef]

34. Pereira, P.; Evangelista, L.; de Brito, J. The Effect of Superplasticisers on the Workability and Compressive Strength of Concrete Made with Fine Recycled Concrete Aggregates. *Constr. Build. Mater.* **2012**, *28*, 722–729. [CrossRef]
35. Fan, C.-C.; Huang, R.; Hwang, H.; Chao, S.-J. Properties of Concrete Incorporating Fine Recycled Aggregates from Crushed Concrete Wastes. *Constr. Build. Mater.* **2016**, *112*, 708–715. [CrossRef]
36. Khatib, J.M. Properties of Concrete Incorporating Fine Recycled Aggregate. *Cem. Concr. Res.* **2005**, *35*, 763–769. [CrossRef]
37. Evangelista, L.; de Brito, J. Durability Performance of Concrete Made with Fine Recycled Concrete Aggregates. *Cem. Concr. Compos.* **2010**, *32*, 9–14. [CrossRef]
38. Lotfy, A.; Al-Fayez, M. Performance Evaluation of Structural Concrete Using Controlled Quality Coarse and Fine Recycled Concrete Aggregate. *Cem. Concr. Compos.* **2015**, *61*, 36–43. [CrossRef]
39. Carro-López, D.; González-Fonteboa, B.; de Brito, J.; Martínez-Abella, F.; González-Taboada, I.; Silva, P. Study of the Rheology of Self-Compacting Concrete with Fine Recycled Concrete Aggregates. *Constr. Build. Mater.* **2015**, *96*, 491–501. [CrossRef]
40. Zhang, H.; Ji, T.; Zeng, X.; Yang, Z.; Lin, X.; Liang, Y. Mechanical Behavior of Ultra-High Performance Concrete (UHPC) Using Recycled Fine Aggregate Cured under Different Conditions and the Mechanism Based on Integrated Microstructural Parameters. *Constr. Build. Mater.* **2018**, *192*, 489–507. [CrossRef]
41. Geng, J.; Sun, J. Characteristics of the Carbonation Resistance of Recycled Fine Aggregate Concrete. *Constr. Build. Mater.* **2013**, *49*, 814–820. [CrossRef]
42. De Juan, M.S.; Gutiérrez, P.A. Study on the Influence of Attached Mortar Content on the Properties of Recycled Concrete Aggregate. *Constr. Build. Mater.* **2009**, *23*, 872–877. [CrossRef]
43. Akbarnezhad, A.; Ong, K.C.G. 10—Separation Processes to Improve the Quality of Recycled Concrete Aggregates (RCA). In *Handbook of Recycled Concrete and Demolition Waste*; Pacheco-Torgal, F., Tam, V.W.Y., Labrincha, J.A., Ding, Y., de Brito, J., Eds.; Woodhead Publishing Series in Civil and Structural Engineering; Woodhead Publishing: Cambridge, UK, 2013; pp. 246–269, ISBN 978-0-85709-682-1.
44. Florea, M.V.A.; Brouwers, H.J.H. Properties of Various Size Fractions of Crushed Concrete Related to Process Conditions and Re-Use. *Cem. Concr. Res.* **2013**, *52*, 11–21. [CrossRef]
45. Evangelista, L.; Guedes, M.; de Brito, J.; Ferro, A.C.; Pereira, M.F. Physical, Chemical and Mineralogical Properties of Fine Recycled Aggregates Made from Concrete Waste. *Constr. Build. Mater.* **2015**, *86*, 178–188. [CrossRef]
46. Evangelista, L.; de Brito, J. Concrete with Fine Recycled Aggregates: A Review. *Eur. J. Environ. Civ. Eng.* **2014**, *18*, 129–172. [CrossRef]
47. Li, Z.; Liu, J.; Tian, Q. Method for Controlling the Absorbed Water Content of Recycled Fine Aggregates by Centrifugation. *Constr. Build. Mater.* **2018**, *160*, 316–325. [CrossRef]
48. Pavlů, T.; Kočí, V.; Hájek, P. Environmental Assessment of Two Use Cycles of Recycled Aggregate Concrete. *Sustainability* **2019**, *11*, 6185. [CrossRef]
49. Singh, R.; Nayak, D.; Pandey, A.; Kumar, R.; Kumar, V. Effects of Recycled Fine Aggregates on Properties of Concrete Containing Natural or Recycled Coarse Aggregates: A Comparative Study. *J. Build. Eng.* **2022**, *45*, 103442. [CrossRef]
50. Kirthika, S.K.; Singh, S.K. Durability Studies on Recycled Fine Aggregate Concrete. *Constr. Build. Mater.* **2020**, *250*, 118850. [CrossRef]
51. Li, T.; Xiao, J.; Zhang, Y.; Chen, B. Fracture Behavior of Recycled Aggregate Concrete under Three-Point Bending. *Cem. Concr. Compos.* **2019**, *104*, 103353. [CrossRef]
52. De Andrade Salgado, F.; de Andrade Silva, F. Recycled Aggregates from Construction and Demolition Waste towards an Application on Structural Concrete: A Review. *J. Build. Eng.* **2022**, *52*, 104452. [CrossRef]
53. Cartuxo, F.; De Brito, J.; Evangelista, L.; Jiménez, J.R.; Ledesma, E.F. Increased Durability of Concrete Made with Fine Recycled Concrete Aggregates Using Superplasticizers. *Materials* **2016**, *9*, 98. [CrossRef]
54. Bravo, M.; de Brito, J.; Evangelista, L.; Pacheco, J. Superplasticizer's Efficiency on the Mechanical Properties of Recycled Aggregates Concrete: Influence of Recycled Aggregates Composition and Incorporation Ratio. *Constr. Build. Mater.* **2017**, *153*, 129–138. [CrossRef]
55. Lovato, P.S.; Possan, E.; Molin, D.C.C.D.; Masuero, Â.B.; Ribeiro, J.L.D. Modeling of Mechanical Properties and Durability of Recycled Aggregate Concretes. *Constr. Build. Mater.* **2012**, *26*, 437–447. [CrossRef]
56. Fan, C.-C.; Huang, R.; Hwang, H.; Chao, S.-J. The Effects of Different Fine Recycled Concrete Aggregates on the Properties of Mortar. *Materials* **2015**, *8*, 2658–2672. [CrossRef]
57. Český Statistický Úřad. Available online: https://www.czso.cz/csu/czso/domov (accessed on 11 June 2022).
58. Sarika, N. *Building Material and Technology Promotion Council*; Newsletter: New Delhi, India, 2018; Volume 7.
59. CPCB Central Pollution Control Board. *Guidelines on Environmental Management of Construction & Demolition (C&D) Wastes*; CPCB Central Pollution Control Board: New Delhi, India, 2017.
60. *BIS: 383-2016*; Specification for Coarse and Fine Aggregates from Natural Sources for Concrete. Bureau of Indian Standards: New Delhi, India, 2016.
61. Alves, A.V.; Vieira, T.F.; de Brito, J.; Correia, J.R. Mechanical Properties of Structural Concrete with Fine Recycled Ceramic Aggregates. *Constr. Build. Mater.* **2014**, *64*, 103–113. [CrossRef]
62. Uddin, M.T.; Mahmood, A.H.; Kamal, M.R.I.; Yashin, S.M.; Zihan, Z.U.A. Effects of Maximum Size of Brick Aggregate on Properties of Concrete. *Constr. Build. Mater.* **2017**, *134*, 713–726. [CrossRef]

63. Leite, M.B.; Figueire do Filho, J.G.L.; Lima, P.R.L. Workability Study of Concretes Made with Recycled Mortar Aggregate. *Mater. Struct.* **2013**, *46*, 1765–1778. [CrossRef]
64. Sri Ravindrarajah, R.; Tam, C.T. Recycling Concrete as Fine Aggregate in Concrete. *Int. J. Cem. Compos. Lightweight Concr.* **1987**, *9*, 235–241. [CrossRef]
65. Evangelista, L.; Brito, J. *Criteria for the Use of Fine Recycled Concrete Aggregates in Concrete Production*; RILEM: Barcelona, Spain, 2004.
66. Kou, S.-C.; Poon, C.-S. Properties of Concrete Prepared with Crushed Fine Stone, Furnace Bottom Ash and Fine Recycled Aggregate as Fine Aggregates. *Constr. Build. Mater.* **2009**, *23*, 2877–2886. [CrossRef]
67. Delobel, F.; Bulteel, D.; Mechling, J.M.; Lecomte, A.; Cyr, M.; Rémond, S. Application of ASR Tests to Recycled Concrete Aggregates: Influence of Water Absorption. *Constr. Build. Mater.* **2016**, *124*, 714–721. [CrossRef]
68. *CSN EN 12620+A1*; Aggregate for Concrete. The Czech Standardization Agency: Prague, Czech Republic, 2008. (In Czech)
69. De Brito, J.; Saikia, N. *Recycled Aggregate in Concrete*; Green Energy and Technology; Springer: London, UK, 2013; ISBN 978-1-4471-4539-4.
70. Dhir, R.K.; de Brito, J.; Silva, R.V.; Lye, C.Q. 10—Recycled Aggregate Concrete: Durability Properties. In *Sustainable Construction Materials*; Dhir, R.K., de Brito, J., Silva, R.V., Lye, C.Q., Eds.; Woodhead Publishing Series in Civil and Structural Engineering; Woodhead Publishing: Cambridge, UK, 2019; pp. 365–418, ISBN 978-0-08-100985-7.
71. Levy, S.M.; Helene, P. Durability of Recycled Aggregates Concrete: A Safe Way to Sustainable Development. *Cem. Concr. Res.* **2004**, *34*, 1975–1980. [CrossRef]
72. Yaprak, H.; Aruntas, H.Y.; Demir, I.; Simsek, O. Effects of the Fine Recycled Concrete Aggregates on the Concrete Properties. *Int. J. Phys. Sci.* **2011**, *6*, 2455–2461.
73. Kou, S.C.; Poon, C.S. Properties of Self-Compacting Concrete Prepared with Coarse and Fine Recycled Concrete Aggregates. *Cem. Concr. Compos.* **2009**, *31*, 622–627. [CrossRef]
74. Kim, S.-W.; Yun, H.-D. Evaluation of the Bond Behavior of Steel Reinforcing Bars in Recycled Fine Aggregate Concrete. *Cem. Concr. Compos.* **2014**, *46*, 8–18. [CrossRef]
75. Kumar, R.; Gurram, S.C.B.; Minocha, A.K. Influence of Recycled Fine Aggregate on Microstructure and Hardened Properties of Concrete. *Mag. Concr. Res.* **2017**, *69*, 1288–1295. [CrossRef]
76. Bogas, J.A.; de Brito, J.; Ramos, D. Freeze–Thaw Resistance of Concrete Produced with Fine Recycled Concrete Aggregates. *J. Clean. Prod.* **2016**, *115*, 294–306. [CrossRef]
77. Sarhat, S.R. An Experimental Investigation on the Viability of Using Fine Concrete Recyled Aggregate in Concrete Production. In *Sustainable Construction Materials and Technologies*; CRC Press: Boca Raton, FL, USA, 2007; ISBN 978-1-00-306102-1.
78. Yildirim, S.T.; Meyer, C.; Herfellner, S. Effects of Internal Curing on the Strength, Drying Shrinkage and Freeze–Thaw Resistance of Concrete Containing Recycled Concrete Aggregates. *Constr. Build. Mater.* **2015**, *91*, 288–296. [CrossRef]
79. Bendimerad, A.Z.; Rozière, E.; Loukili, A. Plastic Shrinkage and Cracking Risk of Recycled Aggregates Concrete. *Constr. Build. Mater.* **2016**, *121*, 733–745. [CrossRef]
80. Wang, Y.; Zhang, H.; Geng, Y.; Wang, Q.; Zhang, S. Prediction of the Elastic Modulus and the Splitting Tensile Strength of Concrete Incorporating Both Fine and Coarse Recycled Aggregate. *Constr. Build. Mater.* **2019**, *215*, 332–346. [CrossRef]
81. Velay-Lizancos, M.; Martinez-Lage, I.; Azenha, M.; Granja, J.; Vazquez-Burgo, P. Concrete with Fine and Coarse Recycled Aggregates: E-Modulus Evolution, Compressive Strength and Non-Destructive Testing at Early Ages. *Constr. Build. Mater.* **2018**, *193*, 323–331. [CrossRef]
82. Omary, S.; Ghorbel, E.; Wardeh, G. Relationships between Recycled Concrete Aggregates Characteristics and Recycled Aggregates Concretes Properties. *Constr. Build. Mater.* **2016**, *108*, 163–174. [CrossRef]
83. Kim, J.; Grabiec, A.M.; Ubysz, A. An Experimental Study on Structural Concrete Containing Recycled Aggregates and Powder from Construction and Demolition Waste. *Materials* **2022**, *15*, 2458. [CrossRef]
84. Mardani-Aghabaglou, A.; Tuyan, M.; Ramyar, K. Mechanical and Durability Performance of Concrete Incorporating Fine Recycled Concrete and Glass Aggregates. *Mater. Struct.* **2015**, *48*, 2629–2640. [CrossRef]
85. Evangelista, L.; de Brito, J. Durability of Crushed Fine Recycled Aggregate Concrete Assessed by Permeability-Related Properties. *Mag. Concr. Res.* **2019**, *71*, 1142–1150. [CrossRef]
86. Ho, H.-L.; Huang, R.; Lin, W.-T.; Cheng, A. Pore-Structures and Durability of Concrete Containing Pre-Coated Fine Recycled Mixed Aggregates Using Pozzolan and Polyvinyl Alcohol Materials. *Constr. Build. Mater.* **2018**, *160*, 278–292. [CrossRef]
87. Pedro, D.; de Brito, J.; Evangelista, L. Structural Concrete with Simultaneous Incorporation of Fine and Coarse Recycled Concrete Aggregates: Mechanical, Durability and Long-Term Properties. *Constr. Build. Mater.* **2017**, *154*, 294–309. [CrossRef]
88. Fumoto, T.; Yamada, M. Durability of Concrete with Recycled Fine Aggregate. In Proceedings of the Seventh CANMET/ACI International Conference on Durability of Concrete 2006, Montreal, Canada, 28 May–3 June 2006; SP-234, pp. 457–472.
89. Zaharieva, R.; Buyle-Bodin, F.; Wirquin, E. Frost Resistance of Recycled Aggregate Concrete. *Cem. Concr. Res.* **2004**, *34*, 1927–1932. [CrossRef]
90. Marinković, S. On the Selection of the Functional Unit in LCA of Structural Concrete. *Int. J. Life Cycle Assess.* **2017**, *22*, 1634–1636. [CrossRef]

91. Pavlů, T.; Fořtová, K.; Řepka, J.; Mariaková, D.; Pazderka, J. Improvement of the Durability of Recycled Masonry Aggregate Concrete. *Materials* **2020**, *13*, 5486. [CrossRef] [PubMed]
92. Pavlu, T.; Pazderka, J.; Fořtová, K.; Řepka, J.; Mariaková, D.; Vlach, T. The Structural Use of Recycled Aggregate Concrete for Renovation of Massive External Walls of Czech Fortification. *Buildings* **2022**, *12*, 671. [CrossRef]

 MDPI

Article

Hydration and Mechanical Properties of High-Volume Fly Ash Concrete with Nano-Silica and Silica Fume

Byung-Jun Kim, Geon-Wook Lee and Young-Cheol Choi *

Department of Civil and Environmental Engineering, Gachon University, Seongnam 13120, Gyeonggi-do, Korea
* Correspondence: zerofe@gachon.ac.kr; Tel.: +82-31-750-5721; Fax: +82-31-754-2772

Abstract: This study investigated the effects of nano-silica (NS) and silica fume (SF) on the hydration reaction of high-volume fly ash cement (HVFC) composites. In order to solve the dispersibility problem caused by the agglomeration of NS powder, NS and NSF solutions were prepared. NS content and SF content were used as main variables, and an HVFC paste was prepared in which 50% of the cement volume was replaced by fly ash (FA). The initial heat of hydration was measured using isothermal calorimetry to analyze the effects of NS and SF on the initial hydration properties of the HVFC. In addition, the compressive strength was analyzed by age. The refinement of the pore structure by the nanomaterial was analyzed using mercury intrusion porosimetry (MIP). The results show that the addition of NS and SF shortened the setting time and induction period by accelerating the initial hydration reaction of HVFC composites and improved the compressive strength during the initial stage of hydration. In addition, the micropore structure was improved by the pozzolanic reaction of NS and SF, thereby increasing the compressive strength during the middle stage of hydration.

Keywords: cement composite; filler effect; microstructure; nano-silica; silica fume

Citation: Kim, B.-J.; Lee, G.-W.; Choi, Y.-C. Hydration and Mechanical Properties of High-Volume Fly Ash Concrete with Nano-Silica and Silica Fume. *Materials* **2022**, *15*, 6599. https://doi.org/10.3390/ma15196599

Academic Editor: Miguel Ángel Sanjuán

Received: 27 August 2022
Accepted: 20 September 2022
Published: 23 September 2022

1. Introduction

With ongoing industrial development and coal consumption, the importance of research on carbon neutrality in the construction industry has increased. Accordingly, studies have been conducted on a variety of construction materials, and the use of supplementary cementing materials (SCMs), such as fly ash (FA), ground granulated blast-furnace slag (GGBS), and silica fume (SF), has attracted attention [1–3]. SCMs are known to contribute to the engineering properties of cement composites through the reaction of such pozzolans with cement [4–7].

Of these SCMs, FA has been widely used as a cement admixture because of its low heat of hydration, shrinkage-reducing effect, and high durability. Roshani et al. [8] predicted the effect of fly ash on the mechanical properties of concrete using artificial neural networks (ANN). Studies have been actively researching high-volume fly ash cement (HVFC) containing large amounts of FA. HVFC reduces greenhouse gas emissions by decreasing cement consumption and has major benefits in terms of the recycling of industrial byproducts. However, it has low initial compressive strength because FA remains inactive during the initial hydration process [4,9–12]. To address this problem, many processes have been studied, including the grinding of FA [13], chemical activation [14], mechanochemical treatment [15], and hydrothermal treatment [16,17]. Yang et al. [18] reported that adding SF to Na_2SO_4-activated high-volume fly ash (HVFA) concrete leads to higher resistance and less strength loss. Bondar et al. [19] reported that the initial strength of HVFA paste was improved by the use of cement kiln dust (CKD) and gypsum as activators.

In recent years, there has been growing interest in research to improve the performance of cementitious materials, including the use of nanomaterials to improve the initial strength of HVFA concrete. In general, nanomaterials accelerate the hydration of cement and improve its microstructure, thereby enhancing its mechanical performance and durability [20–22]. The last ten years has seen rapid development of nanomaterials in various fields. For example, nanomaterials that can transform concrete have appeared, such as nano-SiO_2 (NS) and nano-TiO_2 [23–26]. Onaizi et al. [27] used nano-sized waste glass powder to improve the early strength of HVFA concrete. They reported that the increase in compressive strength decreased when nano glass powder was added more than 10%.

Nano-silica (NS) is a material with very small particle sizes and the characteristics of a pozzolanic material. The dissolved silica component reacts with the calcium hydroxide generated from the cement hydration reaction to form additional calcium silicate hydrate (C-S-H) and contribute to strength development [28–57]. Therefore, many studies on concrete have been conducted using NS. Karakouzian et al. [58] studied the effect of carbon nanotube and NS on the mechanical properties of concrete. They reported that NS can improve not only the early strength of concrete, but also the long-term strength by the pozzolan effect. Kong et al. [35] investigated the effect of NS on the microstructure and mechanical properties of cementitious materials. They found that resistance to calcium-leaching and mechanical performance were improved regardless of the type or amount of NS. Li [28] reported that NS can improve the low initial strength of HVFC by accelerating the cement hydration reaction. The application of NS to HVFC, however, may have a negative impact on long-term strength. Kawashima et al. [38] reported that an NS content of 5% or more decreased the long-term strength of HVFC as the pozzolanic reaction products of NS interfered with the reaction of FA particles.

As NS may form weak zones in the matrix because of its agglomeration characteristics, NS dispersibility is an important matter of concern [59]. An excessive amount of NS may also affect strength negatively by decreasing the fluidity of the concrete. Ghafari et al. [50] investigated the effect of dry NS powder on the fluidity, strength, and transport performance of ultra-high-performance concrete (UHPC). They found that, when the NS content was greater than 3% of the cement by weight, neither strength nor transport performance was improved, owing to the agglomeration of NS particles. Kooshafar and Madani [60] reported that the incorporation of NS into cement composites significantly reduced workability. This reduction was caused by strong agglomeration of NS particles within the cement composite. Wu et al. reported that a dry NS powder content of 1% in UHPC can improve the interfacial bonding between cement matrices, but a higher content may degrade performance instead [61]. As a result of the analysis of existing studies, there are still problems for application to construction products, such as dispersibility and optimal usage of nanomaterials.

The purpose of this study was to investigate the effects of NS and SF on the hydration reaction of HVFC. The main focus of this study is to solve the dispersibility problem caused by the agglomeration of NS powder in previous studies. To this end, NS was prepared in the form of a suspension so that the nano-silica particles were not physically agglomerated with each other. In addition, SF having a larger particle size than NS was mixed with NS to prepare an NSF suspension with a multiple particle size distribution. These two solutions were applied to an HVFC paste in which 50% of the cement volume was replaced with FA. The main variables were the NS and SF content. The hydration properties were evaluated through the setting time, heat of hydration, and compressive strength. In addition, the refinement of the pore structure by NS was analyzed using mercury intrusion porosimetry (MIP).

2. Materials and Methods

2.1. Materials

For this study, ordinary Portland cement (OPC), FA, SF, and NS were used as binders. The chemical compositions of the raw materials were analyzed using an X-ray fluorescence (XRF) spectrometer (ZSX Primus II Rigaku, Tokyo, Japan), with the results as shown in Table 1. The main components of the FA were SiO_2 (50.5%), Al_2O_3 (24.3%), CaO (9.93%), and Fe_2O_3 (5.96%), corresponding to class F according to the ASTM C618 specifications.

The main component of NS is SiO_2 (99.9%). The densities of OPC, FA, SF, and NS are 3.16, 2.32, 2.26, and 2.23 g/cm^3, respectively.

Table 1. Chemical composition of binders.

Binder	Chemical Composition (wt.%)									
	CaO	SiO_2	Al_2O_3	Fe_2O_3	MgO	TiO_2	K_2O	Na_2O	SO_3	LOI
OPC	63.2	20.2	4.1	3.6	2.7	0.2	1.0	0.1	1.4	0.8
FA	9.93	50.5	24.3	6.0	1.5	1.2	1.6	0.9	2.2	1.8
SF	0.2	91.6	0.4	0.9	1.2	–	1.0	0.8	0.4	0.6

Abbreviations: FA, fly ash; OPC, ordinary Portland cement; SF, silica fume.

Figure 1 shows the X-ray diffraction (XRD) patterns for FA, SF, and NS. From the results in Figure 1, it was found that the main minerals constituting FA are quartz and mullite and that SF and NS are mostly amorphous. Both SF and NS have an amorphous structure that causes a centered halo corresponding to the main line of cristobalite.

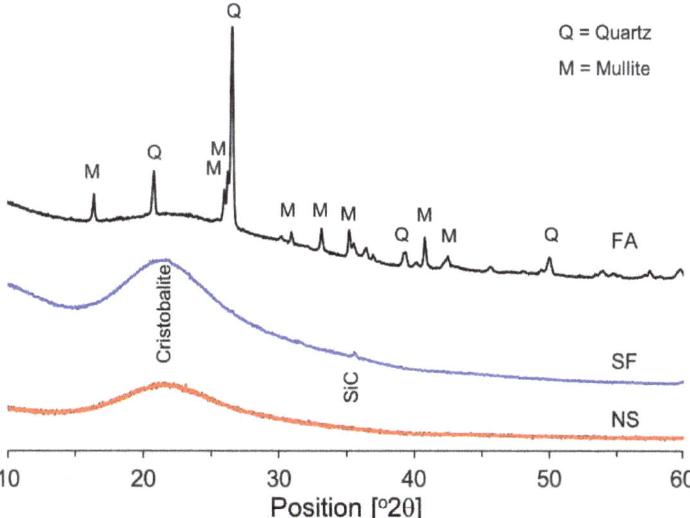

Figure 1. XRD patterns of FA, SF, and NS.

Figure 2 shows the particle size distributions for OPC, FA, and SF. The average particle diameters are 19.1, 26.7, and 0.148 μm, respectively. As can be seen in the figure, the particle size distributions for OPC and SF are unimodal, whereas that for FA is bimodal. In general, the particle size distribution of fly ash varies depending on the type of fuel coal, combustion temperature, and process.

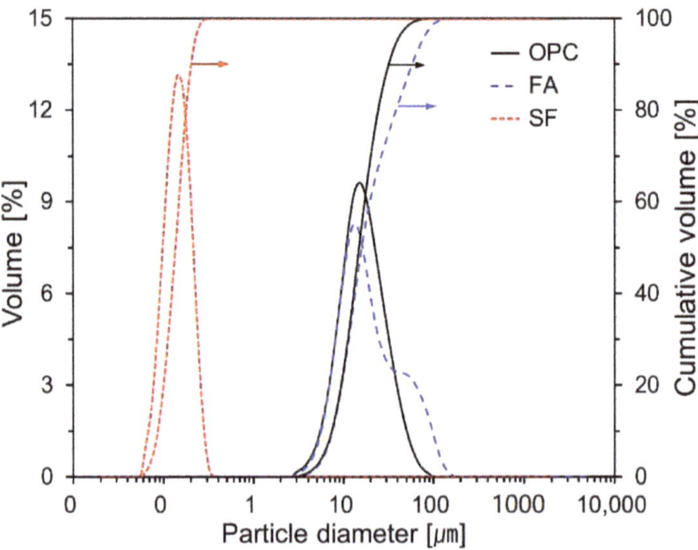

Figure 2. Particle size distributions for OPC, FA, and SF.

For the study, two types of solution were prepared and used to accelerate the hydration of an FA–cement mixture. The first was an NS solution with a solids content of 5%. The other, called the NSF solution, was prepared by mixing NS powder and SF at a weight ratio of 1:8; it had a solids content of 32%. Figure 3 shows SEM images of NS and NSF particles. The average diameter of NS particles measured from Figure 3a is 100 nm. In the case of NSF, as shown in Figure 3b, it was confirmed that the NS particles were evenly distributed around the SF particles.

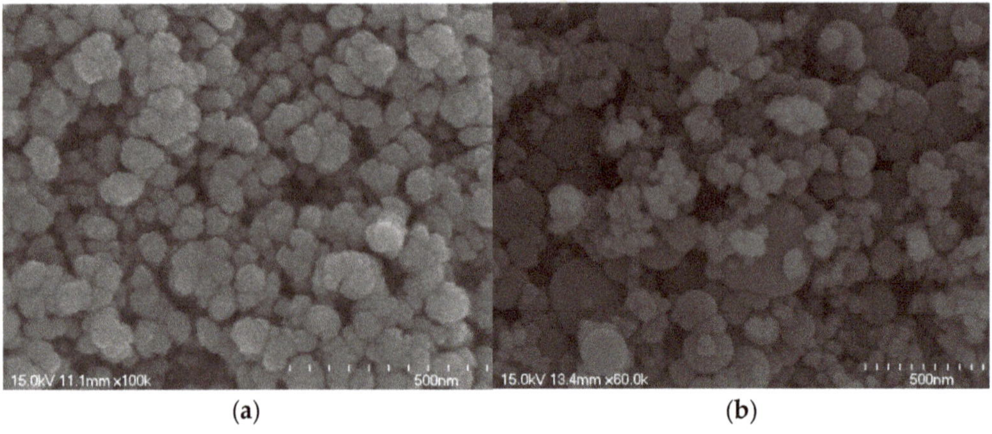

(**a**) (**b**)

Figure 3. SEM images of (**a**) NS solution (×100,000) and (**b**) mixed NS/SF (NSF) solution (×60,000).

2.2. Mix Proportions and Specimen Preparation

Cement pastes were prepared using the NS and NSF solutions, as shown in Table 2. The water-to-binder ratio was fixed at 0.3 for all specimens. For specimen Plain, 50% of the OPC volume was replaced by FA. For specimens NS05 and NS15, NS solution was added such that the NS solids content would be 0.5% and 1.5%, respectively, of the OPC volume. For specimen NS05SF4, NSF solution was added such that the NS and SF content would be 0.5% and 4%, respectively, of the OPC weight. The amounts of water added were determined based on the water content of the NS and NSF solutions.

Table 2. Mix proportions of pastes.

Specimen	W/B	Binder (g)				SP (wt.% by Binder)
		OPC	FA	NS	SF	
Plain	0.3	100	73.42	0	0	0.06
NS05		99	73.42	0.71	0	0.25
NS15		97	73.42	2.12	0	0.62
NS05SF4		91	73.42	0.71	5.72	0.30

Abbreviations: W/B, water-to-binder ratio

After OPC and FA were placed in a forced mixer and mixed for approximately 30 s, NS or NSF solution was added, and the result was mixed at a speed of approximately 100 rpm for approximately 60 s, and then at approximately 200 rpm for 5 min. The paste specimen was then poured into a mold of size 40 mm × 40 mm × 160 mm and cured in a constant temperature and humidity chamber at a temperature of 20 ± 1 °C and a relative humidity of 90% or higher for one day. It was then demolded and subjected to water curing at 20 ± 1 °C until the test measurement date.

2.3. Test Methods

Figure 4 show the schematic diagram of the test process. The compressive strength of each paste specimen was measured at 3, 7, and 28 days of age in accordance with ISO 679. The specimen was split into two pieces for the flexural strength measurement test, and the compressive strength of each piece was measured. For each variable, the compressive strengths of the six samples were measured at each age, and the average values were taken as the results. The setting time was measured using an Acmel PA8 automatic setting time tester (ACMEL LABO, Saint Pierre du Perray, France) in accordance with ISO 9597. The paste for the setting time test was prepared in a constant temperature and humidity chamber at a temperature of 23 °C and a humidity of 60%. Initial setting and final setting were measured using an Acmel PA8 automatic setting time tester.

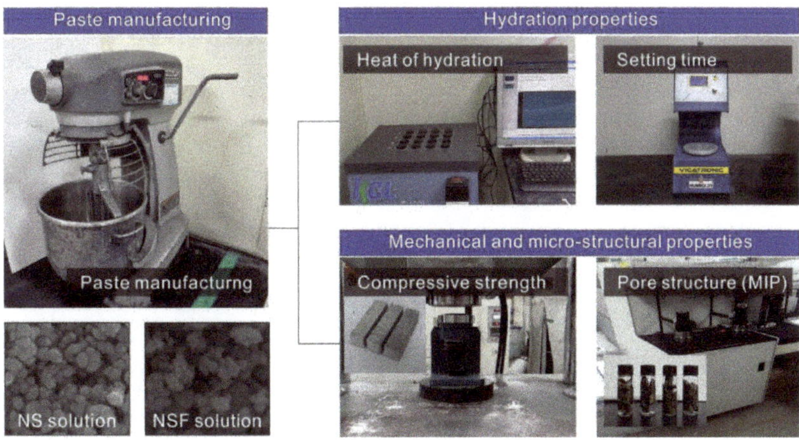

Figure 4. Schematic diagram of the test process.

The hydration heat flow for each variable was measured using a TAM Air isothermal calorimeter (TA instruments, New castle, US). After a paste was prepared for each variable according to the mix proportions in Table 2, approximately 4 g of the paste was inserted into a glass ampule and then placed in the isothermal calorimeter. The temperature of the calorimeter was set to 23 °C, and measurements were performed every minute for 48 h.

MIP analysis was conducted using a Micromeritics AutoPore IV 9500 (Micromeritics, Norcross, US) to measure the pore size distribution and cumulative porosity of each specimen. At each measurement age, a sample was collected from the center of the specimen, and hydration was stopped. The sample was then vacuum-dried and stored to prevent contamination by carbonation. The pressure of mercury intrusion varied from 0 to 30,000 psi.

3. Results and Discussion

3.1. Hydration Properties

Figure 5 shows the penetration depth of the Vicat needle over time. Based on these results, the initial setting times for Plain, NS05, NS15, and NS05SF4 were measured as 5.92, 4.45, 3.97, and 3.82 h, respectively. Thus, the initial setting times for NS05, NS15, and NS05SF4 were reduced by 24.8%, 32.9%, and 35.5%, respectively, from that for Plain. The measured final setting times for Plain, NS05, NS15, and NS05SF4 were 6.89, 5.09, 4.93, and 4.74 h, respectively, similar to the results for the initial setting time. These results are in agreement with the finding by Chithra et al. [62] that NS promotes the setting of cement paste and decreases the dormant period of hydration by accelerating the hydration reaction of cement. Similar results were also confirmed in the study results of Bhatta et al. [63]. They reported that the initial setting time of cement composites decreased with increasing NS content. This result is because NS particles with large specific surface area provided more sites for hydration reaction. The setting time reduction was most obvious in NS05SF4, which contained both NS and SF. This appears to be because NS and SF further improved the filling rate by filling the gaps between cement particles and further accelerated the setting reaction by making the cement microstructure denser [64].

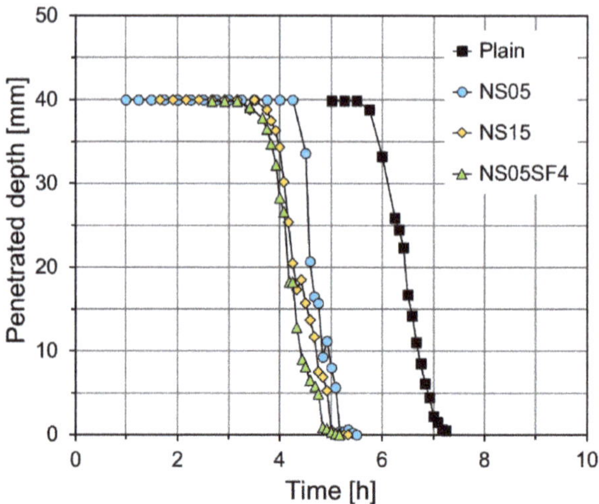

Figure 5. Results of Vicat needle test on the specimens.

Figure 6 shows the hydration heat flow and cumulative heat release for each paste specimen. As can be seen from Figure 6a, the second heat flow peak occurred at 15.85, 11.03, 9.97, and 8.25 h for Plain, NS05, NS15, and NS05SF4, respectively. As the NS content increased, the peak occurred earlier and the magnitude of the peak tended to increase. Similar to the results for setting time (Figure 5), the second heat flow peak occurred earliest for NS05SF4, approximately 7.6 h earlier than that for Plain.

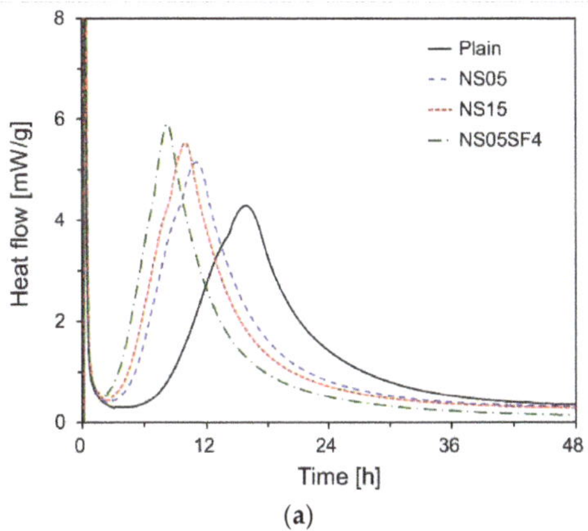

(a)

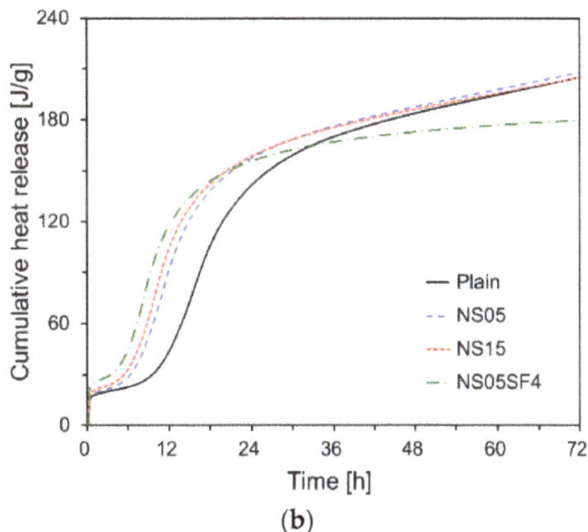

(b)

Figure 6. (**a**) Hydration heat flow and (**b**) cumulative heat release.

The cumulative heat release results (Figure 6b) show that the cumulative heat release for specimens containing NS was substantially greater than that for Plain until 32 h of hydration. This is in agreement with the finding by Xi et al. [65] that NS and SF particles accelerate hydration by acting as additional nucleation sites for cement hydrates during the initial stage of hydration. In particular, initial hydration was accelerated more strongly for NS05SF4 than for NS05 or NS15. For NS05SF4, however, the cumulative heat release became less than that for NS05 or NS15 after approximately 22 h of hydration. These results are consistent with those of Wang et al. [66]. They reported that when 0.5% of NS was added and 5% of SF was added, the cumulative heat release tended to decrease after 24 h.

As shown in Figure 7, NS with nano-sized particles fills the pores between the cement particles during the initial hydration process, thereby causing the structural filler effect inside the cement. This can improve the density of the matrix by effectively refining capillary pores and decreasing the porosity of the matrix [65]. Pacheco-Torgal et al. [67] reported that the addition of NS makes the internal structure of concrete mortar denser.

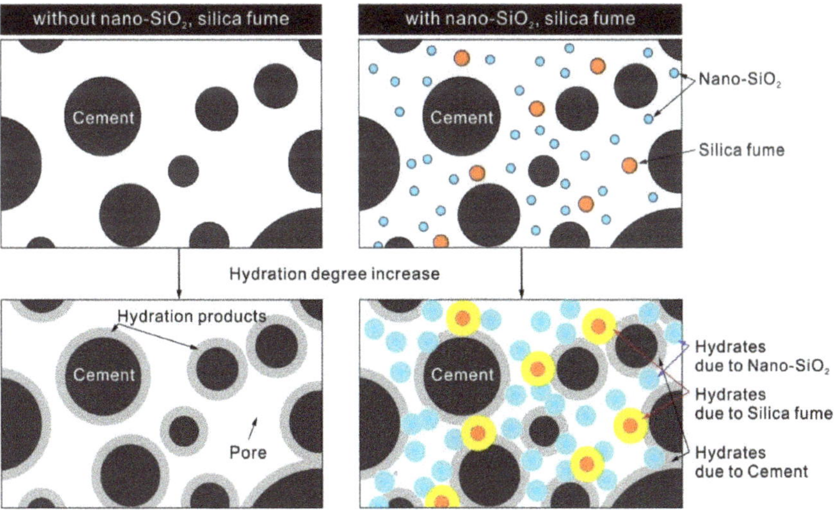

Figure 7. Hydration reaction caused by the NS and SF particles.

In addition, NS particles accelerate the cement hydration reaction by providing more nucleation sites for the precipitation of hydration products during the initial hydration process (Figure 7). NS particles preferentially adsorb the C-S-H gels generated during hydration through their large specific surface area. The C-S-H gels adsorbed on the surface of NS particles act as crystal nuclei as they propagate between the cement particles [65]. This increases strength and promotes setting by further accelerating the cement hydration reaction and forming more C-S-H gels. Silvestre et al. found that NS particles that act as nucleation sites form a larger network through connection with the C-S-H network in the cement system and improve mechanical performance [23].

3.2. Compressive Strength

Figure 8 shows the compressive strengths of the specimens according to age. The compressive strengths of all non-Plain specimens were found to be higher than that of Plain at all ages, except for the compressive strength of NS05 at 7 days. At 3 days, the compressive strengths of NS05, NS15, and NS05SF4 were 105%, 103%, and 116% that of Plain. This is due to the filler effect, in which NS fills empty spaces between the cement particles during the initial stage of hydration, accelerating the hydration reaction by providing nucleation sites [51]. These results are similar to those of Chekravarty et al. [68]. They reported that the compressive strength increased up to 3% of the NS content, which was attributed to the improvement of the microstructure and aggregate–paste interface by NS. At 28 days, the compressive strengths of NS05, NS15, and NS05SF4 were 106%, 110%, and 110% that of Plain. This is because of the acceleration of the hydration reaction by the additional pozzolanic reactions of NS and SF particles during the middle stage of hydration.

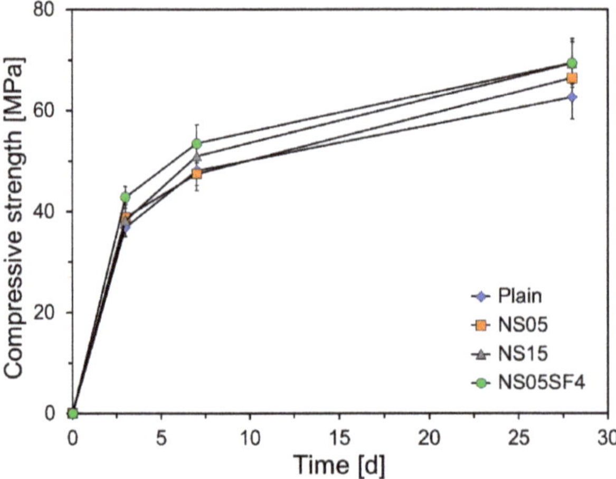

Figure 8. Compressive strengths of specimens.

NS and SF particles can generate C-S-H gels through fast reactions with calcium hydroxide ($Ca(OH)_2$), known as the pozzolanic reaction [65]. NS and SF particles that have interpenetrated the cement particles elute SiO_4^{4-} and AlO_2^{-} ions as the hydration reaction continues. They react with the Ca^{2+} ions eluted from the C_3S particles to form C-S-H gels around the pozzolan particles. This pozzolanic reaction of NS and SF improves long-term strength during the cement hydration process [69]. Hanif et al. reported that the pozzolanic activity can be significantly accelerated by adding NS particles to FA concrete composites [70].

Figure 9 shows the compressive strength increase for the specimens according to age. In general, the strength improvement effect of the C-S-H gels generated by the pozzolanic reaction occurs after 3 or 7 days of age. Thus, the effect of the additional pozzolanic reactions of NS and SF particles during the middle stage of hydration could be assessed by measuring the change in compressive strength after 3 or 7 days. As can be seen from Figure 9, during the interval from 3 to 7 days, the compressive strengths of Plain, NS05, NS15, and NS05SF4 increased by 11.17, 8.58, 12.74, and 10.64 MPa, respectively. Except for NS15, the strength increases of the non-Plain specimens were slightly lower than that of Plain. During the interval from 3 to 28 days, the compressive strengths of Plain, NS05, NS15, and NS05SF4 increased by 25.69, 27.54, 31.15, and 26.57 MPa, respectively, and from 7 to 28 days, the compressive strengths increased by 14.52, 18.96, 18.41, and 15.93 MPa, respectively. After 7 days of age, the strength increase of all non-Plain specimens was greater than that of Plain. The greatest increase was observed during the interval from 7 to 28 days, and the slope of the strength increases for all non-Plain specimens was higher than that for Plain. This is because the pozzolanic reaction was accelerated by NS and SF particles that interpenetrated the cement particles during the hydration process.

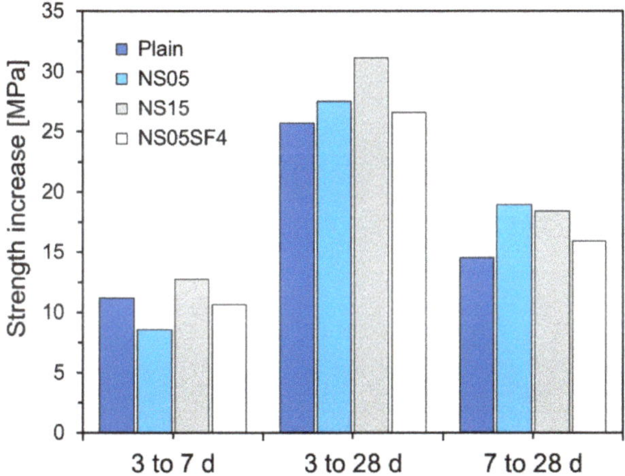

Figure 9. Increases in specimen strengths during different age intervals.

3.3. Microstructure

Figure 10 shows the log differential intrusion for each specimen by pore diameter according to age. Three experiments were conducted for each variable, and the average value was used. The standard deviation was less than 6%.

For all specimens, pore size tended to decrease as age increased. At 1 day, the main pore size of Plain ranged from 500 to 1000 nm. For the other specimens, however, the peaks in pore size occur in approximately the 300–700 nm range. At 3 and 7 days, the pore sizes of all specimens tended to be less than that at 1 day. This appears to be due to the filler effect of NS and SF particles. At 28 days, pore diameters of the non-Plain specimens corresponding to the peaks were substantially less than those of Plain. In particular, the peak in pore size occurs in approximately the 150–200 nm range for NS15 and NS05SF4; this is substantially smaller than for Plain, whose pore sizes peak in the 300–400 nm range. This appears to be due to the improvement in pore structure caused by the pozzolanic activity of NS and SF particles during the middle stage of hydration. Miao et al. [71] reported that NS particles optimize the pore structure and replace large pores with small pores. They reported that NS solution containing PCE (polycarboxylate ether) can effectively improve the strength of cement mortar even with a very small amount.

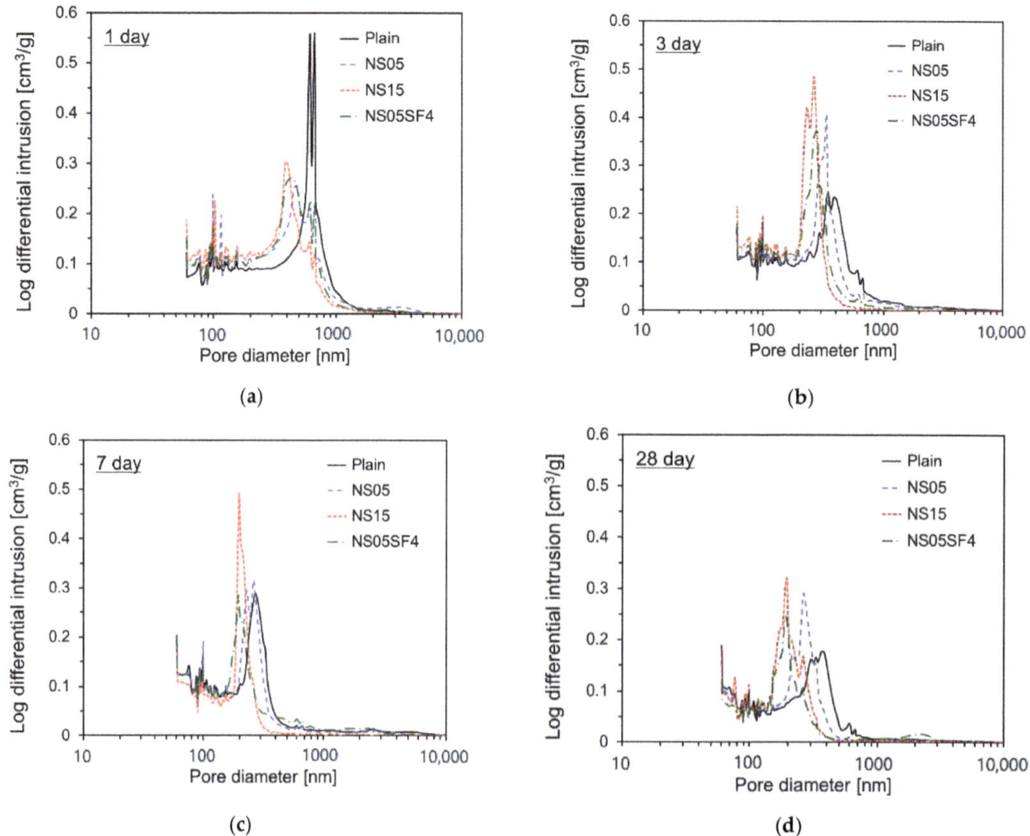

Figure 10. Log differential intrusions for specimens by pore diameter. (**a**) 1 day; (**b**) 3 days; (**c**) 7 days; (**d**) 28 days.

Figure 11 shows the cumulative pore volume of each specimen according to age. As age increased, the pore volumes of all specimens decreased. At 1 day, the cumulative pore volumes of Plain, NS05, NS15, and NS05SF4 were 0.157, 0.165, 0.161, and 0.167 cm^3/g, respectively. Although Plain showed the lowest value, the difference was negligible. At 3 days, however, a different pattern was observed. By 7 days, the difference was pronounced: the cumulative pore volumes of Plain, NS05, NS15, and NS05SF4 were 0.124, 0.121, 0.096, and 0.114 cm^3/g, respectively. Plain showed the highest value, and the value of NS15 was 77.8% that of Plain. These results display a pattern similar to that in the compressive strength increase results for the interval from 3 to 7 days of age. Then, at 28 days, the pore volumes of the non-Plain specimens were less than those of Plain, with the pore volume of NS05SF4 having the lowest value (0.084 cm^3/g). This again is similar to the compressive strength results and appears to be due to the formation of additional reaction products caused by the additional pozzolanic reaction activation of NS and SF particles that penetrated into the cement composite after 7 days. The ratios of the pore volumes at 28 days compared with 7 days for Plain, NS05, NS15, and NS05SF4 were 81.2%, 81.1%, 96.0%, and 73.3%, respectively. NS15 displayed the lowest reduction in pore volume, and NS05SF4 exhibited the highest. This appears to be due to the effect of the pozzolanic reaction for SF with larger particle sizes.

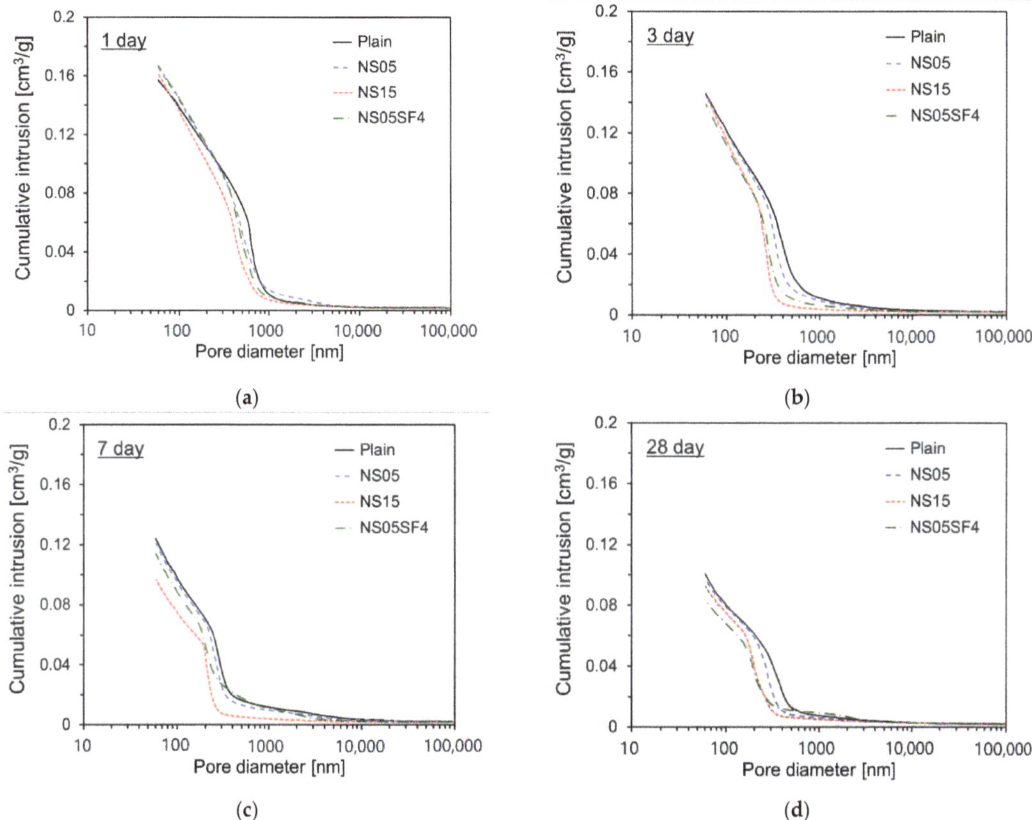

(a) (b) (c) (d)

Figure 11. Cumulative intrusions in specimens. (**a**) 1 day; (**b**) 3 days; (**c**) 7 days; (**d**) 28 days.

4. Conclusions

This study investigated the effects of nano-silica (NS) and silica fume (SF) on the hydration reaction of high-volume fly ash cement (HVFC). The main findings of the study are as follows.

1. The initial and final setting times of HVFC were shortened as the NS and SF particle content increased. The setting time was shorter when NS and SF were added together than when NS was used alone. In particular, in the case of the specimen containing the NSF solution with multiple particle sizes, the initial setting time was reduced by 35.5% from that of the plain specimen. This appears to be the result of the NSF solution with multiple size distribution filling the gaps between the cement particles and improving the filling rate.
2. The peak occurrence time of hydration heat flow decreased and the magnitude of the peak increased as the NS content increased. The cumulative heat release of all non-plain specimens was higher than that of the plain specimen until 22 h of hydration. This appears to be because NS and SF particles acted as additional nucleation sites for cement hydrates during the initial stage of hydration. In the case of the specimen containing the NSF solution, however, the cumulative heat release became less than that for NS05 or NS15 after approximately 22 h of hydration. These results are similar to the previous studies that reported that the cumulative caloric value decreased after 1 day of age when NS and SF were added together. Therefore, future studies on optimal NS and SF content are needed.

3. When NS and SF were mixed, the compressive strength of the cement composite showed a tendency to improve. The compressive strengths of all specimens were higher than that of the Plain at all ages, except for the compressive strength of NS05 at 7 days of age. In addition, the increase in the compressive strength of all non-plain specimens after 7 days was greater than that of the Plain, the largest increase being in the age interval from 7 to 28 days. This appears to be because the pozzolanic reaction was accelerated by NS and SF particles. A quantitative study on the effect of pozzolanic reaction of NS and SF on long-term compressive strength is needed in the future.
4. The pore size of the specimens decreased according to NS and SF particle content. After 28 days of age, the pore diameters of the non-plain specimens were substantially less than those of the Plain. This appears to be due to the improvement in pore structure caused by the pozzolanic activity of NS and SF particles during the middle stage of hydration. In addition, the cumulative pore volumes of all specimens decreased after 7 days of age, with the greatest decrease being that at 28 days in the specimen containing the NSF solution.
5. NS and SF improved the early strength of cement composites and improved the internal pore structure. In particular, the mechanical properties and pore structure of the cement composite were further improved when NS and SF were mixed together than when NS was used alone. It seems that the properties of the cement composite will vary depending on the amount of NS and SF mixed. Therefore, in order to apply nanomaterials such as NS and SF to construction products, research on the optimum mix design suitable for the characteristics of each construction product is required in the future.

Author Contributions: Conceptualization, B.-J.K. and G.-W.L.; data curation, B.-J.K.; formal analysis, B.-J.K.; funding acquisition, Y.-C.C.; investigation, G.-W.L.; methodology, B.-J.K. and G.-W.L.; project administration, Y.-C.C.; supervision, Y.-C.C.; validation, G.-W.L.; visualization, G.-W.L.; writing—original draft, B.-J.K.; writing—review & editing, Y.-C.C. All authors have read and agreed to the published version of the manuscript.

Funding: This work was supported by the National Research Foundation of Korea (NRF) grant funded by the Korea government (MSIT) (NRF-2020R1A2C2008926). This work was also supported by the Gachon University research fund of 2020 (GCU-202008470007).

Institutional Review Board Statement: Not applicable.

Informed Consent Statement: Not applicable.

Data Availability Statement: The raw/processed data will be provided as request.

Conflicts of Interest: The authors declare no conflict of interest.

References

1. Prakash, R.; Raman, S.N.; Subramanian, C.; Divyah, N. 6—Eco-friendly fiber-reinforced concretes. In *Handbook of Sustainable Concrete and Industrial Waste Management*; Woodhead Publishing: Sawston, UK, 2022; pp. 109–145. [CrossRef]
2. Agnihotri, A.; Ramana, P.V. GGBS: Fly-Ash evaluation and mechanical properties within high strength concrete. *Mater. Today Proc.* **2022**, *50*, 2404–2410. [CrossRef]
3. Wang, Q.; Liu, R.; Liu, P.; Liu, C.; Sun, L.; Zhang, H. Effects of silica fume on the abrasion resistance of low-heat Portland cement concrete. *Constr. Build. Mater.* **2022**, *329*, 127165. [CrossRef]
4. Sahoo, S.; Das, B.B.; Mustakim, S. Acid, alkali, and chloride resistance of concrete composed of low-carbonated fly ash. *J. Mater. Civ. Eng.* **2017**, *29*, 04016242. [CrossRef]
5. Bagheri, A.R.; Zanganeh, H.; Moalemi, M.M. Mechanical and durability properties of ternary concretes containing silica fume and low reactivity blast furnace slag. *Cem. Concr. Compos.* **2012**, *34*, 663–670. [CrossRef]
6. De Sensale, G.R. Strength development of concrete with rice husk ash. *Cem. Concr. Compos.* **2006**, *28*, 158–160. [CrossRef]
7. Giner, V.T.; Ivorra, S.; Baeza, F.J.; Zornoza, E.; Ferrer, B. Silica fume admixture effect on the dynamic properties of concrete. *Constr. Build. Mater.* **2011**, *25*, 3272–3277. [CrossRef]
8. Roshani, M.M.; Kargar, S.H.; Farhangi, V.; Karakouzian, M. Predicting the Effect of Fly Ash on Concrete's Mechanical Properties by ANN. *Sustainability* **2021**, *13*, 1469. [CrossRef]

9. Obla, K.H.; Hill, R.L.; Thomas, M.D.A.; Shashiprakash, S.G.; Perebatova, O. Properties of concrete containing ultra-fine fly ash. *ACI Mater. J.* **2003**, *100*, 426–433. [CrossRef]
10. Dai, J.; Wang, Q.; Xie, C.; Xue, Y.; Duan, Y.; Cui, X. The effect of fineness on the hydration activity index of ground granulated blast furnace slag. *Materials* **2019**, *12*, 2984. [CrossRef]
11. Copeland, K.D.; Obla, K.H.; Hill, R.L.; Thomas, M.D.A. Ultra Fine Fly Ash for High Performance Concrete. In Proceedings of the Construction Institute Sessions at ASCE Civil Engineering Conference 2001, Houston, TX, USA, 10–13 October 2001; pp. 166–175. [CrossRef]
12. Das, B.B.; Singh, D.N.; Pandey, S.P. Rapid chloride ion permeability of OPC- and PPC-based carbonated concrete. *J. Mater. Civ. Eng.* **2012**, *24*, 606–611. [CrossRef]
13. Payá, J.; Monzó, J.; Borrachero, M.V.; Peris-Mora, E.; Amahjour, F. Mechanical treatment of fly ashes: Part IV. Strength development of ground fly ash-cement mortars cured at different temperatures. *Cem. Concr. Res.* **2000**, *30*, 543–551. [CrossRef]
14. Qian, J.; Shi, C.; Wang, Z. Activation of blended cements containing fly ash. *Cem. Concr. Res.* **2001**, *31*, 1121–1127. [CrossRef]
15. Babaian, P.M.; Wang, K.; Mishulovich, A.; Bhattacharja, S.; Shah, S.P. Effect of mechanochemical activation on reactivity of cement kiln dust-fly ash systems. *ACI Mater. J.* **2003**, *100*, 55–62. [CrossRef]
16. Goñi, S.; Guerrero, A.; Luxán, M.P.; Macías, A. Activation of the fly ash pozzolanic reaction by hydrothermal conditions. *Cem. Concr. Res.* **2003**, *33*, 1399–1405. [CrossRef]
17. Wang, K.; Shah, S.P.; Mishulovich, A. Effects of curing temperature and NaOH addition on hydration and strength development of clinker-free CKD-fly ash binders. *Cem. Concr. Res.* **2004**, *34*, 299–309. [CrossRef]
18. Yang, G.; Wu, T.; Fu, C.; Ye, H. Effects of activator dosage and silica fume on the properties of Na_2SO_4-activated high-volume fly ash. *Constr. Build. Mater.* **2021**, *278*, 122346. [CrossRef]
19. Bondar, D.; Coakley, E. Use of gypsum and CKD to enhance early age strength of High Volume Fly Ash (HVFA) pastes. *Constr. Build. Mater.* **2014**, *71*, 93–108. [CrossRef]
20. Chuah, S.; Pan, Z.; Sanjayan, J.G.; Wang, C.M.; Duan, W.H. Nano reinforced cement and concrete composites and new perspective from graphene oxide. *Constr. Build. Mater.* **2014**, *73*, 113–124. [CrossRef]
21. Du, H.; Pang, S.D. Enhancement of barrier properties of cement mortar with graphene nanoplatelet. *Cem. Concr. Res.* **2015**, *76*, 10–19. [CrossRef]
22. Li, H.; Xiao, H.; Ou, J. A study on mechanical and pressure-sensitive properties of cement mortar with nanophase materials. *Cem. Concr. Res.* **2004**, *34*, 435–438. [CrossRef]
23. Silvestre, J.; Silvestre, N.; de Brito, J. Review on concrete nanotechnology. *Eur. J. Environ. Civ. Eng.* **2016**, *20*, 455–485. [CrossRef]
24. Wang, Y.; Hughes, P.; Niu, H.; Fan, Y. A new method to improve the properties of recycled aggregate concrete: Composite addition of basalt fiber and nanosilica. *J. Clean. Prod.* **2019**, *236*, 117602. [CrossRef]
25. Yeşilmen, S.; Al-Najjar, Y.; Balav, M.H.; Şahmaran, M.; Yıldırım, G.; Lachemi, M. Nano-modification to improve the ductility of cementitious composites. *Cem. Concr. Res.* **2015**, *76*, 170–179. [CrossRef]
26. Senff, L.; Tobaldi, D.M.; Lucas, S.; Hotza, D.; Ferreira, V.M.; Labrincha, J.A. Formulation of mortars with nano-SiO_2 and nano-TiO_2 for degradation of pollutants in buildings. *Compos. Part B Eng.* **2013**, *44*, 40–47. [CrossRef]
27. Onaizi, A.M.; Lim, N.H.A.S.; Huseien, G.F.; Amran, M.; Ma, C.K. Effect of the addition of nano glass powder on the compressive strength of high volume fly ash modified concrete. *Mater. Today Proc.* **2022**, *48*, 1789–1795. [CrossRef]
28. Li, G. Properties of high-volume fly ash concrete incorporating nano-SiO_2. *Cem. Concr. Res.* **2004**, *34*, 1043–1049. [CrossRef]
29. Ji, T. Preliminary study on the water permeability and microstructure of concrete incorporating nano-SiO_2. *Cem. Concr. Res.* **2005**, *35*, 1943–1947. [CrossRef]
30. Li, H.; Xiao, H.; Yuan, J.; Ou, J. Microstructure of cement mortar with nano-particles. *Compos. Part B Eng.* **2004**, *35*, 185–189. [CrossRef]
31. Jo, B.W.; Kim, C.H.; Lim, J.H. Characteristics of cement mortar with nano-SiO2 particles. *ACI Mater. J.* **2007**, *104*, 404–407. [CrossRef]
32. Liu, R.; Xiao, H.; Li, H.; Sun, L.; Pi, Z.; Waqar, G.Q.; Du, T.; Yu, L. Effects of nano-SiO_2 on the permeability-related properties of cement-based composites with different water/cement ratios. *J. Mater. Sci.* **2017**, *53*, 4974–4986. [CrossRef]
33. Zhang, M.; Li, H. Pore structure and chloride permeability of concrete containing nano-particles for pavement. *Constr. Build. Mater.* **2011**, *25*, 608–616. [CrossRef]
34. Hou, P.; Cheng, X.; Qian, J.; Zhang, R.; Cao, W.; Shah, S.P. Characteristics of surface-treatment of nano-SiO_2 on the transport properties of hardened cement pastes with different water-to-cement ratios. *Cem. Concr. Compos.* **2015**, *55*, 26–33. [CrossRef]
35. Kong, D.; Du, X.; Wei, S.; Zhang, H.; Yang, Y.; Shah, S.P. Influence of nano-silica agglomeration on microstructure and properties of the hardened cement-based materials. *Constr. Build. Mater.* **2012**, *37*, 707–715. [CrossRef]
36. Zhang, M.H.; Islam, J.; Peethamparan, S. Use of nano-silica to increase early strength and reduce setting time of concretes with high volumes of slag. *Cem. Concr. Compos.* **2012**, *34*, 650–662. [CrossRef]
37. Li, X.; Chen, H.; Li, H.; Liu, L.; Lu, Z.; Zhang, T.; Duan, W.H. Integration of form-stable paraffin/nanosilica phase change material composites into vacuum insulation panels for thermal energy storage. *Appl. Energy* **2015**, *159*, 601–609. [CrossRef]
38. Kawashima, S.; Hou, P.; Corr, D.J.; Shah, S.P. Modification of cement-based materials with nanoparticles. *Cem. Concr. Compos.* **2013**, *36*, 8–15. [CrossRef]

39. Hou, P.; Kawashima, S.; Kong, D.; Corr, D.J.; Qian, J.; Shah, S.P. Modification effects of colloidal nanoSiO_2 on cement hydration and its gel property. *Compos. Part B Eng.* **2013**, *45*, 440–448. [CrossRef]
40. Hou, P.; Kawashima, S.; Wang, K.; Corr, D.J.; Qian, J.; Shah, S.P. Effects of colloidal nanosilica on rheological and mechanical properties of fly ash–cement mortar. *Cem. Concr. Compos.* **2013**, *35*, 12–22. [CrossRef]
41. Haruehansapong, S.; Pulngern, T.; Chucheepsakul, S. Effect of the particle size of nanosilica on the compressive strength and the optimum replacement content of cement mortar containing nano-SiO_2. *Constr. Build. Mater.* **2014**, *50*, 471–477. [CrossRef]
42. Quercia, G.; Spiesz, P.; Hüsken, G.; Brouwers, H.J.H. SCC modification by use of amorphous nano-silica. *Cem. Concr. Compos.* **2014**, *45*, 69–81. [CrossRef]
43. Du, H.; Du, S.; Liu, X. Durability performances of concrete with nano-silica. *Constr. Build. Mater.* **2014**, *73*, 705–712. [CrossRef]
44. Du, H.; Du, S.; Liu, X. Effect of nano-silica on the mechanical and transport properties of lightweight concrete. *Constr. Build. Mater.* **2015**, *82*, 114–122. [CrossRef]
45. Ibrahim, R.K.; Hamid, R.; Taha, M.R. Strength and microstructure of mortar containing nanosilica at high temperature. *ACI Mater. J.* **2014**, *111*, 163–170. [CrossRef]
46. Yu, R.; Tang, P.; Spiesz, P.; Brouwers, H.J.H. A study of multiple effects of nanosilica and hybrid fibres on the properties of Ultra-High Performance Fibre Reinforced Concrete (UHPFRC) incorporating waste bottom ash (WBA). *Constr. Build. Mater.* **2014**, *60*, 98–110. [CrossRef]
47. Oertel, T.; Hutter, F.; Tänzer, R.; Helbig, U.; Sextl, G. Primary particle size and agglomerate size effects of amorphous silica in ultra-high performance concrete. *Cem. Concr. Compos.* **2013**, *37*, 61–67. [CrossRef]
48. Oertel, T.; Helbig, U.; Hutter, F.; Kletti, H.; Sextl, G. Influence of amorphous silica on the hydration in ultra-high performance concrete. *Cem. Concr. Res.* **2014**, *58*, 121–130. [CrossRef]
49. Liu, R.; Xiao, H.; Liu, J.; Guo, S.; Pei, Y. Improving the microstructure of ITZ and reducing the permeability of concrete with various water/cement ratios using nano-silica. *J. Mater. Sci.* **2019**, *54*, 444–456. [CrossRef]
50. Ghafari, E.; Costa, H.; Júlio, E.; Portugal, A.; Durães, L. The effect of nanosilica addition on flowability, strength and transport properties of ultra high performance concrete. *Mater. Des.* **2014**, *59*, 1–9. [CrossRef]
51. Yu, R.; Spiesz, P.; Brouwers, H.J.H. Effect of nano-silica on the hydration and microstructure development of Ultra-High Performance Concrete (UHPC) with a low binder amount. *Constr. Build. Mater.* **2014**, *65*, 140–150. [CrossRef]
52. Rong, Z.; Sun, W.; Xiao, H.; Jiang, G. Effect of nano-SiO_2 particles on the mechanical and microstructural properties of ultra-high performance cementitious composites. *Cem. Concr. Compos.* **2015**, *56*, 25–31. [CrossRef]
53. Cai, Y.; Xuan, D.; Poon, C.S. Effects of nano-SiO_2 and glass powder on mitigating alkali-silica reaction of cement glass mortars. *Constr. Build. Mater.* **2019**, *201*, 295–302. [CrossRef]
54. Durgun, M.Y.; Atahan, H.N. Strength, elastic and microstructural properties of SCCs' with colloidal nano silica addition. *Constr. Build. Mater.* **2018**, *158*, 295–307. [CrossRef]
55. Palla, R.; Karade, S.R.; Mishra, G.; Sharma, U.; Singh, L.P. High strength sustainable concrete using silica nanoparticles. *Constr. Build. Mater.* **2017**, *138*, 285–295. [CrossRef]
56. Durgun, M.Y.; Atahan, H.N. Rheological and fresh properties of reduced fine content self-compacting concretes produced with different particle sizes of nano SiO_2. *Constr. Build. Mater.* **2017**, *142*, 431–443. [CrossRef]
57. Lavergne, F.; Belhadi, R.; Carriat, J.; Ben Fraj, A. Effect of nano-silica particles on the hydration, the rheology and the strength development of a blended cement paste. *Cem. Concr. Compos.* **2019**, *95*, 42–55. [CrossRef]
58. Karakouzian, M.; Farhangi, V.; Farani, M.R.; Joshaghani, A.; Zadehmohamad, M.; Ahmadzadeh, M. Mechanical Characteristics of Cement Paste in the Presence of Carbon Nanotubes and Silica Oxide Nanoparticles: An Experimental Study. *Materials* **2021**, *14*, 1347. [CrossRef] [PubMed]
59. Liu, H.; Li, Q.; Ni, S.; Wang, L.; Yue, G.; Guo, Y. Effect of nano-silica dispersed at different temperatures on the properties of cement-based materials. *J. Build. Eng.* **2022**, *46*, 103750. [CrossRef]
60. Kooshafar, M.; Madani, H. An investigation on the influence of nano silica morphology on the characteristics of cement composites. *J. Build. Eng.* **2022**, *30*, 101293. [CrossRef]
61. Wu, Z.; Khayat, K.H.; Shi, C. Effect of nano-SiO_2 particles and curing time on development of fiber-matrix bond properties and microstructure of ultra-high strength concrete. *Cem. Concr. Res.* **2017**, *95*, 247–256. [CrossRef]
62. Chithra, S.; Senthil Kumar, S.R.R.; Chinnaraju, K. The effect of colloidal nano-silica on workability, mechanical and durability properties of high-performance concrete with copper slag as partial fine aggregate. *Constr. Build. Mater.* **2016**, *113*, 794–804. [CrossRef]
63. Bhatta, D.P.; Singla, S.; Garg, R. Experimental investigation on the effect of Nano-silica on the silica fume-based cement composites. *Mater. Today Proc.* **2022**, *57*, 2338–2343. [CrossRef]
64. Nili, M.; Ehsani, A. Investigating the effect of the cement paste and transition zone on strength development of concrete containing nanosilica and silica fume. *Mater. Des.* **2015**, *75*, 174–183. [CrossRef]
65. Xi, B.; Zhou, Y.; Yu, K.; Hu, B.; Huang, X.; Sui, L.; Xing, F. Use of nano-SiO_2 to develop a high performance green lightweight engineered cementitious composites containing fly ash cenospheres. *J. Clean. Prod.* **2020**, *262*, 121274. [CrossRef]
66. Wang, X.; Gong, C.; Lei, J.; Dai, J.; Lu, L.; Cheng, X. Effect of silica fume and nano-silica on hydration behavior and mechanism of high sulfate resistance Portland cement. *Constr. Build. Mater.* **2021**, *279*, 122481. [CrossRef]

67. Pacheco-Torgal, F.; Miraldo, S.; Ding, Y.; Labrincha, J.A. Targeting HPC with the help of nanoparticles: An overview. *Constr. Build. Mater.* **2013**, *38*, 365–370. [CrossRef]
68. Chekravarty, D.S.V.S.M.R.K.; Mallika, A.; Sravana, P.; Rao, S. Effect of using nano silica on mechanical properties of normal strength concrete. *Mater. Today Proc.* **2022**, *51*, 2573–2578. [CrossRef]
69. Moon, G.D.; Oh, S.; Choi, Y.C. Effects of the physicochemical properties of fly ash on the compressive strength of high-volume fly ash mortar. *Constr. Build. Mater.* **2016**, *124*, 1072–1080. [CrossRef]
70. Hanif, A.; Parthasarathy, P.; Ma, H.; Fan, T.; Li, Z. Properties improvement of fly ash cenosphere modified cement pastes using nano silica. *Cem. Concr. Compos.* **2017**, *81*, 35–48. [CrossRef]
71. Miao, J.; Lu, L.; Jiang, J. Reinforcement of ultra-low dosage of polycarboxylate ether (PCE) grafted nano-silica sol to the mechanical and durable properties of cement mortar. *J. Mater. Res. Tech.* **2022**, *19*, 3464–3657. [CrossRef]

 *materials*

Article

Numerical Simulation of the Response of Concrete Structural Elements Containing a Self-Healing Agent

Todor Zhelyazov

Structural Engineering and Composites Laboratory—SEL, Reykjavik University, Menntavegur 1, IS-102 Reykjavik, Iceland; elovar@yahoo.com

Abstract: Self-healing of a crack is a relatively novel technique allowing for the partial recovery of the initial mechanical characteristics of a structural element after some period of exploitation. By a widely accepted convention, self-healing is either autogenous or autonomous. The former is a mechanism inherent for cementitious composites (in particular—concrete), while the latter is an engineered process. Both autogenous and engineered healing have recently been the object of numerous studies. Despite the large amount of research work being carried out, the potential of this technique has not yet been fully realized. The article focuses on the modeling and the finite element simulation of the recovery of the initial material properties resulting from the sealing of cracks. The employed numerical procedure uses a constitutive relation for concrete based on the continuum damage mechanics. It captures both the strain-softening and the inverse process—the crack healing. Finite element simulations of benchmark cases illustrate the effect of self-healing. The numerically obtained constitutive relations for specimens with and without a healing agent are compared.

Keywords: self-healing of cracks; concrete; constitutive relations; damage mechanics; finite element analysis

Citation: Zhelyazov, T. Numerical Simulation of the Response of Concrete Structural Elements Containing a Self-Healing Agent. *Materials* **2022**, *15*, 1233. https://doi.org/10.3390/ma15031233

Academic Editor: Yeonung Jeong

Received: 5 January 2022
Accepted: 4 February 2022
Published: 7 February 2022

1. Introduction

The self-healing of cracks implies the material properties' recovery resulting from the crack sealing [1,2]. Generally, self-healing can be a natural (autogenous) or engineered (autonomous) process. Solutions based on self-healing will possibly contribute to extending the life span of concrete structures. The placement of healing agents (such as encapsulated polymers, minerals, or bacteria) in structural elements programs the subsequent initiation of self-healing upon the realization of specific conditions. More precisely, the release of the self-healing agent, when the cracks unseal the capsules, triggers the process. On the other hand, the introduction of mineral additions, crystalline admixtures, superabsorbent, or other polymers stimulate autogenous self-healing [3].

Using environmentally friendly bacteria such as Bacillus pasteurii can replace the commonly used repair materials for concrete (e.g., epoxy systems and acrylic resins), which are environmentally unfriendly [4]. Moreover, the application of epoxies and acrylic resins is reportedly accompanied by cracking and delamination between concrete and the repair material. Also, it should be noted that an intelligent approach to self-healing is required. Thus, for example, calcium ions can be provided either by internal sources in the cement structure or by adding chemicals such as calcium chloride, calcium nitrate, or calcium lactate [5]. However, the utilization of calcium chloride as a calcium source may cause chloride ion attack and consequently degradation of reinforcement bars [6].

Some authors [7,8] attribute the unexpected longevity of bridges and buildings to autogenous healing, acting simultaneously with other mechanisms. The current understanding of self-healing attributes this phenomenon to an interaction between mechanisms such as continuing hydration, dissolution and crystallization, particle clogging, and carbonation. Presumably, due to the continuing hydration, products of strength comparable to that of calcium silicate hydrate (CSH) gel [9] form. Based on results obtained by microstructural analysis [10], the healed cracks contain CSH, ettringite, and calcium hydroxide ($Ca(OH)_2$).

The partial substitution of cement with fly ash, as well as the addition of crystalline admixture, yields the recovery of both compressive strength and electrical resistivity of concrete up to 94% [11]. According to other sources (see for example [12]), after 30 days of water curing, a damaged specimen containing a high volume of fly ash regained 74% of the lost compressive strength, whereas samples without fly ash—68%. Based on research on the contribution to the healing capacity of crystalline admixture (CA), [13] reports a healing rate of 81–93% for specimens containing CA, whereas the healing rate of specimens without CA was 80–86%. In samples containing CA, cracks of width up to 0.25 mm were healed.

Polymers (polyurethane, superabsorbent polymers (SAP), acrylamide, acrylate, epoxy, poly styrene-divinylbenzene, styrene-butadiene rubber, etc.) improve the rate of the regain in the mechanical properties and contribute to the healing of surface cracks [14].

Reportedly, fibers (either steel or synthetic) in fiber-reinforced concrete control the crack opening what results in ease of application of the self-healing products in terms of adhesion to the crack surface and crack filling. Thus, according to [15], in engineered cementitious composites (ECC) containing synthetic fibers (such as polypropylene (PP), ethylene vinyl alcohol (EVA), and poly-vinyl acrylate (PVA)), the precipitation of calcium carbonate potentially increases due to the presence of synthetic fibers. For concrete specimens containing PVA fibers and SAP, [16] reports a similar effect. Self-healing of ECC samples containing supplementary cementitious material (SCM) after having been exposed to various deterioration triggering processes has been studied by [17]. Damage has been induced in test specimens by exposure to natural weathering and moderate controlled humidity in a laboratory environment. After that, the test specimens remained submerged in water for 90 days to trigger the self-healing mechanisms. The evaluation of the mechanical characteristics showed a regain of stiffness and tensile strength of 60% and 70%, respectively. The presence of water or high humidity appeared as a crucial factor for a rapid and effective healing process. The time required for healing varies in the function of the method employed. Thus, autogenous healing of concrete might continue for up to two years, whereas the healing process based on encapsulated cyanoacrylate is completed in less than a minute [18].

Ultra-high-performance concrete (UHPC) and ultra-high-performance fiber-reinforced concrete (UHPFRC) have the potential to exhibit self-healing capacity because of the high binder content and the low water/binder ratio [19–21]. Also, at the damage state, both UHPC and UHPFRC develop cracks of small width, thus creating favorable conditions for self-healing, either via water emersion or by using healing promoters [22–28]. Based on the available experimental data, the following set of variables can be used in the study and evaluation of the self-healing [18]: the initial opening of the crack and the age of cracking; the curing conditions during the healing period and its duration; the presence of sustained loading along the healing period, which results into through-crack stress states; and the repeatability of the healing action and its effectiveness in consequence of successive repeated cracking phenomena, at the same and/or at different locations.

The modeling of the self-healing of concrete is in close relation with the simulation of the concrete failure behavior. The fracture of concrete has been investigated in numerous research works [29–44].

Among other approaches suitable to model the self-healing phenomena, Yang et al. [45] discuss the possible implementation of the phase-field (PF) methods for self-healing of cementitious materials. A number of literature sources report the application of the PF approach for modeling the fracture behavior of concrete (see among others [46–48]). However, the modeling of the crack-closure effects in the framework of the PF approach still needs researchers' attention. According to [45], possible axes of investigation are: (i) the implementation of cohesive elements along the crack path to prevent overlapping of the crack faces and (ii) a contact scheme, recently proposed by [49].

The deterministic approach to the analysis is often questionable, considering the high scattering in material properties within structural elements and loading. The stochastic perturbation-based finite element method theoretically addresses this issue [50].

Recent studies extend the finite element method by proposing a new finite element formulation with embedded strong discontinuity [51] or by coupling the finite element method (FEM) with the discrete element method (DEM) [52]. These research works provide intuitive decoupling of two different phenomena: damage accumulation in the initially undamaged materials and initiation and propagation of the macroscopic cracks.

In Ref. [51], by hypothesis, the total displacement in the numerical domain is the sum of a component, obtained by linear elastic analysis (of the continuum part), and a jump, associated with the embedded strong discontinuity. The discontinuity presumably follows an elastic-damage cohesive crack constitutive relationship [53]. The proposed finite element formulation accounts for the self-healing by adding a term to account for the healed material. The approach accounts for the healing agent circulation in the material continuum by a modified Lucas–Washburn model [54–57]. The capillary pressure is evaluated by the Young–Laplace equation, defining the dynamic contact angle according to [58].

In [51], by assumption, the shear jump is constant along the discontinuity in each element [59]. Thus, some shear-dominant problems might require mesh refinement to improve accuracy [51]. Equilibrium across the discontinuity and the crack-plane displacements are defined at the element level. Therefore, the lack of continuity across the finite element boundaries should be compensated by adopting the mean values at coincident nodes in the constitutive relations [60]. The formulation of the specialized finite element presumes small rotations—this could restrain its application to certain crack width.

Combining the finite element method (FEM) and the discrete element method (DEM) has drawn considerable research attention [61–64]. Considering that the FEM operates with macroscopic characteristics of the material within equations defined in the framework of continuum mechanics, including failure criteria in the context of the constitutive equations and the DEM is a widely recognized method for modeling the response of granular matter and non-continuum media, combining these two methods appears suitable for simulating the transition from initially undamaged to cracked materials such as geomaterials and concrete.

The combined FEM–DEM formulation implies a multi-scale approach. The frictional contact properties of the interacting discrete particles (for the DEM) are defined at the micro-scale, while the material properties for the FEM generally reflect experimental data obtained at the macroscopic scale. The multi-scale approach involves an identification procedure [52] that generally complicates the overall implementation.

Oñate et al. [52] use the combined FEM–DEM approach to model the concrete degradation within a complex fluid–solid interaction (FSI). The approach requires the generation of a finite element mesh in the computational domain. Damage accumulation in the continuum (using a finite element analysis) is simulated element by element, employing a standard isotropic damage model along with a Mohr–Coulomb failure criterion. Finite elements exceeding a predefined damage threshold are removed from the mesh and replaced by a set of particles modeled with the DEM. These particles overlap the removed finite elements node by node. DEM particles are placed to a maximum close distance, avoiding the initial indentation. The contact interaction forces computed using DEM are transferred to the FE model.

Both aforementioned approaches imply complicated numerical procedures accounting for crack initiation and propagation in the continuum. On the other hand, the complication affects only a part of the initial computational domain.

The article focuses on the numerical simulation of the opening and reclosure of cracks, to model the material properties degradation, due to the former mechanism and the material properties recovery attributed to the latter.

A simplified solution scheme using the finite element method only is applied. Using different methods for different phenomena might be intuitive and consistent. On the other hand, continuum mechanics (specifically, the continuum damage mechanics) describes the microscope manifestation of mechanisms that take place at the microscopic scales. The employed damage model slightly differs from the standard damage model incorporated in the advanced numerical simulations referenced above. The numerical algorithm uses an original procedure that integrates the damage variable into the material constitutive relation. Generally, the concrete fracture is described either by using fracture mechanics or in the context of damage mechanics. Damage mechanics is the preferred option in this study since it allows for modeling the damage accumulation and crack propagation starting with homogeneous isotropic material. Unlike the fracture mechanics approach, there is no need to pre-establish defects in the studied specimen to initiate macroscopic cracking. The present article develops and summarizes some ideas discussed in previously published contributions [65,66].

The underlying hypotheses of the reported study presume that mechanical loading triggers the crack opening while the regain in the mechanical properties results from engineered self-healing. Continuum damage mechanics allows for the modeling of the degradation of the initial mechanical properties via the introduction of a damage variable into the material stress–strain relationship. The damage variable is regarded as an operator acting on the elasticity tensor. Thus, damage variable increase models the damage accumulation in the material, and a critical value of the damage variable corresponds to macroscopic crack initiation in the specified location. Vice versa, the sealing of cracks results in a decrease in the damage variable and a partial regain in the initial mechanical properties. Finite element simulations of some benchmark examples illustrate the self-healing mechanism. Modeled specimens are damaged to introduce multiple cracking patterns (i.e., the load is applied to generate some distributed damage). Then, at a specific moment of the loading history, the activation of the autonomous self-healing leads to a sealing of the formed macroscopic cracks. This mechanism results in a partial regain of the initial material properties.

2. Materials and Methods

The focus of the article is on the finite element analysis of concrete specimens containing a self-healing agent. Within the assumed approach, the material response of concrete is modeled via a coupling between elasticity and damage [33,67].

$$\sigma_{ij} = \frac{\nu}{(1+\nu)(1-2\nu)} E(1-D)\varepsilon_{kk}\delta_{ij} + \frac{1}{(1+\nu)} E(1-D)\varepsilon_{ij} \tag{1}$$

In Equation (1), σ_{ij} denotes the components of the stress tensor, ν is the Poisson's ratio, E is Young's modulus of the material (concrete) before damage occurs, D is the damage variable, ε_{kk} is the trace of the strain tensor $(k = 1, 2, 3)$, δ_{ij} is the Kronecker delta, and ε_{ij} are the components of the strain tensor.

The damage variable acts directly on Young's modulus. Theoretically, it evaluates the mechanical damage induced by the applied load.

$$D = f(C_i, \varepsilon_{D0}, D_c, \varepsilon_{eqv}) \tag{2}$$

As shown in Equation (2), the damage variable depends on a set of material constants $(C_i, \varepsilon_{D0}, D_c)$ that should be identified through comparison with experimental data, and a variable constructed from the positive eigenvalues of the strain tensor [68].

$$\varepsilon_{eqv} = \sqrt{\sum_{j=1}^{3} \langle \varepsilon_j \rangle^2} \tag{3}$$

The constant ε_{D0}, (also referred to as damage threshold) controls the transition between the elastic response and the inelastic behavior. The constants C_i, one set of two constants for tension, and another for compression control the shape of the numerically obtained stress–strain relationship. In Equation (3),

$$\langle \varepsilon_j \rangle = \frac{\varepsilon_j + |\varepsilon_j|}{2} \quad (4)$$

Damage accumulates provided the variable defined in Equation (3) becomes larger than another model parameter, the damage threshold:

$$\frac{dD}{dt} \begin{cases} = 0 & \text{if} \quad \varepsilon_{eqv} \leq \varepsilon_{D0} \\ > 0 & \text{if} \quad \varepsilon_{eqv} > \varepsilon_{D0} \end{cases} \quad (5)$$

Reaching the critical value of the damage variable at a specific location in the continuum ($D = D_c$) means a macroscopic crack initiation or deactivation of the corresponding finite element. The identification of the material constants is performed through curve fitting. For example, one input parameter of the identification procedure is the elasticity modulus of the undamaged material. Material constants are then varied to match the other material characteristics (such as the concrete compressive or tensile strength). In other words, the stress–strain relationship obtained in the numerical simulation of a characterization test should match, as closest as possible, the experimental one.

The activation of the self-healing process leads to the sealing of the newly formed cracks and regaining the initial rigidity in some regions. Therefore, self-healing is modeled by setting the damage variable equal to zero and the Young modulus—equal to its initial value. At the same time, some level of damage remains in regions where the damage variable has not yet reached its critical value. Figure 1 shows a flowchart of the employed numerical procedure.

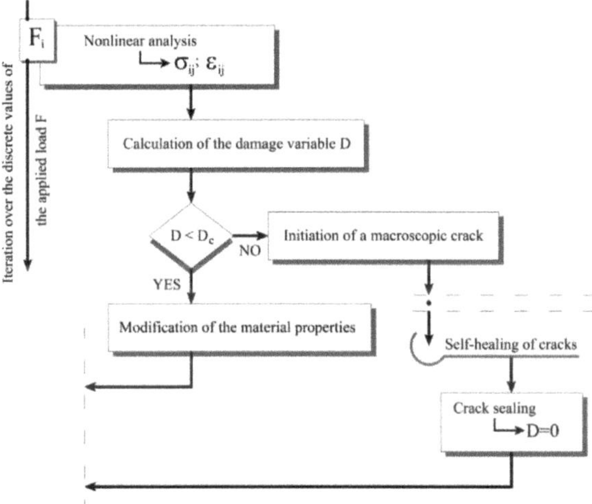

Figure 1. Algorithm for the implementation of the material model taking into account the healing.

Standard compression and tension by flexure tests, as well as a torsion test on concrete specimens, are simulated. Simulations include a preloading phase without reaching the failure load of the specimen. The preloading phase induces some damaged state in the specimen. Also, in some regions, the critical value of the damage variable is reached (i.e., macroscopic cracks are initiated), and the self-healing process is launched (e.g., the healing agent is released from capsules intersected by cracks). The responses of the specimens are obtained by finite element analysis in terms of stress–strain relationships.

The evaluation of the self-healing effectiveness employs an assessment of the recovery of the relevant mechanical properties [18]. Concrete specimens (loaded in compression) exhibit an initial elastic phase. Stress then increases with a decreasing rate, with the increase of the applied quasi-static load up to a maximum value, because of the strain-softening effect. After reaching a specific maximum value, the stress starts to decrease. The mechanical load provokes damage accumulation followed by the initiation of macroscopic crack. By hypothesis, crack initiation (associated with the critical value of the damage variable) triggers autonomous self-healing. The released healing agent seal opened cracks, recovering thus previously degraded stiffness. The stress developed in concrete starts to increase again with the increase of the applied load. In the loading history, including healing, two parameters are to be retained: (i) the maximum stress developed in the specimen during the first loading of the undamaged material (e.g., the compressive strength for concrete loaded in compression) and (ii) the maximum stress developed after the healing. The assessment of the recovery of the material properties consists of a comparison of these two parameters.

3. Results Obtained by Finite Element Modeling

The general-purpose finite element code ANSYS is employed for the numerical simulations. A compression test on a cylindrical concrete specimen, a tension by flexure test on a prismatic concrete specimen, and a torsion test on a cylindrical concrete specimen are simulated. Standard geometries are used for the simulations of the compression and the tension-by-flexure tests.

3.1. Compression Test

The compressive strength of the concrete is measured on cylinders 150 mm in diameter and 300 mm in height, in accordance with ISO 1920-3 [69]. Figure 2 displays the finite element model built. The generated finite element mesh contains a total of 4050 finite elements SOLID185. SOLID185 is a 3-D finite element defined by eight nodes, having three degrees of freedom at each node (specifically, translations in the nodal x-, y-, and z-directions). Nodes on the bottom surface ($z = 0$) are restrained in the z-direction. Additionally, zero displacements in the x- and y-directions are applied to four circumferential nodes on the bottom surface. In the framework of a displacement-controlled simulation, vertical displacements are incrementally applied to top nodes ($z = 300$ mm).

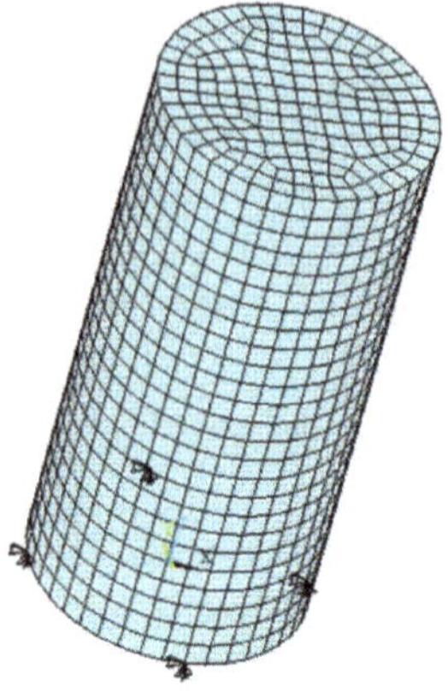

Figure 2. Finite element model built for the compression test.

Figure 3 shows results obtained by finite element analysis. As expected, the stress increases with the increase of the applied displacement, and upon reaching the compressive strength, starts to decrease (the grey line in Figure 3). The concrete specimen is loaded in compression without reaching the value of the imposed displacement corresponding to failure. Despite the accumulated damage, the specimen still has some residual load-carrying capacity. At this point, self-healing is initiated, and the previously formed macroscopic cracks are sealed. On the macroscopic scale, crack sealing corresponds to the axial stress increase with the increase of the applied displacement (the black line in Figure 3).

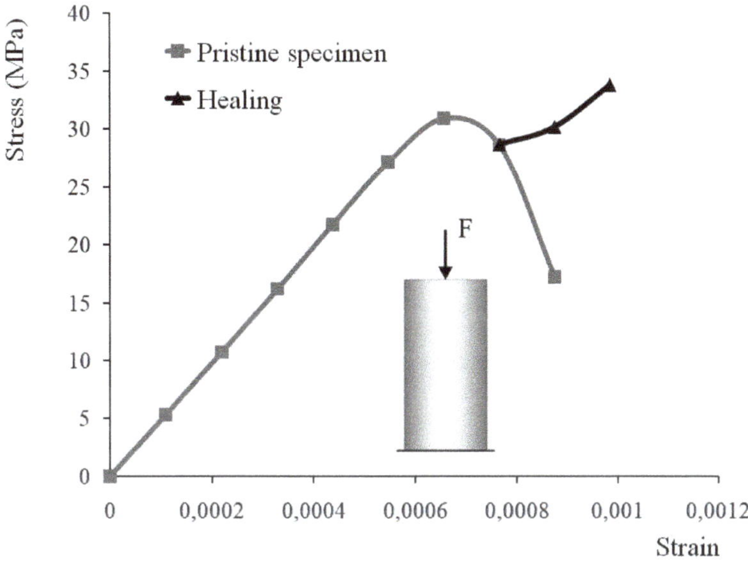

Figure 3. Stress–strain response of a standard cylindrical concrete specimen loaded in compression and then subjected to a self-healing of cracks. The numerically obtained stress–strain relationship for the pristine specimen is shown in grey and the action of the healing agent-in black.

3.2. Tension by Flexure Test

Tension by flexure test commonly performed on a prismatic concrete specimen is simulated. The modeled specimen is 400 mm in length, 300 mm in span, with a rectangular cross-section of 100 mm × 100 mm. The distance between the specimen end and the point load location is 100 mm (denoted by 'a' in Figure 4).

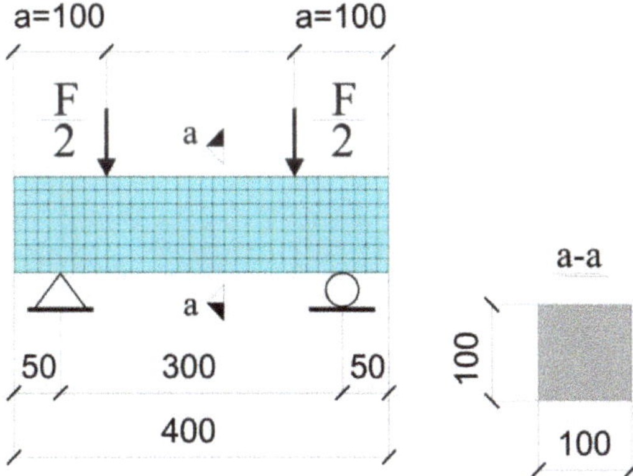

Figure 4. Tension by flexure test: geometry of the modeled specimen and visualization of the generated finite element mesh.

According to the beam theory, between the load application points, the specimen is loaded to flexure only. The absence of reinforcement results in the progressive failure of concrete in the tension face, whereas concrete in the compression face still behaves elastically (the elastic domain is commonly assumed up to 30% of the compressive strength). For the considered experimental setup, unstable crack propagation being a fast process, additional time-stepping, with the definition of a smaller time interval, might be required to capture it.

The finite element analysis simulates a displacement-controlled experimental setup with a constant time step for the entire load history. The solid geometry is meshed with 1568 finite elements SOLID185 (Figure 4). To model supports, vertical displacements in the global y-direction of nodes at the corresponding locations are set to zero. Additional restrictions are defined for two nodes to constrain the model. Displacements are incrementally applied to nodes at the load application points (or lines in the three-dimensional model built).

Figure 5 shows the stress–displacement relationship (the grey line) obtained by finite element analysis. Tensile stress in the specimen is calculated based on the beam theory using the gross moment of inertia up to crack initiation. The black line (in Figure 5) reflects the modification of the overall response after the activation of the self-healing mechanism. Cracks' sealing results in a partial regain of the rigidity and the load-carrying capacity of the damaged specimen. Healing starts after the initiation of macroscopic crack or when the critical value of the damage variable reaches its critical value (in some finite elements). The numerical algorithm implies the deactivation of finite elements with critical damage. Deactivated finite elements do not contribute to the structural element rigidity. The self-healing mechanism reverses this process. With crack reclosure, cracked regions partially restore their initial rigidity.

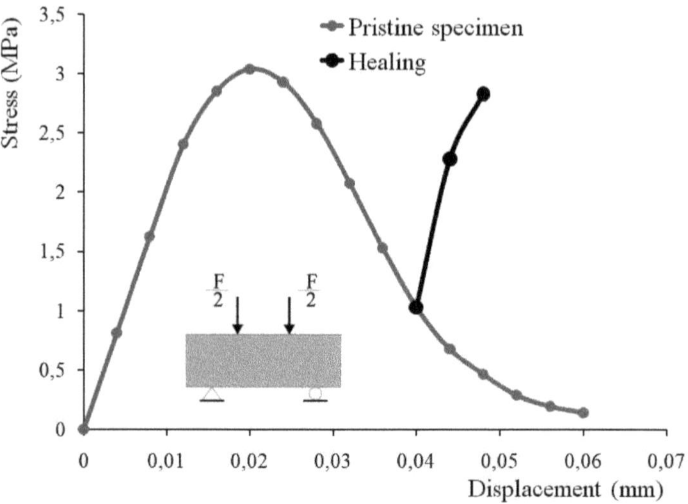

Figure 5. Tension by flexure test: stress–displacement relationship obtained by finite element analysis.

3.3. *Torsion Test*

The finite element simulation reported in this section reproduces the response of a cylindrical concrete specimen subjected to torsion without the ambition to reproduce accurately any experimental setup. Figure 6 displays the finite element model built. It contains a total of 7191 SOLID185 finite elements. Rigid regions are defined where nodes are restrained and where the load is applied to homogenize stress, strain, and damage distributions in the concrete specimen. In the framework of the used cylindrical coordinate system (R, θ, z), the model is constrained by restraining all nodes in the specified region in global z- and tangential θ-directions. Some nodes are also restrained in the radial direction. In another region, displacements are incrementally applied to circumferential nodes, appropriate to model torsional loading.

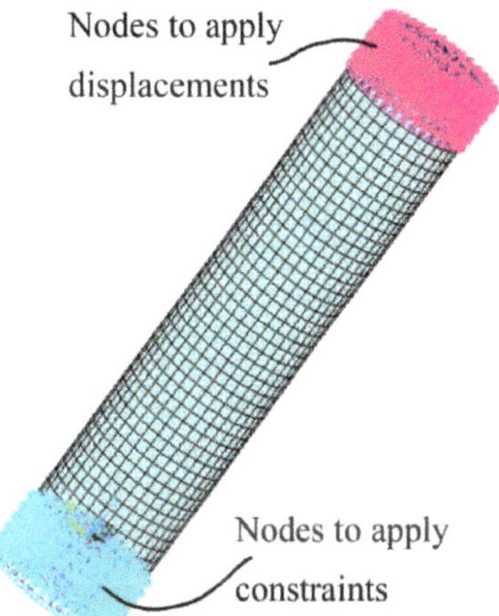

Figure 6. Finite element model built to reproduce the behavior of concrete specimens subjected to torsion: generated finite element mesh and applied boundary conditions.

Figure 7 presents a comparison between the responses of two concrete specimens loaded in torsion as one of them is subjected to healing after having been damaged. The same load sequence for the incrementally imposed displacements applies for both specimens. The provided comparison between the two relationships (angle of the applied rotation-reactive moment) outlines the positive effect of the self-healing of cracks. As a result of the healing, the apparent regain in the specimen rigidity follows the degradation observed in the pre-loading phase.

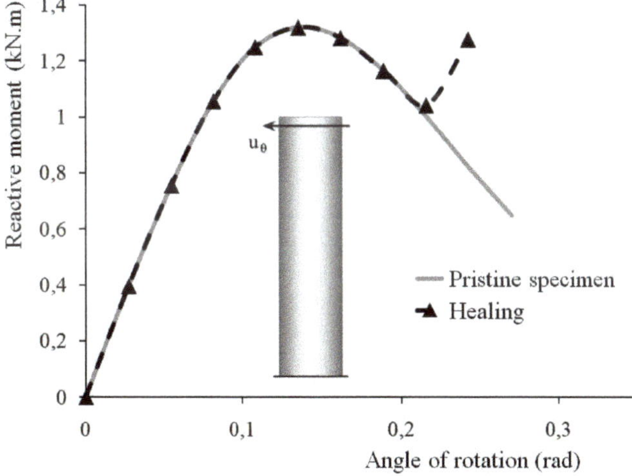

Figure 7. Concrete specimens subjected to torsion: numerically obtained constitutive relations for a specimen containing (dotted black line) and for a specimen without (continuous grey line) healing agent.

4. Discussion

Results obtained by finite element simulation have illustrated the effect of the self-healing of cracks. The strength recovery is 109.2%, 93%, and 96.75% for the specimens loaded in compression, flexure, and torsion, respectively. By assumption, the strength recovery is the ratio of the peak values of the stress or moment in the healed and the pristine specimens.

Further implementation of the employed numerical procedure (as a design tool) requires extensive experimental work. The identification of the model parameters, specifically of the critical value of the damage variable, is inherently empirical. Presumably, the aforementioned critical value of the damage variable is directly related to the crack opening and, thereby, the healing process initiation.

The damage-based constitutive relation slightly differs from that implemented in recent research. For example, in [51], the tensile strength of concrete is involved in the damage evolution law, whereas in the present study, the concrete tensile strength is to be matched via the identification of the model constants (please see equation 2). Furthermore, the specimens (for the direct tension test) studied in [51] contain a predefined notch. The specimens modeled herein are initially homogeneous and isotropic. They do not contain any initially defined precursors of cracking.

Very detailed modeling at the mesoscopic scale has been proposed by [70]. To model the fracture at the mesoscale, cohesive elements are inserted at possible failure locations, appropriate constitutive relations and failure criteria are defined. Investigation of some self-healing concrete parameters (such as volume fractions of capsules and aggregates and core-shell thickness of capsules) on the macroscopic response of a specimen loaded in tension. However, the presented results have been limited to the simulation of specimens containing a self-healing agent. Qualitatively, the obtained results reproduce adequately the macroscopic stress–strain relationship for concrete subjected to tension. In contrast, the modeling employed herein accounts implicitly for the action of the healing agent by increasing the value of the damage variable in the modeled compression, tension by flexure, and torsion tests.

Although the self-healing of cracks emerges as a technique capable of enhancing the durability and sustainability of structures, it still possesses some limitations. The major one is that self-healing results in only a partial regaining of the performances of the undamaged material. According to the vast majority of literature sources, the healing rate remains less than 100%. For comparison, alternative techniques for refurbishment, such as strengthened with externally bonded composite materials, provide a net increase in the load-carrying capacity and stiffness. Also, engineered self-healing implies taking actions during construction (such as the self-healing agent encapsulation). In this context, post-strengthening appears to be more flexible for existing structures, especially not equipped for self-healing. On the other hand, when programmed, the engineered self-healing is expected to run as an autonomous process presuming less maintenance and repairs (during the life cycle of a structure).

Despite some shortcomings, the self-healing of cracks remains a field of active research because of the attractive perspectives. Besides autonomous healing, self-healing includes mechanisms inherent for concrete. Accurate quantification of the natural phenomena in the concrete leading to its longevity will contribute to a more efficient design of sustainable structures.

5. Conclusions

Numerical simulations of the self-healing process in plain concrete specimens have been presented and discussed. Results obtained by finite element simulations of a compression test, tension by flexure test, and torsion test for healed and pristine specimens of the same geometry have been compared to predict the recovery in strength.

Both material degradation and healing are modeled implicitly in the framework of the continuum damage mechanics. The numerical algorithm integrates into the finite element analysis an original procedure designed to simulate the strain-softening concrete response and extended to account for the effects of healing. The computational procedure uses only the finite element method. It does not include any additional computational scheme to decouple effects due to different mechanisms (i.e., the damage accumulation and crack initiation and propagation). The framework of the continuum damage mechanics is preferred to that of fracture mechanics as it enables modeling of crack propagation and initiation in an initially homogeneous and isotropic material (i.e., there is no need to define initial defect that will subsequently induce macroscopic cracking).

Numerical results reproduce the response of the pristine specimens well, whereas the estimated recovery in strength for some tests is unexpectedly high, accounting for the empirical data provided by previous (mainly experimental) research works. In this context, a forthcoming experimental campaign can provide additional empirical background for the model validation and the identification of the material constants.

The presented study considers only the healing of specimens subjected to a quasi-static loading. Simulations of the behavior of structural elements under various loading conditions are forthcoming.

Funding: This research received no external funding.

Institutional Review Board Statement: Not applicable.

Informed Consent Statement: Not applicable.

Data Availability Statement: The study employs data available in publicly accessible repositories as well as data available in publicly accessible repositories that do not issue DOIs.

Conflicts of Interest: The author declares no conflict of interest.

References

1. Lv, Z.; Chen, D. Overview of recent work on self-healing in cementitious materials. *Mater. Constr.* **2014**, *64*, e034. [CrossRef]
2. Mihashi, H.; Nishiwaki, T. Development of engineered self-healing and self-repairing concrete-state-of-the-art report. *J. Adv. Concr. Technol.* **2012**, *10*, 170–184. [CrossRef]
3. De Belie, N.; Gruyaert, E.; Al-Tabbaa, A.; Antonaci, P.; Baera, C.; Bajāre, D.; Darquennes, A.; Davies, R.; Ferrara, L.; Jefferson, A.; et al. A review on self-healing concrete for damage management of structures. *Mater. Interface Sci.* **2018**, *5*, 1800074. [CrossRef]
4. Chen, H.-J.; Peng, C.-F.; Tang, C.-W.; Chen, Y.-T. Self-Healing Concrete by Biological Substrate. *Materials* **2019**, *12*, 4099. [CrossRef]
5. Stuckrath, C.; Serpell, R.; Valenzuela, L.M.; Lopez, M. Quantification of chemical and biological calcium carbonate precipitation: Performance of self-healing in reinforced mortar containing chemical admixtures. *Cem. Concr. Compos.* **2014**, *50*, 10–15. [CrossRef]
6. Seifan, M.; Samani, A.K.; Berenjian, A. Bioconcrete: Next generation of self-healing concrete. *Appl. Microbiol. Biotechnol.* **2016**, *100*, 2591–2602. [CrossRef] [PubMed]
7. Ghosh, S.; Biswas, M.; Chattopadhyay, B.D.; Mandal, S. Microbial activity on the microstructure of bacteria modified mortar. *Cem. Concr. Compos.* **2009**, *31*, 93–98. [CrossRef]
8. Westerbeek, T. *Self-Healing Materials Radio Netherlands*; John Wiley & Sons: Hoboken, NJ, USA, 2005.
9. Hearn, N.; Morley, C.T. Self-sealing property of concrete—Experimental evidence. *Mater. Struct.* **1997**, *30*, 404–411. [CrossRef]
10. Jacobsen, S.; Marchand, J.; Hornain, H. SEM observations of the microstructure of frost deteriorated and self-healed concretes. *Cem. Concr. Res.* **1995**, *25*, 1781–1790. [CrossRef]
11. Li, G.; Liu, S.; Niu, M.; Liu, Q.; Yang, X.; Deng, M. Effect of granulated blast furnace slag on the self-healing capability of mortar incorporating crystalline admixture. *Constr. Build. Mater.* **2020**, *239*, 117818. [CrossRef]
12. Litina, C.; Al-Tabbaa, A. First generation microcapsule-based self-healing cementitious construction repair materials. *Constr. Build. Mater.* **2020**, *255*, 119389. [CrossRef]
13. Roig-Flores, M.; Moscato, S.; Serna, P.; Ferrara, L. Self-healing capability of concrete with crystalline admixtures in different environments. *Constr. Build. Mater.* **2015**, *86*, 1–11. [CrossRef]
14. Medjigbodo, S.; Bendimerad, A.Z.; Rozière, E.; Loukili, A. How do recycled concrete aggregates modify the shrinkage and self-healing properties? *Cem. Concr. Compos.* **2018**, *86*, 72–86. [CrossRef]
15. Nishikawa, T.; Yoshida, J.; Sugiyama, T.; Fujino, Y. Concrete crack detection by multiple sequential image filtering. *Comput. Aided Civ. Infrastruct. Eng.* **2012**, *27*, 29–47. [CrossRef]
16. Snoeck, D.; Van Tittelboom, K.; Steuperaert, S.; Dubruel, P.; De Belie, N. Self-healing cementitious materials by the combination of microfibres and superabsorbent polymers. *J. Intell. Mater. Syst. Struct.* **2014**, *25*, 13–24. [CrossRef]
17. Hung, C.C.; Su, Y.F. Medium-term self-healing evaluation of Engineered Cementitious Composites with varying amounts of fly ash and exposure durations. *Constr. Build. Mater.* **2016**, *118*, 194–203. [CrossRef]

18. Ferrara, L.; Van Mullem, T.; Alonso, M.C.; Antonaci, P.; Borg, R.P.; Cuenca, E.; Jefferson, A.; Ng, P.L.; Peled, A.; Roig-Flores, M.; et al. Experimental characterization of the self-healing capacity of cement based materials and its effects on the material performance: A state of the art report by COST Action SARCOS WG2". *Constr. Build. Mater.* **2018**, *167*, 115–142. [CrossRef]
19. Di Prisco, M.; Ferrara, L.; Lamperti, M.G. Double edge wedge splitting (DEWS): An indirect tension test to identify post-cracking behaviour of fibre reinforced cementitious composites. *Mater. Struct.* **2013**, *46*, 1893–1918. [CrossRef]
20. Ferrara, L.; Ozyurt, N.; Di Prisco, M. High mechanical performance of fibre reinforced cementitious composites: The role of "casting-flow induced" fibre orientation. *Mater. Struct.* **2011**, *44*, 109–128. [CrossRef]
21. Monte, F.L.; Ferrara, L. Tensile behaviour identification in Ultra-High Performance Fibre Reinforced Cementitious Composites: Indirect tension tests and back analysis of flexural test results. *Mater. Struct.* **2020**, *53*, 1–12. [CrossRef]
22. Ferrara, L.; Krelani, V.; Moretti, F. Autogenous healing on the recovery of mechanical performance of High Performance Fibre Reinforced Cementitious Composites (HPFRCCs): Part 2–Correlation between healing of mechanical performance and crack sealing. *Cem. Concr. Compos.* **2016**, *73*, 299–315. [CrossRef]
23. Droval, G.; Feller, J.F.; Salagnac, P.; Glouannec, P. Conductive polymer composites with double percolated architecture of carbon nanoparticles and ceramic microparticles for high heat dissipation and sharp PTC switching. *Smart Mater. Struct.* **2008**, *17*, 025011. [CrossRef]
24. Ferrara, L.; Krelani, V.; Moretti, F.; Flores, M.R.; Ros, P.S. Effects of autogenous healing on the recovery of mechanical performance of High Performance Fibre Reinforced Cementitious Composites (HPFRCCs): Part 1. *Cem. Concr. Compos.* **2017**, *83*, 76–100. [CrossRef]
25. Cuenca, E.; Tejedor, A.; Ferrara, L. A methodology to assess crack-sealing effectiveness of crystalline admixtures under repeated cracking-healing cycles. *Constr. Build. Mater.* **2018**, *179*, 619–632. [CrossRef]
26. Cuenca, E.; Mezzena, A.; Ferrara, L. Synergy between crystalline admixtures and nanoconstituents in enhancing autogenous healing capacity of cementitious composites under cracking and healing cycles in aggressive waters. *Constr. Build. Mater.* **2021**, *266*, 121447. [CrossRef]
27. Cuenca, E.; D'Ambrosio, L.; Lizunov, D.; Tretjakov, A.; Volobujeva, O.; Ferrara, L. Mechanical properties and self-healing capacity of ultra high performance Fibre Reinforced Concrete with alumina nanofibres: Tailoring Ultra High Durability Concrete for aggressive exposure scenarios. *Cem. Concr. Compos.* **2021**, *118*, 103956. [CrossRef]
28. Monte, F.L.; Ferrara, L. Self-Healing Characterization of UHPFRCC with Crystalline Admixture: Experimental Assessment via Multi-Test/Multi-Parameter Approach. *Constr. Build. Mater.* **2021**, *283*, 122579. [CrossRef]
29. Bangert, F.; Kuhl, D.; Meschke, G. Chemo-hygro-mechanical modelling and numerical simulation of concrete deterioration caused by alkali-silica reaction. *Int. J. Numer. Anal. Methods Geomech.* **2004**, *28*, 689–714. [CrossRef]
30. Wittmann, F. Structure of concrete with respect to crack formation. *Fract. Mech. Concr.* **1983**, *43*, 6.
31. Mazars, J.; Pijaudier-Cabot, G. Continuum damage theory—Application to concrete. *J. Eng. Mech.* **1989**, *115*, 345–365. [CrossRef]
32. Schlangen, E.; Van Mier, J. Micromechanical analysis of fracture of concrete. *Int. J. Damage Mech.* **1992**, *1*, 435–454. [CrossRef]
33. Lemaitre, J. *A Course on Damage Mechanics*; Springer: Berlin/Heidelberg, Germany, 1996.
34. Meschke, G.; Lackner, R.; Mang, H.A. An anisotropic elastoplastic-damage model for plain concrete. *Int. J. Numer. Methods Eng.* **1998**, *42*, 703–727. [CrossRef]
35. Wriggers, P.; Moftah, S. Mesoscale models for concrete: Homogenisation and damage behaviour. *Finite Elem. Anal. Des.* **2006**, *42*, 623–636. [CrossRef]
36. Pedersen, R.; Simone, A.; Sluys, L. An analysis of dynamic fracture in concrete with a continuum visco-elastic visco-plastic damage model. *Eng. Fract. Mech.* **2008**, *75*, 3782–3805. [CrossRef]
37. Hain, M.; Wriggers, P. Numerical homogenization of hardened cement paste. *Comput. Mech.* **2008**, *42*, 197–212. [CrossRef]
38. Kim, S.M.; Al-Rub, R.K.A. Meso-scale computational modeling of the plastic-damage response of cementitious composites. *Cem. Concr. Res.* **2011**, *41*, 339–358. [CrossRef]
39. Unger, J.F.; Eckardt, S.; Könke, C. A mesoscale model for concrete to simulate mechanical failure. *Comput. Concr.* **2011**, *8*, 401–423. [CrossRef]
40. Lohaus, L.; Oneschkow, N.; Wefer, M. Design model for the fatigue behaviour of normal-strength, high-strength and ultra-high-strength concrete. *Struct. Concr.* **2012**, *13*, 182–192. [CrossRef]
41. Anderson, T.L. *Fracture Mechanics: Fundamentals and Applications*; CRC Press: Boca Raton, FL, USA, 2017.
42. Zreid, I.; Kaliske, M. A gradient enhanced plasticity–damage microplane model for concrete. *Comput. Mech.* **2018**, *62*, 1239–1257. [CrossRef]
43. Schäfer, N.; Gudžulić, V.; Timothy, J.J.; Breitenbücher, R.; Meschke, G. Fatigue behavior of HPC and FRC under cyclic tensile loading: Experiments and modeling. *Struct. Concr.* **2019**, *20*, 1265–1278. [CrossRef]
44. Gebuhr, G.; Pise, M.; Sarhil, M.; Anders, S.; Brands, D.; Schröder, J. Analysis and evaluation of the pull-out behavior of hooked steel fibers embedded in high and ultra-high performance concrete for calibration of numerical models. *Struct. Concr.* **2019**, *20*, 1254–1264. [CrossRef]
45. Yang, S.; Aldakheel, F.; Caggiano, A.; Wriggers, P.; Koenders, E. A Review on Cementitious Self-Healing and the Potential of Phase-Field Methods for Modeling Crack-Closing and Fracture Recovery. *Materials* **2020**, *13*, 5265. [CrossRef] [PubMed]
46. Ma, R.; Sun, W. FFT-based solver for higher-order and multi-phase-field fracture models applied to strongly anisotropic brittle materials. *Comput. Methods Appl. Mech. Eng.* **2020**, *362*, 112781. [CrossRef]

47. Tarafder, P.; Dan, S.; Ghosh, S. Finite deformation cohesive zone phase field model for crack propagation in multi-phase microstructures. *Comput. Mech.* **2020**, *66*, 723–743. [CrossRef]
48. De Lorenzis, L.; Gerasimov, T. Numerical implementation of phase-field models of brittle fracture. In *Modeling in Engineering Using Innovative Numerical Methods for Solids and Fluids;* Springer: Berlin/Heidelberg, Germany, 2020; pp. 75–101.
49. Aldakheel, F.; Hudobivnik, B.; Artioli, E.; Beirao da Veiga, L.; Wriggers, P. Curvilinear Virtual Elements for Contact Mechanics. *Comput. Methods Appl. Mech. Eng.* **2020**, *372*, 113394. [CrossRef]
50. Kaminski, M. Uncertainty analysis in solid mechanics with uniform and triangular distributions using stochastic perturbation-based Finite Element Method. *Finite Elem. Anal. Des.* **2022**, *200*, 103648. [CrossRef]
51. Freeman, B.L.; Bonilla-Villalba, P.; Mihai, I.C.; Alnaas, W.F.; Jefferson, A.D. A specialised finite element for simulating self-healing quasi-brittle materials. *Adv. Model. Simul. Eng. Sci.* **2020**, *7*, 32. [CrossRef]
52. Oñate, E.; Cornejo, A.; Zárate, F.; Kashiyama, K.; Franci, A. Combination of the finite element method and particle-based methods for predicting the failure of reinforced concrete structures under extreme water forces. *Eng. Struct.* **2022**, *251*, 113510. [CrossRef]
53. Jefferson, A.D.; Mihai, I.C.; Tenchev, R.; Alnaas, W.F.; Cole, G.; Lyons, P. A plastic-damage-contact constitutive model for concrete with smoothed evolution functions. *Comput. Struct.* **2016**, *169*, 40–56. [CrossRef]
54. Gardner, D.; Jefferson, A.D.; Hofman, A. Investigation of capillary fow in discrete cracks in cementitious materials. *Cem. Concr. Res.* **2012**, *42*, 972–981. [CrossRef]
55. Gardner, D.; Jefferson, A.D.; Hofman, A.; Lark, R. Simulation of the capillary fow of an autonomic healing agent in discrete cracks in cementitious materials. *Cem. Concr. Res.* **2014**, *58*, 35–44. [CrossRef]
56. Gardner, D.; Herbert, D.; Jayaprakash, M.; Jefferson, A.D.; Paul, A. Capillary flow characteristics of an autogenic and autonomic healing agent for self-healing concrete. *J. Mater. Civ Eng.* **2017**, *29*, 4017228. [CrossRef]
57. Selvarajoo, T.; Davies, R.E.; Gardner, D.R.; Freeman, B.L.; Jefferson, A.D. Characterisation of a vascular self-healing cementitious materials system: Fow and curing properties. *Constr. Build. Mater.* **2020**, *245*, 118332. [CrossRef]
58. Jiang, T.S.; Soo-Gun, O.H.; Slattery, J.C. Correlation for dynamic contact angle. *J. Colloid Interface Sci.* **1979**, *69*, 74–77. [CrossRef]
59. Dias-da-Costa, D.; Alfaiate, J.; Sluys, L.J.; Julio, E. Towards a generalization of a discrete discontinuity approach. *Comput. Methods Appl. Mech. Eng.* **2009**, *198*, 3670–3681. [CrossRef]
60. Alfaiate, J.; Simone, A.; Sluys, L.J. Non-homogeneous displacement jumps in strong embedded discontinuities. *Int. J. Solids Struct.* **2003**, *40*, 5799–5817. [CrossRef]
61. Zárate, F.; Cornejo, A.; Oñate, E. A three-dimensional FEM–DEM technique for predicting the evolution of fracture in geomaterials and concrete. *Comput. Part. Mech.* **2018**, *5*, 411–420. [CrossRef]
62. Cornejo, A.; Mataix, V.; Zárate, F.; Oñate, E. Combination of an adaptive remeshing technique with a coupled FEM-DEM approach for analysis of crack propagation problems. *Comput. Part. Mech.* **2020**, *7*, 735–752. [CrossRef]
63. Cornejo, A.; Franci, A.; Zárate, F.; Oñate, E. A fully Lagrangian formulation for fluid-structure interaction problems with free-surface flows and fracturing solids. *Comput. Struct.* **2021**, *250*, 106532. [CrossRef]
64. Zárate, F.; Oñate, E. A simple FEM–DEM technique for fracture prediction in materials and structures. *Comput. Part. Mech.* **2015**, *2*, 301–314. [CrossRef]
65. Zhelyazov, T.; Ivanov, R. Modeling of the behavior of concrete element containing a self-healing agent. In Proceedings of the Structural Engineering for Future Societal Needs, IABSE Congress Ghent 2021, Ghent, Belgium, 22–24 September 2021; pp. 79–84.
66. Zhelyazov, T.; Ivanov, R. Numerical simulation of cracking in concrete using damage mechanics. In *Resilient Technologies for Sustainable Infrastructure, Proceedings of the IABSE Congress Christchurch 2020, Christchurch, New Zealand, 2–4 September 2020;* Taylor & Francis: Abingdon, UK, 2020; pp. 881–889.
67. Lemaître, J.; Desmorat, R. *Engineering Damage Mechanics: Ductile, Creep, Fatigue and Brittle Failures;* Springer: Berlin/Heidelberg, Germany, 2004.
68. Mazars, J.J. Damage Mechanics Application to Nonlinear Response and Failure Behavior of Structural Concrete. Ph.D. Thesis, Paris 6 University, Paris, France, 1984.
69. FIB—Fédération Internationale du Béton. *FIB Model Code for Concrete Structures 2010;* FiB-International: Lausanne, Switzerland, 2013.
70. Mauludin, L.M.; Zhuang, X.; Rabczuk, T. Computational modeling of fracture in encapsulation-based self-healing concrete using cohesive elements. *Com. Struct.* **2018**, *196*, 63–75. [CrossRef]

 *materials*

Article

Mechanical and Durability Properties of Cementless Concretes Made Using Three Types of CaO-Activated GGBFS Binders

Woo Sung Yum, Juan Yu, Dongho Jeon, Haemin Song, Sungwon Sim, Do Hoon Kim and Jae Eun Oh *

School of Urban and Environmental Engineering, Ulsan National Institute of Science and Technology (UNIST), UNIST-gil 50, Ulju-gun, Ulsan 44919, Korea; reikoku@nate.com (W.S.Y.); cp9eins@unist.ac.kr (J.Y.); dhjeon0707@gmail.com (D.J.); haemin93@unist.ac.kr (H.S.); conor123@naver.com (S.S.); elvis03@unist.ac.kr (D.H.K.)

* Correspondence: ohjaeeun@unist.ac.kr

Abstract: This study examined the mechanical and durability properties of CaO-activated ground-granulated blast-furnace slag (GGBFS) concretes made with three different additives ($CaCl_2$, $Ca(HCOO)_2$, and $Ca(NO_3)_2$) and compared their properties to the concrete made with 100% Ordinary Portland Cement (OPC). All concrete mixtures satisfied targeted air content and slump ranges but exhibited significantly different mechanical and durability properties. The CaO-activated GGBFS concretes showed different strength levels, depending on the type of additive. The added $CaCl_2$ was the most effective, but $Ca(NO_3)_2$ was the least effective at increasing mechanical strength in the CaO-activated GGBFS system. The OPC concrete showed the most excellent freezing–thawing resistance in the durability test, but only the CaO-activated GGBFS concrete with $CaCl_2$ exhibited relatively similar resistance. In addition, the chemical resistance was significantly dependent on the type of acid solution and the type of binder. The OPC concrete had the best resistance in the HCl solution, while all CaO-activated GGBFS concretes had relatively low resistances. However, in the H_2SO_4 solution, all CaO-activated GGBFS concretes had better resistance than the OPC concrete. All concrete with sulfate ions had ettringite before immersion. However, when they were immersed in HCl solution, ettringite tended to decrease, and gypsum was generated. Meanwhile, the CaO-activated GGBFS concrete with $CaCl_2$ did not change the type of reaction product, possibly due to the absence of ettringite and $Ca(OH)_2$. When immersed in an H_2SO_4 solution, ettringite decreased, and gypsum increased in all concrete. In addition, the CaO-activated concrete with $CaCl_2$ had a considerable amount of gypsum; it seemed that the dissolved C-S-H and calcite, due to the low pH, likely produced Ca^{2+} ions, and gypsum formed from the reaction between Ca^{2+} and H_2SO_4.

Keywords: CaO-activation; auxiliary activator; GGBFS; chemical resistance; freezing; thawing

Citation: Yum, W.S.; Yu, J.; Jeon, D.; Song, H.; Sim, S.; Kim, D.H.; Oh, J.E. Mechanical and Durability Properties of Cementless Concretes Made Using Three Types of CaO-Activated GGBFS Binders. *Materials* **2022**, *15*, 271. https://doi.org/10.3390/ma15010271

Academic Editor: Jean-Marc Tulliani

Received: 8 November 2021
Accepted: 28 December 2021
Published: 30 December 2021

1. Introduction

With global warming, abnormal climates are occurring worldwide, causing many problems [1,2]. The international community agreed to reduce greenhouse gas emissions through the Paris Climate Agreement, to prevent accelerating global warming.

There is increasing demand for materials to replace Portland cement [3,4] because the OPC industry accounts for 8 to 10% of worldwide CO_2 emissions, accelerating global warming [5]. Therefore, many studies have been conducted on developing alternative binders with a low carbon footprint, using various industrial by-products, such as fly ash or ground-granulated blast-furnace slag (GGBFS) to replace OPC [6–8]. Fly ash-based geopolymers using fly ash as the main material use alkaline activators, such as sodium hydroxide (NaOH), potassium hydroxide (KOH), or sodium silicates (e.g., water glass). Palomo et al. [9] reported that geopolymer activated with NaOH 8–12 M cured at 85 °C showed a compressive strength of about 40 MPa, and it reached nearly 90 MPa when water glass was added to the binder. Sun et al. [10] explored the effect of three different aggregate sizes on the strength of geopolymers. M.A.M Ariffin et al. [11] showed that geopolymer concrete exhibited better sulfuric acid resistance than OPC concrete. Although geopolymer has high compressive strength and durability, it has the disadvantage that its performance changes depending on the type of activator and curing temperature [12]. Furthermore, U Yurt et al. [13] showed the performance of alkali-activated GGBFS composite varying with the different alkali activators and activation temperature. In addition, many studies [14,15] reported that alkaline activated GGBFS binders have high compressive strength and chemical resistance. However, similar to geopolymers, GGBFS-based cementless binders have drawbacks in the following usages [16]: (1) high material costs, (2) user safety, and (3) difficulty in producing a one-part binder. Despite these problems, alkali-activated binders produce comparable mechanical properties compared to the OPC binder [17,18]; thus, many studies have been conducted at the concrete level to validate the mechanical and durability properties [19,20].

On the other hand, the CaO-activated binder system [21,22] relatively overcame the disadvantages mentioned above of alkali-activated binders. Kim et al. [23] studied CaO-activated GGBFS binder, and Yum et al. [21,24,25] enhanced mechanical properties of CaO-activated GGBFS binders by adding different types of additional activators. In addition, there are studies to develop cementless binders by activating fly ash using CaO and other chemical additives [22,26]. Although the CaO-activated binders can potentially apply to construction fields, no study has been conducted at the concrete level on mechanical properties and durability.

Therefore, this study compared the mechanical and durability properties of various CaO-activated GGBFS concretes using additives, $CaCl_2$, $Ca(HCOO)_2$, and $Ca(NO_3)_2$. Mechanical properties, such as compressive strength, tensile splitting strength, and elastic modulus were measured. In addition, the durability properties were evaluated by the freezing and thawing cycle test and the chemical resistance test using hydrochloric acid (HCl) and sulfuric acid (H_2SO_4). Changes in hydration products were examined through powder X-ray diffraction (XRD) measurement and thermogravimetry analysis (TGA).

2. Materials and Methods

Commercial OPC (type I, 42.5N) and GGBFS were used in this study. In addition, all used chemicals were analytical grade. CaO (Daejung, Korea) was the main activator, and $CaCl_2$ (Daejung, Korea), $Ca(HCOO)_2$ (Aladdin, China), $Ca(NO_3)_2$ (Sigma Aldrich, Burlington, MA, USA), and $CaSO_4$ (Daejung, Korea) were additives. The maximum size of coarse aggregate was 25 mm. Fine aggregate were classified as aggregates with a size less than 5 mm. All aggregates were used under saturated surface dry (SSD) conditions. In addition, a naphthalene-based commercial plasticizer was used in concrete mixtures.

XRD, XRF, and particle size distribution analysis were performed to identify the characteristics of the raw materials (i.e., OPC and GGBFS). XRD patterns were recorded by means of a D/Max2500 V diffractometer (Rigaku, Japan) with an incident beam of Cu-kα radiation (λ = 1.5418 nm) in a 2θ scanning range from 5° to 60°; X-ray fluorescence was conducted with S8-Tiger spectrometer (Bruker, Billerica, MA, USA).

Particle size distribution results (Figure 1) show that the average particle sizes of OPC and GGBFS were 25 and 30 μm, respectively. Table 1 exhibits the XRF results of OPC and GGBFS, and they mainly consisted of CaO and SiO_2. In Figure 2, the XRD results show that akermanite [$Ca_2Mg(Si_2O_7)$] was identified in GGBFS, and GGBFS primarily consisted of an amorphous phase. On the other hand, dicalcium silicate (C_2S), tricalcium silicate (C_3S), tetracalcium aluminoferrite (C_4AF), tricalcium aluminate (C_3A), and gypsum ($CaSO_4$) were identified in OPC [27].

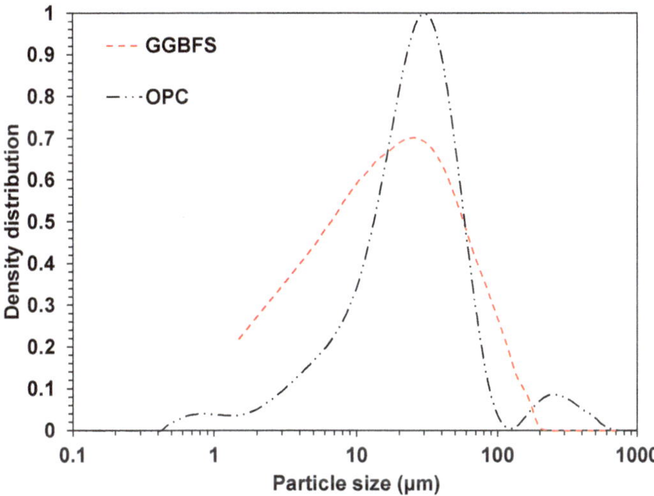

Figure 1. Particle size distributions of OPC and GGBFS.

Table 1. Oxide compositions: OPC and GGBFS (wt%).

Oxide	OPC	GGBFS
CaO	64.05	59.48
SiO_2	18.59	23.05
Al_2O_3	4.45	9.72
MgO	3.14	2.54
SO_3	3.55	1.37
TiO_2	0.29	0.98
MnO	0.19	0.83
Fe_xO_y	3.61	0.68
K_2O	1.23	0.68
Na_2O	0.29	0.28
Others	0.61	0.4

Table 2. Material proportions of binders used in concrete mixture samples (wt%).

Binder Type	OPC	GGBFS	CaO	$CaCl_2$	$Ca(HCOO)_2$	$Ca(NO_3)_2$	$CaSO_4$	Sum
P-binder	100	0	0	0	0	0	0	100
C-binder	0	94	4	2	0	0	0	100
F-binder	0	88	4	0	3	0	5	100
N-binder	0	86	4	0	0	5	5	100

The material proportions of binders used to produce concrete samples are shown in Table 2. The P-binder was 100% OPC. The other binders (C-, F-, and N-binders) were CaO-activated GGBFS binders, which used CaO as a primary activator for activating GGBFS while included different additives. The C-binder, F-binder, and N-binder used $CaCl_2$, $Ca(HCOO)_2$, and $Ca(NO_3)_2$, respectively, as additives, and their material proportions were determined via previous studies [21,24,25]. Note that F-binder and N-binder additionally used five weight percentages of $CaSO_4$ to enhance the early compressive strength of the binders.

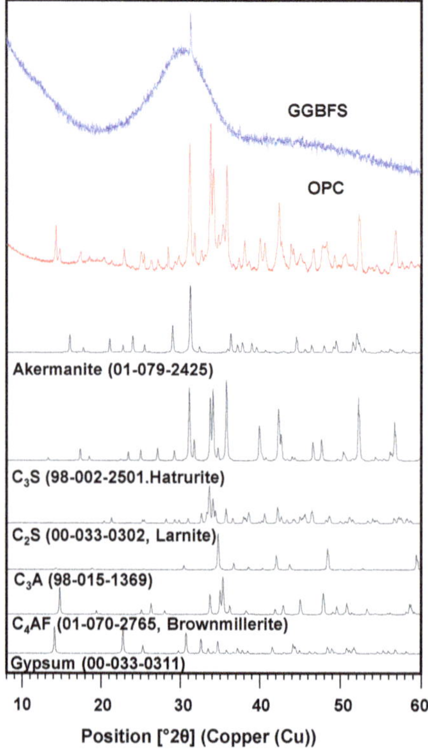

Figure 2. XRD patterns of OPC and GGBFS with identified phases.

The mixture proportions of concrete are given in Table 3, and the samples were labeled similarly after the used binders. According to the Korea Concrete Specification [28], concrete mixture proportions were designed, and the targeted compressive strength was above 45 MPa. Targeted slump and air content ranges were 5–15 cm and 3–6%, respectively.

Table 3. Mixture proportions of concrete samples.

Sample	Binder Type	G_{max} (mm)	w/b (wt%)	s/a (vol%)	Unit Weight (kg/m^3)				
					Water	Binder	FA	CA	Plasticizer
P-CON	P-binder	25	36.73	42.31	176.63	480.82	687.81	955.86	0.14
C-CON	C-binder						676.34	939.93	0.14
F-CON	F-binder						673.87	936.48	0.14
N-CON	N-binder						673.87	936.48	0.14

Note: FA denotes fine aggregate; CA denotes coarse aggregate; G_{max} denotes the ratios of the maximum size of coarse aggregate; w/b denotes water to binder weight ratio, and s/a denotes sand to aggregate volume.

For concrete production, all mixing procedures followed the Korea Standard (KS) F 2403 [29]. The fresh concrete samples were cast in cylindrical molds (φ 10 cm × 20 cm^3) to test determine compressive strength, split strength, elastic modulus, and chemical resistance, and samples in prism molds (10 × 10 × 40 cm^3) for freezing and thawing cycle tests. The fresh concrete samples were compacted and cured at 23 °C and 99% relative humidity.

Slump and air content were examined according to the KS F 2402 and KS F 2421 [30,31]. The slump and air content tests were conducted once for each mixture proportion. Triplicate samples were tested for compressive and split strength of each mixture at 3, 7, and 28 days of curing. Chord elastic modulus was measured for the 28-day cured samples, and triplicate specimens were tested for each mixture. The elastic modulus was calculated using the chord modulus using stress-strain curves obtained from three linear variable differential transformers (LVDT) installed in longitudinal directions of concrete [27]. The measured elastic modulus was compared with models proposed by the American Concrete Institute (ACI) 318 and the Euro-International Committee for Concrete—the International Federation for Pre-stressing (CEB-FIP); the CEP-FIP equations are as follows [32,33]:

$$\text{ACI Model}: \ E_c = {W_c}^{1.5} \times 0.043\sqrt{f_{28}}$$

$$\text{CEB}-\text{FIP}: \ E_c = 2.15 \times 10^4 {\left(\frac{f_{28}}{10}\right)}^{1/3}$$

where E_c is the elastic modulus (GPa), f_{28} is the 28-day compressive strength (MPa), and w_c is density (kg/m^3).

Freezing and thawing cycle resistance tests were performed using 14-day cured samples, and all experimental procedures followed by KS F 2456 [34]. In addition, three identical samples were used on each measurement date. The initial weight and dynamic modulus of elasticity were recorded for 14 days after curing, and then the samples were tested for freezing and thawing cycles. Each cycle consisted of −20 °C to 4 °C; the time of each cycle was less than 4 h, and the relative dynamic modulus of elasticity and weight change were measured every 30 cycles. Three identical samples were used for each concrete, and the experiments were conducted up to 300 cycles.

Chemical resistance tests were carried out according to ASTM C 267 [35]. The 28-day cured samples were immersed in a container with 5% HCl and 5% H_2SO_4 solutions, and the solutions were replaced once in 4 weeks. The weight changes were measured at 3, 7, 14, 28, 56, and 91 days after immersion to identify the evolution of reaction products. On each weight measurement, three identical specimens were weighed. Analysis of XRD and TGA identified reaction products after 28, 56, and 91 days of immersion. The exposed samples were fractured and finely ground using a mortar and pestle. In addition, XRD and TGA were conducted once for each measurement day and mixture proportion. In addition, SDT Q600 (TA Instruments, New Castle, DE, USA) was used to measure the reaction products of each sample with a heating rate of 30 °C/min from 25 to 1000 °C in a nitrogen atmosphere in an alumina pan.

3. Results

3.1. Slump and Air Content

Figure 3 displays the air content and slump measurement test results. The air content and the slump values satisfied the criteria specified in KS F 2421 (air content = 3–6%) and KS F 2402 (slump = 5–15 cm) [30,31], respectively.

3.2. Compressive and Split Strength

Compressive strengths of hardened concretes are shown in Figure 4, and the greatest compressive strength was measured from P-CON at all curing days. In addition, all mixtures showed increased strengths with increasing curing days. All cementless concrete samples exhibited significantly lower compressive strengths than P-CON at three days, mainly due to the lower reaction rate of CaO-activated GGBFS binders than OPC binder [36]. The type of used additives influenced the strength development of CaO-activated GGBFS concretes. $CaCl_2$ was the most effective in increasing strength, while $Ca(NO_3)_2$ was the least effective.

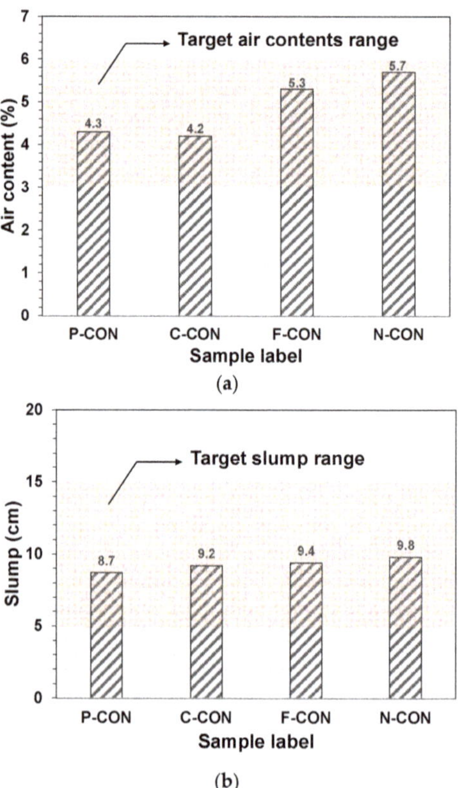

Figure 3. The test results: (**a**) air content and (**b**) slump of fresh concrete samples.

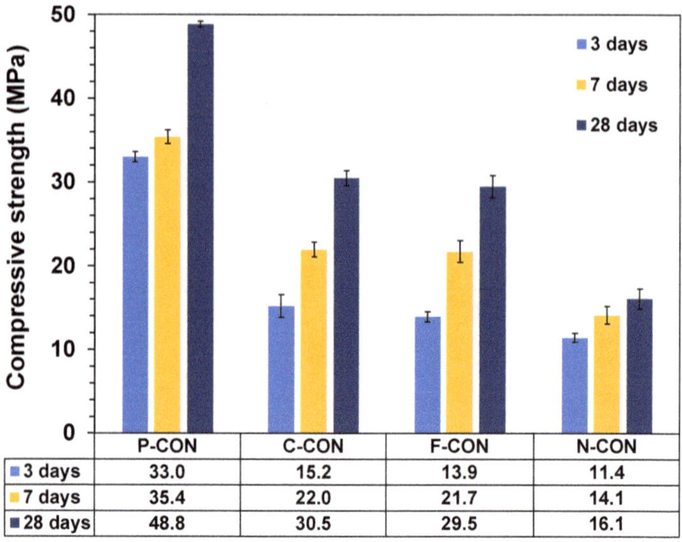

	P-CON	C-CON	F-CON	N-CON
3 days	33.0	15.2	13.9	11.4
7 days	35.4	22.0	21.7	14.1
28 days	48.8	30.5	29.5	16.1

Figure 4. Compressive strengths of hardened concrete samples.

The mixture proportion of concrete was designed following the Concrete Standard Specification in Korea [37], and the designed 28-day compressive strength (f_{cr}) was 45 MPa. The compressive strength was 48 MPa for P-CON, but the CaO-activated GGBFS concretes gained only about 10 to 30 MPa, significantly lower compressive strength than f_{cr}.

The split strengths are presented in Figure 5. The tensile strength of OPC concrete was reported as about 10 to 15% of the compressive strength [27]. The concrete specimens made with CaO-activated GGBFS binders having no OPC also exhibited a similar trend in this study.

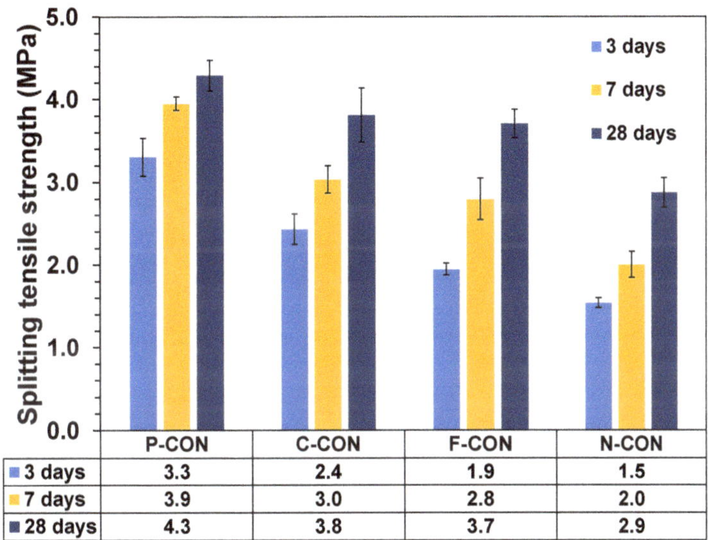

Figure 5. Split strengths of hardened concrete samples.

3.3. Chord Elastic Modulus

The measured elastic modulus values of concrete samples are given in Table 4 with the calculated values from ACI 318 and CEB-FIP model codes. It is known that the strength and elastic modulus had a proportional relationship in OPC concrete [27], and the same tendency was observed for all binder types of concrete in this study.

Table 4. The results of elastic modulus using experiments and model codes.

Sample Label	Compressive Strength at 28 Days (MPa)	Density (kg/m^3)	Elastic Modulus (GPa)		
			Exp.	ACI 318	CEB-FIP
P-CON	44.8	2415.5	28.8	34.1	41.5
C-CON	30.5	2372.6	20.9	27.4	35.7
F-CON	29.5	2349.7	19.1	26.6	34.7
N-CON	20.7	2360.2	17.2	19.7	29.1

When comparing the experimental values with the calculated values from the ACI and CEB-FIP models, the models showed significantly higher elastic modulus than the measured values. Previous studies similarly reported that the ACI and CEB-FIP models tended to overestimate the elastic modulus of OPC concrete, and it was observed for all types of concretes in this study [38–40].

3.4. Freezing and Thawing Cycle Tests

Figure 6 presents the freezing and thawing cycle test results for changing relative dynamic modulus of elasticity and weight loss with increasing cycles.

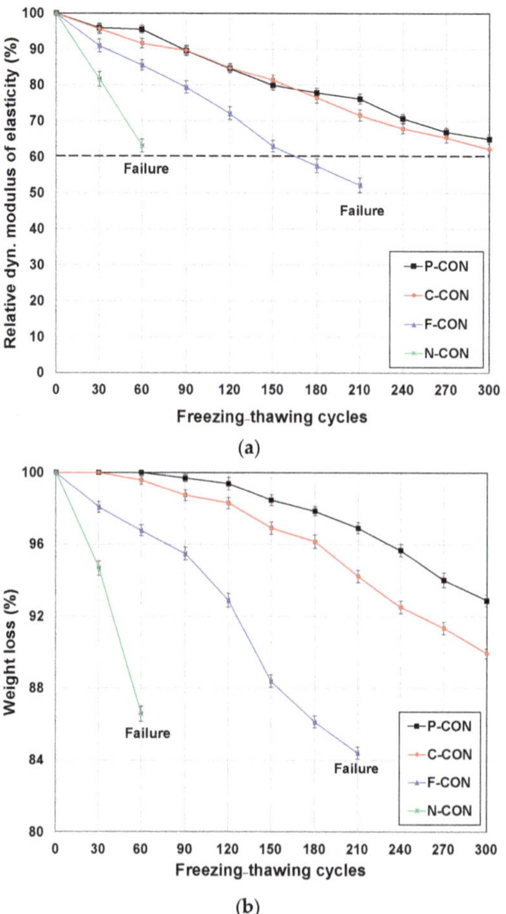

Figure 6. Freezing and thawing test results: (**a**) relative dynamic modulus of elasticity and (**b**) weight loss.

The dynamic elastic modulus of P-CON and C-CON decreased down to about 60% up to 300 cycles, while those of F-CON and N-CON were not measured because they were damaged significantly before reaching 300 cycles. Thus, F-CON and N-CON are vulnerable to freezing and thawing cycles.

It is worth noting that although C-CON and F-CON had similar compressive strengths and elastic modulus, they showed significantly different resistances to freezing and thawing cycles. However, a sample with a higher drop of the dynamic modulus tended to show a steeper decrease in the weight. Thus, the strength itself was a less accurate indicator than the dynamic modulus for estimating the ability to resist freezing and thawing cycles in this study.

For OPC concretes, KS F 2456 [34] determines that OPC concrete is susceptible to freezing and thawing cycles when OPC concrete samples are destroyed before 300 cycles or when the dynamic elastic modulus decreases below 60% after 300 cycles. Thus, this study implies that KS F 2456 can also be applied to the CaO-activated GGBFS system.

3.5. Chemical Resistance Tests (5% HCl and 5% H_2SO_4)

The results of chemical resistance tests using 5% HCl and 5% H_2SO_4 solutions are shown in Figure 7. Significantly different results were obtained, depending on the type of solutions. In 5% HCl aqueous solutions, although weight loss was observed in all immersed samples, P-CON had the most excellent resistance, while N-CON was the most vulnerable. However, the samples immersed in 5% H_2SO_4 aqueous solutions yielded different results; the most considerable weight loss occurred in P-CON, followed by N-CON, F-CON, and C-CON, in order.

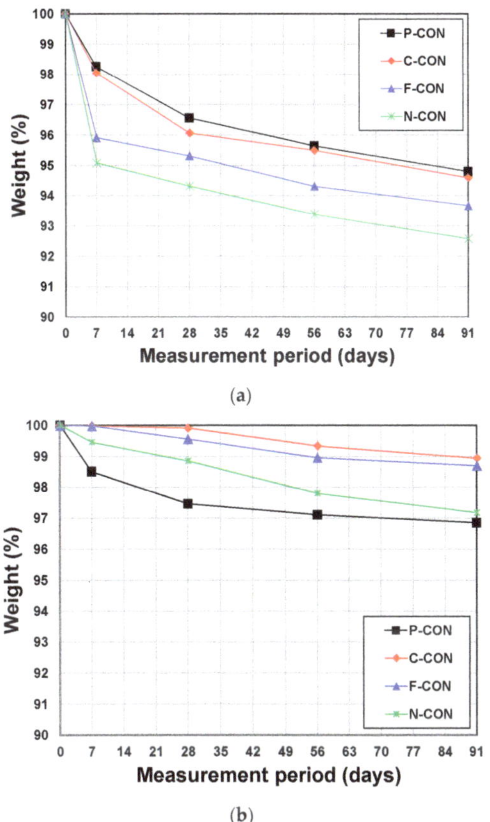

Figure 7. Chemical resistance test results: (**a**) 5% HCl solution and (**b**) 5% H_2SO_4 solution.

XRD and TG/DTG Analysis in Chemical Resistance Tests

When concrete samples were immersed in acid solutions (i.e., HCl and H_2SO_4), the pH of the concrete was lowered, resulting in the change of hydration products.

In the HCl solution, the change of the reaction products was investigated using the XRD and TG/DTG analysis, shown in Figures 8 and 9, respectively. After immersion, $Ca(OH)_2$ disappeared in the P-CON sample, and Friedel's salt ($Ca_2Al(OH)_6Cl(H_2O)_2$) was found. The HCl solution likely lowered the pH of P-CON, resulting in removing OH^- of $Ca(OH)_2$ [41,42], and Friedel's salt was produced by the reaction of Cl^- ions from HCl and Ca^{2+} ions from $Ca(OH)_2$. In addition, regeneration of $Ca(OH)_2$ was observed at the 91-day sample. It seemed that as the chemical resistance test proceeded, $Ca(OH)_2$ was regenerated by a complex chemical reaction; more research will likely be needed for the regeneration of $Ca(OH)_2$.

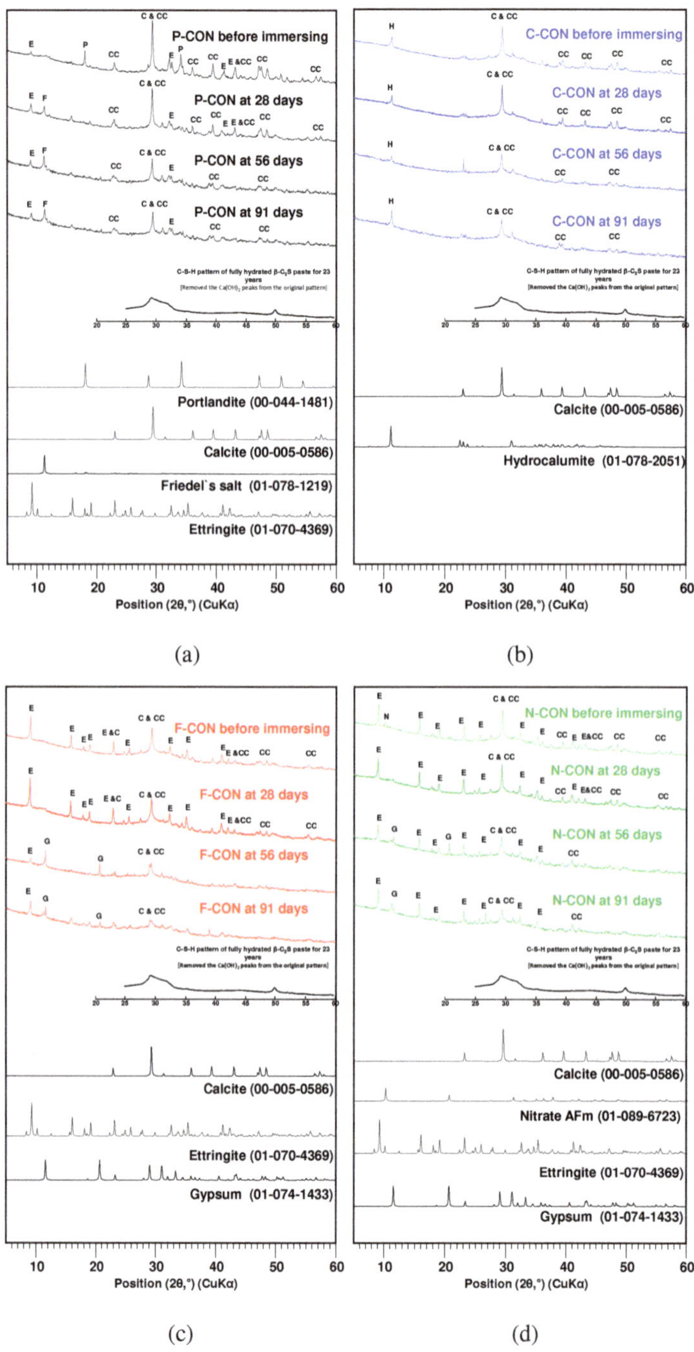

Figure 8. XRD patterns of hardened samples immersed in 5% HCl solution: (**a**) P-CON, (**b**) C-CON, (**c**) F-CON, and (**d**) N-CON. (Note. E: ettringite, C: C-S-H, CC: calcite, G: gypsum, H: hydrocalumite, F: Friedel's salt, P: portlandite, and N: nitrate AFm, respectively).

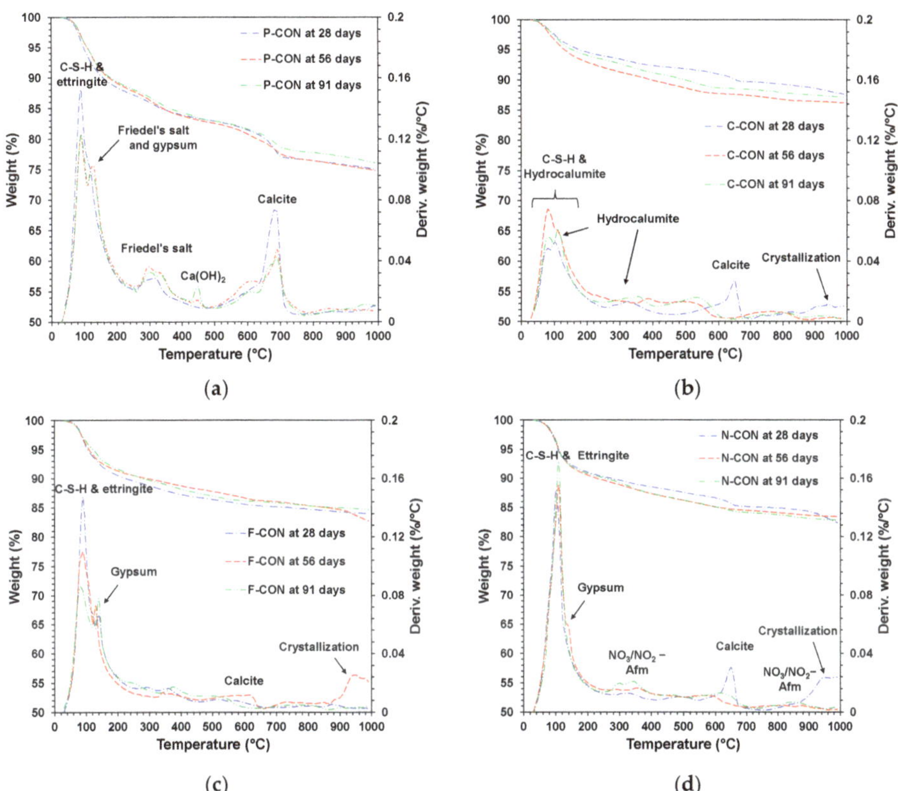

Figure 9. TG/DTG curves of hardened samples immersed in 5% HCl solution: (**a**) P-CON, (**b**) C-CON, (**c**) F-CON, and (**d**) N-CON.

The XRD showed that all samples with sulfate ion (SO_4^{2-}) (i.e., P-, F-, and N-CON) had ettringite before immersion. When immersed in HCl solution, the DTG analysis showed that ettringite tended to decrease, and gypsum was generated; F-CON exhibited the most prominent decrease of ettringite, while N-CON displayed the slightest change in ettringite peak. Previous studies [41,43,44] reported that, below pH value of 10.6, ettringite was no longer stable, and gypsum formed. Unlike the other samples, C-CON did not change the type of reaction product in the HCl solution, possibly due to the absence of ettringite and $Ca(OH)_2$ before immersion. In addition, the weight loss occurred in the vicinity of 800 to 900 °C in CaO-activated concrete, which might be the recrystallization of C-S-H [45].

Figures 10 and 11 show the results of XRD and TG/DTG tests, respectively, for the samples in the H_2SO_4 solution. In addition, similar to the HCl results, regeneration of $Ca(OH)_2$ at the P-CON sample and crystallization were observed. It is worth noting that ettringite is relatively more susceptible to the drying process of sample preparation than other cementitious reaction products and, thus, ettringite can often be underestimated in quantity in XRD and TG/DTG [46,47]. Unfortunately, in this study, the result for ettringite in the H_2SO_4 solution in TG/DTG was not precisely consistent with that of XRD; however, careful interpretation enabled us to recognize the approximate tendency of the product change.

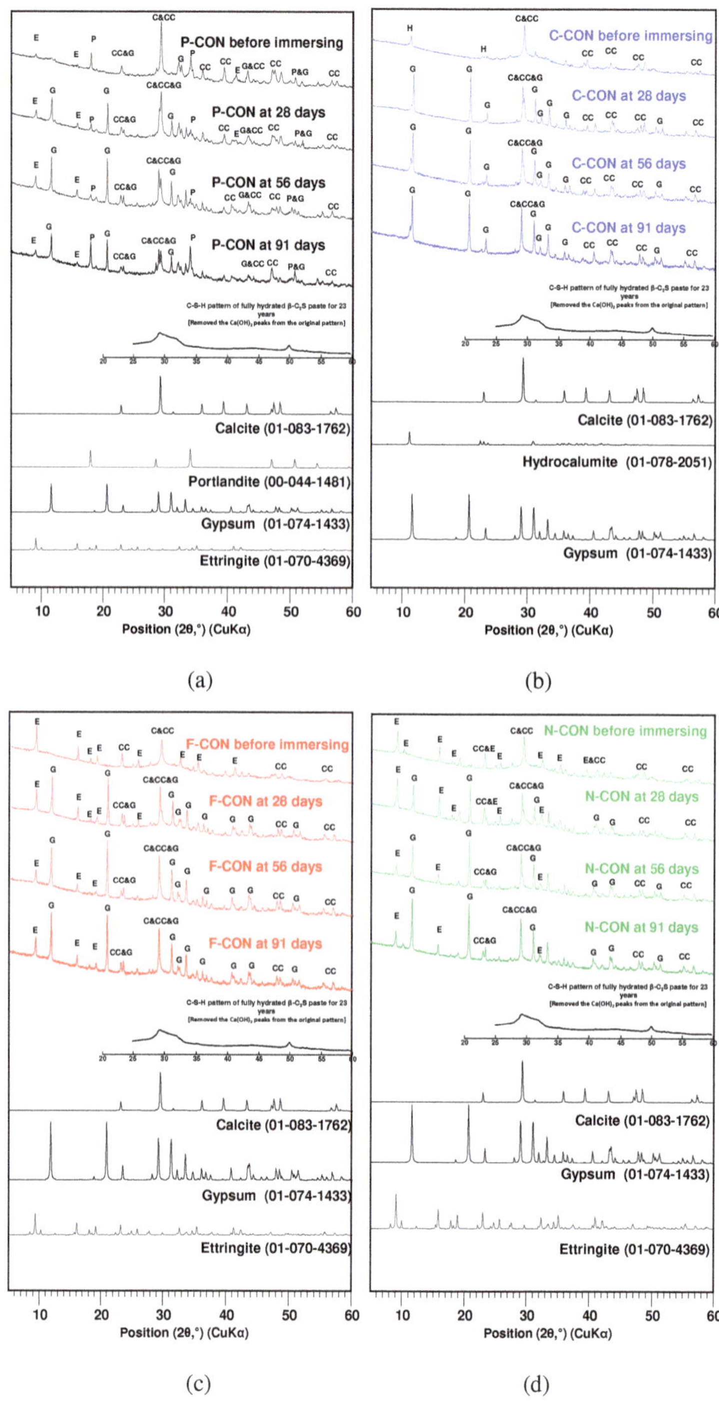

Figure 10. XRD patterns of hardened samples immersed in 5% H_2SO_4 solution: (**a**) P-CON, (**b**) C-CON, (**c**) F-CON, and (**d**) N-CON. (Note. E: ettringite, C: C-S-H, CC: calcite, G: gypsum, H: hydrocalumite, and P: portlandite, respectively).

In the samples that had ettringite before immersion (i.e., P-, F-, and N-CON in Figure 10), as the DTG peaks of ettringite decreased, the gypsum DTG peaks tended to increase, as shown in Figure 11. In particular, the F-CON and N-CON samples showed relatively strong gypsum XRD and DTG peaks, and the peaks were even stronger than those of samples in the HCl solution. The cause of gypsum formation was similar to the case of HCl immersion; as the pH of the concrete was likely lowered in the H_2SO_4 solution, ettringite became unstable and decomposed, producing gypsum.

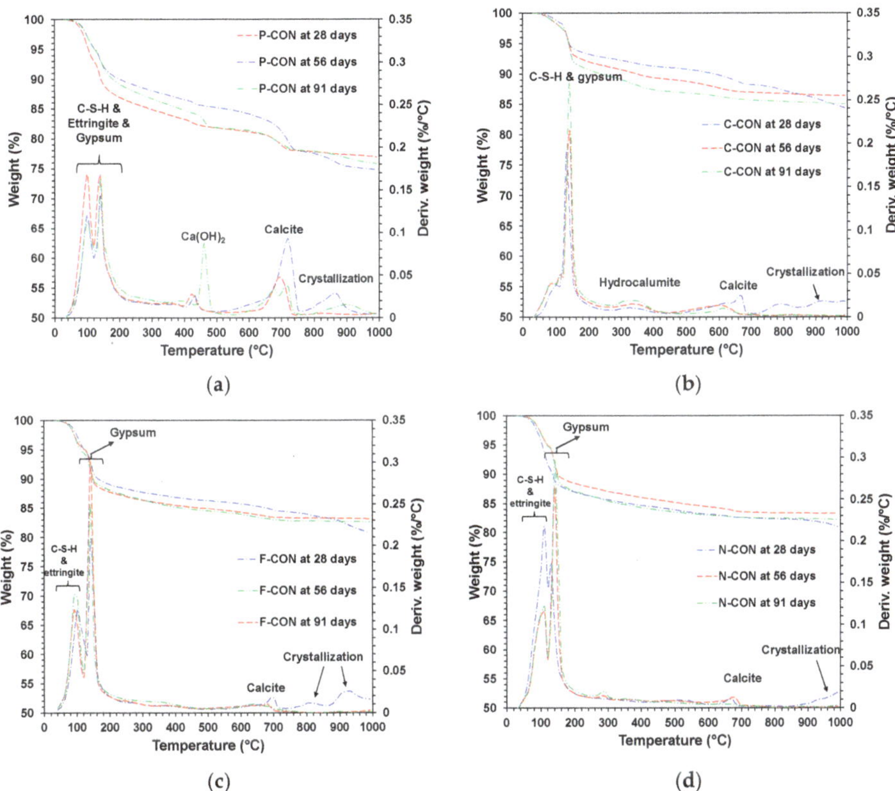

Figure 11. TG/DTG curves of hardened samples immersed in 5% H_2SO_4 solution: (**a**) P-CON, (**b**) C-CON, (**c**) F-CON, and (**d**) N-CON.

However, note that gypsum can also form from the reaction between $Ca(OH)_2$ and H_2SO_4 without ettringite decomposition [27,41]. For instance, in P-CON, because $Ca(OH)_2$ was present before immersion, this reaction surely accounted for a portion of gypsum formation. However, this reaction was also likely responsible for the gypsum formation in C-CON despite no $Ca(OH)_2$ before immersion (Figures 10b and 11b). Previous studies [27,41,48–51] reported that C-S-H and calcite can dissolve, producing Ca^{2+} ions (or $Ca(OH)_2$) when the pH is lowered. In the TG/DTG results, the decrease in calcite peak was—relatively—clearly seen, but the decrease in C-S-H peak was difficult to observe because the C-S-H peak below 200 °C overlapped with the peaks of ettringite and gypsum, making it challenging to analyze. However, when referring to the other study results [49–51], C-S-H likely dissolved to some extent similar to calcite. Thus, the dissolved Ca^{2+} ions can produce gypsum from the reaction with H_2SO_4 in C-CON. On the other hand, in the case of P-CON, the coexistence of $Ca(OH)_2$ generally prevents chemical deterioration of C-S-H [49,52]; thus, a relatively smaller amount of C-S-H likely dissolved than the other samples.

4. Conclusions

This study examined the mechanical and durability properties of three types of CaO-activated GGBFS concretes made with three different additives ($CaCl_2$, $Ca(HCCO)_2$, and $Ca(NO_3)_2$), and compared their properties with those of Portland cement concrete. Detailed conclusions are summarized as follows:

1. All CaO-activated GGBFS concretes satisfied the targeted air content and slump ranges. However, they exhibited lower mechanical properties (compressive strength, split strength, and elastic modulus) than OPC concrete.
2. In the CaO-activated GGBFS system, the additive $CaCl_2$ was the most effective at increasing strength and elastic modulus, while the additive $Ca(NO_3)_2$ was the least effective.
3. In the freezing and thawing cycle test, the CaO-activated GGBFS concretes with added $Ca(HCOO)_2$ and $Ca(NO_3)_2$ were destroyed before 300 cycles; however, the one made with $CaCl_2$ had a relatively comparable resistance in the freezing and thawing cycles test to OPC concrete.
4. The strength was a less accurate indicator than the dynamic modulus for estimating the ability to resist freezing and thawing cycles in this study.
5. The chemical resistance tests using 5% HCl and 5% H_2SO_4 solutions showed significantly different results depending on the type of acid solutions. In a 5% HCl solution, although weight loss was observed in all immersed samples, the OPC concrete had the most excellent resistance, while the CaO-activated GGBFS concrete with $Ca(NO_3)_2$ was the most vulnerable. However, in 5% H_2SO_4 aqueous, the most considerable weight loss occurred in the OPC concrete.
6. In XRD, all concrete with added sulfate ion (SO_4^{2-}) (i.e., Portland cement concrete and CaO-activated GGBFS concretes with $Ca(HCOO)_2$ and $Ca(NO_3)_2$) had ettringite before immersion. When they were immersed in HCl solution, their DTG results showed that ettringite tended to decrease, and gypsum was generated; however, the CaO-activated GGBFS concrete with $CaCl_2$ did not change the type of reaction product in the HCl solution, possibly due to the absence of ettringite and $Ca(OH)_2$ before immersion.
7. When immersed in H_2SO_4 solution, similar to the case of HCl, the DTG peaks of ettringite decreased, and the gypsum DTG peaks tended to increase. However, in the H_2SO_4 solution, gypsum was synthesized from the reaction between $Ca(OH)_2$ and H_2SO_4 in all concrete samples. In particular, the concrete without ettringite (i.e., CaO-activated GGBFS concrete with $CaCl_2$) had a considerable amount of gypsum formation; in this concrete, dissolved C-S-H and calcite, due to the low pH, likely produced Ca^{2+} ions, and gypsum formed from the reaction between Ca^{2+} and H_2SO_4.

Author Contributions: Conceptualization, W.S.Y. and J.Y.; methodology, W.S.Y. and D.J.; software, W.S.Y. and H.S.; investigation, W.S.Y. and S.S.; resources, W.S.Y. and D.H.K.; writing—original draft preparation, W.S.Y. and J.E.O.; writing—review and editing, W.S.Y. and J.E.O.; visualization, W.S.Y. and J.E.O.; supervision, J.E.O.; project administration, J.E.O.; funding acquisition, J.E.O. All authors have read and agreed to the published version of the manuscript.

Funding: This research was supported by the Basic Research Program through the National Research Foundation of Korea (NRF) funded by the MSIT (grant number: 2021R1A4A1030867), and Carbon Neutral Institute Project Fund (Project Number: 1.210109.01) of UNIST(Ulsan National Institute of Science & Technology).

Institutional Review Board Statement: Not applicable.

Informed Consent Statement: Not applicable.

Data Availability Statement: Not applicable.

Conflicts of Interest: The authors declare no conflict of interest.

References

1. Kerr, R.A. Global warming is changing the world. *Science* **2007**, *316*, 188–190. [CrossRef]
2. Drake, F. *Global Warming*; Routledge: Abingdon, UK, 2014.
3. Aïtcin, P.-C. Cements of yesterday and today: Concrete of tomorrow. *Cem. Concr. Res.* **2000**, *30*, 1349–1359. [CrossRef]
4. Aydin, E. Novel coal bottom ash waste composites for sustainable construction. *Constr. Build. Mater.* **2016**, *124*, 582–588. [CrossRef]
5. Poudyal, L.; Adhikari, K. Environmental sustainability in cement industry: An integrated approach for green and economical cement production. *Resour. Environ. Sustain.* **2021**, *4*, 100024. [CrossRef]
6. Hwang, C.-L.; Huynh, T.-P. Engineering, Evaluation of the performance and microstructure of ecofriendly construction bricks made with fly ash and residual rice husk ash. *Adv. Mater. Sci. Eng.* **2015**, *2015*, 891412. [CrossRef]
7. Karim, M.; Zain, M.F.M.; Jamil, M.; Lai, F. Fabrication of a non-cement binder using slag, palm oil fuel ash and rice husk ash with sodium hydroxide. *Constr. Build. Mater.* **2013**, *49*, 894–902. [CrossRef]
8. Billong, N.; Melo, U.; Kamseu, E.; Kinuthia, J.; Njopwouo, D.J.C. Improving hydraulic properties of lime–rice husk ash (RHA) binders with metakaolin (MK). *Constr. Build. Mater.* **2011**, *25*, 2157–2161. [CrossRef]
9. Palomo, A.; Grutzeck, M.; Blanco, M. Alkali-activated fly ashes: A cement for the future. *Cem. Concr. Res.* **1999**, *29*, 1323–1329. [CrossRef]
10. Sun, Z.; Lin, X.; Vollpracht, A. Pervious concrete made of alkali activated slag and geopolymers. *Constr. Build. Mater.* **2018**, *189*, 797–803. [CrossRef]
11. Ariffin, M.; Bhutta, M.; Hussin, M.; Tahir, M.M.; Aziah, N. Sulfuric acid resistance of blended ash geopolymer concrete. *Constr. Build. Mater.* **2013**, *43*, 80–86. [CrossRef]
12. Ryu, G.S.; Lee, Y.B.; Koh, K.T.; Chung, Y.S. The mechanical properties of fly ash-based geopolymer concrete with alkaline activators. *Constr. Build. Mater.* **2013**, *47*, 409–418. [CrossRef]
13. Yurt, Ü. High performance cementless composites from alkali activated GGBFS. *Constr. Build. Mater.* **2020**, *264*, 120222. [CrossRef]
14. Rostami, M.; Behfarnia, K. The effect of silica fume on durability of alkali activated slag concrete. *Constr. Build. Mater.* **2017**, *134*, 262–268. [CrossRef]
15. Messina; Ferone; Colangelo; Roviello; Cioffi. Alkali activated waste fly ash as sustainable composite: Influence of curing and pozzolanic admixtures on the early-age physico-mechanical properties and residual strength after exposure at elevated temperature. *Compos. Part B Eng.* **2018**, *132*, 161–169. [CrossRef]
16. Van Deventer, J.S.; Provis, J.L.; Duxson, P. Technical and commercial progress in the adoption of geopolymer cement. *Miner. Eng.* **2012**, *29*, 89–104. [CrossRef]
17. Yang, K.-H.; Song, J.-K.; Song, K.-I. Assessment of CO2 reduction of alkali-activated concrete. *J. Clean. Prod.* **2013**, *39*, 265–272. [CrossRef]
18. Fernandez-Jimenez, A.M.; Palomo, A.; Lopez-Hombrados, C. Engineering properties of alkali-activated fly ash concrete. *ACI Mater. J.* **2006**, *103*, 106.
19. Atiş, C.D.; Bilim, C.; Çelik, Ö.; Karahan, O. Influence of activator on the strength and drying shrinkage of alkali-activated slag mortar. *Constr. Build. Mater.* **2009**, *23*, 548–555. [CrossRef]
20. Palacios, M.; Puertas, F. Effect of carbonation on alkali-activated slag paste. *J. Am. Ceram. Soc.* **2006**, *89*, 3211–3221. [CrossRef]
21. Yum, W.S.; Suh, J.-I.; Sim, S.; Yoon, S.; Jun, Y.; Oh, J.E. Influence of calcium and sodium nitrate on the strength and reaction products of the CaO-activated GGBFS system. *Constr. Build. Mater.* **2019**, *215*, 839–848. [CrossRef]
22. Suh, J.-I.; Yum, W.S.; Jeong, Y.; Park, H.-G.; Oh, J.E. The cation-dependent effects of formate salt additives on the strength and microstructure of CaO-activated fly ash binders. *Constr. Build. Mater.* **2019**, *194*, 92–101. [CrossRef]
23. Kim, M.S.; Jun, Y.; Lee, C.; Oh, J.E. Use of CaO as an activator for producing a price-competitive non-cement structural binder using ground granulated blast furnace slag. *Cem. Concr. Res.* **2013**, *54*, 208–214. [CrossRef]
24. Yum, W.S.; Jeong, Y.; Yoon, S.; Jeon, D.; Jun, Y.; Oh, J.E. Effects of CaCl2 on hydration and properties of lime (CaO)-activated slag/fly ash binder. *Cem. Concr. Compos.* **2017**, *84*, 111–123. [CrossRef]
25. Yum, W.S.; Suh, J.-I.; Jeon, D.; Oh, J.E. Strength enhancement of CaO-activated slag system through addition of calcium formate as a new auxiliary activator. *Cem. Concr. Compos.* **2020**, *109*, 103572. [CrossRef]
26. Jeon, D.; Yum, W.S.; Jeong, Y.; Oh, J.E. Properties of quicklime (CaO)-activated Class F fly ash with the use of CaCl2. *Cem. Concr. Res.* **2018**, *111*, 147–156. [CrossRef]
27. Mehta, P.K.; Monteiro, P.J. *Concrete Microstructure, Properties and Materials*; McGraw-Hill Education: New York, NY, USA, 2017.
28. Ministry of Land, I.T. *Korea, Korea Concrete Specification*; Ministry of Land: Daejeon, Korea, 2017.
29. Korean Standards Association. *Standard Test Method for Making Concrete Specimens*; KS F 2403; Korean Standards Association: Seoul, Korea, 2019.
30. Korean Standards Association. *Method of Test for Slump of Concrete*; KS F 2402; Korean Standards Association: Seoul, Korea, 2012.
31. Korean Standards Association. *Standard Test Method for Air Content of Fresh Concrete by the Pressure Method (Air Receiver Method)*; KS F 2421; Korean Standards Association: Seoul, Korea, 2011.
32. ACI Committee. *Building Code Requirements for Structural Concrete (ACI 318-08) and Commentary*; American Concrete Institute: Farmington Hills, MI, USA, 2008.

33. Comite Euro-International Du Beton. *CEB-FIP Model Code 1990, Design Code*; Institution of Civil Engineers Publishing: London, UK, 1993.
34. Korean Standards Association. *Standard Test Method for Resistance of Concrete to Rapid Freezing and Thawing*; KS F 2456; Korean Standards Association: Seoul, Korea, 2013.
35. ASTM. *Standard Test Methods for Chemical Resistance of Mortars, Grouts, and Monolithic Surfacings and Polymer Concretes*; ASTM C 267; ASTM: West Conshohocken, PA, USA, 2020.
36. Jeong, Y.; Park, H.; Jun, Y.; Jeong, J.-H.; Oh, J.E. Microstructural verification of the strength performance of ternary blended cement systems with high volumes of fly ash and GGBFS. *Constr. Build. Mater.* **2015**, *95*, 96–107. [CrossRef]
37. Ministry of Land, Infrastructure and Transport. *Korea, Concrete Standard Specification*; Korean Standards Association: Seoul, Korea, 2016.
38. Jo, B.-W.; Shon, Y.-H.; Kim, Y.-J. The evalution of elastic modulus for steel fiber reinforced concrete. *Russ. J. Nondestruct. Test.* **2001**, *37*, 152–161. [CrossRef]
39. Kim, J.-K.; Han, S.H.; Song, Y.C. Effect of temperature and aging on the mechanical properties of concrete: Part I. Experimental results. *Cem. Concr. Res.* **2002**, *32*, 1087–1094. [CrossRef]
40. Alsalman, A.; Dang, C.N.; Prinz, G.S.; Hale, W.M. Evaluation of modulus of elasticity of ultra-high performance concrete. *Constr. Build. Mater.* **2017**, *153*, 918–928. [CrossRef]
41. Allahverdi, A.; Škvára, F. Acidic corrosion of hydrated cement based materials. *Ceram. Silikáty* **2000**, *44*, 152–160.
42. Shi, Z.; Geiker, M.R.; Lothenbach, B.; De Weerdt, K.; Garzón, S.F.; Enemark-Rasmussen, K.; Skibsted, J. Friedel's salt profiles from thermogravimetric analysis and thermodynamic modelling of Portland cement-based mortars exposed to sodium chloride solution. *Cem. Concr. Compos.* **2017**, *78*, 73–83. [CrossRef]
43. Goyal, S.; Kumar, M.; Sidhu, D.S.; Bhattacharjee, B. Resistance of mineral admixture concrete to acid attack. *J. Adv. Concr. Technol.* **2009**, *7*, 273–283. [CrossRef]
44. Roy, D.; Arjunan, P.; Silsbee, M. Effect of silica fume, metakaolin, and low-calcium fly ash on chemical resistance of concrete. *Cem. Concr. Res.* **2001**, *31*, 1809–1813. [CrossRef]
45. Taylor, H.F. *Cement Chemistry*; Thomas Telford: London, UK, 1997; Volume 2.
46. Scrivener, K.; Snellings, R.; Lothenbach, B. *A Practical Guide to Microstructural Analysis of Cementitious Materials*; CRC Press: Boca Raton, FL, USA, 2018.
47. Song, H.; Jeong, Y.; Bae, S.; Jun, Y.; Yoon, S.; Eun Oh, J. A study of thermal decomposition of phases in cementitious systems using HT-XRD and TG. *Constr. Build. Mater.* **2018**, *169*, 648–661. [CrossRef]
48. Bakharev, T.; Sanjayan, J.G.; Cheng, Y.-B. Resistance of alkali-activated slag concrete to acid attack. *Cem. Concr. Res.* **2003**, *33*, 1607–1611. [CrossRef]
49. Liu, X.; Feng, P.; Li, W.; Geng, G.; Huang, J.; Gao, Y.; Mu, S.; Hong, J. Effects of pH on the nano/micro structure of calcium silicate hydrate (CSH) under sulfate attack. *Cem. Concr. Res.* **2021**, *140*, 106306. [CrossRef]
50. Sverdrup, H.U. Calcite dissolution and acidification mitigation strategies. *Lake Reserv. Manag.* **1984**, *1*, 345–355. [CrossRef]
51. Pedrosa, E.T.; Fischer, C.; Morales, L.F.; Rohlfs, R.D.; Luttge, A. Influence of chemical zoning on sandstone calcite cement dissolution: The case of manganese and iron. *Chem. Geol.* **2021**, *559*, 119952. [CrossRef]
52. Feng, P.; Miao, C.; Bullard, J.W. A model of phase stability, microstructure and properties during leaching of portland cement binders. *Cem. Concr. Compos.* **2014**, *49*, 9–19. [CrossRef]

Article

Modeling the Effect of Alternative Cementitious Binders in Ultra-High-Performance Concrete

Solmoi Park [1], Namkon Lee [2], Gi-Hong An [2], Kyeong-Taek Koh [3] and Gum-Sung Ryu [2,*]

[1] Department of Civil Engineering, Pukyong National University, 45 Yongso-ro, Nam-gu, Busan 48513, Korea; solmoi.park@pknu.ac.kr

[2] Department of Structural Engineering Research, Korea Institute of Civil Engineering and Building Technology, 283 Goyangdae-ro, Ilsanseo-gu, Goyang-si 10223, Korea; nklee@kict.re.kr (N.L.); agh0530@kict.re.kr (G.-H.A.)

[3] Korean Peninsula Infrastructure Special Committee, Korea Institute of Civil Engineering and Building Technology, 283 Goyangdae-ro, Ilsanseo-gu, Goyang-si 10223, Korea; ktgo@kict.re.kr

* Correspondence: ryu0505@kict.re.kr

Abstract: The use of alternative cementitious binders is necessary for producing sustainable concrete. Herein, we study the effect of using alternative cementitious binders in ultra-high-performance concrete (UPHC) by calculating the phase assemblages of UHPC in which Portland cement is replaced with calcium aluminate cement, calcium sulfoaluminate cement, metakaolin or blast furnace slag. The calculation result shows that replacing Portland cement with calcium aluminate cement or calcium sulfoaluminate cement reduces the volume of C-S-H but increases the overall solid volume due to the formation of other phases, such as strätlingite or ettringite. The modeling result predicts that using calcium aluminate cement or calcium sulfoaluminate cement may require more water than it would for plain UHPC, while a similar or lower amount of water is needed for chemical reactions when using blast furnace slag or metakaolin.

Keywords: UHPC; thermodynamic modeling; phase assemblage; alternative cementitious binders; supplementary cementitious materials

Citation: Park, S.; Lee, N.; An, G.-H.; Koh, K.-T.; Ryu, G.-S. Modeling the Effect of Alternative Cementitious Binders in Ultra-High-Performance Concrete. *Materials* **2021**, *14*, 7333. https://doi.org/10.3390/ma14237333

Academic Editor: F. Pacheco Torgal

Received: 19 October 2021
Accepted: 22 November 2021
Published: 30 November 2021

1. Introduction

Ultra-high-performance concrete (UHPC) is considered one of the most promising construction materials in terms of performance. It is typically produced at low water-to-binder (w/b) ratios (0.15–0.25) and exhibits outstanding mechanical properties (compressive and tensile strengths exceeding 120 and 5 MPa after 28 days of curing) [1,2]. Due to the use of low w/b ratios, UHPC generally possesses extremely low porosity and excellent resistance to various chemical degradations. UHPC has been a topic of numerous studies which focused on various aspects of UHPC, such as its fresh and hardened state properties [3–7], effects of mix designs [8–11] and its performance at a structural level [12–15]. Due to the fact that the mixture proportioning and production methods of UHPC became mostly standardized in many countries, Portland cement (PC) is used as a cementitious binder along with silica fume in most cases. This is particularly important in recent years, where the high CO_2 footprint of PC has been a global concern and the demand for more sustainable cementitious binders is dramatically increasing.

Some attempts have been made to investigate the effect of using alternative cements or supplementary cementitious materials in the replacement of PC in UHPC. Song et al. [16] studied the effect of calcium sulfoaluminate cement (CSA) addition on the microstructure of UHPC, and showed that CSA tends to reduce the autogenous shrinkage of UHPC, concluding that a 5–15% addition is optimal. It is also reported that a denser matrix is formed when CSA is added to UHPC [16]. The use of calcium aluminate cement (CAC) as a cementitious binder in UHPC was intensively investigated by Lee et al. [17], who reported that CSA-based UHPC is capable of withstanding heat exposure and is free from spalling, whereas typical UHPC is expected to completely lose its form at an exposure temperature ~400 °C.

The effect of blast furnace slag (BFS) incorporation in UHPC can vary depending on the curing age. The incorporation of BFS in UHPC is generally known to decrease the compressive strength at early age (1–3 days) by 18–39% [18], while the strength can be higher in the BFS-containing UHPC in comparison with the plain one after 28 days [19,20]. The effect of BFS on the properties of UHPC can also vary depending on its replacement ratios. A study by Abdulkareem et al. [21] suggests that incorporation of BFS accelerates the hydration and refines the pore structure, while a high content of BFS may reduce the compressive strength by decreasing the amount of C-S-H in UHPC. The use of metakaolin (MK) in UHPC has been found to improve the early-age strength while decreasing the later-age strength [22]. The loss in the strength when MK is used to replace PC can be 11.8% relative to the strength of the control sample after 28 days of curing [22]. On the other hand, the use of very fine MK (so-called nano MK) is reported to increase the strength by 7.9% at the sample age of 28 days at a dosage as low as 1% [23]. Materials obtained from other sources can also be used to make UHPC (i.e., demolition waste [24], mine tailings [11], cement kiln dust, rice husk ash [25]) and achieve performance comparable to the ordinary system.

Despite an increasing number of studies having investigated the mechanical properties of UHPC incorporating cementitious binders other than PC, their microstructural information, which dictates the evolution of their properties, and performance is rarely available in the literature. Therefore, this study conducts thermodynamic simulations to predict the phase assemblages of UHPC in which PC is replaced with other cementitious binders, including CAC, CSA, MK or BFS.

2. Materials and Methods

The mix proportion of UHPC used in this work is shown in Table 1. The water-to-cement and water-to-binder ratios of the modeled UHPC were 0.2 and 0.16, respectively. The cement denotes PC, which was gradually replaced by either CAC, CSA, MK or BFS in the simulation. The compositions (Table 2) and reaction degrees of the binders reported in previous studies were used (PC [26], CAC [27], CSA [26], MK [28] and BFS [29]). The phase assemblages of UHPC mixtures incorporating PC, CAC, CSA, MK or BFS were predicted using GEM-Selektor v3.7 [30,31] and Cemdata18 [32].

Table 1. Mix proportion of UHPC.

Materials	Cement	Silica Fume	Silica Powder	Water
Mass Ratio	1.00	0.25	0.25	0.20

Table 2. Oxide compositions (mass-%) of raw materials obtained from other studies.

Oxides	**PC [26]**	**CAC [27]**	**CSA [26]**	**MK [28]**	**BFS [29]**
CaO	60.7	36.6	41.8		41.6
SiO_2	20.6	4.1	8.5	52.0	36.6
Al_2O_3	5.0	40.3	30.4	43.8	12.2
Fe_2O_3	3.4	16.3	2.1	0.3	0.9
SO_3	2.4	0.3	12.0	0.1	0.6
Na_2O	0.2		0.1	0.3	0.2
K_2O	1.0		0.3	0.1	0.3
MgO		0.1	2.2		7.1
SrO	0.1		0.1		
TiO_2		1.8	1.5	1.5	
P_2O_5		0.2		0.2	
MnO					0.1

3. Results

3.1. CAC-Containing UHPC

The predicted phase assemblage of UHPC containing CAC is shown in Figure 1. The phases which were predicted as stable in neat UHPC are C-S-H, ettringite, Fe-hydrogarnet and calcite. It is noted that portlandite was found to be unstable in all mixtures. Replacing PC with CAC in UHPC of up to 10% by mass resulted in the formation of more ettringite and Fe-hydrogarnet but reduced the amount of C-S-H and the overall volume of solid phases. The modeling result suggested that the overall volume of solid phases would increase when the replacement ratio is above 10%, due to the formation of strätlingite. The formation of strätlingite and the increase in the solid volume continued up to a 42% replacement ratio, where formation of $Al(OH)_3$ begins; the increase in the solid volume continues, but at a much lower rate, due to the gradual formation of $Al(OH)_3$ and the reduced strätlingite formation.

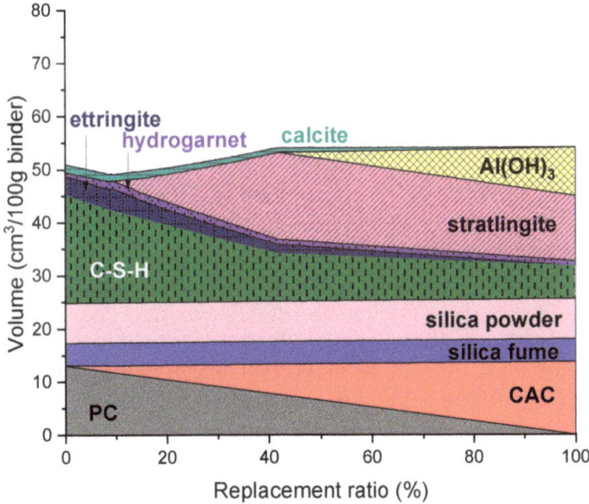

Figure 1. Predicted phase assemblage of UHPC containing CAC.

3.2. CSA-Containing UHPC

The predicted phase assemblage of UHPC containing CSA is shown in Figure 2. The phases which are predictedas stable in the CSA-containing UHPC were similar in the case of the CAC-containing mixture, while their volumes varied. In general, the volume of ettringite was higher, and accordingly the volume of C-S-H was predicted to be lower in the CSA-containing UHPC. The formation of strätlingite and $Al(OH)_3$ in the CSA-containing UHPC was observed at similar replacement ratios that resulted in formation of those phases in the CAC-containing UHPC. The overall solid volume sharply increased at a 10% CSA replacement ratio, which coincides with the formation of strätlingite, similar to the CAC-containing UHPC. The increase in the solid volume continued even after the formation of $Al(OH)_3$ started in the CSA-containing UHPC, unlike the case of the CAC-containing UHPC, which can be associated with the continued formation of ettringite.

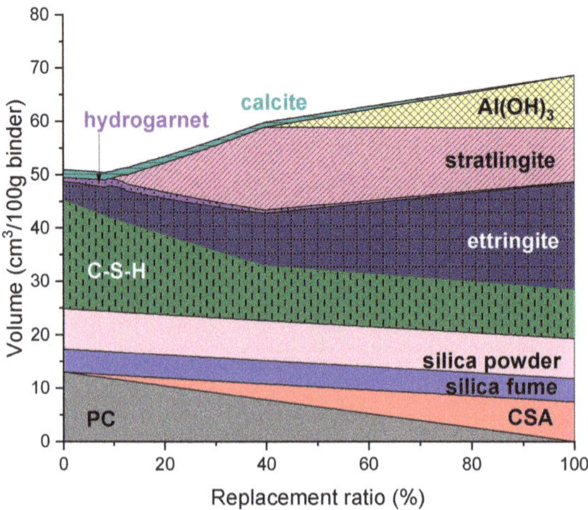

Figure 2. Predicted phase assemblage of UHPC containing CSA.

3.3. MK-Containing UHPC

The predicted phase assemblage of UHPC containing MK is shown in Figure 3. Replacing PC with MK in UHPC initially decreased the amount of C-S-H that forms in the mixture. Unlike those containing CSA or CAC, ettringite was found stable up to 25% replacement. In addition, the amount of Fe-hydrogarnet gradually decreased as less PC was used. The formation of $Al(OH)_3$ and amorphous silica was initiated at 17% and 31% replacement ratios, respectively. The formation of amorphous silica in MK-containing UHPC indicates that the proportion of SiO_2 in the mixture can be excessive; thus, replacing MK beyond this ratio (31%) may not be beneficial. The modeling result suggested that gypsum may precipitate at the replacement ratio of 25–90%.

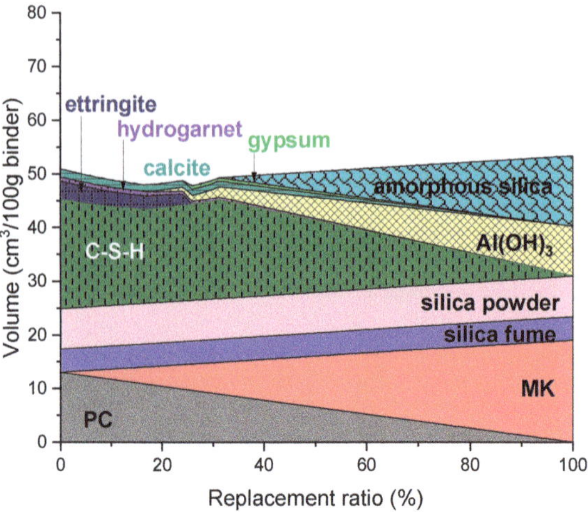

Figure 3. Predicted phase assemblage of UHPC containing MK.

The predicted volume change in the MK-containing UHPC was relatively lower than what was expected for the CSA- and CAC-containing UHPC. The overall solid volume was predicted to gradually decrease up to ~25% replacement and start increasing beyond this replacement level.

3.4. BFS-Containing UHPC

The predicted phase assemblage of UHPC containing BFS is shown in Figure 4. It is noticed that replacing PC with BFS brought the fewest changes in the phase assemblage in comparison with the other mixture combinations. Replacing PC with BFS in UHPC resulted in a marginal decrease in the volume of C-S-H and a more notable decrease in ettringite.

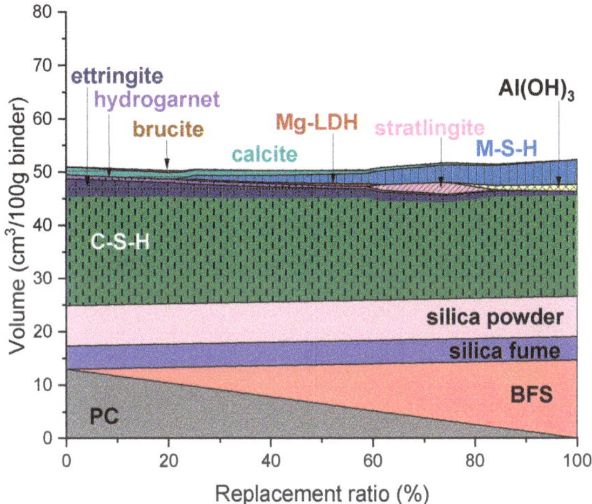

Figure 4. Predicted phase assemblage of UHPC containing BFS.

There were a number of Mg-bearing phases observed in the simulation. Brucite and Mg-LDH were predicted to form as transient phases, which are stable at 1–25% and 23–62% replacement ratios, respectively. These Mg-bearing phases were expected to convert into M-S-H at higher replacement ratios. Strätlingite was found stable at 59–84% replacement and expected to destabilize to $Al(OH)_3$ and M-S-H at higher replacement levels.

The predicted volume change throughout all replacement ratios was the lowest of the prediction results presented in this work. This may be associated with the phase assemblage in the BFS-containing UHPC, which remains almost unchanged upon replacement of PC.

4. Discussion

The thermodynamic modeling results imply that the porosity evolution would vary dramatically according to the cementitious binders that were used to replace PC and their dosages (Figure 5). The prediction results suggest that when the dosage of CAC and CSA is greater than 24 and 12%, respectively, the mixtures are expected to have lower porosity compared to the UHPC solely consisting of PC. Replacing PC with CSA is expected to have the lowest porosity among all simulated mixtures and can induce expansion problems when the replacement ratio exceeds 35%, according to the simulation.

The water demand of the mixtures is simulated in Figure 6 as a function of replacement ratios. Note that the predicted water demand is the amount of water needed for chemical reactions during hydration as predicted by thermodynamic calculations; thus, it is not related to the amount of water needed to ensure homogeneous mixing. The obtained results showed some correlations with the predicted porosity, suggesting that higher water demand leads to generating lower porosity in the matrix. It is advisable that more water is added to the mixtures incorporating CAC or CSA in replacement of PC, because these binders are predicted to consume more water; otherwise, they are known to cause expansion at later ages due to the hydration of anhydrous clinkers at hardened states [26,33,34].

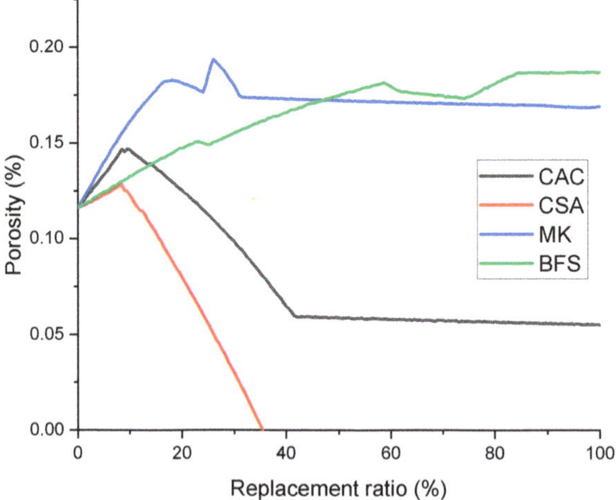

Figure 5. Predicted porosity as a function of replacement ratios.

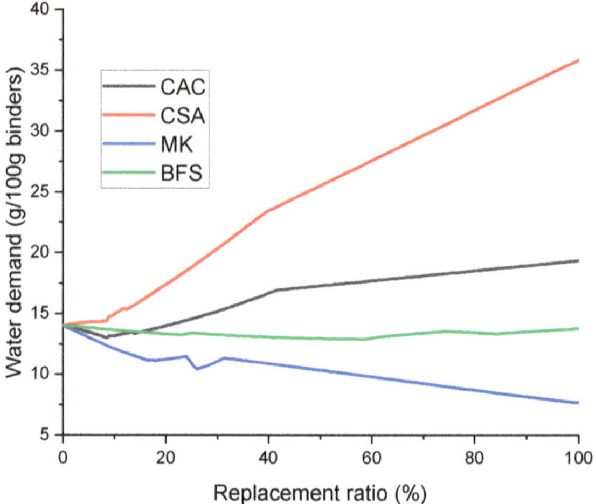

Figure 6. Predicted water demand of UHPC mixtures as a function of replacement ratios.

It is important that the results reported in this work should be cross-checked with the test results before being implemented in practice. The experimental tests for UHPC consisting of alternative cementitious binders will therefore be conducted in forthcoming studies.

5. Conclusions

This study investigated the effect of using alternative cementitious binders in UHPC by adopting thermodynamic calculations. Replacing PC in UHPC with CAC, CSA, MK or BFS is expected to result in significant variations in the phase assemblages. The main outcomes of this study can be summarized as follows.

(1) Strätlingite is predicted as a predominant phase in both CAC- and CSA-containing UHPC. Ettringite would increasingly form as PC is replaced with CSA.
(2) The volume of C-S-H is expected to notably decrease when replacing PC with MK, while this was not the case with BFS-containing UHPC. C-S-H remained as a predominant phase in BFS-containing UHPC throughout all replacement ratios.

(3) The predicted water demand per binder mass suggests that more water is needed for chemical reactions when using CSA and CAC, while a similar or lower amount of water is needed when using BFS and MK as a replacement of PC.
(4) It is expected that use of CSA or CAC would lead to decreasing the porosity in UHPC, while using BFS or MK may increase the porosity in comparison with that solely consisting of PC.

Author Contributions: Conceptualization, G.-S.R.; Data curation, S.P., G.-H.A. and K.-T.K.; Formal analysis, S.P., N.L., G.-S.R., G.-H.A. and K.-T.K.; Funding acquisition, G.-S.R.; Investigation, S.P., N.L., G.-S.R., G.-H.A. and K.-T.K.; Methodology, S.P., N.L., G.-S.R., G.-H.A. and K.-T.K.; Project administration, G.-S.R.; Resources, G.-S.R.; Validation, G.-H.A.; Visualization, G.-S.R. and K.-T.K.; Writing—original draft, S.P.; Writing—review and editing, N.L., G.-S.R., G.-H.A. and K.-T.K. All authors have read and agreed to the published version of the manuscript.

Funding: This work was financially supported by the Korea Institute of Civil Engineering and Building Technology (project no. 20210120).

Institutional Review Board Statement: Not applicable.

Informed Consent Statement: Not applicable.

Data Availability Statement: The data presented in this study are available on request.

Conflicts of Interest: The authors declare no conflict of interest.

References

1. Shi, C.; Wu, Z.; Xiao, J.; Wang, D.; Huang, Z.; Fang, Z. A review on ultra high performance concrete: Part I—Raw materials and mixture design. *Constr. Build. Mater.* **2015**, *101*, 741–751. [CrossRef]
2. Du, J.; Meng, W.; Khayat, K.H.; Bao, Y.; Guo, P.; Lyu, Z.; Abu-obeidah, A.; Nassif, H.; Wang, H. New development of ultra-high-performance concrete (UHPC). *Compos. Part B Eng.* **2021**, *224*, 109220. [CrossRef]
3. Dils, J.; Boel, V.; De Schutter, G. Influence of cement type and mixing pressure on air content, rheology and mechanical properties of UHPC. *Constr. Build. Mater.* **2013**, *41*, 455–463. [CrossRef]
4. Khayat, K.H.; Meng, W.; Vallurupalli, K.; Teng, L. Rheological properties of ultra-high-performance concrete—An overview. *Cem. Concr. Res.* **2019**, *124*, 105828. [CrossRef]
5. Choi, M.S.; Lee, J.S.; Ryu, K.S.; Koh, K.-T.; Kwon, S.H. Estimation of rheological properties of UHPC using mini slump test. *Constr. Build. Mater.* **2016**, *106*, 632–639. [CrossRef]
6. Pyo, S.; Wille, K.; El-Tawil, S.; Naaman, A.E. Strain rate dependent properties of ultra high performance fiber reinforced concrete (UHP-FRC) under tension. *Cem. Concr. Compos.* **2015**, *56*, 15–24. [CrossRef]
7. Wang, R.; Gao, X.; Huang, H.; Han, G. Influence of rheological properties of cement mortar on steel fiber distribution in UHPC. *Constr. Build. Mater.* **2017**, *144*, 65–73. [CrossRef]
8. Richard, P.; Cheyrezy, M. Composition of reactive powder concretes. *Cem. Concr. Res.* **1995**, *25*, 1501–1511. [CrossRef]
9. Kim, H.; Koh, T.; Pyo, S. Enhancing flowability and sustainability of ultra high performance concrete incorporating high replacement levels of industrial slags. *Constr. Build. Mater.* **2016**, *123*, 153–160. [CrossRef]
10. Pyo, S.; Kim, H.-K.; Lee, B.Y. Effects of coarser fine aggregate on tensile properties of ultra high performance concrete. *Cem. Concr. Compos.* **2017**, *84*, 28–35. [CrossRef]
11. Pyo, S.; Tafesse, M.; Kim, B.-J.; Kim, H.-K. Effects of quartz-based mine tailings on characteristics and leaching behavior of ultra-high performance concrete. *Constr. Build. Mater.* **2018**, *166*, 110–117. [CrossRef]
12. Bajaber, M.; Hakeem, I. UHPC evolution, development, and utilization in construction: A review. *J. Mater. Res. Technol.* **2020**, *10*, 1058–1074. [CrossRef]
13. Xue, J.; Briseghella, B.; Huang, F.; Nuti, C.; Tabatabai, H.; Chen, B. Review of ultra-high performance concrete and its application in bridge engineering. *Constr. Build. Mater.* **2020**, *260*, 119844. [CrossRef]
14. Bae, Y.; Pyo, S. Ultra high performance concrete (UHPC) sleeper: Structural design and performance. *Eng. Struct.* **2020**, *210*, 110374. [CrossRef]
15. Bae, Y.; Pyo, S. Effect of steel fiber content on structural and electrical properties of ultra high performance concrete (UHPC) sleepers. *Eng. Struct.* **2020**, *222*, 111131. [CrossRef]
16. Song, M.; Wang, C.; Cui, Y.; Li, Q.; Gao, Z. Mechanical Performance and Microstructure of Ultra-High-Performance Concrete Modified by Calcium Sulfoaluminate Cement. *Adv. Civ. Eng.* **2021**, *2021*, 4002536. [CrossRef]
17. Lee, N.; Koh, K.; Park, S.; Ryu, G. Microstructural investigation of calcium aluminate cement-based ultra-high performance concrete (UHPC) exposed to high temperatures. *Cem. Concr. Res.* **2017**, *102*, 109–118. [CrossRef]

18. Pyo, S.; Kim, H.-K. Fresh and hardened properties of ultra-high performance concrete incorporating coal bottom ash and slag powder. *Constr. Build. Mater.* **2017**, *131*, 459–466. [CrossRef]
19. Peng, Y.Z.; Huang, J.; Ke, J. Preparation of ultra-high performance concrete using phosphorous slag powder. *Appl. Mech. Mater.* **2013**, *357–360*, 588–591. [CrossRef]
20. He, Z.-H.; Du, S.-G.; Chen, D. Microstructure of ultra high performance concrete containing lithium slag. *J. Hazard. Mater.* **2018**, *353*, 35–43. [CrossRef] [PubMed]
21. Abdulkareem, O.M.; Fraj, A.B.; Bouasker, M.; Khouchaf, L.; Khelidj, A. Microstructural investigation of slag-blended UHPC: The effects of slag content and chemical/thermal activation. *Constr. Build. Mater.* **2021**, *292*, 123455. [CrossRef]
22. Rangaraju, P.R.; Li, Z. Development of UHPC Using Ternary Blends of Ultra-Fine Class F Fly Ash, Meta-kaolin and Portland Cement. In Proceedings of the First International Interactive Symposium on Ultra-High Performance Concrete, Des Moines, IA, USA, 18–20 July 2016.
23. Norhasri, M.M.; Hamidah, M.; Fadzil, A.M.; Megawati, O. Inclusion of nano metakaolin as additive in ultra high performance concrete (UHPC). *Constr. Build. Mater.* **2016**, *127*, 167–175. [CrossRef]
24. He, Z.-H.; Zhu, H.-N.; Zhang, M.-Y.; Shi, J.-Y.; Du, S.-G.; Liu, B. Autogenous shrinkage and nano-mechanical properties of UHPC containing waste brick powder derived from construction and demolition waste. *Constr. Build. Mater.* **2021**, *306*, 124869. [CrossRef]
25. Park, S.; Wu, S.; Liu, Z.; Pyo, S. The role of supplementary cementitious materials (SCMs) in ultra high performance concrete (UHPC): A review. *Materials* **2021**, *14*, 1472. [CrossRef] [PubMed]
26. Park, S.; Jeong, Y.; Moon, J.; Lee, N. Hydration characteristics of calcium sulfoaluminate (CSA) cement/Portland cement blended pastes. *J. Build. Eng.* **2020**, *34*, 101880. [CrossRef]
27. Bizzozero, J.; Scrivener, K.L. Limestone reaction in calcium aluminate cement–calcium sulfate systems. *Cem. Concr. Res.* **2015**, *76*, 159–169. [CrossRef]
28. Avet, F.; Li, X.; Scrivener, K. Determination of the amount of reacted metakaolin in calcined clay blends. *Cem. Concr. Res.* **2018**, *106*, 40–48. [CrossRef]
29. Kocaba, V.; Gallucci, E.; Scrivener, K.L. Methods for determination of degree of reaction of slag in blended cement pastes. *Cem. Concr. Res.* **2012**, *42*, 511–525. [CrossRef]
30. Kulik, D.; Berner, U.; Curti, E. *Modelling Chemical Equilibrium Partitioning with the GEMS-PSI Code*; Paul Scherrer Institute: Villigen, Switzerland, 2004.
31. Kulik, D.A.; Wagner, T.; Dmytrieva, S.V.; Kosakowski, G.; Hingerl, F.F.; Chudnenko, K.V.; Berner, U.R. GEM-Selektor geochemical modeling package: Revised algorithm and GEMS3K numerical kernel for coupled simulation codes. *Comput. Geosci.* **2013**, *17*, 1–24. [CrossRef]
32. Lothenbach, B.; Kulik, D.A.; Matschei, T.; Balonis, M.; Baquerizo, L.; Dilnesa, B.; Miron, G.D.; Myers, R.J. Cemdata18: A chemical thermodynamic database for hydrated Portland cements and alkali-activated materials. *Cem. Concr. Res.* **2019**, *115*, 472–506. [CrossRef]
33. Sorensen, A.; Thomas, R.; Maguire, M.; Quezada, I. Calcium Sulfoaluminate (CSA) Cement: Benefits and Applications. *Concr. Int.* **2018**, *40*, 65–69.
34. Singh, B.K.; Hafeez, M.A.; Kim, H.; Hong, S.; Kang, J.; Um, W. Inorganic waste forms for efficient immobilization of radionuclides. *ACS EST Eng.* **2021**, *1*, 1149–1170. [CrossRef]

Article

Effect of the Addition of Agribusiness and Industrial Wastes as a Partial Substitution of Portland Cement for the Carbonation of Mortars

Wilfrido Martinez-Molina [1,*], Hugo L. Chavez-Garcia [1,*], Tezozomoc Perez-Lopez [2], Elia M. Alonso-Guzman [1,3], Mauricio Arreola-Sanchez [1], Marco A. Navarrete-Seras [1], Jorge A. Borrego-Perez [1], Adria Sanchez-Calvillo [3], Jose A. Guzman-Torres [1] and Jose T. Perez-Quiroz [4]

1 Faculty of Civil Engineering, Universidad Michoacana San Nicolas de Hidalgo, Morelia 58070, Mexico; elia.alonso@umich.mx (E.M.A.-G.); mauricio.arreola@umich.mx (M.A.-S.); mnavarrete@umich.mx (M.A.N.-S.); jorge.borrego@umich.mx (J.A.B.-P.); jaguzman@umich.mx (J.A.G.-T.)
2 Centro de Investigación en Corrosión, Universidad Autónoma de Campeche, Campeche 24070, Mexico; tezperez@uacam.mx
3 Faculty of Architecture, Universidad Michoacana San Nicolas de Hidalgo, Morelia 58070, Mexico; adria.sanchez@umich.mx
4 Coordinación de Ingeniería Vehicular e Integridad Estructural, Mexican Institute of Transportation (IMT), Queretaro 76703, Mexico; jtperez@imt.mx
* Correspondence: wilfrido.martinez@umich.mx (W.M.-M.); luis.chavez@umich.mx (H.L.C.-G.)

Citation: Martinez-Molina, W.; Chavez-Garcia, H.L.; Perez-Lopez, T.; Alonso-Guzman, E.M.; Arreola-Sanchez, M.; Navarrete-Seras, M.A.; Borrego-Perez, J.A.; Sanchez-Calvillo, A.; Guzman-Torres, J.A.; Perez-Quiroz, J.T. Effect of the Addition of Agribusiness and Industrial Wastes as a Partial Substitution of Portland Cement for the Carbonation of Mortars. *Materials* **2021**, *14*, 7276. https://doi.org/10.3390/ma14237276

Academic Editors: Panagiotis G. Asteris and Yeonung Jeong

Received: 18 September 2021
Accepted: 23 November 2021
Published: 28 November 2021

Abstract: The present research work shows the effect on the carbonation of Portland cement-based mortars (PC) with the addition of green materials, specifically residues from two groups: agricultural and industrial wastes, and minerals and fibres. These materials have the purpose of helping with the waste disposal, recycling, and improving the durability of concrete structures. The specimens used for the research were elaborated with CPC 30R RS, according to the Mexican standard NMX-C-414, which is equivalent to the international ASTM C150. The aggregates were taken from the rivers Lerma and Huajumbaro, in the State of Michoacan, Mexico, and the water/cement relation was 1:1 in weight. The carbonation analyses were performed with cylinder specimens in an accelerated carbonation test chamber with conditions of 65 +/− 5% of humidity and 25 +/− 2 °C temperature. The results showed that depending on the PC substitutions, the carbonation front advance of the specimens can increase or decrease. It is highlighted that the charcoal ashes, blast-furnace slags, and natural perlite helped to reduce the carbonation advance compared to the control samples, consequently, they contributed to the durability of concrete structures. Conversely, the sugarcane bagasse ash, brick manufacturing ash, bottom ash, coal, expanded perlite, metakaolin, and opuntia ficus-indica dehydrated fibres additions increased the velocity of carbonation front, helping with the sequestration of greenhouse gases, such as CO_2, and reducing environmental pollution.

Keywords: durability; residues; carbon dioxide; pollution; porosity; pozzolanic activity

1. Introduction

In the world, the most employed construction material is the hydraulic concrete Portland cement (PC). Due to its great mechanical performance, low cost, durability and versatility, PC is used for all types of structures and construction purposes. In the plastic state, it can take any geometric form or design, adapting to the formworks or shoring, later curing and acquiring the desired mechanical resistance.

Concrete consists of coarse aggregate (gravel), fine aggregate (sand), water, PC, and eventually different admixtures. After the curing process, concrete transforms into an artificial rock material. If it is reinforced with steel cores, it is named reinforced concrete; this composite material combines the major uniaxial compression resistance of concrete with the elasticity, resiliency and ductility of the steel reinforcements, allowing to build structures with great durability properties.

Nevertheless, the PC materials, such as concrete and construction mortars, generate a big environmental impact, especially during the producing of the clinker. For each 1000 kg of clinker produced, 600 to 800 kg of CO_2 are emitted, making the cement industry one of the most pollutant over the world, contributing to the 8% of the total CO_2 worldwide emissions [1]. The extraction and processing of the prime materials for the elaboration of PC also contributes to this pollution issue. In addition, the petrous aggregates are non-renewable materials, and the exploitation of the quarries causes a negative environmental impact. Many studies have warned from the importance of diminish the effects of the PC industry on the natural environment [2,3].

Besides all this situation, the durability and preservation of the concrete is severely affected by the adverse atmospheric conditions. The durability is the capacity of a construction material, element or structure to resists the physical, chemical, biological, environmental and global warming actions during a determined and designed period of time, while preserving its original form, mechanical properties and service conditions [4].

An aggressive ambient, which contains chloride ions, can be found in constructions such as the docks or bridges built in coastal regions, due to the high salinity of the water and the sea breeze [5]. In addition, the structures built over soils with greater sulphate content can be prone to suffer damages which affect their durability [6]. The cities or industrial zones where the CO_2 content in the atmosphere is elevated suffer from similar problems. When the gases penetrate the concrete through the pores of the material, the CO_2 chemically react with the calcium hydroxide and the calcium silicates, which are products of the reaction of the PC with the water during the hydration process of concrete, forming calcium carbonate. This chemical reaction reduces the pH of the material, which contributes to a faster degradation by corrosion of the reinforced steel, compromising the structural integrity of the constructions [4]. Other environmental factors such as the CO_X are precursors of the carbonation of the reinforced steel of concrete structures [3].

This phenomenon may affect the stability of the structural reinforcements and produces the oxidation and depassivation processes [7,8], starting major damages in civil and industrial structures. The inclusion and research of additives and substitutions to suppress the carbonation allows us to increase the durability of reinforced concrete constructions and helps to diminish the exploitation of the petrous aggregate quarries. The simplest way to achieve this is to maintain a low water/cement ratio and to comply with good construction practices and regulations with correct quality management.

The prevention of corrosion can be achieved mainly during the design phase using high quality concrete with the adequate covering; this approach has been standardized in the Eurocode 2 and the standard EN 206 [7–9]. It is important, because the majority of the corrosion damages are caused by poor design and execution of the concrete (positioning, compaction and curing). Regarding the quality of the concrete, the benefits from and low water-binder ratio and its permeability are well known [7,9,10].

Nevertheless, it is possible to benefit from the carbonation process of the materials and learn from the previous research which has obtained encouraging results. Chen and Gao (2019), for example, found that the water loss of the PC pastes from 30% to 40% is optimal for the CO_2 absorption [11]. The combination of a correct precuring and carbonation period can also increase the uniaxial compressive strength of the PC mortars effectively, especially at early ages; the PC pastes can improve their mechanical resistance and microstructure with this method [11]. Other papers point with thermogravimetric analysis that the carbonation process can delay the hydration of the cement pastes and with higher CO_2 concentrations the crystallinity degree of the carbonates increases. Conversely, with energy-dispersive X-ray spectroscopy (EDS), it is possible to determine if an excessive carbonation of the material can cause a decalcification of the calcium silicate hydrate (C-S-H), which is the core of the resistance and durability of the PC mixtures.

Different works have studied and documented the mechanical properties and the porosity of PC pastes with high resistance to sulphates (HS SR PC) subdued to curing by carbonation at early ages. One study case submitted two pastes to different carbonation ages (1 and 24 h) [12]. It was found that the sample with 1 h improved its mechanical properties, while the other sample reduced them in comparison with the reference samples. Besides, it was found that the increase of the carbonation time from 1 to 24 h enhanced substantially the absorption properties. Notwithstanding, the higher mechanical resistance of the one-hour sample also entailed a higher content of absorbed water compared with the reference test.

Currently, the employment of PC presents some challenges to diminish the carbon footprint. Some of the approaches to achieve it include the incorporation of materials which act as partial replacements of PC [13], to reinforce circular economy, improve the physical, mechanical and chemical properties, reduce the relation water/cement of concrete, and the retardation of the carbonation [14,15]. Several research works have explored these strategies with encouraging results and publications relying on materials with pozzolanic activity [16–22]. The origin of the pozzolanic activity materials lies in the locality of Pozzuoli, Italy, where the volcanic activity of the region produced these materials which present great cementitious properties only after reacting with calcium hydroxide.

Another strategy is the utilization of waste materials such as glass [23], PET bottles or containers [24], ceramic tiles and coverings, ceramic sanitary products [25], ceramic clay bricks [26], tyres and rubber [27], concrete residues [28], agricultural wastes [29,30], metallurgy slags [31–33], industrial wastes [34–36], or coconut husks [37]. All these additions proportionate a positive environmental impact due to the reutilization of materials and products which had completed their service life. These waste materials are a viable alternative for construction purposes, as many investigations have proven improvements in the mechanical resistance, durability and elasticity while reducing the exploitation economic costs [38–44]. Conversely, many of the substitutions may reduce the workability of the concrete casting and diminish the tensile strength [45,46].

In addition, the diverse mineral additions have been researched with binary combinations of PC (silica fume, metakaolin, fly ash, among others). They have been studied by means of the monitoring and analysis of thermodynamic and kinetic parameters, and the evaluation of the corrosion process in reinforced concrete specimens subdued to prolonged chloride attack (44 weekly moistening cycles with a NaCl solution and air-dried, the equivalent of 308 days' chloride attack). The electrochemical techniques: Ecorr, Rp and EIS (Electrochemical Impedance Spectroscopy), in combination with the electrical resistivity, have helped to evaluate the protection capabilities of the studied concretes, regarding the strong environmental attacks.

The mineral additions have produced meaningful gains in the resistivity of the concretes, mainly due to the physical alterations of the pore structure of PC pastes. The additives also cause an enhancement of the interphase paste-aggregate and the conductivity of the porous solution.

The employed mineral additions have significantly increased the resistivity of concretes, due to the physical alterations in the porous structures of the PC pastes, the improvement of the paste-aggregate interphase, and the conductivity changes of the porous solution [47].

The carbonation process absorbs the CO_2 in the atmosphere, compensating partially the polluting emissions generated during the cement production [48,49]. Jacobsen and Jahren [50] estimated that the 16% of the CO_2 emissions of PC production are reabsorbed during the service life of the material and this carbonation process. Recent works and additions, such as glass powder wastes from organic light-emitting diodes (OLED), have demonstrated a greater CO_2 absorption while carbonation process [51].

Recently, the study of using waste materials to enhance the carbonation resistance in mortars has attracted attention in order to improve the properties of mortars [52–57]. Recent studies have analysed the acceleration of carbonation of two types of lime mortars (standard sand and extra ceramic dust) [58]; the mortar with extra ceramic dust showed advanced calcite precipitation and early improvement in various properties. Dvender Sherma and Shweta Goyal used cement kiln dust as partial replacement of cement and an accelerated carbonation curing process to improve the compressive strength, and they studied the porosity and pH of the mortar mix [59]; the results showed an increase of 20% in strength and a reduction of the porosity. Ioannis Rigopoulos et al. studied the carbonation of air lime mortars modified with quarry waste [60], rich in Ca, Mg and Fe silicate minerals. It has been demonstrated how the incorporation of quarry wastes increases the compressive strength and density the microstructure of the mortar. Shan Liu et al. analysed the adding of different additives (calcined hydrotalcite, calcium silicate, gypsum and silica fume) [60], on alkali-activated mortar to enhance the carbonation resistance; the results showed that the incorporation of calcined hydrotalcite improved the carbonation resistance and compressive strength.

2. Materials and Methods

The study aim is to analyse the carbonation behaviour of mortars by adding different waste materials and configurations with cementitious and/or pozzolanic properties as substitutions of PC. It was determined if such additions can limit or enhance the carbonation of the mixtures when hardened.

2.1. Sampling Materials and Selection

A regional study was made in the state of Michoacan, Mexico, to find the most abundant and suitable solid wastes and the market necessities and opportunities. The residues were characterized in the laboratory of the Faculty of Civil Engineering of the university "Universidad Michoacana San Nicolas de Hidalgo" to determine their cementitious capacity and their feasibility as construction materials. The additions were studied as partial substitution of PC with different percentages; the effects on the mechanical properties of the mortars were monitored.

The PC substitutions were considered among two main groups: wastes from agricultural and industrial processes, and mineral origin materials and fibres. All the additions were compared with control samples elaborated with cement CPC 30 R RS according to the Mexican code NMX-C-414 [61], directly related with cement Type 4 of the international standard ASTM C-150 [62]. Table 1 shows the list of materials analysed for mortar mixtures in Mexico. The table displays the origin of each addition and the percentage added as PC substitution (%).

Sands from two rivers located in the Mexican state of Michoacan were used for the preparation of the different mixtures in order to compare the influence that each type of sand has on the different mortars studied. C1 are core mortars made with sand from the Lerma river and C2 are core mortars made with sand from the Huajumbaro river. The mortars made with sand from the Lerma river are: BMA (Brick manufacturing ash), BA (Bottom ash), C (Coal), OF (Opuntia ficus-indica dehydrated fibres). Conversely, the mortars made with sand from the Huajumbaro river are: SBA (Sugarcane bagasse ash), CA (Charcoal ashes), BMA (Brick manufacturing ash), EP (Expanded perlite), NP (Natural perlite), MK (Metakaolin).

2.2. Elaboration of the Specimens

The mortars were designed with a 1:2.5 cement/aggregate proportion and a 1:1 water/cement ratio in weight. The purpose was to achieve good workability and fluidity (including the specimens with incorporation of the substitutions). The studied materials were added as a partial replacement of the PC.

The mortars were elaborated under lab conditions, dosed by weight, mechanically mixed, with safe water, and with the dosages and substitutions shown in Table 1. The specimens designed for the accelerated carbonation analysis were cylinders of 5 cm diameter and 10 cm height. All the specimens were subdued to curing process by immersion after the uncasing, according to the standard ASTM C31 [63].

Table 1. Classification of the Portland cement substitutions.

Classification	Description	Code	(%)	Origin
Agricultural and industrial wastes	Sugarcane bagasse ash	SBA5	5	Sugar cane mill of Taretan, Michoacan, Mexico
		SBA20	20	
	Charcoal ashes	CA	5	Cuitzeo lake margins
	Brick manufacturing ash	BMA	5	Santiago Undameo, Michoacan, Mexico
	Blast-furnace slag	BFS	15	Arcelor Mittal Industry
	Bottom ash	BA	15	Arcelor Mittal Industry
	Coal	C	15	Arcelor Mittal Industry
Minerals and fibres	Expanded perlite	EP	20	Construction products with high-purity standards and quality
	Natural perlite	NP	5	
	Metakaolin	MK	20	
	Opuntia ficus-indica dehydrated fibers	OF	4	Food grade product
Fine Aggregates	Control 1, ARL and ARL2	C1	-	Lerma river margins
	Control 2, HUAJ	C2	-	Huajumbaro river margins

All the materials proposed as substitutions and the aggregates were characterized with X-Ray fluorescence spectroscopy to determine their composition and relate it with the pozzolanic activity.

2.3. Carbonation Analysis

After that, they were exposed in series of four elements to the carbonation chamber with a relative humidity of 65 +/− 5%, temperature of 25 +/− 2 °C and 3% of CO_2, (See Figure 1). Cylinders were covered with vinyl paint in their both cross sections, as is shown in Figure 2, to prevent the carbonation advance through the longitudinal axis. After being subjected to the accelerated carbonation, each series of cylinders was cut into 5 mm slices and sprinkled in the exposed side with the indicators (phenolphthalein and thymolphthalein-based), to measure the carbonation front, as shown in Figure 3. The measurements were taken at 30, 60, 90, 120, and 180 days.

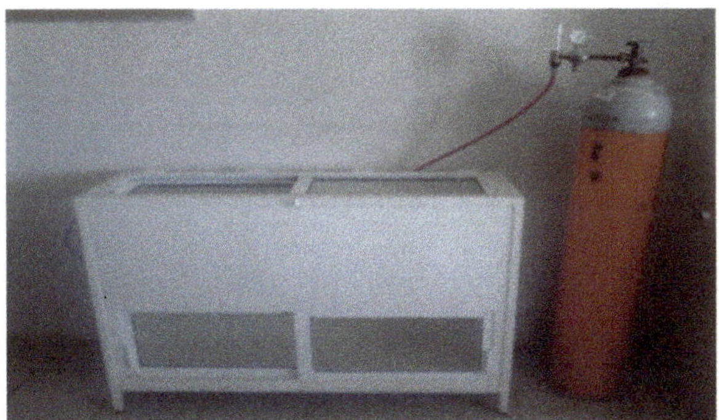

(a)

(b)

Figure 1. (**a**) Accelerated carbonation chamber, ACC. (**b**) Samples distribution in ACC.

Figure 2. Vinyl paint applied in cross sections of cylinders.

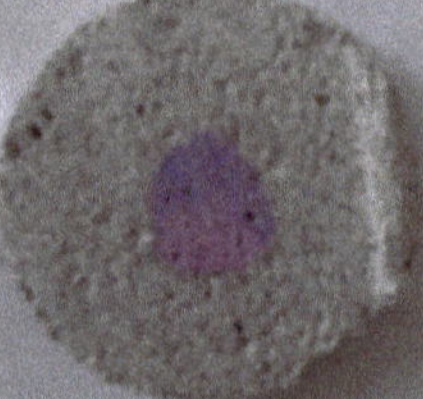
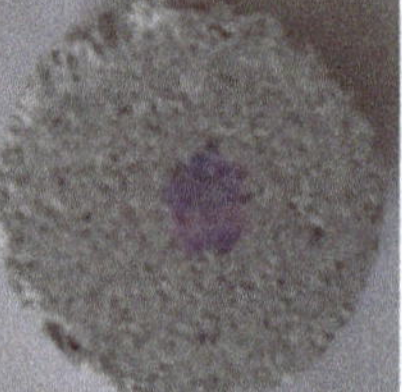

Figure 3. Carbonation monitoring by using phenolphthalein and thymolphthaleine.

2.4. Compressive Strength

To complement the study of the effects on the properties of the researched materials, the uniaxial compressive strength was applied to the specimens, as is shown in Figure 4. This test, as well as the elaboration of the samples, were performed following the standards ASTM C109/109M [64].

Figure 4. Illustration of the compressive strength test.

2.5. Electrical Resistivity Test

In addition to the compressive resistance, the electrical resistivity test was applied to the samples, as can be seen in Figure 5. The test was performed with the same specimens used in the mechanical analyses following the standards ASTM C1876 and the DURAR manual [44,65].

Figure 5. Illustration of the electrical resistivity test and equipment.

3. Results

3.1. Characterization of the Aggregates

Table 2 presents the mechanical properties of the sand aggregates employed in the concrete mixtures. Regarding the absorption percentage and the sand equivalent value we find significant differences; nevertheless, the rest of the properties remain equivalent.

Table 2. Results of the mechanical properties of the aggregates used.

Test	Standard	Lerma Sand	Huajumbaro Sand
Sampling	ASTM D-75-03 [66]	250 kg	250 kg
Reducing sampling	ASTM C-702 [67]	0.500 kg	0.500 kg
Bulk density (unit weight and voids)	ASTM C-29/C-29M [68]	1.353	1.226
Bulk density (unit weight)	ASTM C-29/C-29M [68]	1.444	1.331
Relative density	ASTM C-128 [69]	2–40	2.31
Specific gravity	ASTM C-128 [69]	2.39–2.48	2.24–2.36
Surface moisture (%)	ASTM C-128 [69] ASTM C-70 [70]	0.748	0.741
Absorption percentage (%)	ASTM C-128 [69] ASTM C-566 [71]	1.89	3.18
Sand equivalent value (%)	ASTM D-2419 [72]	86.97	98.25
Clay lumps and friable particles (%)	ASTM C-142 [73]	8.165	2.498

In the characterization of the sands of the Huajumbaro and Lerma rivers, it was found that both materials are composed of silica and have similar physical properties. Table 2 shows significant differences in the absorption percentage and the clay content of both samples, which probably has a repercussion on the carbonation process and evolution and the k constant, later presented in this research work. In addition, their mechanical behaviour in compression is very similar, where the mortars made with both sands reached the same strength (13.3 MPa).

3.2. Characterization of the Additions

The PC substitution materials studied in this research were analysed with SEM microscopy to observe the granulometry and particle size of each one of them. In Figures 6a–f and 7a–f several morphologies and particle shapes are displayed, with well-defined particles with diameter measurements between 40 and 60 µm. The BMA and SBA samples show a composition of morphologies with great surface area, while conversely, the CA sample presents more scattered conglomerates, with smaller size particles. This variety of size, morphology and surface area provides important information to later analyse the mixtures with PC.

Figure 6. SEM images of the substitution materials. (**a**) SBA; (**b**) CA; (**c**) BMA; (**d**) BFA; (**e**) BA; (**f**) C.

The samples were characterized by scanning electron microscopy—energy dispersive X-ray spectroscopy (SEM-EDS). The images in Figure 8a–c show the morphology of the mortars with a considerable surface area and conglomerate compounds of petrous materials and cement. Most of the samples present porous surfaces and voids filled with air during the solidification. Overall, the samples show good homogenization and coupling regarding the proposed design and additions.

In Table 3, the results of the EDS mapping and the elemental composition of the substitution samples are shown. They are expressed in percentage by total weight.

A variety in the composition of the substitutions can be noted. SiO_2, Al_2O_3 and Fe_2O_3 are chemical compounds with high pozzolanic activity, and their presence in all the addition materials is significant. Table 4 presents the pozzolanic activity value of the materials, being MK (94.969), EP (88.186), NP (86.791) and SBA (69.474) the ones with major values. This composition of the additions allows subsequent reactions to the hydration of the cement compounds, providing improvements in the mechanical resistance and a reduction of the porosity. Nevertheless, these secondary reactions are conducted with the alkalises generated during the cement hydration, with the disadvantage of reducing the pH and the concentration of alkaline species which confer high pH values to the fresh concrete.

Figure 7. SEM microscopy images of the substitution materials. (**a**) PE; (**b**) NP; (**c**) MK; (**d**) OF; (**e**) C1; (**f**) C2.

Table 3. EDS XRF results of the chemical composition of the materials.

	SiO_2	TiO_2	Al_2O_3	Fe_2O_3	MnO	MgO	CaO	Na_2O	K_2O	P_2O_5	SO_3	PXC	Total
SBA	60.04	0.43	6.289	3.145	0.13	1.825	1.64	0.446	1.856	0.786	-	23.6	100
OF	14.58	0.83	4.287	26.77	2.423	5.995	37.456	0.032	0.024	0.814	0.8	4.79	98
CA	32.52	0.76	13.55	5.371	0.112	2.108	18.759	0.67	1.027	0.541	-	22.2	97.6
BA	27.93	0.2	6.437	2.217	0.083	1.301	49.773	0.669	1.255	0.118	3.37	5.12	95.1
BMA	19.10	0.32	8.776	2.008	0.538	4.243	27.874	0.545	6.051	1.763	0.8	27.3	98.5
BFS	36.38	0.56	10.63	0.335	0.417	10.107	37.551	0.298	0.424	0.053	1.93	0.72	96
MK	49.75	1.53	44.71	0.509	0.013	0.159	0.044	0.23	0.141	0.033	-	0.77	97.9
EP	73.59	0.13	13.43	1.166	0.045	0.197	1.333	2.918	5.013	0.02	-	1.18	99
NP	72.20	0.12	13.58	1.011	0.073	0.538	1.052	3.216	4.285	0.025	-	3.99	100
HUAJ	78.19	0.2	11.56	1.567	0.03	0.239	1.015	2.666	3.577	0.036	-	1.19	100
ARL	77.57	0.28	10.67	2.384	0.025	0.479	1.338	2.232	3.083	0.075	-	2.2	100.3
ARL2	75.25	0.32	11.92	2.689	0.026	0.568	1.482	2.219	3.124	0.077	-	2.5	100

As expected, the mortars with additions reduced their porosity and increased their uniaxial compression strength. Nevertheless, some authors mention that, in few cases, the results can be opposed [13,14]. Besides, the referred secondary reactions of the pozzolanic compounds consume the alkalis existing in the pores dissolution and can generate a decrease of the pH. Regarding this alkalis generation, BFS and BA present higher contents of MgO and CaO; and NP has 3.216% Na_2O and 4.285% K_2O. The presence of these compounds underwater could be a great source of alkalis generation, which would keep the pH over 12, even after the pozzolanic reactions.

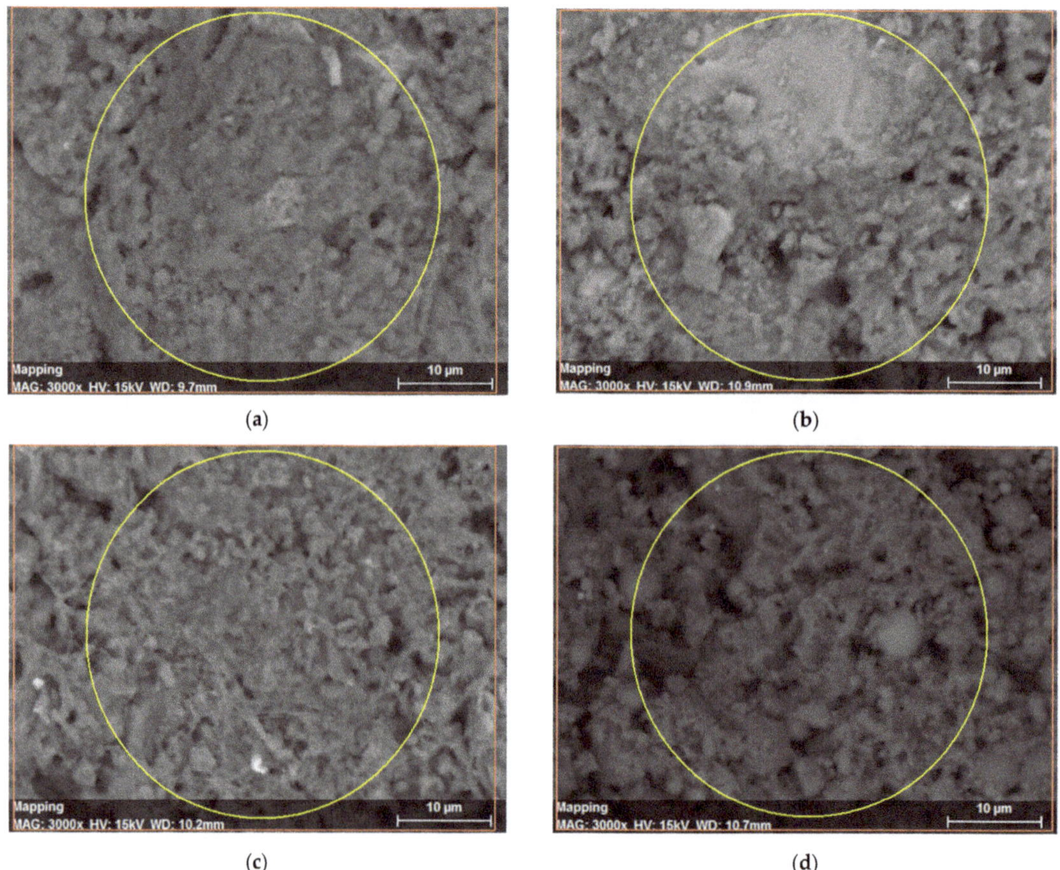

(a) (b) (c) (d)

Figure 8. SEM images of the mortar samples with different substitution materials. (**a**) BMA 5%; (**b**) SBA 5%; (**c**) CA 5%; (**d**) BFS 15%.

In respect of the aggregates, their composition with great siliceous content makes them susceptible to experiment alkali-silica reactions, which is one of the main causes of severe damages in concrete structures. However, the silica of these materials is present in quartzes; consequently, they do not react, but otherwise, they neither achieve the pozzolanic activity due to the crystallinity properties. Then, the aggregates work as inert materials.

3.3. Carbonation Analysis

For a proper analysis of the figures and tables presented in this section, it is important to make clear that each substitution sample is compared with the control samples.

Figures 9 and 10 display the advance of the carbonation front of the mortar samples of this research. It can be noted that all the samples met 25 mm of carbonation front in the maximum period of 180, although some of them achieved it earlier; this measurement is important since 25 mm is the most common depth of the reinforced steel in Mexican concrete structures.

Table 4. Pozzolanic activity.

Sample	SiO_2	Al_2O_3	Fe_2O_3	Pozzolanic Activity	Acid Character
SBA	60.04	6.289	3.145	69.47	66.32
OF	14.58	4.287	26.77	45.64	19.37
CA	32.52	13.55	5.371	51.44	46.07
BA	27.93	6.437	2.217	36.58	34.41
BMA	19.1	8.776	2.008	29.88	27.87
BFS	36.38	10.63	0.335	47.35	47.01
MK	49.75	44.71	0.509	94.97	94.46
EP	73.59	13.43	1.166	88.19	87.03
NP	72.2	13.58	1.011	86.79	85.78
HUAJ	78.19	11.56	1.567	91.32	89.75
ARL	77.57	10.67	2.384	90.62	88.24
ARL2	75.25	11.92	2.689	89.86	87.16

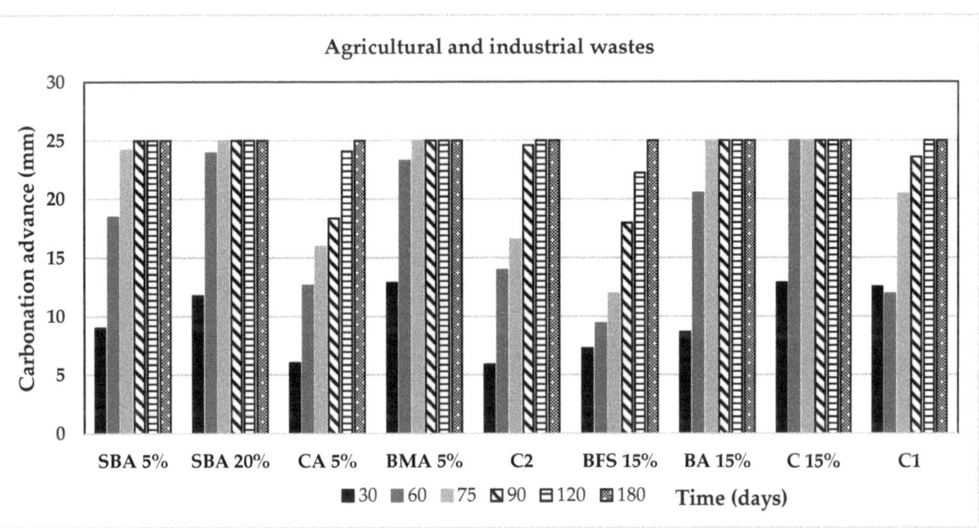

Figure 9. Advance of the carbonation front as a function of the time for the agricultural and industrial wastes group of additions.

Figures 11 and 12 present the required time to reach out the mentioned 25 mm of carbonation front. The EP and OF samples achieved the mark within the initial period of 30 days; the MK and C samples reached it at 60 days; SBA5, BMA, and BA substitutions lasted 75 days to carbonate; at 90 days the SBA20 met the mark; C1 and C2, the control samples needed 120 days; and at last, BFS, CA and NP reached the total carbonation at 180 days. Taking this parameter as an indicator of the concrete quality, the carbonation sequence indicates which additions are the better to increase the durability of structures. Conversely, the samples that reached carbonation in lesser time periods could be useful for carbon dioxide reduction and sequestration, helping to absorb these greenhouse gas emissions.

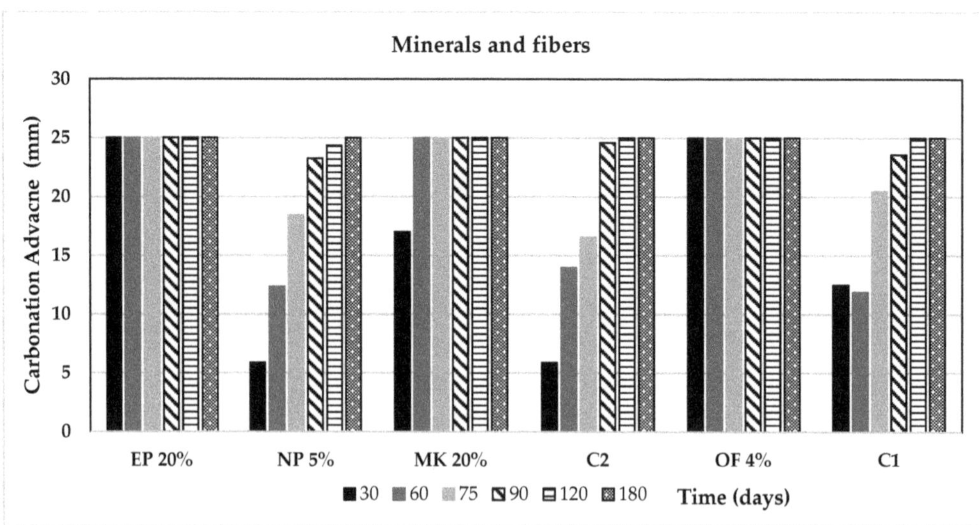

Figure 10. Advance of the carbonation front as a function of the time for the minerals and fibres group of additions.

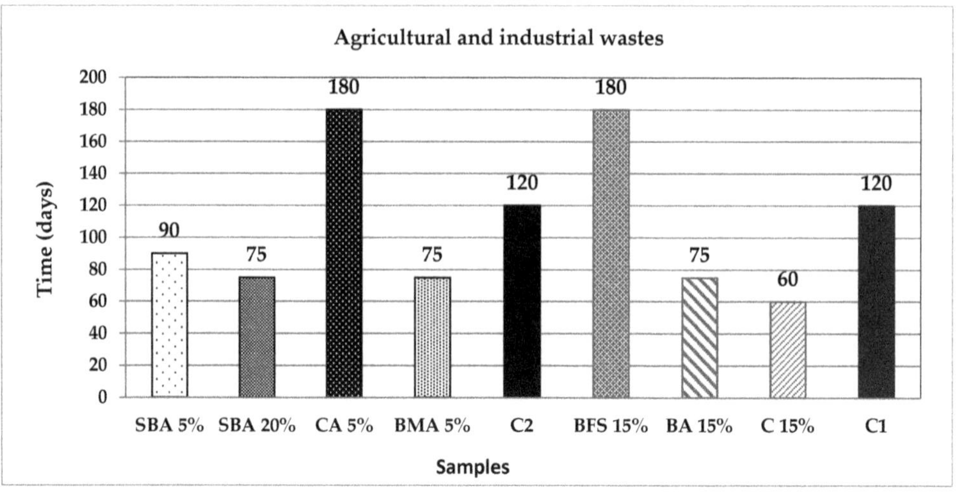

Figure 11. Critical time to reach out 25 mm of carbonation front, for the agricultural and industrial wastes group of additions.

Another way to quantify and correlate the carbonation advance in concrete structures is the following equation:

$$e = kt^{1/2} \quad (1)$$

where:

e = carbonation front depth, in mm;

k = proportionality constant of the carbonation process;

t = exposure time, in years.

Figures 13 and 14 display the value of constant k (in mm/year^2) for all the specimens. According to the equation, it can be noted that the higher k values correspond to the mortars with faster carbonation times. CA, BFS and NP reduced the carbonation velocity in comparison with the control samples, C1 and C2; therefore, these substitutions are suitable to increase the durability of reinforced concrete structures. It can also be noted that there is not a definite trend regarding the carbonation rapidity depending on the incorporated additions.

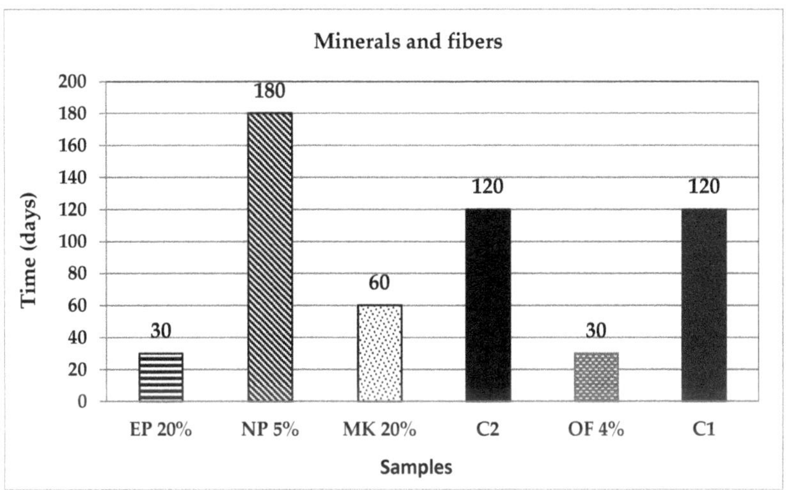

Figure 12. Critical time to reach out 25 mm of carbonation front, for the minerals and fibres group of additions.

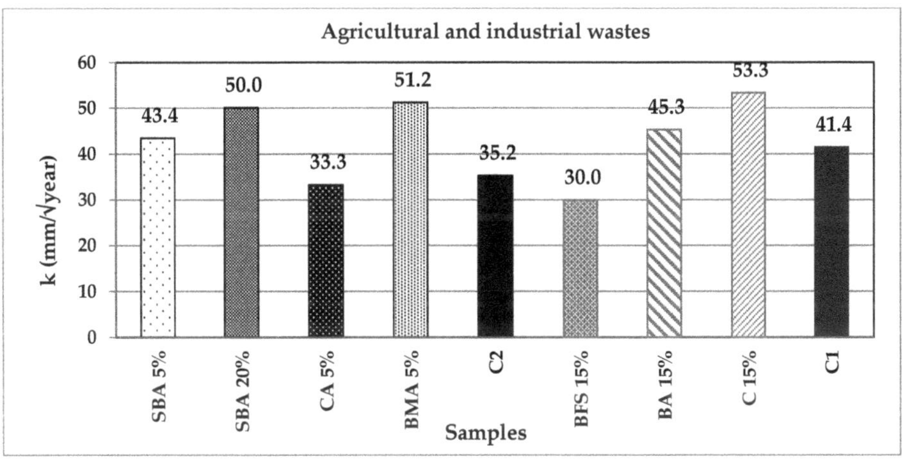

Figure 13. Constant k for the agricultural and industrial wastes group of additions.

According to the manual of the DURAR network [44,74], in natural conditions, the carbonation coefficients above 6 mm/year$^{1/2}$ are representative of bad quality concretes, and below m mm/year$^{1/2}$ are representative of high quality concretes. In the present study case, the CO_2 experimental concentration was 3%, which cannot be compared with the natural exposure of concrete structures.

In the case of the samples SBA, BMA, BA, C, EP, MK and OF, it is apparent that the incorporation of these substitutes furthers the carbonation of the mortar mixtures to a greater or lesser extent. The EP and OF samples, which reached out the 25 mm carbonation front in only 30 days (three times faster), standing out from the rest; MK and C also achieve the carbonation two times faster than the control samples. This quick carbonation may be due to the high porosity of the mixtures and the low reactivity of the pozzolanic compounds with the alkalis existing after the hydration of the PC.

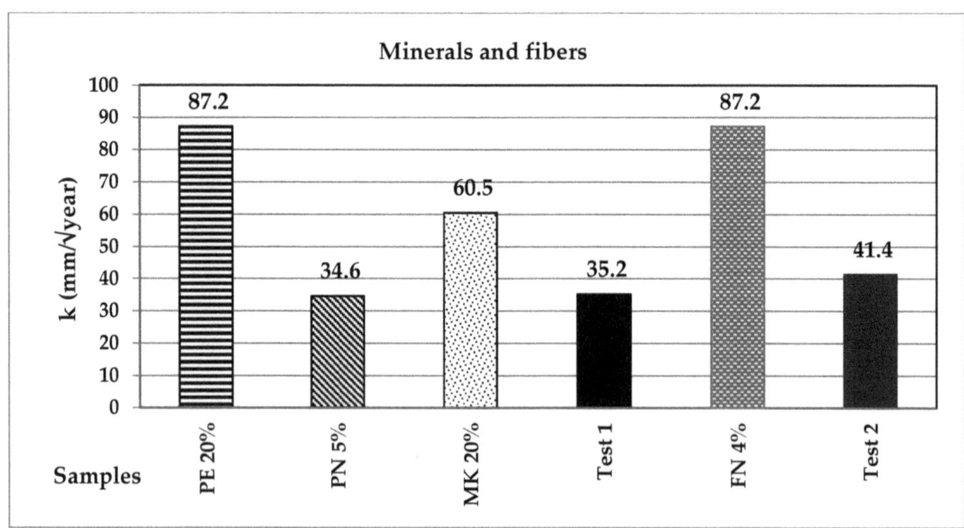

Figure 14. Constant k for the minerals and fibres group of additions.

Figures 15 and 16 present the uniaxial compressive strength values of the analysed specimens; it is notable that the mixtures with higher k values and lower carbonation times also present lower mechanical resistances compared to the control samples. Two samples stand out from the rest: SBA and MK; they present higher mechanical properties while at the same time promote carbonation; in the case of SBA, it can be observed how the greater is the substitution percentage, the higher is the carbonation velocity and the lower is the mechanical resistance.

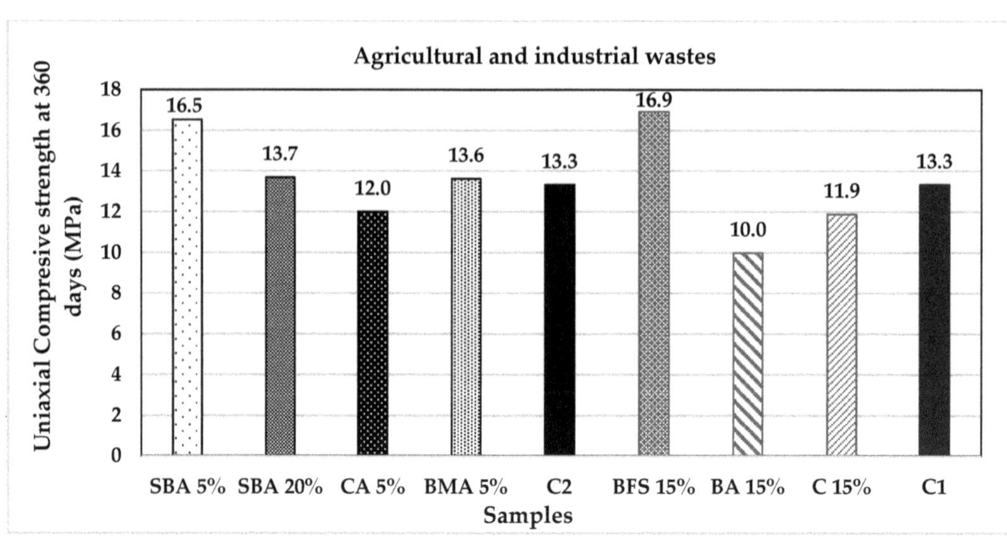

Figure 15. Uniaxial compressive strength of agricultural and industrial wastes group of additions.

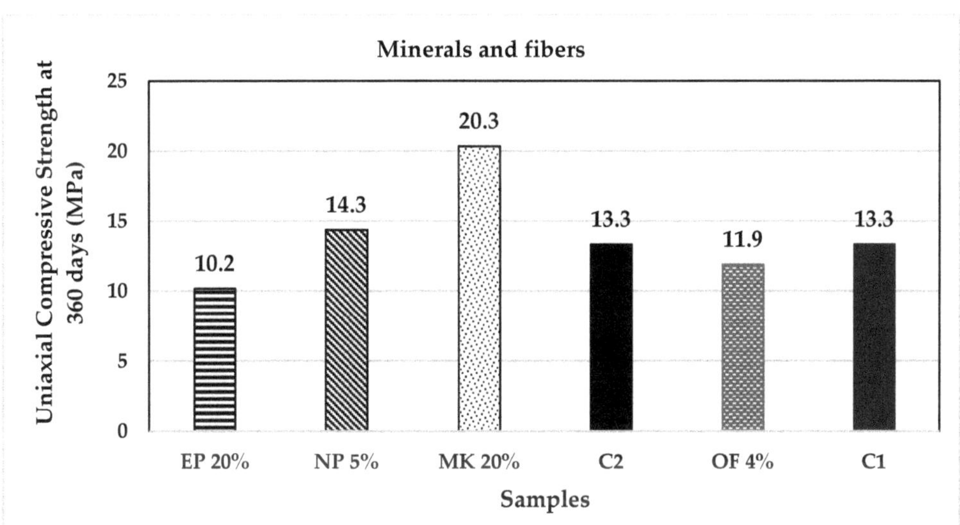

Figure 16. Uniaxial compressive strength of minerals and fibres group of additions.

Other substitution material with an unusual behaviour was CA, showing lower mechanical values than control samples but with a low carbonation velocity, being one of the three materials that achieve in reducing it. CA may act as a filler and not as a pozzolan, allowing to limit the interconnection of the cementitious matrix pores and increase its tortuosity without actually densifying the matrix enough to bring more mechanical resistance.

At last, the results of the electrical resistivity test are shown in Figures 17 and 18. We can observe that all the substitutions proposed present better behaviours than the control samples. Similar than Figures 15 and 16, SBA and MK have higher values of electrical resistivity and uniaxial compressive strength while they favour the faster carbonation. It is known that SBA contain amorphous SiO_2, which acts as a puzzolana reacting with $Ca(OH)_2$ produced during the hydration process of cement and water, acting as a retarding reaction, resulting into the generation of a saturated zone of calcium silicate hydrate (CSH) gel, selling porous and increasing compression resistance. This CSH gel reduces the amount of calcium hydroxide $Ca(OH)_2$ and consequently the pH of the cementitious paste [75]. Metakaolin (MK), $Al_2Si_2O_7$, is a highly amorphous dehydration product of kaolinite, $Al_2(OH)_4Si_2O_5$; it contains silica and alumina in an active form which will react with CH. For MK, similar pozzolanic reaction occurs through its interaction with the calcium hydroxide present in the cement paste, forming hydrated calcium silicates (C-S-H) and aluminates (C_2ASH_8, C_4AH_{13} and C_3AH_6) [76]. Both material samples were exposed in the accelerated carbonation chamber under experimental temperature, relative humidity and CO_2 concentration. This favoured the $Ca(OH)_2$ reactions, reducing their pH and increasing their mechanical resistance. This demonstrates that both materials have a high potential as CO_2 environmental recruiters without compromising the mechanical properties of the concrete mixtures.

The results can be correlated with the origin of the substitution materials. The materials with organic and mineral origin, as well as the agricultural and industrial wastes, can increase or decrease the carbonation process and with that modify the quality of the concrete in terms of durability. Only NP, CA and BFS improved the carbonation resistance in comparison with the control samples; the rest of additions presented higher k values, not being recommendable for high quality structures which require great durability.

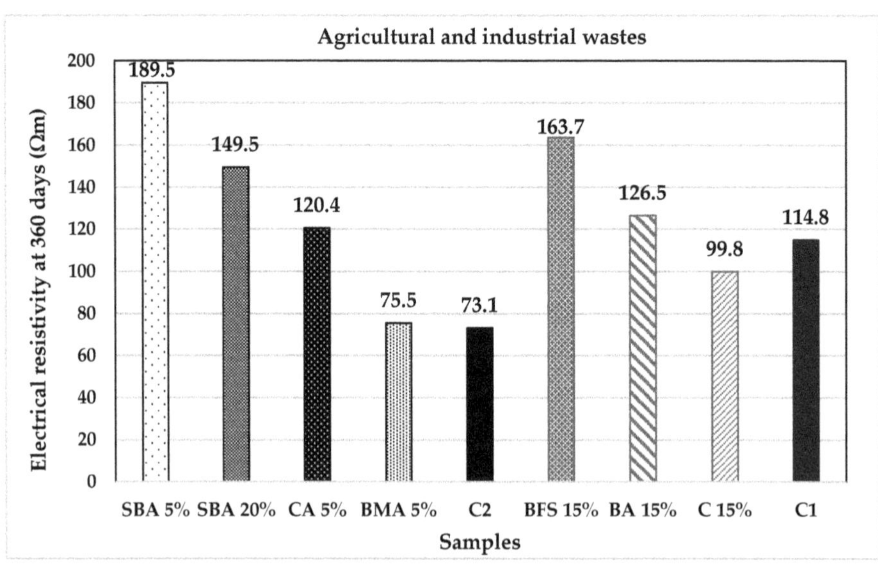

Figure 17. Electrical resistivity of the agricultural and industrial wastes group of additions.

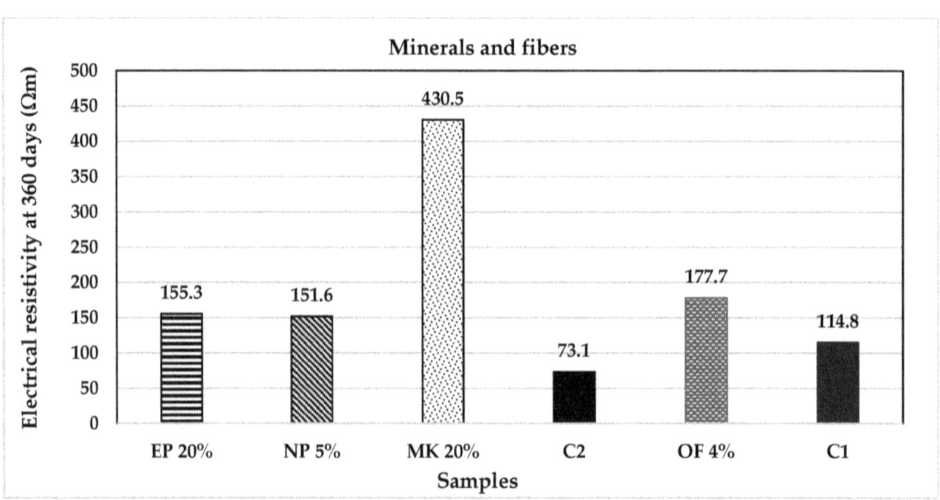

Figure 18. Electrical resistivity of the mineral and fibres group of additions.

According to the elemental composition of samples and the pozzolanic activity, SBA, MK and EP react with the alkalis of the cement pastes, while $Ca(OH)_2$, NaOH and KOH reduce the capability to resist the carbonation advance. In the case of NP, the alkalis quantity, 3.216 Na_2O and 4.285 K_2O, have the capacity to generate NaOH and KOH when they come into contact with water and then provide chemical reactions to the concrete to reduce the carbonation velocity. Regarding the pozzolanic activity of CA (51.44) and BFS (47.35), which are the other two materials that improved the carbonation resistance, they do not consume great amounts of alkalis; furthermore, they present higher concentrations of NaOH (2.108 and 10.107) and KOH (18.759 and 37.551), respectively.

It is important to mention that the criteria for the selection of substitutions and mixtures is not subdued to the carbonation velocity. A designed mixture could achieve great mechanical properties but suffer carbonation easily; in that case, it is recommended to apply superficial refurbishments and protect the structure.

The numerical analysis is an important task employed in this kind of studies in order to establish or find some relation between the features analysed. The present study establishes a multivariate analysis employing the method of least squares to estimate the compressive strength of the samples, as it is shown in Figure 19.

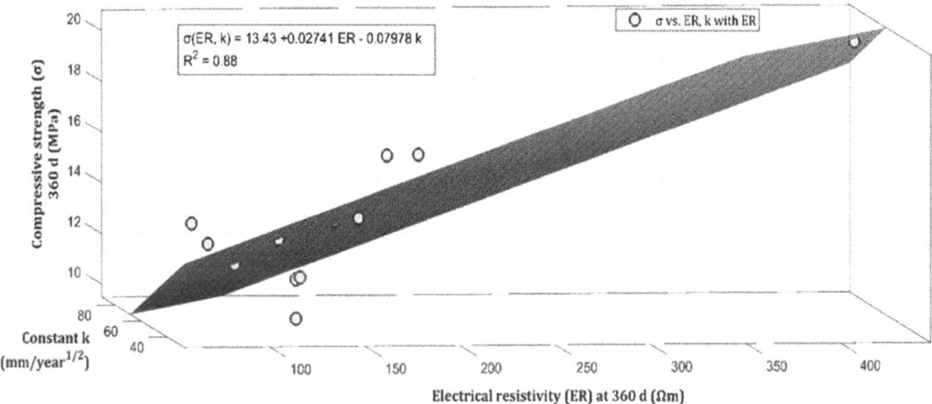

Figure 19. Multi-variable linear regression estimating the compressive strength d.

The figure displays the performance of the linear regression model depicted in a plane that fits all the data analysed. The electrical resistivity (ER) and the constant k are the two predictive variables used to build the multivariate regression model. This model yields a linear equation which is described in Equation (2).

$$\sigma\,(\mathrm{ER}, \mathrm{k}) = 13.43 + 0.02741\mathrm{ER} - 0.07978\mathrm{k} \tag{2}$$

One of the most common metrics for evaluating the accuracy of a regression model is the determination coefficient R^2. This coefficient provides the proportion of the entire variance involved by the regression task and is related with the accuracy achieved by the model [77]. The coefficient of determination can be used to evaluate the goodness-of-fit of the model and it always varies between 0 and 1. When R^2 is equal to 1, it means that there is a perfect correlation and that the sample data lie exactly on the regression line [78]. For this case the coefficient value is equal to 0.88, which means that the plane captures 88% of the information correctly. Furthermore, it is crucial to establish that the compressive strength in mortars is a value that is possible to estimate with acceptable accuracy (88%) just using the electrical resistivity value and a constant k with a simple linear multivariate regression. Nevertheless, the equation can be improved by exploring a superior order regression, a task which will be completed in further research projects.

4. Conclusions

From the eleven total materials studied in this research as partial substitutions of PC for concrete mixtures; three of them showed to reduce the carbonation velocity (CA, BFS, and NP) presenting lower k carbonation coefficients than the control samples without additions. The other eight substitution materials presented higher carbonation values; however, the whole of the specimens partially substituted the PC, allowing to decrease the emission of CO_2 to the atmosphere. In the case of the agricultural and industrial wastes materials, they presented minor reduction of the polluting emissions.

Other disclosure of the research was that the mixtures with SBA and MK acted as collectors of the environmental CO_2 without provoking a decrease of the mechanical properties. This can be extremely useful for the implementation in the construction industry, helping to reduce the PC consumption and improving the disposal of waste. Additionally, the mixtures elaborated with BMA, BA, C, EP and OF also acted as CO_2 fixers, but they did present lower mechanical resistance values. These substitutions could be applied for non-structural uses inside the construction industry, for example, acting as refurbishments or partition walls.

It is important to continue the research on potential PC substitutions from the perspective of the green economy, ecology and durability. One of the main objectives was to determine how different additions react to the carbonation process. Furthermore, these additions can proportionate other features to the concrete mixtures, which are desired for different construction processes and conditions. The materials also help to reduce the CO_2 footprint, and because of their origins, diminish the quantity of demolitions and repairs in old buildings.

Regarding the current situation with global warming and to reduce the greenhouse emissions produced by the construction industry, it is essential to find efficient and feasible ways to diminish them. This research work demonstrates the suitability of the incorporation of waste materials from diverse industries which could be used as construction materials. The employment of these materials as substitutions helps to increase the durability of concrete structures and to passively absorb the CO_2 of the environment.

Author Contributions: Conceptualization, W.M.-M., H.L.C.-G., T.P.-L. and E.M.A.-G.; methodology, M.A.-S.; software, M.A.N.-S., J.A.G.-T. and M.A.-S.; validation, M.A.-S., M.A.N.-S., J.T.P.-Q. and J.A.G.-T.; formal analysis, W.M.-M., H.L.C.-G., M.A.-S. and J.A.B.-P.; investigation, W.M.-M., T.P.-L., H.L.C.-G., J.T.P.-Q. and E.M.A.-G.; resources, W.M.-M., H.L.C.-G. and E.M.A.-G.; writing—original draft preparation, W.M.-M., M.A.-S., H.L.C.-G., A.S.-C., E.M.A.-G., J.A.B.-P., M.A.N.-S., J.A.G.-T., T.P.-L. and J.T.P.-Q.; writing—review and editing, W.M.-M., M.A.-S., H.L.C.-G. and A.S.-C.; supervision, W.M.-M., H.L.C.-G., M.A.-S. and E.M.A.-G.; project administration, W.M.-M. and H.L.C.-G.; funding acquisition, W.M.-M. and H.LC.-G. All authors have read and agreed to the published version of the manuscript.

Funding: This research received no external funding.

Institutional Review Board Statement: Not applicable.

Informed Consent Statement: Not applicable.

Data Availability Statement: Data sharing is not applicable for this article.

Acknowledgments: Authors would like to thank the financial support of: Coordinación de Investigación Científica, CIC, de Universidad Michoacana de San Nicolas de Hidalgo, UMSNH; Secretaría de Educación Pública, SEP, Programa Prodep; Consejo Nacional de Ciencia y Tecnología, CONACYT with 315660 and 315680 PRONACES Proposals and Programa Estancias Posdoctorales 460429; Universidad Autónoma de Campeche; and Programa de Becas de Movilidad Académica entre Instituciones Asociadas a la Asociación Universitaria Iberoamericana de Posgrado. In addition, the technical support received from the "Ing. Luis Silva Ruelas" Materials laboratory staff of the UMSNH Faculty of Civil Engineering, the Mexican Institute of Transportation of the SCT, and Jairo Valentín Tamayo Zapata for their technical support during the carbonation tests.

Conflicts of Interest: The authors declare no conflict of interest. The funders had no role in the design of the study; in the collection, analyses, or interpretation of data; in the writing of the manuscript, or in the decision to publish the results.

References

1. Rosewitz, J.A.; Wang, S.; Scarlata, S.F.; Rahbar, N. An enzymatic self-healing cementitious material. *Appl. Mater. Today* **2021**, *23*, 101035. [CrossRef]
2. International Energy Agency (IEA). World Energy Outlook. 2019. Available online: https://iea.blob.core.windows.net/assets/98909c1b-aabc-4797-9926-35307b418cdb/WEO2019-free.pdf (accessed on 22 November 2021).
3. USGS Mineral Commodity Summaries (Commodity statistics and information); United States Geological Survey (USGS). *Mineral Yearbooks*; USGS: Washington, DC, USA, 2019.

4. NMX-C-530-ONNCCE. *Durabilidad-Norma General de Durabilidad de Estructuras de Concreto Reforzado-Criterios y Especificaciones*; Industria de la Normalización y Certificación de la Construccion y Edificación: Mexico City, Mexico, 2018.
5. Noushini, A.; Castel, A. Performance-based criteria to assess the suitability of geopolymer concrete in marine environments using modified ASTM C1202 and ASTM C1556 methods. *Mater. Struc.* **2018**, *51*, 146. [CrossRef]
6. Ziane, S.; Khelifa, M.-R.; Mezhoud, S.; Medaoud, S. Durability of concrete reinforced with alfa fibres exposed to external sulphate attack and thermal stresses. *Asian J. Civ. Eng.* **2020**, *21*, 555–567. [CrossRef]
7. Collepardi, M. *The New Concrete*, 1st ed.; Titoretto: Villorba, Italy, 2006.
8. European Committee for Standardization. *Concrete—Specification, Performance, Production and Conformity*; European Committee for Standardization: Brussels, Belgium, 2013.
9. European Committee for Standardization. *Eurocode 2: Design of Concrete Structures—Part 1-1: General Rules and Rules for Buildings*; European Committee for Standardization: Brussels, Belgium, 2004.
10. Bertolini, L.; Elsenes, B.; Pedeferri, P.; Redaelli, E.; Polder, R. *Corrosion of Steel in Concrete: Prevention, Diagnosis, Repair*, 2nd ed.; Wiley-VCH: Weinheim, Germany, 2013.
11. Chen, T.; Gao, X. Effect of carbonation curing regime on strength and microstructure of Portland cement paste. *J. CO2 Util.* **2019**, *34*, 74–86. [CrossRef]
12. Neves Junior, A.; Toledo Filho, R.D.; de Moraes Rego Fairbairn, E.; Dweck, J. The effects of the early carbonation curing on the mechanical and porosity properties of high initial strength Portland cement pastes. *Constr. Build. Mater.* **2015**, *77*, 448–454. [CrossRef]
13. Zhutovsky, S.; Shishkin, A. Recycling of hydrated Portland cement paste into new clinker. *Constr. Build. Mater.* **2021**, *280*, 122510. [CrossRef]
14. Talukdar, S.; Banthia, N. Carbonation in concrete infrastructure in the context of global climate change: Development of a service lifespan model. *Constr. Build. Mater.* **2013**, *40*, 775–782. [CrossRef]
15. Chávez-Ulloa, E.; López, T.P.; Trujeque, J.R.; Pérez, F.C. Deterioration of concrete structures due to carbonation in tropical marine environment and accelerated carbonation chamber. *Rev. Tec. Fac. Ing. Univ. Zulia* **2013**, *36*, 104–113.
16. Malhotra, V.M.; Metha, P. *Pozzolanic and Cementitious Materials*, 1st ed.; CRC Press: Ottawa, ON, Canada, 1996.
17. Escalante, J.I.; Navarro, A.; Gomez, L.Y. Caracterización de morteros de cemento portland substituido por metacaolín de baja pureza. *Rev. ALCONPAT* **2012**, *1*, 186–199. [CrossRef]
18. Memon, M.J.; Jhatial, A.A.; Murtaza, A.; Raza, M.S.; Phulpoto, K.B. Production of eco-friendly concrete incorporating rice husk ash and polypropylene fibres. *Environ. Sci. Pollut. Res.* **2021**, *28*, 39168–39184. [CrossRef]
19. Mustakim, S.M.; Das, S.K.; Mishra, J.; Aftab, A.; Alomayri, T.S.; Assaedi, H.S.; Kaze, C.R. Improvement in Fresh, Mechanical and Microstructural Properties of Fly Ash- Blast Furnace Slag Based Geopolymer Concrete by Addition of Nano and Micro Silica. *Silicon* **2021**, *13*, 2415–2428. [CrossRef]
20. Kannur, B.; Chore, H.S. Utilization of sugarcane bagasse ash as cement-replacing materials for concrete pavement: An overview. *Innov. Infrastruct. Solut.* **2021**, *6*, 184. [CrossRef]
21. Chandrasekhar Reddy, K. Investigation of mechanical and durable studies on concrete using waste materials as hybrid reinforcements: Novel approach to minimize material cost. *Innov. Infrastruct. Solut.* **2021**, *6*, 197. [CrossRef]
22. Martinez-Molina, W.; Torres-Acosta, A.A.; Martínez-Peña, G.E.; Alonso-Guzmán, E.; Mendoza-Pérez, I.N. Cement-based, materials-enhanced durability from opuntia ficus indica mucilage additions. *ACI Mater. J.* **2015**, *112*, 165–172. [CrossRef]
23. Islam, G.M.S.; Rahman, M.H.; Kazi, N. Waste glass powder as partial replacement of cement for sustainable concrete practice. *Int. J. Sustain. Built Environ.* **2017**, *6*, 37–44. [CrossRef]
24. Janfeshan Araghi, H.; Nikbin, I.M.; Rahimi Reskati, S.; Rahmani, E.; Allahyari, H. An experimental investigation on the erosion resistance of concrete containing various PET particles percentages against sulfuric acid attack. *Constr. Build. Mater.* **2015**, *77*, 461–471. [CrossRef]
25. Anderson, D.J.; Smith, S.T.; Au, F.T.K. Mechanical properties of concrete utilizing waste ceramic as coarse aggregate. *Constr. Build. Mater.* **2016**, *117*, 20–28. [CrossRef]
26. Tavakoli, D.; Heidari, A.; Pilehrood, S.H. Properties of concrete made with waste clay brick as sand incorporating nano SiO2. *Indian J. Sci. Technol.* **2014**, *7*, 1899–1905. [CrossRef]
27. Gupta, T.; Chaudhary, S.; Sharma, R.K. Assessment of mechanical and durability properties of concrete containing waste rubber tire as fine aggregate. *Constr. Build. Mater.* **2014**, *73*, 562–574. [CrossRef]
28. Cartuxo, F.; de Brito, J.; Evangelista, L.; Jiménez, J.R.; Ledesma, E.F. Rheological behaviour of concrete made with fine recycled concrete aggregates—Influence of the superplasticizer. *Constr. Build. Mater.* **2015**, *89*, 36–47. [CrossRef]
29. Mo, K.H.; Alengaram, U.J.; Jumaat, M.Z.; Yap, S.P.; Lee, S.C. Green concrete partially comprised of farming waste residues: A review. *J. Clean. Prod.* **2016**, *117*, 122–138. [CrossRef]
30. Devadiga, D.G.; Bhat, K.S.; Mahesha, G.T. Sugarcane bagasse fiber reinforced composites: Recent advances and applications. *Cogent Eng.* **2020**, *7*, 1823159. [CrossRef]
31. Han-Seung, L.; Wang, X.-Y. Evaluation of compressive strength development and carbonation depth of high-volume slag-blended concrete. *Constr. Build. Mater.* **2016**, *124*, 45–54. [CrossRef]
32. Miah, M.J.; Hossain Patoary, M.M.; Paul, S.C.; Babafemi, A.J.; Panda, B. Enhancement of mechanical properties and porosity of concrete using steel slag coarse aggregate. *Materials* **2020**, *13*, 2865. [CrossRef] [PubMed]

33. Landa-Sánchez, A.; Bosch, J.; Baltazar-Zamora, M.A.; Croche, R.; Landa-Ruiz, L.; Santiago-Hurtado, G.; Moreno-Landeros, V.M.; Olguín-Coca, J.; López-Léon, L.; Bastidas, J.M.; et al. Corrosion behavior of steel-reinforced green concrete containing recycled coarse aggregate additions in sulfate media. *Materials* **2020**, *13*, 4345. [CrossRef]
34. Irshidat, M.R.; Al-Nuaimi, N. Industrial waste utilization of carbon dust in sustainable cementitious composites production. *Materials* **2020**, *13*, 3295. [CrossRef]
35. Srimahachota, T.; Yokota, H.; Akira, Y. Recycled nylon fiber from waste fishing nets as reinforcement in polymer cement mortar for the repair of corroded RC beams. *Materials* **2020**, *13*, 4276. [CrossRef]
36. Ulewicz, M.; Pietrzak, A. Properties and structure of concretes doped with production waste of thermoplastic elastomers from the production of car floor mats. *Materials* **2021**, *14*, 872. [CrossRef]
37. Kanojia, A.; Jain, S.K. Performance of coconut shell as coarse aggregate in concrete. *Constr. Build. Mater.* **2017**, *140*, 150–156. [CrossRef]
38. Morandeau, A.; Thiéry, M.; Dangla, P. Impact of accelerated carbonation on OPC cement paste blended with fly ash. *Cem. Concr. Res.* **2015**, *67*, 226–236. [CrossRef]
39. Tavakoli, D.; Hashempour, M.; Heidari, A. Use of waste materials in concrete: A review. *Pertanika J. Sci. Technol.* **2018**, *26*, 499–522.
40. Criado, M.C.; Vera, C.; Downey, P.; Soto, M.C. Influences of the fiber glass in physical-mechanical properties of the concrete. *Rev. Ing. Construcción* **2005**, *20*, 201–212.
41. Page, J.; Khadraoui, F.; Gomina, M.; Boutouil, M. Influence of different surface treatments on the water absorption capacity of flax fibres: Rheology of fresh reinforced-mortars and mechanical properties in the hardened state. *Constr. Build. Mater.* **2019**, *199*, 424–434. [CrossRef]
42. El-Nadoury, W.W. Applicability of Using Natural Fibers for Reinforcing Concrete. In *IOP Conference Series: Materials Science and Engineering*; IOP Publishing: Bristol, UK, 2020; Volume 809. [CrossRef]
43. He, J.; Kawasaki, S.; Achal, V. The utilization of agricultural waste as agro-cement in concrete: A review. *Sustainability* **2020**, *12*, 6971. [CrossRef]
44. Trocónis de Rincón, O. *DURAR Network Members, Manual for Inspecting, Evaluating and Diagnosing Corrosion in Reinforced Concrete Structures*, 1st ed.; CYTED: Maracaibo, Venezuela, 2000.
45. Gómez-Zamorano, L.Y.; Escalante-García, J.I.; Mendoza-Suárez, G. Geothermal waste: An alternative replacement material of portland cement. *J. Mater. Sci.* **2004**, *39*, 4021–4025. [CrossRef]
46. Morandeau, A.; Thiéry, M.; Dangla, P. Investigation of the carbonation mechanism of CH and C-S-H in terms of kinetics, microstructure changes and moisture properties. *Cem. Concr. Res.* **2014**, *56*, 153–170. [CrossRef]
47. de Oliveira, A.M.; Cascudo, O. Effect of mineral additions incorporated in concrete on thermodynamic and kinetic parameters of chloride-induced reinforcement corrosion. *Constr. Build. Mater.* **2018**, *192*, 467–477. [CrossRef]
48. Galan, I.; Andrade, C.; Mora, P.; Sanjuan, M.A. Sequestration of CO_2 by Concrete Carbonation. *Environ. Sci. Technol.* **2010**, *44*, 3181–3186. [CrossRef] [PubMed]
49. Possan, E.; Cristiano Fogaça, J.; Catiussa, E.; Pazuch, M. Sequestro de CO_2 Devido à Carbonatação do Concreto: Potencialidades da Barragem de Itaipu. *Rev. Estud. Ambient.* **2012**, *14*, 28–38.
50. Jacobsen, S.; Jahren, P. Binding of CO_2 by carbonation of Norwegian Concrete. In Proceedings of the CANMET/ACI International Conference on Sustainability and Concrete Technology, Lyon, France, 1 September 2001; pp. 329–338.
51. Kim, S.-K.; Jang, I.-Y.; Yang, H.-J. An Empirical Study on the Absorption of Carbon Dioxide in OLED-Mixed Concrete through Carbonation Reaction. *KSCE J. Civ. Eng.* **2020**, *24*, 2495–2504. [CrossRef]
52. Guedes de Paiva, F.F.; Tamashiro, J.R.; Pereira Silva, L.H.; Kinoshita, A. Utilization of inorganic solid wastes in cementitious materials—A systematic literature review. *Const. Build. Mater.* **2021**, *285*, 122833. [CrossRef]
53. Berenguer, R.; Lima, N.; Cruz, R.; Pinto, L.; Lima, N. Thermodynamic, microstructural and chemometric analyses of the reuse of sugarcane ashes in cement manufacturing. *J. Environ. Chem. Eng.* **2021**, *9*, 105350. [CrossRef]
54. Liu, G.; Florea, M.; Brouwers, H. The role of recycled waste glass incorporation on the carbonation behaviour of sodium carbonate activated slag mortar. *J. Clean. Prod.* **2021**, *292*, 126050. [CrossRef]
55. Mi, R.; Pan, G.; Li, Y.; Kuang, T.; Lu, X. Distinguishing between new and old mortars in recycled aggregate concrete under carbonation using iron oxide red. *Const. Build. Mater.* **2019**, *222*, 601–609. [CrossRef]
56. Zhang, S.; Ghouleh, Z.; Shao, Y. Green concrete made from MSWI residues derived eco-cement and bottom ash aggregates. *Const. Build. Mater.* **2021**, *297*, 123818. [CrossRef]
57. Duygu, E.; Fort, R. Accelerating carbonation in lime-based mortar in high CO_2 environments. *Const. Build. Mater.* **2018**, *188*, 314–325. [CrossRef]
58. Sharma, D.; Goyal, S. Accelerated carbonation curing of cement mortars containing cement kiln dust: An effective way of CO_2 sequestration and carbon footprint reduction. *J. Clean. Prod.* **2018**, *192*, 844–854. [CrossRef]
59. Rigopoulos, I.; Kyriakou, L.; Vasiliades, M.A.; Kyratsi, T.; Efstathiou, A.M.; Ioannou, I. Improving the carbonation of air lime mortars at ambient conditions via the incorporation of ball-milled quarry waste. *Const. Build. Mater.* **2021**, *301*, 124703. [CrossRef]
60. Liu, S.; Hao, Y.; Ma, G. Approaches to enhance the carbonation resistance of fly ash and slag-based alkali-activated mortar-experimental evaluations. *J. Clean. Prod.* **2021**, *280*, 124321. [CrossRef]
61. NMX-C-414-ONNCCE-2017. *Cementantes Hidraulicos. Especificaciones y Métodos de Ensayo*; ONNCCE: Mexico City, Mexico, 2017.
62. ASTM C150/C150M-21. *Standard Specification for Portland Cement*; ASTM International: West Conshohocken, PA, USA, 2021.

63. ASTM C31/C31M-21a. *Standard Practice for Making and Curing Concrete Test Specimens in the Field*; ASTM International: West Conshohocken, PA, USA, 2021.
64. ASTM C109/C109M-21. *Standard Test Method for Compressive Strength of Hydraulic Cement Mortars (Using 2-in. or [50 mm] Cube Specimens)*; ASTM International: West Conshohocken, PA, USA, 2021.
65. ASTM C1876-19. *Standard Test Method for Bulk Electrical Resistivity or Bulk Conductivity of Concrete*; ASTM International: West Conshohocken, PA, USA, 2019.
66. ASTM D75-03. *Standard Practice for Sampling Aggregates*; ASTM International: West Conshohocken, PA, USA, 2003.
67. ASTM C702 / C702M-18. *Standard Practice for Reducing Samples of Aggregate to Testing Size*; ASTM International: West Conshohocken, PA, USA, 2018.
68. ASTM C29 / C29M-17a. *Standard Test Method for Bulk Density ("Unit Weight") and Voids in Aggregate*; ASTM International: West Conshohocken, PA, USA, 2017.
69. ASTM C128-15. *Standard Test Method for Relative Density (Specific Gravity) and Absorption of Fine Aggregate*; ASTM International: West Conshohocken, PA, USA, 2015.
70. ASTM C70-20. *Standard Test Method for Surface Moisture in Fine Aggregate*; ASTM International: West Conshohocken, PA, USA, 2020.
71. ASTM C566-19. *Standard Test Method for Total Evaporable Moisture Content of Aggregate by Drying*; ASTM International: West Conshohocken, PA, USA, 2019.
72. ASTM D2419-14. *Standard Test Method for Sand Equivalent Value of Soils and Fine Aggregate*; ASTM International: West Conshohocken, PA, USA, 2014.
73. ASTM C142 / C142M-17. *Standard Test Method for Clay Lumps and Friable Particles in Aggregates*; ASTM International: West Conshohocken, PA, USA, 2017.
74. NMX-C-515-ONNCCE-2016. *Industria de la Construcción, Concreto Hidráulico, Durabilidad, Determinación de la Profundidad de Carbonatación en Concreto Hidráulico, Especificaciones y Método de Ensayo*; Industria de la Normalización y Certificación de la Construccion y Edificación: Mexico City, Mexico, 2016.
75. Ali, S.; Kumar, A.; Rizvi, S.H.; Ali, M.; Ahmed, I. Effect of Sugarcane Bagasse Ash as Partial Cement Replacement on the Compressive Strength of Concrete. *Quaid-E-Awam Univ. Res. J. Eng. Sci. Technol.* **2018**, *18*, 44–49. [CrossRef]
76. El-Diadamony, H.; Amer, A.A.; Sokkary, T.M.; El-Hoseny, S. Hydration and characteristics of metakaolin pozzolanic cement pastes. *HBRC J.* **2018**, *14*, 150–158. [CrossRef]
77. Guzmán-Torres, J.A.; Domínguez-Mota, F.J.; Alonso-Guzmán, E.M. A multi-layer approach to classify the risk of corrosion in concrete specimens that contain different additives. *Case Stud. Constr. Mater.* **2021**, *15*, e00719. [CrossRef]
78. Silva, A.; de Brito, J.; Lima Gaspar, P. *Methodologies for Service Life Prediction of Buildings. With a Focus on Façade Claddings*, 1st ed.; Springer: Cham, Switzerland, 2016. [CrossRef]

 materials MDPI

Review

A Review of Carbon Footprint Reduction in Construction Industry, from Design to Operation

Banu Sizirici [1], Yohanna Fseha [1], Chung-Suk Cho [1,*], Ibrahim Yildiz [2] and Young-Ji Byon [1]

1 Department of Civil Infrastructure and Environmental Engineering, Khalifa University, Abu Dhabi P.O. Box 127788, United Arab Emirates; banu.yildiz@ku.ac.ae (B.S.); 100049403@ku.ac.ae (Y.F.); youngji.byon@ku.ac.ae (Y.-J.B.)
2 Department of Chemistry, Khalifa University, Abu Dhabi P.O. Box 127788, United Arab Emirates; ibrahim.yildiz@ku.ac.ae
* Correspondence: chung.cho@ku.ac.ae

Abstract: Construction is among the leading industries/activities contributing the largest carbon footprint. This review paper aims to promote awareness of the sources of carbon footprint in the construction industry, from design to operation and management during manufacturing, transportation, construction, operations, maintenance and management, and end-of-life deconstruction phases. In addition, it summarizes the latest studies on carbon footprint reduction strategies in different phases of construction by the use of alternative additives in building materials, improvements in design, recycling construction waste, promoting the utility of alternative water resources, and increasing efficiencies of water technologies and other building systems. It was reported that the application of alternative additives/materials or techniques/systems can reduce up to 90% of CO_2 emissions at different stages in the construction and building operations. Therefore, this review can be beneficial at the stage of conceptualization, design, and construction to assist clients and stakeholders in selecting materials and systems; consequently, it promotes consciousness of the environmental impacts of fabrication, transportation, and operation.

Keywords: embodied carbon; recycled asphalt; recycled aggregate; construction waste materials; alternative additives; alternative water resources

Citation: Sizirici, B.; Fseha, Y.; Cho, C.-S.; Yildiz, I.; Byon, Y.-J. A Review of Carbon Footprint Reduction in Construction Industry, from Design to Operation. *Materials* **2021**, *14*, 6094. https://doi.org/10.3390/ma14206094

Academic Editor: Alessandro P. Fantilli

Received: 9 September 2021
Accepted: 12 October 2021
Published: 15 October 2021

1. Introduction

This paper aims to bring attention to the carbon footprint in the construction industry (building, maintaining, and deconstructing the structures), since the construction industry is listed as the single largest global consumer of resources [1,2]. In the European Union, building construction consumes 40% of materials and 40% of primary energy, and generates 40% of waste annually [1]. Globally, in developed and developing countries, buildings contributes to 33% of the greenhouse gas (GHG) emissions and 40% of the global energy consumption which stem from the usage of the equipment, the manufacturing of building materials and transportation [3,4]. The total CO_2 emission of the construction sector was 5.7 billion tons which made up 23% of the emissions of global economic activity in 2009 [5]. Globally, the urban population is predicted to exceed six billion in 2045, and this could lead to more construction in the future.

According to the 4th Assessment Report of the Intergovernmental Panel on Climate Change (IPCC), GHG emissions from buildings contributed 8.6 billion t-CO_2-e in 2004. It is predicted that it could reach up to 15.6 billion t-CO_2-e by 2030, creating an increase of 26% CO_2 which accounts for 30–40% of the total GHG emissions [6]. It is necessary to take action to reduce GHGs resulted from construction activities. Hence, it is vital to implement policies that focus on GHG emissions mitigation. Such schemes are broadly classified into two approaches: (1) indirect pricing such as regulations and (2) direct pricing such as carbon taxes and emission trading schemes (ETS) [7].

Regulations such as building codes can effectively reduce GHG emissions if enforced well enough, and can ensure new buildings incorporate designs that are both cost and energy effective [8]. Required codes, including the European Union's zero energy mandate by 2021, Australia's NatHERS 5-star standard, volunteer certificates such as Leadership in Energy and Environmental Design (LEED) which is required for all new federal government construction projects and renovations in the USA but voluntary for private construction, and the Building Research Establishment Environmental Assessment Method (BREEAM), would force designers and contractors to reconsider material usage that has a high embodied carbon content and also to rethink way they conduct their operations [9,10].

Another instrument for the mitigation of GHG emission is the carbon tax. Carbon taxes are simpler to design, have relatively low administration costs, and are attractive to stakeholders in the building sector due to their familiarity with the tax mechanism [11,12]. Carbon taxes encourage industry and the general public to help reduce GHG emissions by using energy efficiently and opting for cleaner, renewable sources of energy which in turn leads to innovations in technology and processes [13]. In terms of ETS, the cumulative amount of GHG emissions mitigated can be quantified with ETS and emission permits can be distributed for free or auctioned off [7,14]. As both energy supply and demand have equal weights, an ETS can be especially useful in the construction industry, thereby, encouraging the use of technologies that are energy efficient [15].

Studies have shown that a variety of factors slow down the move towards a carbon neutral construction industry. A study conducted in Singapore and Hong Kong found that lack of awareness, education, incentives, and high initial costs are the obstacles to such a move [16]. In another study that focused on commercial buildings in the Chinese cities of Beijing and Shanghai, the barriers were identified to be lack of regulations and financial incentives, ineffective monitoring, and lack of awareness around energy saving [17]. Therefore, this paper aims to bring attention/awareness where carbon footprint resulting from design to operation/management phases, such as manufacturing, transportation, construction, operation and maintenance, and end-of-life deconstruction in construction industry. If these sources are well identified, it will be helpful to reduce GHGs at the stage of conceptualization, design, construction, and management via selecting material, system, operation and management having less carbon footprint, which will promote environmental consciousness in whole construction operations.

There are many studies focused on CO_2 reduction at different phases in the construction industry. However, there is no other study focusing on carbon reduction in all stages from design to operation and management phases with emphasis on manufacturing, transportation, construction, operation and maintenance, and end-of-life deconstruction comprehensively. Therefore, this paper reviewed a variety of the latest techniques for reducing the carbon footprint of each phase such as the use of alternative additives in building materials, improvements in the design, recycling of construction waste, promoting the use of alternative water resources, increasing the efficiency of water technologies, and building novel systems to improve the sustainability of the construction industry.

2. Carbon Footprint of Mining, Manufacturing, and Materials Transporting in the Construction Industry and GHG Reduction

Construction process undergoes several phases, starting with production of materials (non-metallic minerals, oil, cement mortar, iron, steel, concrete) and material transportation which contributes 82–96% of the total CO_2 emissions through the construction period as shown in Figure 1 [18–21].

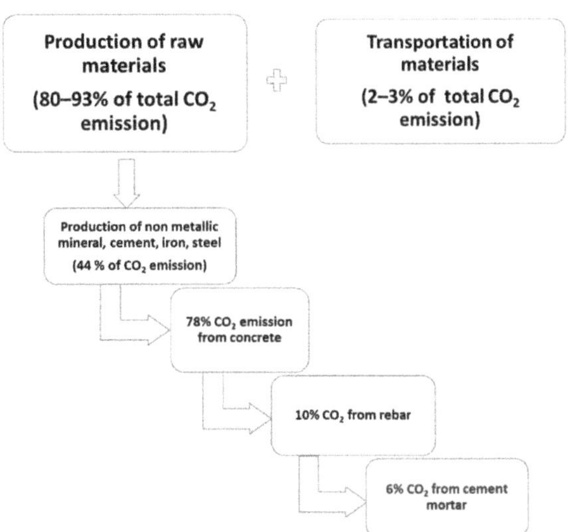

Figure 1. CO_2 emission from different phases in the construction industry.

A study showed that carbon footprint of urban buildings increased from 8.95 million tons in 2005 to 13.57 million tons in 2009, and that 45% of CO_2 resulted from building material production whereas 40% of CO_2 resulted from building energy in Xiamen, China [22]. Another study indicated that life-cycle carbon emission of a five-story brick-concrete residential building in Nanjing city of PR China was 1807.31 t, and 90% of CO_2 were emitted at the stage of construction materials preparation and the stage of building operation [23].

2.1. Carbon Footprint of Limestone Quarrying

Limestone is one of the largest produced crushed rocks which is the basic component of construction materials, such as aggregate, lime, cement, and building stones for the construction industry [24]. The energy required for lime quarrying is associated with the machine fuel, diesel, and electricity that are needed for the limestone processing. The machines used together with their energy requirements and CO_2 emissions are listed in Figure 2. A study found that the main cause of resource depletion in limestone quarrying was the use of diesel fuel in the transportation process, and that based on the GHG Protocol the GHGs emission was found to be 3.13 kg CO_2 eq. per ton crushed rock product. This study suggested the adoption of alternative renewable energies such as solar, thermal, and biodiesel which will have significant impact on the reduction of GHG emissions (0.21 Mt-CO_2 eq. annually) [25].

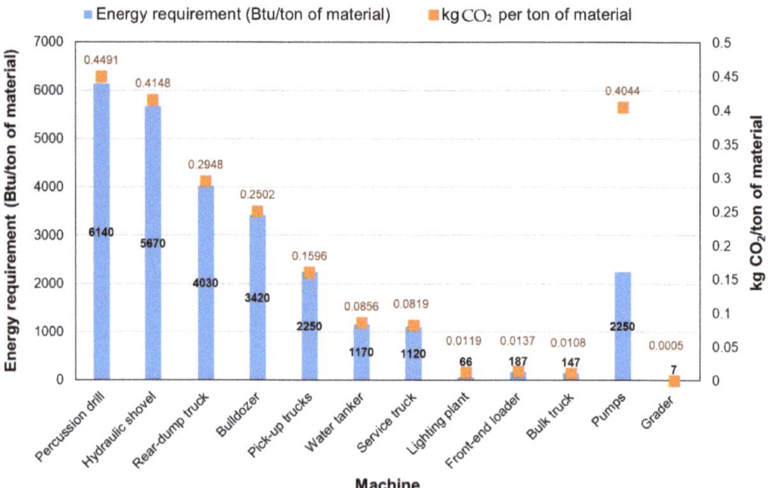

Figure 2. The machines used together with their energy requirements and CO_2 emissions in limestone processing. Adapted from [24].

2.2. *Carbon Footprint of Cement and Concrete Manufacturing*

Globally, cement manufacturing accounts for 5% of CO_2 emissions [26]. It has been reported that manufacturing of 1 kg of Portland clinker releases nearly 1 kg of CO_2 to the atmosphere. The calcination process that takes place in the cement kiln contributes nearly 0.55 kg CO_2 per kg of cement clinker [27]. Concrete's typical composition is 34% sand, 12% Portland cement, 48% crushed stone, and 6% water. Since the cement percentage is relatively small in concrete, it is considered non-energy intensive compared to other construction materials [24]. CO_2 emission rate during the production of concrete is between 347 and 351 kg of CO_2-e/m^3 [28]. According to Solís-Guzmán, Cement II/AL 32.5 N in two four-story blocks of flats (a total of 107 dwellings with total area of 10,243.69 m^2) gives 148,180 kg CO_2 eq/year and concrete HA25/B/40 gives 312,596.55 kg CO_2 eq/year during one year of construction process [1]. A study in China reported that 1 km Portland cement concrete pavement construction gives 8215.31 CO_2e tons in which raw material production accounts for 92.7%, concrete manufacturing phase accounts for 7.2% and onsite pavement construction phase accounts for 0.1% of the total GHG emissions [26]. The energy consumption on-site and CO_2 emissions from the production of cement/concrete annually are listed in Table 1. The United States was the third largest producer of cement globally with 50–55 million metric tons (Mt) of CO_2eq emissions which is equivalent to 4% of the total GHG emissions in the country in 2012. These numbers are expected to increase further as the production of cement grows [29].

Table 1. On-site energy consumption and CO_2 emissions from cement/concrete production annually. Adapted from [24,26].

Activity		Cement		Concrete	
		Energy Use/Ton (Btu)	CO_2 Emissions (Ton)/Ton of Material	Energy Use/Ton (Btu)	CO_2 Emissions (Ton)/Ton of Material
Quarrying and crushing		4.29×10^4	4.05×10^{-3}	1.61×10^5	1.44×10^{-2}
Cement manufacturing	Raw grinding	9.39×10^4	1.69×10^{-2}		
	Kiln fuels	4.62×10^6	4.33×10^{-1}		
	Reactions		5.44×10^{-1}		
	Finish milling	2.71×10^5	4.86×10^{-2}		
Concrete production	Blending/mixing			3.54×10^5	6.36×10^{-2}
	Transportation			6.97×10^5	5.10×10^{-2}

2.3. Carbon Footprint of Asphalt Production and Construction

Asphalt is a substance used as binder for pavement materials. The energy consumption for asphalt binder production includes the extraction of crude oil, transportation, and the refining process. The energy consumption for asphalt binders is 4900 MJ per ton, and the corresponding GHG emissions is 285 kg CO_2 per ton [30]. Heating aggregates account for 67% CO_2 emission, asphalt heating accounts for 14% CO_2, and mixing process accounts for 12% of total carbon emissions [31]. A case study in China reported that 20 km long asphalt pavement construction emitted 52,264,916.06 kg CO_2-e, which includes raw materials production accounting for 43% of total GHG emissions, mixing accounting for 54% of total GHG emissions, and transportation, laying, compacting, and curing phase accounting for 3% of total GHG emission [32].

2.4. Carbon Footprint of Steel Production

Steel production starts with the reaction between iron ore and a reducing agent, coking coal, in the blast furnaces producing melted iron which is converted to steel in a later stage. The reaction of iron ore with carbon is the major contributor of CO_2 emission in the steel production corresponding to 70–80% of the total CO_2 emissions [25]. Globally, steel manufacturing accounts for 6% of CO_2 emissions [33]. Globally, steel manufacturing accounts for 6% of CO_2 emissions [26]. According to Solis Guzmán [1], Steel B 500S in two four-story blocks of flats (a total of 107 dwellings with total area of 10,243.69 m^2) gives 281,898.38 kg CO_2 eq/year during one year of construction process. Table 2 shows the energy type and consumption, and CO_2 emissions associated with steel production in the integrated steel making and secondary steel making stages.

Table 2. The energy type and consumption and the CO_2 emissions associated with steel production. Adapted from [34].

Activity	Energy Type and Consumption	Final Energy (MBtu/Ton)	CO_2 Emission (Ton)/Ton of Material
	Primary steel making		
Sinter making	26 PJ fuel and 2 PJ electricity	0.264	0.009
Coke making	74 PJ fuel and 2 PJ electricity	0.718	0.007
Iron making	676 PJ fuel and 4 PJ electricity	6.421	0.120
Steel making (Basic oxygen furnace)	19 PJ fuel and 6 PJ electricity	0.236	0.005
Casting	15 PJ fuel and 11 PJ electricity	0.236	0.010
Hot rolling	157 PJ fuel and 34 PJ electricity	1.803	0.041
Cold rolling and finishing	43 PJ fuel and 15 PJ electricity	0.548	0.014
Boilers	167 PJ fuel and 0 electricity	1.577	0.085
Co-generation (integrated steel making)	101 PJ fuel and -22 PJ electricity	0.746	0.004
	Secondary steel making		
Steel making using electric arc furnace	6 PJ fuel and 62 PJ electricity	0.642	0.031
Casting	1 PJ fuel and 4 PJ electricity	0.028	0.002
Hot rolling	102 PJ fuel and 22 PJ electricity	1.171	0.026
Cold rolling and finishing	Not required	none	-
Boilers	42 PJ fuel and 0 PJ electricity	0.397	0.026
Co-generation	11 PJ fuel and -2 PJ electricity	0.085	0.0004

2.5. GHG Reduction in Materials and Chemicals

Alternative additives or recycled concrete waste materials can be used in common construction materials such as cement, concrete, asphalt, and clay to reduce environmental impact in the construction industry.

2.5.1. Cement and Concrete Additives

Cement manufacturing requires energy; therefore, it is recommended to substitute the clinker content partially with industrial by-products. It is safe to substitute the clinker content by 30% (by weight of total binder) without compromising the strength or performance [35–38]. Also, high energy milling can be done to blend constituents to increase their reactivity and to increase their surface area, both of which can help improve the compressive strength development [39]. Recent studies have shown that regular Portland Cement can be replaced with alkali-activated slag mortars. These alkali-activated slags (AAS) can help reduce environmental impacts greatly since the production of AAS results in low energy consumption and lower energy consumption leading to lower CO_2 emissions [40].

The admixtures used for alkali-activated slag were Peramin SRA 40 (SRA), polymer polyethylene glycol (PEG), and polypropylene glycol (PPG) [41]. In order to reduce the CO_2 emissions, alternative clinker chemistries can be used as well as changing cement production methods in favor of more energy efficient technologies which result in reduction of 374 kg CO_2/t clinker and totaling annual 224,540 t-CO_2 emission release [42]. Table 3 reports CO_2 emission reductions from some alternative technologies and materials in cement manufacturing. For instance, fluidized bed kilns can be used instead of conventional rotary kilns to burn raw materials into powder using a new technology called granulation control/hot self-granulation; such a change can lower energy consumption by 10–15% and reduce NO_x emissions to 0.77 kg per ton of clinker as compared to 2.1–2.6 kg per ton of clinker for conventional kilns [43]. Oxy-fuel technologies have emerged as a promising candidate for CO_2 capture in new cement kilns by using pure oxygen for fuel burning. Due to reduced fuel combustion, oxy-fuel technology reduces CO_2 emissions by 454–726 kg CO_2 per ton of cement. However, due to increased electricity usage, CO_2 emissions increase slightly by 50–68 kg of CO_2 per ton of cement [44].

Table 3. Reduction in CO_2 emissions from alternative technology/materials in cement production. Adapted from [42].

Technology/Material	Alternative	Reduction in CO_2 Emissions
Cement production methods	Fluidized bed kiln; high activation grinding	20 to 30 kg CO_2/ ton product
Changes in raw material	Calcareous oil shale, steel slag	60 kg CO_2/ton of clinker
	Carbide slag	374 kg CO_2/ton of clinker
Emerging alternative cement products	Novacem cement	750 kg CO_2/ton product
	Geopolymer cement	300 kg CO_2/ton product
	Calera cement manufacturing	500 kg CO_2/ton of product
Carbon capture technologies	Concrete curing	120 kg CO_2/ton product
	Carbonate looping	370 to 500 kg CO_2/ ton product
Fuel technologies	Oxygen enrichment and Oxy-fuel	404 to 676 kg CO_2/ton cement
Post-combustion carbon capture	Absorption	690 to 725 CO_2/ton clinker
Industrial recycling	CO_2 from cement process into high-energy algal biomass	1800 kg of CO_2 will be utilized per ton of dry algal biomass produced

In order to mitigate the impact of concrete on the environment, its physical and mechanical properties such as strength, durability and light weight can be enhanced. For instance, lightweight concretes (LWCs) with high volume of additives such as fly ash or silica fume, which reduces the overall structural volume to withstand load, reduces CO_2 emissions by 30–50% as compared to conventional concrete and improves mechanical properties of LWC [45]. Demolition waste such as old tires, crushed glass, and various materials from the incineration process can be granulated and cast into concrete as fillers [27]. According to a study, a sustainable Ultra High-Performance Fiber Reinforced Cement Composite (UHPFRCC) was produced using silica flour, blast-furnace slag cement, silica fume, superplasticizers, wollastonite, and steel fibers [46].

Another study stated that pulverized fuel ash (PFA) and high calcium wood ash (HCWA) were reused as concrete materials and HCWA:PFA of 50:50 and 40:60 provide the optimal flexural and compressive strength [47]. Titanium dioxide (TiO_2) can be used as both an additive and as a coating layer. It was found that photocatalytic concrete containing TiO_2 was effective to remove NO_x in urban streets [48]. According to a study, the carbon footprint of a building with 4020 m^2 gross area and 5633 tons of total weight was 14,229 tons of CO_2-e; in particular, this building contributed to 42% of the total emissions during both productions of material and construction stages [49]. A fractional replacement of cement in concrete with fly-ash together with the use of ground granulated blast furnace slag and the use of natural aggregates with recycle crushed aggregate can reduce up to 3.8% (10.5 kg

CO_2-e) in comparison to the conventional concrete mixture during the life cycle of the building [50].

2.5.2. Asphalt Additives

Asphalt is used in most road and pavement construction, and it is a considerable contributor to GHGs in construction industry [31]. There are several additives that can assist in reducing GHG emissions, such as Sasobit, which also can reduce mixture viscosity and lower conventional mix temperature. Recent studies compared Warm Mix Asphalt (WMA) and Hot Mix Asphalt (HMA) in terms of their emission profiles. It was determined that mixture containing Sasobit additives with WMA produces the lowest CO_2 emissions which ranges from 450 ppm to 550 ppm while HMA produces 700 ppm to 750 ppm of CO_2 [41]. Another additive for the production of WMA is synthetic zeolite. It reduces the viscosity and increases asphalt mixtures' workability. Furthermore, by allowing stronger coatings of bitumen on aggregates, it improves the bonding [51]. To improve the bonding of aggregates with bitumen at low temperature, zeolite can be doped with Ca $(OH)_2$ which would also control the emission of CO_2 [52]. Studies have shown that with the addition of 6% of additive by weight, mixing temperatures of asphalt mixtures reduced from 180 °C HMA to 120 °C WMA which in turn reduces the CO_2 emissions from 7500 ppm to 500 ppm [53].

2.5.3. Clay Additives

Fired and unfired clay bricks are used in the construction industry. However, fired clay bricks require a large amount of energy for their production [54]. In order to lessen the environmental impacts and achieve sustainable building industry development, unfired clay bricks are more suitable than fired bricks. Unfired clay bricks are composed of clay soils and a binder such as lime or cement [55]. Calcium-based binder such as lime and cement increases carbon in the air, due to high energy consumption during manufacturing; furthermore, the rocks naturally change $CaCO_3$ into CaO which further releases CO_2 [56]. Various additives have been tested, and it was found that MgO can be a potential alternative to calcium-based binders. MgO has some similar attributes of CaO, however MgO has the ability to immobilize heavy metals in contaminated soil. In addition, magnesite is used in manufacturing refractory products [57]. The reduction of CO_2 emissions for unfired clay bricks were estimated 9.96 kg CO_2-e per fu (functional unit) [50].

2.5.4. Recycled Aggregate Concrete

In addition to crushed concrete, recycled aggregate concrete (RAC) consists of materials such as bricks, metals, tiles, and other materials including plastic, wood, glass, and paper [4]. RAC has inferior durability and mechanical properties as compared to conventional concrete. However, desirable RAC properties can be obtained by using admixtures such as silica fume, GGBS, fly ash, and meta-kaolin, and by modifying mixing procedures [4]. In a study, it was found that RAC together with industrial wastes such as fly ash, silica fume, blast slag, etc. can improve concrete's carbon footprint and provide great benefits [58]. Chetumal Institute of Technology in Mexico studied the influence of the fine and normal/recycled coarse aggregates on carbon footprint [59]. The result showed that recycled coarse aggregates contributes 39% of CO_2-e, fine aggregate contributes 19% of CO_2-e and normal coarse aggregate contributes 42% of CO_2-e. The study concluded that increasing recycled aggregates may help reduce 22,343 tons of CO_2-e annually in the region of Mexico alone. A study found that 100% reclaimed and recycled aggregates, which is called Pixelcrete, reduced the content of Portland cement (60% less than the conventional concrete) in office building, and led to 73.8 t-CO_2-e reduction in GHG emission [60].

2.5.5. Recycled Asphalt

Reclaimed asphalt pavement (RAP) is used to describe re-used asphalt containing pavement. In 2010, 62.1 million tons of RAP was used in asphalt pavements [61]. The

RAP could be used in three different categories of production: either as hot mix asphalt, or cold mix asphalt, or as aggregates. The RAP is generated through removal of asphalt pavement by either milling the surface using a milling machine or full depth removal. The recycling process includes both hot and cold mix asphalt and can be done in recycling plant or in place [62]. In a study, it was found that virgin asphalt produces 132 kg CO_2-e per t. In addition, 1/3 of this value was contributed by the energy intensive processes such as heating and drying; therefore, several studies were carried out to determine the factors that affect the reduction of CO_2 emissions including the RAP [28]. It was reported that RAP mix resulted in 5.5% reduction of carbon content, and it enhanced the reduction by 14% when larger aggregates sizes were used. By using RAP, embodied carbon content dropped to an average of 84.35 kg CO_2-e per t [3].

3. Carbon Footprint of On-Site Construction Processes and GHGs Reduction

Carbon footprints are resulted during manufacturing, transportation, and installation processes of ground foundation, wood/steel/concrete framed construction at on-site construction activities. The amount of CO_2 released from a concrete-steel residential tower in the Tehran Metropolitan City was 13,076,390,236 kg CO_2-e, and the amount of CO_2 emissions in 1 m^2 of Gross Floor Area (GFA) was 435,879.67 kg CO_2e/m^2, of which 83% was related to the emissions from transportation of materials and 14% was related to construction wastes and 3% was related to on-site construction process [20]. A prefabricated wood-frame multi-story building in Quebec City produced a total of embodied carbon emissions of 275 kg CO_2-e, which was 25% less than traditional buildings built with steel or concrete. The fabrication phase of building material contributed the most (75%) to the carbon emissions, while transportation (13%), construction (1%), and waste management (11%) contributed 25% [50].

A study found that the embodied carbon of a 3-bedroom semi-detached house constructed using offsite panelized timber frame was approximately 35 t-CO_2 (82% of the total embodied carbon is embodied in the materials incorporated in the building, 2% of the total embodied carbon resulted from transportation of the materials from point of distribution to site and the rest resulted from waste materials exported from the site and energy used onsite), and an equivalent home constructed using traditional masonry construction was 52 t-CO_2. Using modern methods in construction resulted in a 34% reduction in embodied carbon [63]. The overall CO_2 emissions from the 1008 m wastewater pipeline project in China were calculated in tons through the entire construction period; the results were found to be 452.81 tons, 61.32 tons, and 6.59 tons from transportation phase, material manufacturing phase, and installation phase, respectively [64]. The global warming and energy consumption of 1 m^2 of hoarding construction using large amounts of steel products and concrete in the construction site resulted in 3 tons of CO_2 eq GWP and 39 GJ of non-renewable energy consumption [65]. Another study showed that home building with ready mix concrete results in 40% less CO_2 emissions and less fuel consumption per lot by changing concrete slab size from 3000 ft^2 to 1500 ft^2. In addition, choosing the closest ready mix concrete plant saves 46 gallons of diesel and eliminates 1020 lb of CO_2 emissions per lot in Greater Phoenix Arizona area [66].

Enhancement of energy efficiency and optimization of construction machines can reduce direct carbon emissions in construction industry [67]. Oil and electricity consumption during the on-site construction contribute to carbon footprint of construction industry. According to this study, the sources of CO_2 emission from the on-site construction are as follows: reinforced concrete work produced 44.1 t-CO_2 (23.9% of the total CO_2 emissions), earthworks produced 39.1 t-CO_2 (21.2% of the total CO_2 emissions), ground heat construction (close loop) produced 31.9 t-CO_2 (16.7% of the total CO_2 emissions), foundation work (PHC PILE) produced 26.7 t-CO_2 (14.4% of the total CO_2 emissions), and ground heat construction (open loop) produced 16.6 t-CO_2 (8.5% of the total CO_2 emissions) of 84.6% of the total CO_2 at the on-site construction phase. Furthermore, electricity consumption of concrete works on-site accounts for 41.9% of the total electricity used during the construction, resulting 14.1% (13,279 kWh) of the total electricity usage during building operations [21]. A case study has shown that on an average 99.8% of carbon present in the fossil fuel consumed by an excavator is released into the atmosphere as CO_2 [5,67]. Additionally, emission factors during idling times contribute to overall average emission factors.

A study showed that the total CO_2 emission increased during engine idling of non-road diesel construction equipment was considered although during the idle the time fuel use and CO_2 emissions are between 1/3 to 1/5 of the non-idle time. During idling time, 2.7 kg CO_2/liter was produced at a diesel fuel consumption rate of 03.7 L/h [68]. According to the EPA (2005), operators should take the equipment needs into consideration, including the time required for warm-ups and cool-downs. An operational efficiency system that is commonly accepted and used to estimate equipment productivity is 50 min = h (83%), which indicates 50 min of non-idle time and 10 min of idle time per hour. Equipment such as backhoes and bulldozers have equipment productivity ranging from 80% to 85%. However, off-road trucks have equipment productivity of 41% considering that a large part of their time is spent cycle idling, mainly loading and offloading of cargo. If off-road truck average operational efficiency increased from 40% to 50% by reducing idle time by only 6 min/h, the hourly fuel use and CO_2 emissions can be reduced 10% [68].

A case study of a construction project in USA involved a roadway construction of an 18.8-mile highway requiring 184 pieces of machinery categorized into 35 equipment types, with idle time assumed to be 6 h per day for 7 days per week for this machinery. It was shown that the net total emission was 179,055 Mt-CO_2-e during a period of 2.5 years (71,609 Mt-CO_2-e per year), of which 40,023 Mt-CO_2-e/km was contributed by the constructed roadway [69]. Amount of CO_2 resulted from idling time can be reduced using different technologies such as direct-fire heaters, auxiliary power units (APU), thermal storage systems, on-board batteries, and automatic engine shut-off devices [70]. According to a study, direct fired heaters can reduce NO_x and CO_2 emissions by 99% and 94–96%, respectively, since heat is transferred directly to the heat exchanger from the combustion flame resulting in less fuel usage than diesel engines [71].

4. Carbon Footprint of Construction and Demolition Waste Generation and GHG Reduction

Construction demolition waste (CDW) stems from construction, renovation, and demolition workplaces which include (i) excavation materials, (ii) road building and maintenance materials, (iii) demolition materials, and (iv) other worksite waste materials, (e.g., unpainted, non-treated wood scrap, unpainted, non-treated wood pallets, plastic, packaging), land clearing, and development activities [72]. Construction waste is increasing in volume and affecting the environment adversely [73]. Over 80% of CDW is composed of excavated earth in construction works. Mixed CDW contains the remaining of materials and packaging. [74] A 3-bedroom modular timber frame semi-detached house with 83 m^2 internal floor area produced 17 m^3 of waste (excavated inert materials, waste and unused construction materials, and other waste) totaling 4.9 t-CO_2 equating to 109 kgCO_2 per m^2. Timber and packaging contributed to 33% and 31% of the total waste, respectively [63].

When a building reaches the end of its service life, it is demolished; the process is responsible for an emission of 0.004 to 0.01 kg CO_2 per kg of the concrete material. This figure depends on the type of reinforcement and structure used, in addition to the general working conditions on the site during demolition [27]. A situ-concrete type building was being demolished in Korea; it required total energy consumption of 51.5 MJ/m^2 from diesel fuel to demolish it; thus, the level of CO_2 emitted during demolition was 10.3 kg-CO_2/10 m^2. In consideration of the CO_2 that is emitted during the transportation of the demolition debris, 24.4 Kg-CO_2/10 m^2, 26.3 kg-CO_2, and 17.6 kg-CO_2 were obtained for a single-family house, a flat, and a multi-family house, respectively [75]. Waste transportation consumes energy which leads to CO_2 emission. According to study, during the construction period, 530 tons of waste generated and during the transportation of this waste 527 L of diesel oil consumed totaling 1.4 t-CO_2 emission from the waste transportation phase [21].

Waste materials generated from the construction industry (concrete and concrete rubble, construction ceramics, timber and wood, glass, plastics, steel, iron, aluminum, excavated soil, and Styrofoam) or from general life can be recycled as alternative construction materials [61]. During demolition, interior finishing from buildings can be reused or recycled. To look after the environment and determine the recycling and reuse values of CDW, the waste management must be planned via volume and composition determination [76]. Concrete blocks can be crushed so that they can be used for landscaping or landfilled. The fiber generated from the carpet waste can be used in fiber reinforced concrete (FCR) and fiber reinforced soil as well. The fiber improved several mechanical properties of the concrete such as toughness, strength in tension, fatigue strength, and durability, while it reduced possible cracks and defects [77]. Waste materials can act as substitutes of concrete components; it is estimated that plastic and glass can replace fine aggregates in concrete mixes by up to 20%, while waste concrete could make up for 20% of the coarse aggregate mixes in concrete [78].

Recycling one kg of aluminum as building demolition waste can contribute to emission reduction of 20.07 kg CO_2-e [79]. Demolition debris that contains steel is separated so that the steel can be sold to scrap dealers. The economically not valuable waste can be sent to dump sites [80]. When the waste steel from hoarding construction is recycled as steel scraps, 281 kg CO_2-e/m GHGs emission can be reduced [65]. New asphalt can be used from asphalt removed from road that is refurbished. The landscaping clearing wastes can be used as well. A portion of waste glass can be used in place of fine aggregate in asphalt paving mixtures (glassphalt) [81]. Reusing wood waste in production of particleboard reduced embodied carbon emissions up to 14.6% (−28.6 kg CO_2-e/m^2) [50].

5. Carbon Footprint during Operational Stage and GHGs Reduction

Over the full cycle, building operations contribute to the CO_2 balance when in service [82]. Carbon emission during operational stage of a building was a major contributor, accounting for 85.4% of the total emission followed by the construction stage which accounted for 12.6% of total emissions [83]. A high-rise residential housing block in Hong Kong demonstrated that GHG emission was estimated to about 213.03 t-CO_2-e/flat and 4980 kg CO_2-e/m^2, of which 85.82% was stemming from the operating energy, 12.69% from materials, 1.14% from renovation, 0.28% from end-of-life of the building, and 0.07% from other factors [84]. The energy consumption per area of the buildings from urban, national, and global scales are 3.03 GJ/m^2, 4.27 GJ/m^2 and 0.44 GJ/m^2 which correspond to 0.40 t-CO_2-e/m^2, 0.14 t-CO_2-e/m^2 and 0.04 t-CO_2-e/m^2 greenhouse gas emissions, respectively, based on hybrid systems analysis combining input–output analysis and process analysis in China [85].

In order to contribute to CO_2 reduction, new technologies were implemented in buildings. According to a study, low-carbon strategies, such as increased energy efficiency design for new buildings and energy-saving retrofit for existing buildings would decrease energy consumption by 2.98% with a carbon emission reduction of 3.15 million t-CO_2-e [22]. Choosing correct materials, systems, and technologies which are listed in following sections at the phase of design and materials selection, will reduce energy consumption and CO_2 emissions during operational stage of the buildings.

5.1. *Alternate Water Resources for Water Reuses*

Reusing water in a typical office building is estimated to conserve about 75% of the indoor potable water [86]. The average water saving of a green building was estimated to reach 37.6% with applying water efficiency technologies [87]. The rise of the water savings will reduce energy consumption and CO_2 emissions [88]. The passive irrigation system has two stages: collecting water when it rains and supplying water in drought conditions [89]. Water flow in the system is natural under gravity or capillarization method [90]. A 250-room hotel in Birmingham, UK, with the rainwater recycling system saved up to 780 m^3 of potable water per year [91]. According a comparative simulation model, gravity fed rainwater harvesting system for a high rise building in Mexico saved up to 8.5% of GHG [92]. Graywater is the water produced by bathroom, laundry machines, sinks, showers, and bathtubs [93]. Treated graywater can be reused for landscape irrigation and toilets [86].

Efficiency of water use can be improved by graywater recycling systems for flushing of toilets by dual piping, which will contribute to reducing urban water demand from 10% to 25% [94]. NH Campo de Gibraltar hotel substitutes 20% of potable water with filtered and treated grey water from showers, which resulted in a 20% reduction in annual water bill [91]. Blackwater comes from toilets and kitchens. Blackwater reuse showed a positive response from people who used automated or remotely controlled systems by the installer. Another study reported that it is costly and has poor process design [95]. Condensate recovery reuses water produced by air conditioning (AC) systems [95]. AC condensate can be used in flushing toilets, irrigation, cooling towers, roof cleaning, green roofs, and spray cooling [96].

Examples of water reuse and alternative water supplies include water conserving toilets, waterless toilets, waterless urinals, alternative shower and faucet fixtures (alternative controls, self-powering, low flow), water efficient appliances, and alternative landscaping (high efficiency irrigation, water conserving plant selection) [95]. Some statistical studies showed that water technologies increase water efficiency. For instance, urinals and commercial dishwashers showed the greatest reductions of water use, while showers and commercial toilets showed the least savings [88]. In the same manner, wastewater centralized reuse system (WWCRS) require more energy for treatment which leads to higher CO_2 emissions, while the greywater decentralized reuse system (GWDRS) requires less energy (11.8–37.5%) than WWCRS consumed [97]. A constructed wetlands system treats wastewater in a building so that it can be used in low-flow toilets and urinals, which reduces the water use in total by a percentage higher than 60% [86].

5.2. Heating, Ventilation and Air Conditioning

Heating, ventilation, and air conditioning (HVAC) systems of buildings consume about 40–60% of total energy taking into consideration the embodied energy which stems from the production of the building [98]. Owing to their large thermal mass, concrete and other heavy weight materials positively impact the energy consumption of buildings; for a heavy weight building (based on concrete frame), energy needed for heating/cooling/ventilation is 10 MJ/m^2 resulting in 1.3 CO_2/m^2; and for light-weight building, (based on plaster boards stud walls), it is 20 MJ/m^2 resulting in 2.6 CO_2/m^2 in Northern Europe [27].

Equipment sizing and selection, pipe/duct sizing, energy performance analysis, system optimization, real-time performance optimization, control analysis, control optimization, and simulation and programming for HVAC systems can reduce energy consumption and increase the comfort of residents [99,100]. According to a study, using a high energy performance air conditioner resulted in 7664.4 t-CO_2-e reduction in an office building in Nanhaiyiki 3, China; 451.5 t-CO_2-e reduction in a Pixel building in Australia during the life cycle of the buildings. In the same fashion, using natural ventilation and lighting resulted in a 5687.6 t-CO_2-e reduction in Nanhaiyiki 3, China; 4649.8 t-CO_2-e reduction in the Pixel building in Australia during the life cycle of the buildings [60].

5.3. Other Building Systems and Technologies

There are various technologies and systems that can be applied to enhance the efficiency of buildings and decrease CO_2 emissions. Such innovations include: windows and building surfaces with tunable optical properties; high-efficiency heat pumps; highly efficient lighting devices; thin insulating materials; improved software for analyzing building design and operations; inexpensive, energy harvesting sensors and controls; optimized control strategies; and interoperable building communication systems [101]. A study was conducted to compare different systems in a building, and it found that systems like interior lights (−150%), mechanical ventilation (−25%), and pumps (−11%) had the least energy savings whereas systems like interior fans (100%), heat rejection units (56%) and receptacle equipment (33%) had the highest energy savings. The negative values show that the systems are less efficient when compared to the baseline [82]. In another study, it was found that using renewable energies such as a solar photovoltaic system, wind turbine, and anaerobic digester resulted in 1204.1 t-CO_2-e reduction in an office building in Australia, and using renewable energy such as a solar photovoltaic system, a solar thermal water system, and a ground source heat pump resulted in 2871.6 t-CO_2-e reduction in an office building in China during the life cycle of the buildings [60]. Expanded polystyrene (EPS), cellulose, and elastomer as insulation and sealing materials resulted in an average 3.5 kg CO_2-e/kg emission, some insulation materials such as sheep's wool could reduce its impact up to 98% [50].

6. Discussion

Globally, in the developed and developing countries, the whole process of construction and building operations contributes to 33% of greenhouse gas (GHG) emissions and 40% of global energy consumption, stemming from the usage of the equipment, transportation, and the manufacturing of building materials. The urban population is increasing, which leads to more construction in the future and increased GHGs emissions [6]. Therefore, new policies are required for mitigation of GHG emissions. Regulations such as building codes can effectively reduce GHG emissions if enforced well enough and can ensure new buildings incorporate designs that are both cost and energy effective. However, regulations alone can result in extra costs for the governments, and they should be designed to cover all aspects of GHG emission activities [7]. Moreover, this policy instrument has been widely criticized for being inflexible, complex and for not taking into consideration differences in technology and geography [102].

On the other hand, a carbon tax is simpler to design, has relatively low administration costs and is attractive to stakeholders in the building sector due to their familiarity with the tax mechanism. The revenues earned from carbon tax can be redistributed to other policy instruments such as incentives [7]. However, establishing an appropriate tax rate can be a challenging task for governments as it involves complete knowledge of costs of mitigation, the growth of the economy, progresses in technology and other factors which need to be taken into consideration. Moreover, due to opposition from the public and also to avoid pressuring the construction industry intensively, governments could also face problems in establishing a deterring tax rate that can reduce GHG emissions [102].

The cumulative amount of GHG emissions mitigated can be quantified with ETS and emission permits can be distributed for free or auctioned off. However, there are concerns of market failures and regulatory based loop holes because the construction sectors lacks proper GHG accounting [12]. It is necessary to move beyond the debate of policy instruments in order to be able to pinpoint the factors that are actually slowing the move to a carbon neutral construction industry. One of the common cited barriers to carbon reduction schemes in the construction industry is the incremental cost associated with it [103,104]. Studies have shown that building contractors and developers often overestimate the cost associated with energy efficiency [105]. For example, in Germany, new buildings with very little heating requirements can be constructed with an extra cost of no more than 5–12%, while, in Northern China, a building project was able to achieve reductions of 65% in heating consumption with an extra cost of no more than 8% without compromising thermal comfort [103,106]. Therefore, correct estimations are important for cost estimations.

Other cited barriers to carbon reduction in construction industry were the skills and knowledge gap of not only the designers and contractors but also of the end users, i.e., the occupants of the buildings [104,107]. As a conclusion, each instrument has some limitations; therefore, a variety of economic, environmental, political, and social factors need to be taken into consideration [7]. Training and education should be emphasized as important ways to reduce GHG emissions in the construction industry by enabling behavioral changes within organizations. In this context, identifying sources of the carbon footprint at the different stages during construction and showing possible carbon reduction technologies/systems and techniques as summarized in Table 4 will be helpful for awareness and to fulfill the knowledge gap at the design stage from clients to designers and contractors.

Table 4. Summary of findings.

Building Operations	CO_2 Emission	Reduction Material/Techniques	CO_2 Reduction	References
Limestone quarrying	3.13 kg CO_2-e per ton crushed rock product	Application of alternative/ renewable energy such as solar thermal and biodiesel as compared to acquiring energy needs for quarrying from the grid or natural gas	More than 81% reduction in GHG emissions annually	[24]
Portland clinker manufacturing	nearly 1 kg of CO_2 per one kg of Portland clinker [(b)]	Alternative clinker substitution—use of calcium carbide residue in replacement of limestone partially	374 kg CO_2/ton of clinker annually, or more than 37% reduction in CO_2 emissions per ton of clinker	[26,39]
Asphalt	0.05 ppm of CO_2 per ton per year for conventional asphalt production	Sasobit additives with Warm Mix Asphalt	0.003 ppm to 0.004 ppm of CO_2 per ton or more than 94% reduction in CO_2 emissions	[38,51]
		Sasobit additives with Hot Mix Asphalt	0.005 ppm to 0.0054 ppm of CO_2/ton, or more than 90% reduction in CO_2 emissions	[38]
	132 kg CO_2 equivalent /ton of virgin asphalt produced	Reclaimed asphalt pavement	Dropped to average of 84.35 kg CO_2 equivalent/ton, or more than 36% reduction in CO_2 emissions	[3,56]
Concrete	5 w/c were between 347 and 351 kg of CO_2-e/m^3	Recycled coarse aggregates	Reduce 0.03 tons of CO_2-e/m^3	[56,58]
	293 kg of CO_2-e/m^3	Fractional replacement of cement in concrete with fly-ash and ground granulated blast furnace slag and natural aggregates with recycle crushed aggregate	Reductions of up to 3.8% (10.5 kg CO_2-e/m^3)	[47]
Onsite construction process	(a) During idling, at a fuel consumption rate of 0.84 gal/hour, 2.7 kg CO_2/liter was produced	Reducing idling time by using direct fired heaters instead of diesel engines	Direct fired heaters can reduce NOx and CO_2 emissions by 99% and 94–96% respectively during idling time	[70]
	Traditional building with steel products or concrete produces 366 kg CO_2-e/m^2 total embodied carbon emissions	Using prefabricated wood instead of steel or concrete	25% reduction in total GHG emissions	[47,64]
	3-bedroom semi-detached house constructed using traditional masonry construction produces 405 kg CO_2/m^2	using offsite panelized timber frame and modern methods of construction	34% reduction in total embodied carbon emissions	[61]
Construction, demolition waste	0.004 to 0.01 kg CO_2 per kg of the demolition waste	Recycling building demolition waste such as aluminum	20.07 kg CO_2-e per kg of aluminum recycled	[26,78]
		Recycling waste steel from hoarding construction as steel scraps	281 kg CO_2-e per m, or about 8% reduction in CO_2 emissions	[64]
		Reusing wood waste into production-use of particleboard	14.6% reduction in CO_2 emissions (−28.6 kg CO_2-e/m^2)	[47]
Building's operations when in service	Account for 85.4% of the total emissions of a building's life cycle	High energy performance air-conditioner	19 kg CO_2-eq/m^2	[47,48,82]
		Utilization of renewable energy such as a solar photovoltaic system, solar thermal water system, and a ground source heat pump	4.6 kg CO_2-eq/m^2	
		Use of natural ventilation and lighting	9.1 kg CO_2-eq/m^2	
		Use of sheep's wool as insulation material in buildings	98% reduction in GHG emissions	
		Applying large thermal mass, concrete, and other heavy weight materials for reduction of HVAC energy	50% reduction in CO_2 emissions/m^2	[26]
		Rainwater harvesting system	8.5% reduction of GHG	[91]

7. Conclusions

GHG emissions mitigation can be achieved by indirect pricing such as regulations and direct pricing such as carbon tax and emission trading schemes (ETS). However, regulations can be inflexible, complex, and may not take into consideration differences in technology and geography. In addition, ETS can be complex because the construction sector lacks proper GHG accounting. Therefore, increasing the awareness, education, and incentives can lessen the carbon footprint of construction industry. Consequently, we aimed to increase awareness of the carbon footprint sources in construction and building operations during manufacturing, transportation, construction, operations/management, and end-of-life deconstruction. As a result, various carbon reduction techniques/systems were identified. It was found that mining and manufacturing of materials and chemicals contributed to high energy usage and 90% of the total CO_2 emissions. Therefore,

- Testing different blends of cement with addition of alternative additives such as alkali-activated slag mortars or fly ash in concrete;
- Changing cement production methods;
- Addition of Sasobit or reclaimed asphalt pavement in asphalt mixtures;
- Recycling building wastes such as concrete aggregate and recycled asphalt in common construction materials;
- Conversion from the wet process to the dry process in concrete manufacturing;
- Substitution of lower carbon content fuels for coal, coke, and petroleum coke;
- Alternate options in terms of vehicle type, engine power, truck capacity, and fuel type to improve the fuel efficiency in the construction vehicles;
- Reducing idle time by using direct fired heaters, auxiliary power units (APU), thermal storage systems, on-board batteries, and automatic engine shut-off devices;
- Applications of alternate water resources for water reuse purposes;
- Switching to efficient HVAC systems; and
- Utilization of different building operations/systems will lessen energy consumption and reduce GHG emissions up to 90% in different stages in construction industry.

This review can be useful at the stage of conceptualization, design, and construction to assist clients and stakeholders in selecting materials and systems. There is large scope for further research on how to decrease carbon footprint in construction. Some of the areas that require attention include:

- improving recyclable waste materials such as glass, rubber crumbs, etc., as construction materials;
- developing decision making tools for effective carbon footprinting;
- creating inventory databases for Life Cycle Assessment for each alternative material's embodied carbon value.

Author Contributions: Conceptualization, B.S. and Y.F.; formal analysis, B.S. and Y.F.; resources, Y.F., C.-S.C. and I.Y.; writing—original draft preparation, B.S. and Y.F.; writing—review and editing, C.-S.C., I.Y. and Y.-J.B.; project administration, B.S.; funding acquisition, B.S. All authors have read and agreed to the published version of the manuscript.

Funding: This research was funded by the Khalifa University, KUIRF L1 210045 grant.

Institutional Review Board Statement: Not applicable.

Informed Consent Statement: Not applicable.

Data Availability Statement: Not applicable.

Acknowledgments: The authors would like to extend gratitude to the following students for their assistance: Tethkar Alhammadi, Madeya Al Mehairbi, Fatmah Aldhanhani, and Fatima Almusharrekh.

Conflicts of Interest: The authors declare no conflict of interest.

References

1. Solís-Guzmán, J.; Martínez-Rocamora, A.; Marrero, M. Methodology for determining the carbon footprint of the construction of residential buildings. In *Assessment of Carbon Footprint in Different Industrial Sectors*; Springer: Singapore, 2014; Volume 1, pp. 49–83.
2. WEF. *Shaping the Future of Construction: A Breakthrough in Mindset and Technology*; WEF Cologny: Geneva, Switzerland, 2016.
3. Yan, H.; Shen, Q.; Fan, L.C.H.; Wang, Y.; Zhang, L. Greenhouse gas emissions in building construction: A case study of One Peking in Hong Kong. *Build. Environ.* **2010**, *45*, 949–955. [CrossRef]
4. Kisku, N.; Joshi, H.; Ansari, M.; Panda, S.; Nayak, S.; Dutta, S.C. A critical review and assessment for usage of recycled aggregate as sustainable construction material. *Constr. Build. Mater.* **2017**, *131*, 721–740. [CrossRef]
5. Huang, L.; Krigsvoll, G.; Johansen, F.; Liu, Y.; Zhang, X. Carbon emission of global construction sector. *Renew. Sustain. Energy Rev.* **2018**, *81*, 1906–1916. [CrossRef]
6. Mardiana, A.; Riffat, S. Building energy consumption and carbon dioxide emissions: Threat to climate change. *J. Earth Sci. Clim. Chang.* **2015**, 1–3.
7. Wang, T.; Foliente, G.; Song, X.; Xue, J.; Fang, D. Implications and future direction of greenhouse gas emission mitigation policies in the building sector of China. *Renew. Sustain. Energy Rev.* **2014**, *31*, 520–530. [CrossRef]
8. Iwaro, J.; Mwasha, A. A review of building energy regulation and policy for energy conservation in developing countries. *Energy Policy* **2010**, *38*, 7744–7755. [CrossRef]
9. Wong, P.S.; Ng, S.T.; Shahidi, M. Towards understanding the contractor's response to carbon reduction policies in the construction projects. *Int. J. Proj. Manag.* **2013**, *31*, 1042–1056. [CrossRef]
10. King, D. *Engineering a Low Carbon Built Environment: The Discipline of Building Engineering Physics*; Royal Academy of Engineering: London, UK, 2010.
11. Hahn, R.W. Greenhouse gas auctions and taxes: Some political economy considerations. *Rev. Environ. Econ. Policy* **2009**, *3*, 167–188. [CrossRef]
12. Spash, C.L. The brave new world of carbon trading. *New Political Econ.* **2010**, *15*, 169–195. [CrossRef]
13. Lu, Y.; Zhu, X.; Cui, Q. Effectiveness and equity implications of carbon policies in the United States construction industry. *Build. Environ.* **2012**, *49*, 259–269. [CrossRef]
14. Burtraw, D.; Palmer, K. Compensation rules for climate policy in the electricity sector. *J. Policy Anal. Manag. J. Assoc. Public Policy Anal. Manag.* **2008**, *27*, 819–847. [CrossRef]
15. Figueres, C.; Bosi, M. *Achieving Greenhouse Gas Emission Reductions in Developing Countries through Energy Efficient Lighting Projects in the Clean Development Mechanism (CDM)*; Unit, T.C.F., Ed.; World Bank: Washington, DC, USA, 2006.
16. Chan, E.H.; Qian, Q.K.; Lam, P.T. The market for green building in developed Asian cities—The perspectives of building designers. *Energy Policy* **2009**, *37*, 3061–3070. [CrossRef]
17. Jiang, P.; Tovey, N.K. Opportunities for low carbon sustainability in large commercial buildings in China. *Energy Policy* **2009**, *37*, 4949–4958. [CrossRef]
18. Wang, X.; Duan, Z.; Wu, L.; Yang, D. Estimation of carbon dioxide emission in highway construction: A case study in southwest region of China. *J. Clean. Prod.* **2015**, *103*, 705–714. [CrossRef]
19. Fischedick, M.; Roy, J.; Acquaye, A.; Allwood, J.; Ceron, J.-P.; Geng, Y.; Kheshgi, H.; Lanza, A.; Perczyk, D.; Price, L. *Industry In: Climate Change 2014: Mitigation of Climate Change*; Intergovernmental Panel on Climate Change: Cambridge, United Kingdom; Cambridge University Press: New York, NY, USA, 2014.
20. Jafary Nasab, T.; Monavari, S.M.; Jozi, S.A.; Majedi, H. Assessment of carbon footprint in the construction phase of high-rise constructions in Tehran. *Int. J. Environ. Sci. Technol.* **2020**, *17*, 3153–3164. [CrossRef]
21. Seo, M.-S.; Kim, T.; Hong, G.; Kim, H. On-Site Measurements of CO2 Emissions during the Construction Phase of a Building Complex. *Energies* **2016**, *9*, 599. [CrossRef]
22. Huang, W.; Li, F.; Cui, S.-H.; Li, F.; Huang, L.; Lin, J.-Y. Carbon Footprint and Carbon Emission Reduction of Urban Buildings: A Case in Xiamen City, China. *Procedia Eng.* **2017**, *198*, 1007–1017. [CrossRef]
23. Li, D.Z.; Chen, H.X.; Hui, E.C.M.; Zhang, J.B.; Li, Q.M. A methodology for estimating the life-cycle carbon efficiency of a residential building. *Build. Environ.* **2013**, *59*, 448–455. [CrossRef]
24. Choate, W.T. *Energy and Emission Reduction Opportunities for the Cement Industry*; BCS Inc.: Laurel, MD, USA, 2003; p. 1218753.
25. Kittipongvises, S. Assessment of environmental impacts of limestone quarrying operations in Thailand. *Environ. Clim. Technol.* **2017**, *20*, 67–83. [CrossRef]
26. Ma, F.; Sha, A.; Yang, P.; Huang, Y. The Greenhouse Gas Emission from Portland Cement Concrete Pavement Construction in China. *Int. J. Environ. Res. Public Health* **2016**, *13*, 632. [CrossRef] [PubMed]
27. Nielsen, C.V. Carbon footprint of concrete buildings seen in the life cycle perspective. In Proceedings of the NRMCA 2008 Concrete Technology Forum: Focus on Sustainable Development, Denver, CO, USA, 20–22 May 2008; pp. 1–14.
28. Gibson, S.; Strachan, P. *Reducing the Embodied Carbon Content of Asphalt*; University of Strathclyde: Glasgow, Scotland, 2011.
29. Miller, S.A.; Horvath, A.; Monteiro, P.J.; Ostertag, C.P. Greenhouse gas emissions from concrete can be reduced by using mix proportions, geometric aspects, and age as design factors. *Environ. Res. Lett.* **2015**, *10*, 114017. [CrossRef]

30. Chehovits, J.; Galehouse, L. Energy usage and greenhouse gas emissions of pavement preservation processes for asphalt concrete pavements. In Proceedings of the 1st International Conference of Pavement Preservation, Newport Beach, CA, USA, 13–15 April 2010; pp. 27–42.
31. Peng, B.; Cai, C.; Yin, G.; Li, W.; Zhan, Y. Evaluation system for CO_2 emission of hot asphalt mixture. *J. Traffic Transp. Eng. (Engl. Ed.)* **2015**, *2*, 116–124. [CrossRef]
32. Ma, F.; Sha, A.; Lin, R.; Huang, Y.; Wang, C. Greenhouse gas emissions from asphalt pavement construction: A case study in China. *Int. J. Environ. Res. Public Health* **2016**, *13*, 351. [CrossRef] [PubMed]
33. International Energy Agency. *Greenhouse Gas Emissions from Major Industrial Sources III-Iron and Steel Production Report PH3/30*; International Energy Agency Greenhouse Gas R&D Programme; Stoke Orchard: Cheltenham, UK, 2000.
34. Worrell, E.; Blinde, P.; Neelis, M.; Blomen, E.; Masanet, E. *Energy Efficiency Improvement and Cost Saving Opportunities for the US Iron and Steel Industry an ENERGY STAR (R) Guide for Energy and Plant. Managers*; Lawrence Berkeley National Lab. (LBNL): Berkeley, CA, USA, 2010.
35. Danielsen, S.; Gränne, F.; Hólmgeirsdóttir, Þ.; Jonsson, G.; Krage, G.; Mathiesen, D.; Nielsen, C.; Wigum, B. Best Available Technology Report for the Aggregate and Concrete Industries in Europe. *ECO-SERVE Netw.* **2006**, *3*, 99–108.
36. Naqi, A.; Jang, J.G. Recent progress in green cement technology utilizing low-carbon emission fuels and raw materials: A review. *Sustainability* **2019**, *11*, 537. [CrossRef]
37. Deolalkar, S.; Shah, A.; Davergave, N. *Designing Green Cement Plants*; Butterworth-Heinemann: Oxford, UK, 2015.
38. Abdul-Wahab, S.A.; Hassan, E.M.; Al-Jabri, K.S.; Yetilmezsoy, K. Application of zeolite/kaolin combination for replacement of partial cement clinker to manufacture environmentally sustainable cement in Oman. *Environ. Eng. Res.* **2019**, *24*, 246–253. [CrossRef]
39. Kumar, R.; Kumar, S.; Mehrotra, S. Towards sustainable solutions for fly ash through mechanical activation. *Resour. Conserv. Recycl.* **2007**, *52*, 157–179. [CrossRef]
40. Mikhailova, O.; Šimonová, H.; Topolář, L.; Rovnaník, P. Influence of Polymer Additives on Mechanical Fracture Properties and on Shrinkage of Alkali Activated Slag Mortars. *Key Eng. Mater.* **2018**, *761*, 39–44. [CrossRef]
41. LeBlanc, A.; Keches, C.M. *Reducing Greenhouse Gas Emissions from Asphalt Materials*; Worcester Polytechnic Institute: Worcester, MA, USA, 2007.
42. Hasanbeigi, A.; Price, L.; Lin, E. Emerging energy-efficiency and CO_2 emission-reduction technologies for cement and concrete production: A technical review. *Renew. Sustain. Energy Rev.* **2012**, *16*, 6220–6238. [CrossRef]
43. Worrell, E.; Galitsky, C. *Energy Efficiency Improvement and Cost Saving Opportunities for Cement Making*; Ernest Orlando Lawrence Berkeley National Laboratory, University of California: Berkeley, CA, USA, 2008.
44. Barnett, K.; Torres, E. *Available and Emerging Technologies for Reducing Greenhouse Gas Emissions from the Portland Cement Industry*; United States Environmental Protection Agency: Research Triangle Park, NC, USA, 2010.
45. D'Alessandro, A.; Pisello, A.; Fabiani, C.; Ubertini, F.; Cabeza, L.; Cotana, F.; Materazzi, A. Innovative Structural Concretes with Phase Change Materials for Sustainable Constructions: Mechanical and Thermal Characterization. In Proceedings of the Conference on Italian Concrete Days, Lecco, Italy, 14–15 June 2018; pp. 172–183.
46. Toledo Filho, R.; Koenders, E.; Formagini, S.; Fairbairn, E. Performance assessment of ultra high performance fiber reinforced cementitious composites in view of sustainability. *Mater. Des.* **2012**, *36*, 880–888. [CrossRef]
47. Cheah, C.B.; Part, W.K.; Ramli, M. The long term engineering properties of cementless building block work containing large volume of wood ash and coal fly ash. *Constr. Build. Mater.* **2017**, *143*, 522–536. [CrossRef]
48. Ballari, M.M.; Hunger, M.; Hüsken, G.; Brouwers, H. NOx photocatalytic degradation employing concrete pavement containing titanium dioxide. *Appl. Catal. B Environ.* **2010**, *95*, 245–254. [CrossRef]
49. Biswas, W.K. Carbon footprint and embodied energy consumption assessment of building construction works in Western Australia. *Int. J. Sustain. Built Environ.* **2014**, *3*, 179–186. [CrossRef]
50. Padilla-Rivera, A.; Amor, B.; Blanchet, P. Evaluating the Link between Low Carbon Reductions Strategies and Its Performance in the Context of Climate Change: A Carbon Footprint of a Wood-Frame Residential Building in Quebec, Canada. *Sustainability* **2018**, *10*, 2715. [CrossRef]
51. Shafabakhsh, G.; Taghipoor, M.; Sadeghnejad, M.; Tahami, S. Evaluating the effect of additives on improving asphalt mixtures fatigue behavior. *Constr. Build. Mater.* **2015**, *90*, 59–67. [CrossRef]
52. Koga, N.; Tsuru, K.; Takahashi, I.; Ishikawa, K. Effects of humidity on calcite block fabrication using calcium hydroxide compact. *Ceram. Int.* **2015**, *41*, 9482–9487. [CrossRef]
53. Sharma, A.; Lee, B.-K. Energy savings and reduction of CO_2 emission using Ca $(OH)_2$ incorporated zeolite as an additive for warm and hot mix asphalt production. *Energy* **2017**, *136*, 142–150. [CrossRef]
54. BDA. Brick Industry Sustainability. Available online: http://www.brick.org.uk/resources/brick-industry/sustainability/2010 (accessed on 12 July 2020).
55. Seco, A.; Urmeneta, P.; Prieto, E.; Marcelino, S.; García, B.; Miqueleiz, L. Estimated and real durability of unfired clay bricks: Determining factors and representativeness of the laboratory tests. *Constr. Build. Mater.* **2017**, *131*, 600–605. [CrossRef]
56. Espuelas, S.; Omer, J.; Marcelino, S.; Echeverría, A.M.; Seco, A. Magnesium oxide as alternative binder for unfired clay bricks manufacturing. *Appl. Clay Sci.* **2017**, *146*, 23–26. [CrossRef]

57. Miqueleiz, L.; Ramírez, F.; Seco, A.; Nidzam, R.; Kinuthia, J.; Tair, A.A.; Garcia, R. The use of stabilised Spanish clay soil for sustainable construction materials. *Eng. Geol.* **2012**, *133*, 9–15. [CrossRef]
58. Meyer, C. Concrete and sustainable development. *ACI Spec. Publ.* **2002**, *206*, 501–512.
59. Heidari, B.; Marr, L.C. Real-time emissions from construction equipment compared with model predictions. *J. Air Waste Manag. Assoc.* **2015**, *65*, 115–125. [CrossRef]
60. Wang, T.; Seo, S.; Liao, P.-C.; Fang, D. GHG emission reduction performance of state-of-the-art green buildings: Review of two case studies. *Renew. Sustain. Energy Rev.* **2016**, *56*, 484–493. [CrossRef]
61. Bolden, J.; Abu-Lebdeh, T.; Fini, E. Utilization of recycled and waste materials in various construction applications. *Am. J. Environ. Sci.* **2013**, *9*, 14–24. [CrossRef]
62. FHWA. *User Guidelines for Waste and Byproduct Materials in Pavement Construction*; FHWA: Washington, DC, USA, 2012.
63. Monahan, J.; Powell, J.C. An embodied carbon and energy analysis of modern methods of construction in housing: A case study using a lifecycle assessment framework. *Energy Build.* **2011**, *43*, 179–188. [CrossRef]
64. Zhang, B.; Ariaratnam, S.T.; Wu, J. Estimation of CO_2 Emissions in a Wastewater Pipeline Project. In Proceedings of the International Conference on Pipelines and Trenchless Technology, Wuhan, China, 19–22 October 2012; pp. 521–531.
65. Hossain, M.U.; Poon, C.S. Global warming potential and energy consumption of temporary works in building construction: A case study in Hong Kong. *Build. Environ.* **2018**, *142*, 171–179. [CrossRef]
66. Palaniappan, S.; Bashford, H.; Fafitis, A.; Li, K.; Stecker, L. Carbon emissions based on ready-mix concrete transportation: A production home building case study in the Greater Phoenix Arizona area. In Proceedings of the Associated Schools of Construction 45th Annual International Conference, Gainesville, FL, USA, 1–4 April 2009.
67. Abolhasani, S.; Frey, H.C.; Kim, K.; Rasdorf, W.; Lewis, P.; Pang, S.-H. Real-World In-Use Activity, Fuel Use, and Emissions for Nonroad Construction Vehicles: A Case Study for Excavators. *J. Air Waste Manag. Assoc.* **2008**, *58*, 1033–1046. [CrossRef]
68. Lewis, P.; Leming, M.; Rasdorf, W. Impact of Engine Idling on Fuel Use and CO_2 Emissions of Nonroad Diesel Construction Equipment. *J. Manag. Eng.* **2012**, *28*, 31–38. [CrossRef]
69. Melanta, S.; Miller-Hooks, E.; Avetisyan, H.G. Carbon Footprint Estimation Tool for Transportation Construction Projects. *J. Constr. Eng. Manag.* **2013**, *139*, 547–555. [CrossRef]
70. Kolpakov, A.; Reich, S.L. *Synthesis of Research on the Use of Idle Reduction Technologies in Transit*; University of South Florida, Center for Urban Transportation Research: Tampa, FL, USA, 2015.
71. Shancita, I.; Masjuki, H.; Kalam, M.; Fattah, I.R.; Rashed, M.; Rashedul, H. A review on idling reduction strategies to improve fuel economy and reduce exhaust emissions of transport vehicles. *Energy Convers. Manag.* **2014**, *88*, 794–807. [CrossRef]
72. Cochran, K.; Townsend, T.G. Estimating construction and demolition debris generation using a materials flow analysis approach. *Waste Manag. Res.* **2010**, *30*, 2247–2254. [CrossRef]
73. Jiménez, L.F.; Domínguez, J.A.; Vega-Azamar, R.E. Carbon Footprint of Recycled Aggregate Concrete. *Adv. Civil. Eng.* **2018**, *2018*, 1–6. [CrossRef]
74. Solis-Guzman, J.; Marrero, M. *Ecological Footprint Assessment of Building Construction*; Bentham Science Publishers: Sharjah, United Arab Emirates, 2015.
75. Seo, S.; Hwang, Y. Estimation of CO_2 Emissions in Life Cycle of Residential Buildings. *J. Constr. Eng. Manag.* **2001**, *127*, 414–418. [CrossRef]
76. Lage, I.M.; Abella, F.M.; Herrero, C.V.; Ordóñez, J.L.P. Estimation of the annual production and composition of C&D Debris in Galicia (Spain). *Waste Manag.* **2010**, *30*, 636–645.
77. Wang, Y. Chapter 14: Utilization of Recycled Carpet Waste Fibers for Reinforcement of Concrete and Soil. In *Recycling in Textiles*; Woodhead Publishing: Cambridge, UK, 2006.
78. Batayneh, M.; Marie, I.; Asi, I. Use of selected waste materials in concrete mixes. *Waste Manag. Res.* **2007**, *27*, 1870–1876. [CrossRef] [PubMed]
79. Wang, J.; Wu, H.; Duan, H.; Zillante, G.; Zuo, J.; Yuan, H. Combining life cycle assessment and Building Information Modelling to account for carbon emission of building demolition waste: A case study. *J. Clean. Prod.* **2018**, *172*, 3154–3166. [CrossRef]
80. Islam, R.; Nazifa, T.H.; Yuniarto, A.; Uddin, A.S.; Salmiati, S.; Shahid, S. An empirical study of construction and demolition waste generation and implication of recycling. *Waste Manag.* **2019**, *95*, 10–21. [CrossRef] [PubMed]
81. Salem, Z.T.A.; Khedawi, T.S.; Baker, M.B.; Abendeh, R. Effect of waste glass on properties of asphalt concrete mixtures. *Jordan J. Civil. Eng.* **2017**, *11*, 117–131.
82. Ramachanderan, S.S.; Venkiteswaran, V.K.; Chuen, Y.T. Carbon (CO_2) Footprint Reduction Analysis for Buildings through Green Rating Tools in Malaysia. *Energy Procedia* **2017**, *105*, 3648–3655. [CrossRef]
83. Peng, C.; Wu, X. Case study of carbon emissions from a building's life cycle based on BIM and Ecotect. *Adv. Mater. Science. Eng.* **2015**, 1–15. [CrossRef]
84. Yim, S.Y.C.; Ng, S.T.; Hossain, M.U.; Wong, J.M.W. Comprehensive Evaluation of Carbon Emissions for the Development of High-Rise Residential Building. *Buildings* **2018**, *8*, 147. [CrossRef]
85. Li, Y.L.; Han, M.Y.; Liu, S.Y.; Chen, G.Q. Energy consumption and greenhouse gas emissions by buildings: A multi-scale perspective. *Build. Environ.* **2019**, *151*, 240–250. [CrossRef]
86. Kehoe, P.; Rhodes, S. Pushing the conservation envelope through the use of alternate water sources. *J.-Am. Water Work. Assoc.* **2013**, *105*, 46–50. [CrossRef]

87. Cheng, C.-L. Evaluation of Water Efficiency in Green Building in Taiwan. *Water* **2016**, *8*, 236. [CrossRef]
88. Dyballa, C.; Hoffman, H.W.B. The Role of Water Efficiency in Future Water Supply. *J.-Am. Water Work. Assoc.* **2015**, *107*, 35–44. [CrossRef]
89. Moise, G. Passive irrigation system for green roofs. *Sci. Pap. Ser. Manag. Econ. Eng. Agric. Rural. Dev.* **2016**, *16*, 331–334.
90. Köhler, M. Long-term vegetation research on two extensive green roofs in Berlin. *Urban. Habitats* **2006**, *4*, 3–26.
91. Styles, D.; Schönberger, H.; Galvez Martos, J. *Best environmental management practice in the tourism sector*; Publication Office of the European Union: Luxembourg, 2013. [CrossRef]
92. Valdez, M.C.; Adler, I.; Barrett, M.; Ochoa, R.; Pérez, A. The Water-Energy-Carbon Nexus: Optimising Rainwater Harvesting in Mexico City. *Environ. Process.* **2016**, *3*, 307–323. [CrossRef]
93. Paulo, P.L.; Azevedo, C.; Begosso, L.; Galbiati, A.F.; Boncz, M.A. Natural systems treating greywater and blackwater on-site: Integrating treatment, reuse and landscaping. *Ecol. Eng.* **2013**, *50*, 95–100. [CrossRef]
94. March, J.G.; Gual, M. Breakpoint Chlorination Curves of Greywater. *Water Environ. Res.* **2007**, *79*, 828–832. [CrossRef] [PubMed]
95. Chambers, B.D.; Pearce, A.R.; Edwards, M.A.; Dymond, R.L. Experiences of Green Building Professionals With Water-Related Systems. *J.-Am. Water Work. Assoc.* **2017**, *109*, 37–46. [CrossRef]
96. Algarni, S.; Saleel, C.A.; Mujeebu, M.A. Air-conditioning condensate recovery and applications—Current developments and challenges ahead. *Sustain. Cities Soc.* **2018**, *37*, 263–274. [CrossRef]
97. Matos, C.; Pereira, S.; Amorim, E.V.; Bentes, I.; Briga-Sá, A. Wastewater and greywater reuse on irrigation in centralized and decentralized systems—An integrated approach on water quality, energy consumption and CO_2 emissions. *Sci. Total Environ.* **2014**, *493*, 463–471. [CrossRef] [PubMed]
98. Yüksek, I.; Karadayi, T.T. Energy-efficient building design in the context of building life cycle. In *Energy Efficient Buildings*; IntechOpen: London, UK, 2017; pp. 93–123.
99. Fong, K.F.; Hanby, V.I.; Chow, T.-T. HVAC system optimization for energy management by evolutionary programming. *Energy Build.* **2006**, *38*, 220–231. [CrossRef]
100. Trcka, M.; Hensen, J. HVAC system simulation: Overview, issues and some solutions. In Proceedings of the 23rd IIR International Congress of Refrigeration, Prague, Czech Republic, 21–26 August 2011.
101. DOE, U. *An Assessment of Energy Technologies and Research Opportunities*; United States Department of Energy: Washington, DC, USA, 2015.
102. Johnson, M.; Harfoot, M.; Musser, C.; Wiley, T. *Cap and Share. Phase 1: Policy Options for Reducing Greenhouse Gas Emissions*; AEA Energy and Environment: Didcot, UK, 2008.
103. Li, J.; Colombier, M. Managing carbon emissions in China through building energy efficiency. *J. Environ. Manag.* **2009**, *90*, 2436–2447. [CrossRef]
104. Harvey, L.D. *A Handbook on Low-Energy Buildings and District-Energy Systems: Fundamentals, Techniques and Examples*; Earthscan: London, United Kingdom; Sterling, VA, USA, 2012.
105. Liu, F. *Economic Analysis of Residential Building Energy-Efficient Design Standards in Northern China: A Case Study of Tianjin City*; World Bank: Washington, DC, USA, 2006.
106. Heffernan, E.; Pan, W.; Liang, X.; De Wilde, P. Zero carbon homes: Perceptions from the UK construction industry. *Energy Policy* **2015**, *79*, 23–36. [CrossRef]
107. Zuo, J.; Read, B.; Pullen, S.; Shi, Q. Achieving carbon neutrality in commercial building developments–Perceptions of the construction industry. *Habitat Int.* **2012**, *36*, 278–286. [CrossRef]

MDPI

Article

Empirical NOx Removal Analysis of Photocatalytic Construction Materials at Real-Scale

Miyeon Kim [1,2], Hyunggeun Kim [1] and Jinchul Park [2,*]

1 SH Urban Research Center, Seoul Housing and Communities Corporation, Seoul 06336, Korea; mykim1983@naver.com (M.K.); hgkim@i-sh.co.kr (H.K.)
2 Department of Architectural Engineering, Chung-Aug University, Seoul 06974, Korea
* Correspondence: jincpark@cau.ac.kr; Tel.: +82-2-820-5261; Fax: +82-2-816-7740

Abstract: The NOx removal performance of photocatalytic construction materials is demonstrated using two experiments under indoor and outdoor environments: (1) A photoreactor test was conducted to assess the NO removal performance of construction materials (e.g., coatings, paints and shotcrete) using a modified ISO 22197-1 method; (2) A water washing test was conducted using two specimens enlarged to the size of actual building materials and artificially exposed to NOx in a laboratory to analyze NOx removal performance. For (1), the UV irradiation of the outdoor environment was analyzed and the experiment was conducted in an indoor laboratory under UV irradiation identical to that of the outdoor condition. Photoreactor tests were conducted on construction materials applied to actual buildings located in Seoul, South Korea. In (2), the enlarged specimen was used for a field experiment by applying a modified method from the ISO 22197-1 standard. On sunny days, the NOx removal performance (3.12–4.76 μmol/150 cm^2·5 h) was twice as much as that of the ISO 22197-1 standard specification (2.03 μmol/150 cm^2·5 h) in the real-world. The washing water test results indicated that general aqueous paint achieved a NOx removal of 3.88 μmol, whereas photocatalytic paint was superior to 14.13 μmol.

Keywords: photocatalysis; construction materials; ISO22197-1; titanium dioxide (TiO_2); nitrogen oxides (NOx)

Citation: Kim, M.; Kim, H.; Park, J. Empirical NOx Removal Analysis of Photocatalytic Construction Materials at Real-Scale. *Materials* **2021**, *14*, 5717. https://doi.org/10.3390/ma14195717

Academic Editor: Yeonung Jeong

Received: 25 August 2021
Accepted: 23 September 2021
Published: 30 September 2021

1. Introduction

Recently, the emission of nitrogen oxide (NOx) has been considered a major environmental concern. NOx not only causes respiratory disorders in the human body, but also destroys the ozone layer in the stratosphere, thereby promoting climate change [1,2]. Further, NOx is a precursor substance of particulate matter generated via chemical reactions with other precursors [3–5]. Thus, NOx abatement regulation based on various air pollution policies established by the Gothenburg and Kyoto Protocols has contributed to a decrease in the worldwide emissions of NOx [6]. In addition to regulating NOx generation from automobiles, power plants and manufacturing facilities [7], technologies that remove NOx in the atmosphere play a vital role in NOx abatement [8–12]. Heterogeneous photocatalysis is a promising NOx removal technology that has been applied as a construction material [13,14]. Titanium dioxide (TiO_2) is among the most widely used photocatalytic nanoparticles for such applications [15]. The NOx removal process using TiO_2 involves the illumination of the surface TiO_2, which produces two types of carriers: an electron (e^-) and a hole (h^+), followed by oxidation of a donor molecule adsorbed on TiO_2 by the photo-induced hole. The strong oxidation power of the holes enables the production of hydroxyl radicals (·OH) via the oxidation of water. Then, the adsorbed oxygen can be reduced by the promoted electron to form a superoxide ion ($\cdot O_2^-$). NOx oxidation is a complex process involving several steps and intermediate species including the reduction of previously oxidized species. The NOx oxidation mechanism is summarized in Table 1 and Figure 1 (where hv: ultraviolet radiation, Site^(**): surface of TiO_2 and OH: hydroxyl radical). Extensive studies have been conducted to analyze the oxidation mechanism of TiO_2 under various experimental conditions [16–21].

Table 1. Mechanism for oxidation of NOx on the surface TiO_2.

Process	Formula
Activation	$TiO_2 + hv \rightarrow h^+ + e^-$ $H_2O(g) + Site^{**} \rightarrow H_2O_{ads}$ $O_2(g) + Site^{**} \rightarrow O_{2ads}$ $NO(g) + Site^{**} \rightarrow NO_{ads}$
Hole trapping	$H_2O + h^+ \rightarrow \cdot OH + H^+$
Electron trapping	$O_2(g) + e^- \rightarrow O_2^-$
Hydroxyl attack	$NO_{ads} + 2 \cdot OH \rightarrow NO_{2ads} + H_2O$ $NO_{ads} + \cdot OH \rightarrow HNO_3$

Site **: surface of TiO_2 and OH hydroxyl radical.

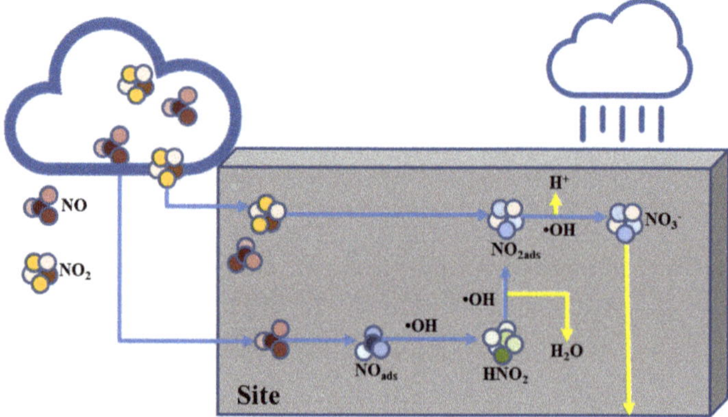

Figure 1. Mechanism for oxidation of NOx on the wall coated with TiO_2.

In the past few decades, TiO_2 has been used as a building material for atmospheric NOx removal by exploiting the NOx oxidation mechanism of TiO_2 [22–25]. Chen and Poon [26] provided an overview of the development of photocatalytic construction materials in terms of academic achievements and practical applications. They determined that photocatalytic substances can be successfully applied to commercialized construction materials such as concrete, glass, paints and various types of cementitious materials. Seo and Yun [27] studied the nitric oxide (NO) removal performance of TiO_2 cementitious materials under wet conditions. They found that the NO removal and absorption rates under dry conditions is higher as TiO_2 particles come smaller. The recovery rate changed during evaporation under wet conditions, and three distinct phases (rapid recovery, stationary reaction and final recovery) were observed. In addition, they determined that the rate of NO removal was reduced when the photocatalytic cementitious material was rewetted. Song et al. [28] studied the NOx removal performance of TiO_2-based coating materials used as paint. They demonstrated that sufficient NOx removal can be achieved although the NO gas concentration and UV-A irradiance are smaller than the experimental conditions of the ISO 22197-1: 2007 standard when their TiO_2 photocatalyst-infused coating is used. Further, they determined that sufficient NOx removal can be achieved under adverse climatic conditions such as winter and gloomy days using TiO_2 photocatalyst-infused coating materials. Based on the ISO 22197-1: 2007 standard, empirical studies on NOx removal efficiency and durability settings were conducted comprehensively in an indoor laboratory.

Recently, the extensive studies on the NOx removal efficiency and durability of TiO_2-based construction materials have resulted in the development of methods for deriving quantitative results in outdoor environments for more practical purposes [29–34]. Guerrini [35] analyzed the NOx removal performance of photocatalytic cement-based paint practically by demonstrating the findings of two environmental monitoring campaigns involving the study of NOx levels measured before and after the reconstruction of Rome's "Umberto I" tunnel. The results obtained by Guerrini indicate that the photocatalytic treatment of the Umberto I tunnel vault with cementitious paint resulted in effective pollution abatement, as shown by the lower concentrations found after the renovation. Yu et al. [36] investigated the performance of a mineral-based transparent air-purifying paint. The efficiency of this photocatalytic paint for removing air pollutants was determined in a laboratory setting using the ISO 22197-1 procedure. Subsequently, outdoor monitoring over a 20-month period was conducted to determine the air pollution removal efficiency under realistic conditions. They proposed a new protocol using the measured nitrate nitrogen (NO_3^-) produced by the photocatalytic oxidation of NO to monitor the air pollutant removal efficiency. Cordero et al. [37] evaluated the NOx removal efficiency of 10 selected materials using two pilot-scale demonstration platforms installed at two separate locations. The materials were exposed outdoors for measuring ground-level NO and nitrogen dioxide (NO_2) concentrations over a one-year period. The pollutant removal efficiency of the materials was determined by comparing them to simultaneously measured concentrations of reference, nonactive materials.

Even though current studies on practical photocatalytic materials on a real scale have proposed their own distinctive methodology, there remain several challenges to providing an accurate assessment of NOx removal efficiency. The modified methodology based on the ISO 22197-1 standard addresses the following challenges. It is necessary to prove that there is a similar trend of quantitative NOx outcomes derived from indoor and outdoor experiments to confirm the NOx removal performance of photocatalytic materials in an outdoor environment. Further, the size of a real construction material, which is a trend indicating an increase in the amount of NOx observed as the size changes, should be demonstrated to certify the NOx removal efficiency of photocatalytic materials.

To this end, the NOx removal performance of photocatalytic construction materials is demonstrated in this study using two different types of experiments in both indoor and outdoor environments.

(1) A photoreactor test was conducted to assess the NO removal performance of construction materials such as coatings, paints and shotcrete using a modified ISO 22197-1 method. The UV irradiation of the outdoor environment was analyzed, and the experiment was conducted in an indoor laboratory under UV irradiation identical to that of the outdoor condition. Subsequently, photoreactor tests were conducted on construction materials applied to actual buildings located in Seoul, South Korea. The NO removal performance at the real scale was demonstrated by confirming that the trends of the indoor and outdoor environments showed comparable results.
(2) In this experiment, the NOx removal performance was analyzed by assessing the amount of NOx ions remaining in the water after washing the surface of the specimen artificially exposed to NOx in the laboratory. The preliminary test used two same-sized specimens according to the specification in the ISO 22197-1 standard; the specimens were enlarged to the size of building materials. The experimental conditions were confirmed when the NOx removal performance was found to increase similarly to the tendency of the specimen to increase in size. The enlarged specimen was used for a field experiment by applying a modified method from the ISO 22197-1 standard.

2. Preliminary Test for Measuring NOx Removal

2.1. Experimental Setup

Various types of photocatalytic materials were used to analyze NOx removal performance (e.g., paint, coating material and shotcrete). Each photocatalytic material was prepared by mixing anatase-based TiO_2 powder (NP-400 (Product Overview of Photoctalytic

Powder NP-100. http://www.btfgreen.com/bbs/board.php?bo_table=product1&wr_id=4 (accessed 23 May 2021))from Bentech Frontier, Jeollanam-do, South Korea) with a commercial material. The mixing ratio of anatase-based TiO_2 powder for each photocatalytic material is summarized in Table 2. In the table, composition of each photocatalytic material is specified in order to enhance the understanding of the results of this study. TiO_2 powder was added to each commercial material. A fluid ceramic binder was used in advance to prevent substances in the paint from interfering with the photocatalytic reaction by chemically affecting the TiO_2 powder. Other components of the paints are similar to those of common paint. The optimal photocatalytic reaction was elicited by adjusting the mixing ratio. Synthetic resin emulsions are used as a binder to increase the photocatalytic performance of shotcrete.

Table 2. Composition of photocatalytic materials.

Material no.	Photocatalytic Coating		Photocatalytic Paint		Photocatalytic Shotcrete	
	Contained Chemicals	**Proportion (%)**	**Contained Chemicals**	**Proportion (%)**	**Contained Chemicals**	**Proportion (%)**
1	TiO_2	1.75	TiO_2	3.5	TiO_2	3.5
2	Silicone compound	5.6	Fluid ceramic binder	7.2	Synthetic resins emulsions	2.2
3	Water	51.0	Others	89.3	Cements	17.2
4	Others	41.65	-	-	Others (Aggregates and water, etc.)	77.1

2.2. Photoreactor Test

We performed a preliminary test in the laboratory to measure the amount of NO removal from various types of construction materials such as coatings, paints and shotcrete using the photocatalyst before conducting the field application experiment. A preliminary test was conducted following ISO 22197-1 to verify the NO removal performance of the photocatalytic product specimens quantitatively. The NO injections ranged from 37.8–38.6 micromole (μmol) on a 5 h basis and 8.61 μmol on a 1 h basis in shotcrete measurement. A test condition of high NO gas concentration was prepared by connecting a hose line for gas injection and a bag containing NO gas while the flow rate of 1 ppm of NO gas was set to 1 L/min. Temperature was set to be 25 ± 2.5 °C and relative humidity was set to be 50% according to the ISO 22197-1 standard.

Figure 2 shows the apparatus used for NO removal analysis (CM2041, Casella, London, UK). In the ISO standard, the requirements of the analysis apparatus are clearly stated, and the apparatus used in this study is suitable for the ISO standard. For this test, UV-A with a wavelength range of 300 nm to 400 nm was used as the light source. This study followed the ISO standard's content to use black light lamps with wavelength ranges of up to 351 nm. For the test, the ultraviolet light irradiance (UV irradiation) was kept constant at 10 W/m^2. The NO removal performance of the photocatalytic coating and photocatalytic paint was measured for 5 h each, and that of photocatalytic shotcrete for 1 h each.

Table 3 lists the NO removal results for each photocatalytic construction material. Even with a 1-h measurement, the performance level of NO removal from photocatalytic shotcrete was higher than that of other materials.

Even though the ISO 22197-1 standard officially recommends setting UV irradiation as 10 W/m^2, this study performed an additional test to determine UV irradiation identical to outdoor UV conditions for a more empirical experiment. We intended to determine the level of actual UV light incident on a vertical wall where the photocatalyst is applied because the UV light for photocatalytic action varies in the field whereas it is constant in the laboratory. As the photocatalyst is applied to the wall to which the reactor is attached, UV light on the southern wall, eastern wall and floor is measured on clear winter days. In the preliminary test, the NO removal amount on the eastern wall was measured in the field to determine the amount of NO injection (see Figure 3). A self-made photoreactor with a length of 300 mm and a width of 500 mm was used for the field measurements; this photoreactor is three times larger than the reactor prescribed in the ISO-22197-1 test method. The transparent window material was composed of quartz to facilitate the passage of ultraviolet light. The photoreactor was installed on the rooftop of the building with an NO injection of 3 ppm. Further, the NO removal amount was measured in the preliminary test to determine the appropriate level of NO injection for use in the field test.

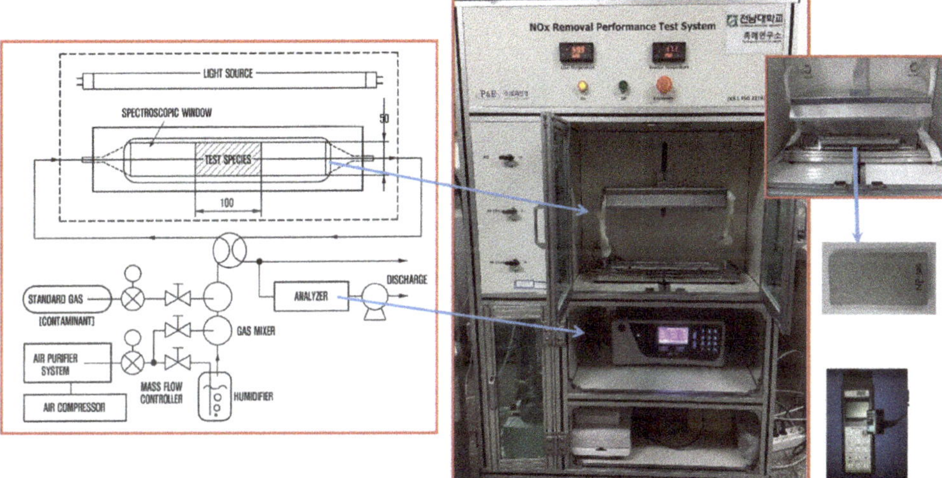

Figure 2. Apparatus used to measure NO removal as per ISO 22197-1 specifications.

Table 3. Photoreactor test results for NO removal rate at the preliminary test.

Materials	NO Injection (μmol)	NO Removal (μmol/50 cm^2)	Measuring Time (Hour)	UV Irradiation (W/m^2)
Photocatalytic coating	38.57	7.29	5	10
Photocatalytic paint	37.78	5.76	5	10
Photocatalytic shotcrete	8.61	11.28	1	10

Figure 4 shows the UV irradiation analysis results performed on the southern, eastern and floor surfaces between 9 a.m. and 4 p.m. at on 21 March. On the southern wall, the UV irradiation reached the highest level at noon and decreased gradually. On the eastern wall, the UV irradiation was relatively lower than that on the southern wall and floor surface, which indicates that the highest irradiation level was 6 W/m^2. The UV irradiation is the highest on the floor because the construction materials receive sunlight when they are attached horizontally towards the sky. Thus, the installation strategy for photocatalytic materials needs to consider solar insolation efficiency.

2.3. Washing Water Experiment

This experiment involves the quantitative analysis of NO_2 and NO_3^- ions contained in washing water after washing the surface coated with photocatalytic paint with distilled water after the specimen is exposed to UV light for a certain period.

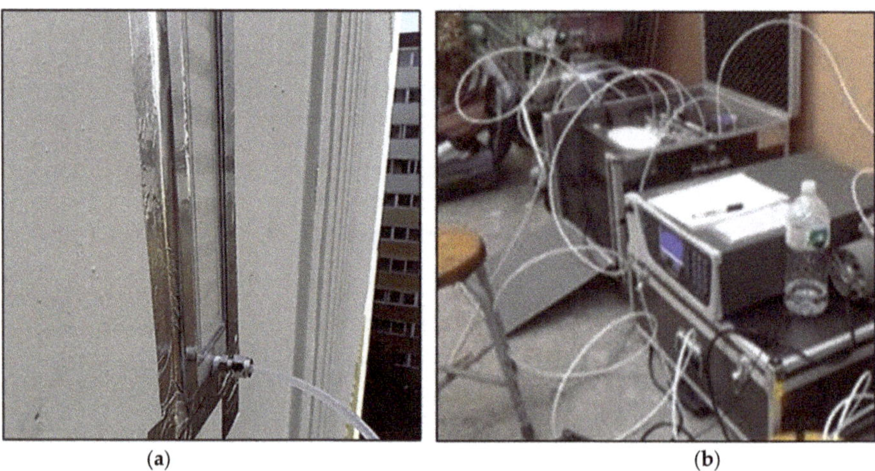

Figure 3. The experimental setup of photoreactor experiment on the wall. (**a**) Photoreactor with NO gas injection hoses on the eastern wall; (**b**) portable analysis device for NOx removal.

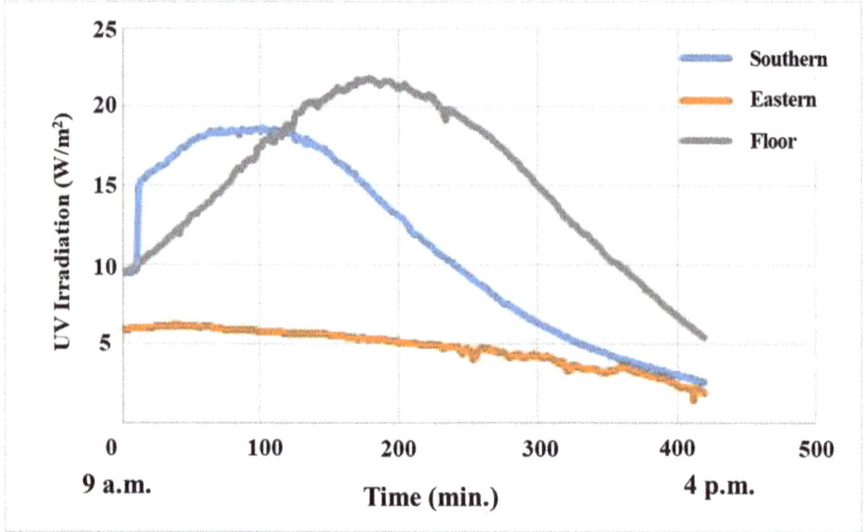

Figure 4. UV irradiation analysis results for southern, eastern and roof surfaces.

This quantitative experiment is conducted to determine the amount of NOx removed by the actual photocatalytic reaction. The experiment based on the washing water recovery method specified in Item 8.3 of ISO 22197-1 was conducted (Figure 5). Distilled water (50 mL) was used, and the washing time was set to 1 h. Elution operations were performed twice, and ions were measured via chromatography (Figure 6). Following ion measurement, the removed amount of NOx was calculated based on the number of ions eluted by the formulas specified in Item 9.7 of the ISO 22197-1 test method.

$$n = n_{w1} + n_{w2} \tag{1}$$

$$n_{w1} + n_{w2} = V_{w1}\left(p_{NO_3^-,w1} \div 62 + p_{NO_2^-,w1} \div 46\right) + V_{w2}\left(p_{NO_3^-,w2} \div 62 + p_{NO_2^-,w2} \div 46\right) \tag{2}$$

where n, v, $p_{NO_3^-}$ and $p_{NO_2^-}$ denote the number of moles of nitric acid eluted from the specimen, amount of distilled water recovered for washing, concentration of nitric acid eluted from the specimen and concentration of nitrous acid eluted from the specimen, respectively.

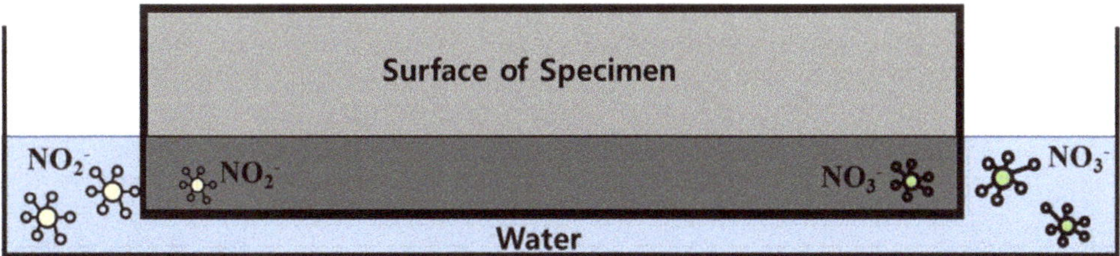

Figure 5. Overview of the washing water experiment of the specimen specified in ISO 22197-1.

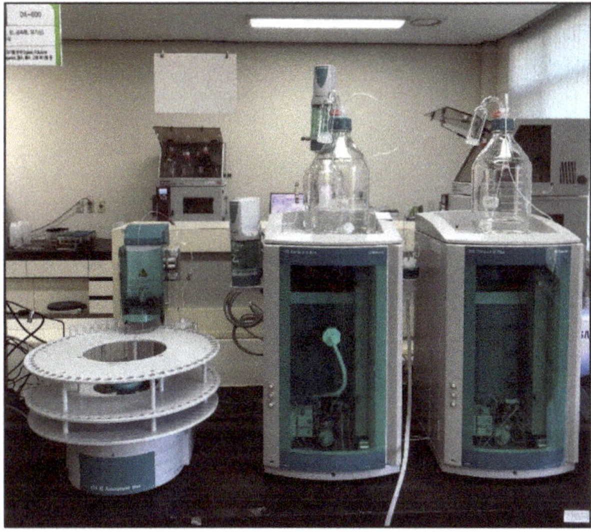

Figure 6. 930 Compact IC Flex ion chromatography (940 Compact IC Flex, Metrohm, FL, USA).

In this study, several experimental conditions were reviewed to modify the existing test methods of ISO 22197-1 for outdoor field test conditions. First, the amount of water washing was analyzed. If more washing water than necessary is used, it can dilute the ion concentration and make the measurement difficult; separate enrichment work is required to solve this problem. Second, a long washing time can increase the time required to process the specimen, which can result in an excessively long experimental schedule. The ISO-based specimens were washed following the test method specified in ISO 22197-1.

Before the field test, a few experimental variables were modified to examine that-the impact of the simplification of ISO 22197-1 method may have a significant effect on deriving the experimental results. Thus, the amount of washing water was adjusted to 20 mL for each washing considering the quantity of samples to be measured via ion chromatography; the washing time was limited to 5 min. The specific experimental conditions are addressed in Table 4.

Table 4. The conditions of chromatographic ionic analysis.

Model	930 Compact IC Flex
Column	Metrosep A Supp 5, 250 × 4 mm
Eluent	3.2 mM Sodium carbonate, 1.0 M Sodium bicarbonate
Flow rate	0.7 mL/min
Inj. Volume	20 μL
Detection	Conductivity Detector

The outdoor preliminary tests were conducted to check whether the test results showed a similar tendency to that observed in the laboratory experiments. A wall with photocatalytic paint was presumed, and common aqueous paint and photocatalytic paint were applied to ordinary concrete specimens. These specimens were then ex-posed to the same outdoor conditions and washed under the same conditions as the ones used in the preceding experiments to check the state of ion detection. For the pre-liminary outdoor experiments, specimens were installed on the roof of an apartment in Seoul, Korea. The specimens were exposed to natural conditions for a certain period, collected and washed by dipping in 20 mL of washing water for 5 min. The washing water was then collected to analyze the ion number using ion chromatography The size of the specimen is 49.5 mm in width and 99.0 mm in length, and the thickness of the specimen is 5 mm. The specimen is made by cutting the large plate with photocatalytic paint to the size specified above. The Specimen A and B were made by applying photocatalytic paint (containing about 3.5% of TiO_2 component) to the surface of a plate made of concrete. In this study, the following pre-process is conducted on the specimens. First, in order to remove organic matter remaining on the surface, UV rays are irradiated for 16 h. The irradiated UV rays are10 W/m^2 which can sufficiently decompose organic matter on the surface.

Two photocatalytic specimens, A and B, were used to conduct the ISO 22197-1 experiment to verify the activity for the preliminary test. Figure 7 shows the test results for the two types of specimens. In order to sufficiently supply NO gas in the reactor and react sufficiently with UV-A, lag occurred at the beginning. Since the experiment time is 5 h, it shows constant removal performance up to 300 min and shows the following results after 300 min. Specimen A showed an active photocatalytic reaction, with an NO removal rate of approximately 12%; specimen B showed a removal rate of approximately 5%, which is approximately half of that of specimen A.

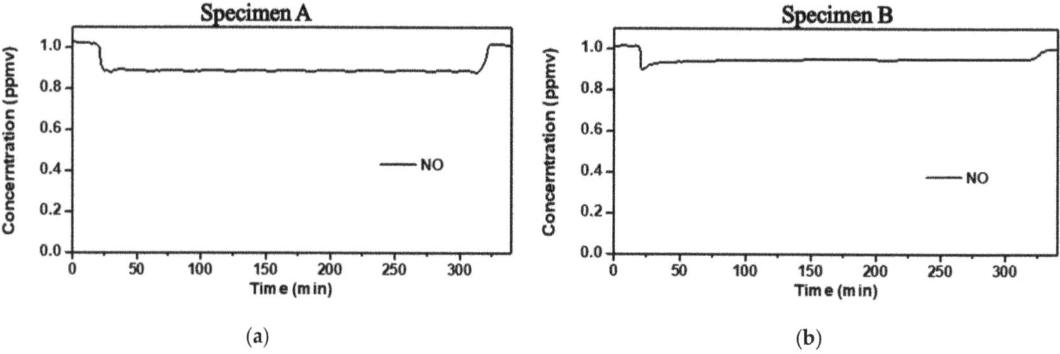

Figure 7. NO removal rate results for 6 h 30 min while UV-A is illuminated. (**a**) and (**b**), Ion chromatography result of specimen A and specimen B.

Table 5 shows the results of analyzing NO removal of the NO attached to the surface during the test performed using the ISO 22197-1 measurement method and the nitric acid ionic weight detected in the washing water. For this experiment, the amount of washing water was 20 mL, and specimens were eluted five times for 5 min.

Table 5. NO removal for each specimen, nitric acid ion weight and ratio analysis results.

Division	Removed NO (μmol)	Nitric Acid Ionic Weight (μmol)	Ratio (%)
Specimen A	4.74	0.96	48
Specimen B	2.03	0.34	41

NO removal results and recovered nitric acid ionic weights from each result showed different trends. However, the recovery rate was found to be less than 50%, which differs from the calculated NO removal amount. Based on this measurement method, it can be concluded that the conditions for washing the specimen presented above are not correlated with the amount of NOx attached to the surface by photocatalytic activity. Therefore, only performing the test method specified in ISO 22197-1 can produce an overall error. In the above experiment, we used cured concrete specimens as the base materials to be painted; however, it is difficult to use them for panels exposed to outdoor environment. Even in the case of ISO specimens, it is expected that using mass-produced base materials for various experiments will yield more objective results. To sum up, the amount of washing water was adjusted from 20 mL to 50 mL for each washing considering the quantity of samples to be measured via ion chromatography; the washing time was also re-adjusted from 5 min to 1 h.

Considering the experimental conditions above, a cellulose fiber reinforced cement (CRC) board (from Byuksan corporation (Product Overview of CRC board used in this study. http://www.byucksan.com/01_product/product.asp?cate=001004002, accessed on 22 September 2021) Korea) was selected as the base material that satisfies the above conditions. The CRC board was fabricated by mixing natural pulp such as cellulose fiber, Portland cement and silica sand with water. The CRC board is a non-flammable (flame-retardant grade 1) construction material produced through an autoclave curing process after pressurizing to 10,000 tons. It is an eco-friendly asbestos-free construction material that exhibits only slight changes in length due to temperature variation, water resistance and noncombustibility (flame retardant grade 1). The experiment was conducted according to the method specified in ISO 22197-1 because washing the specimen under the conditions set previously does fully reflect the NO amount on the surface. The results are summarized in Table 6. The CRC-A and CRC-B were made by applying photocatalytic paint (containing about 3.5% of TiO_2 component) to the surface of CRC board.

Table 6. NO removal results for CRC board panel experiments.

Division	NO Amount with Washing Once (μmol)	Accumulated Removed NO Amount (μmol)
CRC-A	0.81	3.57
CRC-B	1.11	2.84

CRC boards are supplied through various processes such as manufacturing and storage, and therefore, it is difficult to verify if a significant amount of nitrogen oxides is already present. Therefore, removing nitrogen oxides contained in the base material is deemed necessary to use the CRC board as the base material. For the preliminary experiment, the experimental method with the modified specimens was conducted based on ISO standard methods established for outdoor field experiments. The experimental conditions and methods were as follows:

1. Preparation of specimen
    - The base material of the specimen was unified as the CRC board.

- The CRC board used as the base material was washed sufficiently to remove pre-exiting nitrogen oxides.
- The test specimens are prepared with CRC boards; the photocatalytic paint was applied to these boards, and the general aqueous paint specimens were prepared as a control group.

2. Methods of analyzing NO removal and washing water.
   - NO removal was conducted in accordance with the ISO 22197-1 method.
   - Concentrations other than those specified in ISO 22197-1 were not significant in the experiment; hence, the experiments were conducted only at the prescribed concentration of 1 ppm.
   - The washing of the specimens that have undergone the ISO 22197-1 experiment are in accordance with the washing method specified in ISO 22197-1.
   - The washing of ISO specimens exposed to the outdoor environment should be performed according to the washing method specified in ISO 22197-1.

3. Methodology for the Field Test

3.1. Photoreactor Field Test Method

In this study, photocatalytic materials applied to a real building were used to conduct field tests (Figure 8). After applying primers all over the southern wall of Building A, the photocatalytic coating agent was applied using rollers; the coating area was 297 m^2. In Building B, we applied the photocatalytic paint by brushing and spraying onto the eastern wall, and the area of application was 889 m^2. In Building C, the photocatalytic shotcrete was applied to a site of 61 m^2, up to the second floor on the east side of the building. In the same manner as the preliminary test method, a reactor was installed in the area with the photocatalytic material applied to each building to measure the NO amount of injection and removal for more than 5 h. We simultaneously measured the UV irradiation. In the case of outdoor experiments, since it is impossible to adjust experimental variables such as illuminance, temperature and humidity, the experimental variables were carried out under natural experimental conditions.

(a)

(b)

(c)

Figure 8. Photocatalytic materials applied on the real buildings in the field assessment. (**a**) Building A, photocatalyst coating; (**b**) Building B, photocatalyst paint; (**c**) Building C, photocatalyst shotcrete.

Figure 9 shows installed photoreactor and UV radiometer used in the field experiments. The size of the UV radiometer is 50 × 300 mm, which is three times larger than the specimen defined in the ISO method. Considering that the gas could evenly reach the surface of specimen, the space between the transparent plate and the inner specimen was set to 5 mm. The photoreactor was set to 3 ppm as the specimen size was increased by a factor of three. We conducted a comparative evaluation of NO removal using three photocatalytic products based on the minimum criteria for photocatalytic product certification set by the Photocatalysis Industry Association of Japan.

Figure 9. Installed photoreactor and UV radiometer on the rooftop.

3.2. ISO-Based Washing Water Field Test Method for Real-Scale Construction Materials

As the ISO-based specimen size is smaller than the size of the actual building material, a modified experimental method is used to perform practical and empirical analyses. To this end, the established methods are summarized as follows.

It is difficult to fabricate panels made for outdoor exposure using concrete. Hence, the CRC board reviewed in the ISO specimen tests was used as the base material. The size of the CRC board was set to 800 × 900 mm. The panel simulates the photocatalyst-coated surface for the convenience of the experiment because collecting washing water from the wall where photocatalytic paint is applied is difficult; this facilitates the identification of the amount of NOx removed from the surface by exposing it to the outdoor environment. Therefore, considering the field situation, we applied a washing method before measuring instead of dipping the specimens as prescribed in the ISO method. Figure 10 shows that the surface was evenly washed using a sprayer containing 2 L of distilled water. The quantity of distilled water is determined based on the empirical experiment, which indicates that 2 L of distilled water can fully wash the ions attached on the surface of CRC board.

(a)

(b)

Figure 10. Spraying and collecting process of distilled water. (**a**) Spraying on the CRC panel; (**b**) collecting process of washing water.

The experiment was conducted from 25 October to 6 December. The reason why the outdoor experiment of this study was conducted during above period is to prevent the experimental results from being biased due to weather conditions such as excessive rainfall or insufficient rainfall. This experiment was conducted in Seoul, the capital of Korea. Seoul's meteorological characteristics are that the rainfall is higher than other periods due to the influence of rainy seasons and typhoons in summer season (June to September). In winter season (December to February), snow falls and the temperature is low, making the walls coated with photocatalytic paint to freeze. These seasonal characteristics are expected to have an impact on the experimental results of this experiment. Therefore, in order to prevent bias in the experimental results due to the previous seasonal characteristics, the experimental period was set from 25 October to 6 December.

We performed washing up to five times as in the preliminary test to check the concentration of ions contained in the collected washing water; the results showed that the concentration was significantly lowered as the number of washings increased and that the concentration after the second washing was not significant. Therefore, the panels were only washed twice in the actual experiment. The experiments were conducted on the rooftop of a building adjacent to the roadway because measurements immediately next to the roadway were not feasible. The following field experimental conditions determined through preliminary experiments were used.

1. Panels with photocatalytic paint applied
   - The material of the panel was CRC board.
   - The paint was applied to the panel after washing and drying the CRC board.
   - The control group was prepared with a CRC board with general aqueous paint applied.
2. Methods of washing and analyzing the panels
   - A total of 2 L of washing water was used for each panel.
   - The number of washes was limited to two; the washing intervals were determined by considering the weather conditions.
   - The panel was washed by spraying.
   - A separate holder and recovery bin was designed to collect rainwater.
   - The number of ions in the washing water were measured via ion chromatography.

4. Results of NO Removal Analysis Experiment

4.1. Photoreactor Test Results

The results for the three experimental construction materials in each field are listed in Table 7. The NO removal amount on a clear day was more than twice that on the cloudy day for all three buildings. Assuming that the NO removal was converted to the ISO test reactor size of 50 cm^2, Building C with photocatalytic shotcrete applied had the lowest NO removal amount of 0.87 $\mu mol/50\ cm^2$. However, it is still higher than that specified by the Photocatalysis Industry Association of Japan performance standard (0.5 μmol). Building C received less UV irradiation compared to buildings A and B because of the shade formed for a certain period of time. The preliminary test showed the highest NO removal performance of the photocatalytic shotcrete; however, the field measurement showed the lowest performance. As the NO removal performance is highly influenced by weather conditions, conditions such as the bearing of the building and its relationship with adjacent buildings need to be considered when applying photocatalytic products as exterior construction materials. Compared to preliminary test results measured in the laboratory, the field test results of Buildings A and B showed higher NO removal performance. For Building C, lower NO removal occurred in the field test than in the laboratory, which can be attributed to the difference in UV irradiation, which was 10 W/m^2 in the preliminary test and less than 10 W/m^2 in the field measurement.

4.2. Washing Water Field Test Results of Specimens

A specimen experiment was conducted according to the ISO 22197-1 method to determine how the amount of NOx removed by the photocatalyst is reflected in the washing water. The experiment was conducted with the specimen applied with aqueous paint as a comparison group and four types of specimens applied with photocatalytic paint. The specimens were classified into aqueous and photocatalytic paint specimens to facilitate the distinction of experimental specimens.

Table 7. Outdoor NO removal results for construction materials.

Materials		NO Injection Rate (μmol)	NO Removal (μmol/50 cm^2)	NO Removal (μmol/50 cm^2)	UV Irradiation (W/m^2)	Measurement Date (Weather)
Building A: Photocatalytic coating		118.97	11.06	3.68	2–13	25 October 2018 (Sunny)
Building B: Photocatalytic paint	1st	107.82	10.03	3.34	6–14	14 November 2018 (Sunny)
	2nd	122.79	4.87	1.62	2–16	5 December 2018 (Cloudy)
Building C: Photocatalytic shotcrete	1st	95.84	6.78	2.26	1	13 November 2018 (Sunny)
	2nd	114.52	2.62	0.87	2–9	6 December 2018 (Cloudy)

The size of the specimen here is 800 mm in width and 900 mm in length, and the thickness of the specimen is 5 mm. Photocatalytic paint containing the identical amount of TiO_2 powder (about 3.5%) was applied to the surface of CRC boards to create Specimen A, B, C and D. For the photocatalytic paint specimens, newly prepared specimens A and B for this experiment and specimens C and D with photocatalytic paint applied approximately a month ago were used. These CRC specimens also went through the pre-process to remove organic matter remaining on the surface of CRC boards, UV rays are irradiated for 16 h. The irradiated UV rays are10 W/m^2 which can sufficiently decompose organic matter on the surface. Each specimen was washed four times; however, only the ions in the first and second washing water were added and analyzed as prescribed in ISO 22197-1. Table 8 summarizes the results of the ISO 22197-1 experiment for the five specimens used for NOx removal with the results of washing twice.

Table 8. Washing water field test results of five specimens.

Specimen Type	NOx Removal Amount (μmol)	Eluted NOx Amount (μmol)
Aqueous paint specimen	0.08	0.53
Photocatalytic paint specimen A	7.90	7.86
Photocatalytic paint specimen B	10.80	10.38
Photocatalytic paint specimen C	2.40	3.62
Photocatalytic paint specimen D	2.45	3.25

Refer to the preliminary test calculation formula for analyzing and calculating NO elution from ISO specimens.

Experimental results with the aqueous paint showed that the NOx removal amount was 0.08 μmol, which resulted in a weak removal effect; the washing water test results showed a small NOx removal amount of 0.53 μmol. The results obtained using specimens A and B for this experiment show that the calculated NOx removal amount and the amount of eluted nitrogen oxides are similar; in the case of existing specimens C and D, the results show a slightly more eluted NOx than the removed NOx. Thus, the results indicate that the amount removed by the ISO evaluation method is similar to the amount of elution measured from the analysis of the washing water. However, for existing specimens previously exposed to the outdoor environment, the NOx contained—whether through adsorption or reaction—was not completely removed during the pretreatment process. This is believed to have a partial effect on the elution into the washing water. When the specimen was washed sufficiently before the experiment, the error shown in the existing specimen is expected to decrease. In addition, it can be inferred that nitrogen oxides were sufficiently removed through the washing water experiment.

The specimen was placed outside the building from 3 June 2019 to 12 June 2019; during this period, the specimen was moved indoors on rainy days to prevent the surface from being washed by rainwater. The comparison results of the values obtained from experiments and the specimens exposed under the above conditions according to the ISO test method are listed in Table 9.

Table 9. Washing water field test results of specimens for which the surface was washed sufficiently.

Specimen Type	ISO NOx Removal Performance (μmol/50 cm^2·5 h)	Elution Amount after ISO Experiment (μmol)	Elution Amount after Outdoor Exposure for Five Days (μmol)
Aqueous paint	0.08	0.53	3.88
Photocatalytic paint sample C	2.40	3.62	14.13

The specimens were exposed outdoors for five days in the apartment after washing was completed for the ISO experiment. For the general aqueous paints, 3.88 μmol of NOx was removed, whereas photocatalytic paints showed significantly better NOx elution. Some elution was found in the case of general aqueous paints; however, this was attributed to the result from surface adsorption rather than photocatalytic reaction. Further, for the photocatalytic paint with photocatalytic activity, we detected a large amount of elution in a definite contrast. Thus, we confirmed that the ISO NOx removal amount was equal to the value of the nitrogen oxide elution and that the photocatalytic performance showed a significant difference when the photocatalytic specimens were exposed in the field.

4.3. *Washing Water Field Test Results of Photocatalytic Panel*

An ISO-based washing water field experiment was conducted for real-scale construction materials under the same conditions as the specimen experiments.

Figure 11 shows panels installed on the roof of an apartment. The installation period is the same as that for which the ISO specimens were exposed outdoors; further, the panels were moved and kept indoors on rainy days to prevent the surface from being washed by rainwater. All panels were washed together under the same conditions, and all washing water was collected thoroughly and labeled (see Figure 12).

Figure 11. Specimens placed on the rooftop of the apartment.

Table 10 summarizes the results of measuring the ionic value of the washing water of panels exposed on the roof of apartments in Seoul, Korea. In this study, three different types of panels are used for field test. First one is panel coated with aqueous paint. S1 is the panel that is newly coated with photocatalytic paint before the experiment day. S2 and S3 photocatalytic panels are created before two weeks before the experiment day. After exposing the panel to the outside, the surface was washed with water and reused.

(a)

(b)

Figure 12. Collecting process and storage procedure of washing water. (**a**) collecting process of washing water; (**b**) collected washing water in seperate bottles.

Table 10. Washing water field test results of photocatalytic panels.

Photocatalytic Panels	Eluted NO_2^-, NO_3^- Ions (μmol)
Aqueous paint	210
Photocatalytic panel (S1)	1469
Photocatalytic panel (S2)	475
Photocatalytic panel (S3)	504

A clear difference is observed between general aqueous paint and photocatalytic paint in terms of washing water for the CRC panels installed in the apartment. NOx removed by photocatalytic effects remains in an ionic state on the surface of the photocatalyst paint, and they are eluted by water washing and released into ionic nitric acid and nitrous acid. The amount of NOx ions eluted by water washing, which was investigated based on the NOx removal performance in the ISO test method, was almost the same as the amount of NOx removed by photocatalytic action. These results indicate that the amount of NOx ions in the washing water can be analyzed to quantify the amount of NOx removed by the photocatalyst. The NOx ions were observed in the washing water of general aqueous paint; however, only a few were detected in the washing water of photocatalytic paints. When tested and exposed to a real outdoor environment, the results indicated that photocatalysts are effective for removing NOx in the real world.

The results of s1 and s2 have the following implications for the NOx reduction performance of the photocatalytic paint used in this study. First, it shows that contaminants adhering to the surface are sufficiently removed just by washing the surface of the wall coated with photocatalytic paint with rainwater. Photocatalytic paint can maintain its NOx removal performance even after a certain period of time has passed since it is actually installed on the exterior wall of a building. Through this experiment, the durability of the performance when the photocatalytic paint was applied to an actual building is demonstrated.

Table 11 shows the amount of NOx that can be removed per day; it is calculated from the panel test results above. The average daily removal of the panel was calculated by averaging the installation days, excluding the rainy days when there was almost no UV light irradiation. Further, the estimated annual removal of the photocatalytic panel was calculated by excluding the average annual rainfall days in Korea. Moreover, the annual removal amount corresponds to the amount of NOx that can be removed per photocatalytic panel unit area.

Table 11. Washing water field test results of photocatalytic panels.

Division	Daily Average NOx Removal Amount of Panel [(1)] (μmol)	Estimated Annual NOx Removal Amount of Panel [(2)] (μmol)	Estimated Annual NOx Removal Amount [(3)] (g/m^2)
Aqueous paint	42.00	10,920	0.46
Photocatalytic panel (S1)	293.80	76,388	3.19
Photocatalytic panel (S2)	95.00	24,700	1.03
Photocatalytic panel (S3)	100.80	26,208	1.09

[(1)] Total nitrogen oxides washed off from the panel/5 (number of days the panel was installed). [(2)] Panel daily average removal amount × 260 days (excluding the average number of rainfall days per year). [(3)] Estimated annual removal amount (μmol)/106 (calculated as mol) × 1.39 (area multiplier, m^2) × 30 g/mol (nitrogen molecular weight).

5. Conclusions

A modified ISO 22197-1 standard method to analyze the NOx removal performance of photocatalytic construction materials in both indoor and outdoor environments was presented. The first experiment involved a photoreactor test conducted to assess the NO removal performance of construction materials such as coatings, paints and shotcrete. In the preliminary analysis, the impact of UV irradiation according to the direction at the outdoors was analyzed. For sunny days, the NOx removal performance (3.12–4.76 μmol/150 $cm^2 \cdot 5$ h) was twice as much as the ISO 22197-1 standard specification (2.03 μmol/150 $cm^2 \cdot 5$ h) in the real world. Even on cloudy days, it is slightly lower than that of the ISO 22197-1 standard specification; the NOx removal performance ranging from 0.68–1.89 μmol/5 $cm^2 \cdot 5$ h is shown, which is 63% of the indoor experiment results.

The second experiment was a washing water experiment, wherein the NOx removal performance was analyzed by assessing the amount of NOx ions remaining in the water after washing the surface of the specimen artificially exposed to NOx in the laboratory. The preliminary test employed two specimens, one of which was of the same size as that defined in the ISO 22197-1 standard, and the other was expanded to a size comparable to that of the construction materials. The experimental conditions were validated when it was discovered that the NOx removal efficiency improved in tandem with the specimen's propensity to grow. The expanded specimen was then used in a field experiment using an adapted approach from the ISO 22197-1 standard. The washing water test was performed after sufficient washing of the ISO test specimen and exposure to the outdoors for 5 days in natural light. The washing test results indicated that general aqueous paint showed a NOx removal of 3.88 μmol, whereas that for photocatalytic paint was higher at 14.13 μmol. The amount of NOx removed by the washing water test was estimated to be 3.19 g/m^2. This was demonstrated by the results of outdoor exposure, which is almost similar to the ISO standard test. Thus, it is possible to prove the direct correlation between the photocatalytic activity according to the ISO standard and the NOx removal performance.

The results of this study are expected to encourage the application of photocatalytic construction materials to real buildings with obvious NOx removal effectiveness. Empirically demonstrating that the real NOx removal effect increases when the small-sized specimen stated in the ISO standard is expanded to a size close to that of the actual construction material is a significant contribution from both academic and practical standpoints. Future work will include more extensive analysis using various types of photocatalytic construction materials at various sites for a longer experiment time.

Author Contributions: Conceptualization, M.K. and H.K.; methodology, M.K. and H.K.; software, M.K.; validation, M.K. and H.K.; formal analysis, M.K.; resources, M.K. and H.K.; data curation, M.K.; writing—original draft preparation, M.K.; writing—review and editing, H.K.; visualization, M.K.; supervision, J.P.; project administration, H.K.; funding acquisition, H.K. All authors have read and agreed to the published version of the manuscript.

Funding: This research was supported by a grant (21SCIP-B146966-04) from construction technology research project funded by Ministry of Land, Infrastructure and Transport (MOLIT) of Korea government and Korea Agency for Infrastructure Technology Advancement (KAIA).

Conflicts of Interest: The authors declare no conflict of interest.

References

1. Sillman, S. The relation between ozone, NO_x and hydrocarbons in urban and polluted rural environments. *Atmos. Environ.* **1999**, *33*, 1821–1845. [CrossRef]
2. Sun, Y.; Zwolińska, E.; Chmielewski, A.G. Abatement technologies for high concentrations of NOx and SO_2 removal from exhaust gases: A review. *Crit. Rev. Environ. Sci. Technol.* **2016**, *46*, 119–142. [CrossRef]
3. Yoshida, K.; Makino, S.; Sumiya, S.; Muramatsu, G.; Helferich, R. Simultaneous reduction of NOx and particulate emissions from diesel engine exhaust. *SAE Trans.* **1989**, 1994–2005. [CrossRef]
4. Hodan, W.M.; Barnard, W.R. *Evaluating the Contribution of PM2.5 Precursor Gases and Re-Entrained Road Emissions to Mobile Source PM2.5 Particulate Matter Emissions*; MACTEC Federal Programs: Research Triangle Park, NC, USA, 2004.
5. Kim, J.-Y.; Lee, E.-Y.; Choi, I.; Kim, J.; Cho, K.-H. Effects of the particulate matter2.5 (PM2.5) on lipoprotein metabolism, uptake and degradation, and embryo toxicity. *Mol. Cells* **2015**, *38*, 1096.
6. Skalska, K.; Miller, J.S.; Ledakowicz, S. Trends in NO_x abatement: A review. *Sci. Total Environ.* **2010**, *408*, 3976–3989. [CrossRef]
7. Bowman, C.T. Control of combustion-generated nitrogen oxide emissions: Technology driven by regulation. *Symp. (Int.) Combust.* **1992**, *24*, 859–878. [CrossRef]
8. Heck, R.M. Catalytic abatement of nitrogen oxides–stationary applications. *Catal. Today* **1999**, *53*, 519–523. [CrossRef]
9. Chang, M.B.; Lee, H.M.; Wu, F.; Lai, C.R. Simultaneous removal of nitrogen oxide/nitrogen dioxide/sulfur dioxide from gas streams by combined plasma scrubbing technology. *J. Air Waste Manag. Assoc.* **2004**, *54*, 941–949. [CrossRef]
10. Guo, M.-Z.; Ling, T.-C.; Poon, C.-S. TiO_2-based self-compacting glass mortar: Comparison of photocatalytic nitrogen oxide removal and bacteria inactivation. *Build. Environ.* **2012**, *53*, 1–6. [CrossRef]
11. Brogren, C.; Karlsson, H.T.; Bjerle, I. Absorption of NO in an alkaline solution of $KMnO_4$. *Chem. Eng. Technol.* **1997**, *20*, 396–402. [CrossRef]
12. Guseva, T.; Potapova, E.; Tichonova, I.; Shchelchkov, K. Nitrogen oxide emissions reducing in cement production. *IOP Conf. Ser. Mater. Sci. Eng.* **2021**, *1083*, 012083. [CrossRef]
13. Rodríguez, M.H.; Melián, E.P.; Díaz, O.G.; Araña, J.; Macías, M.; Orive, A.G.; Rodríguez, J.D. Comparison of supported TiO_2 catalysts in the photocatalytic degradation of NO_x. *J. Mol. Catal. A Chem.* **2016**, *413*, 56–66. [CrossRef]
14. Bengtsson, N.; Castellote, M. Heterogeneous photocatalysis on construction materials: Effect of catalyst properties on the efficiency for degrading NO_x and self cleaning. *Mater. Constr.* **2014**, *64*, e013. [CrossRef]
15. Cárdenas, C.; Tobón, J.I.; García, C.; Vila, J. Functionalized building materials: Photocatalytic abatement of NO_x by cement pastes blended with TiO_2 nanoparticles. *Constr. Build. Mater.* **2012**, *36*, 820–825. [CrossRef]
16. Devahasdin, S.; Fan Jr, C.; Li, K.; Chen, D.H. TiO_2 photocatalytic oxidation of nitric oxide: Transient behavior and reaction kinetics. *J. Photochem. Photobiol. A Chem.* **2003**, *156*, 161–170. [CrossRef]
17. Ichiura, H.; Kitaoka, T.; Tanaka, H. Photocatalytic oxidation of NOx using composite sheets containing TiO_2 and a metal compound. *Chemosphere* **2003**, *51*, 855–860. [CrossRef]
18. Yu, Q.; Brouwers, H. Indoor air purification using heterogeneous photocatalytic oxidation. Part I: Experimental study. *Appl. Catal. B Environ.* **2009**, *92*, 454–461. [CrossRef]
19. Luévano-Hipólito, E.; Martínez-De La Cruz, A.; López-Cuellar, E.; Yu, Q.; Brouwers, H. Synthesis, characterization and photocatalytic activity of WO_3/TiO_2 for NO removal under UV and visible light irradiation. *Mater. Chem. Phys.* **2014**, *148*, 208–213. [CrossRef]
20. Ballari, M.; Carballada, J.; Minen, R.I.; Salvadores, F.; Brouwers, H.; Alfano, O.M.; Cassano, A.E. Visible light TiO_2 photocatalysts assessment for air decontamination. *Process Saf. Environ. Prot.* **2016**, *101*, 124–133. [CrossRef]
21. Kuo, C.-S.; Tseng, Y.-H.; Huang, C.-H.; Li, Y.-Y. Carbon-containing nano-titania prepared by chemical vapor deposition and its visible-light-responsive photocatalytic activity. *J. Mol. Catal. A Chem.* **2007**, *270*, 93–100. [CrossRef]
22. Maggos, T.; Plassais, A.; Bartzis, J.; Vasilakos, C.; Moussiopoulos, N.; Bonafous, L. Photocatalytic degradation of NO_x in a pilot street canyon configuration using TiO_2-mortar panels. *Environ. Monit. Assess.* **2008**, *136*, 35–44. [CrossRef] [PubMed]
23. Liu, J.; Jee, H.; Lim, M.; Kim, J.H.; Kwon, S.J.; Lee, K.M.; Nezhad, E.Z.; Bae, S. Photocatalytic performance evaluation of titanium dioxide nanotube-reinforced cement paste. *Materials* **2020**, *13*, 5423. [CrossRef] [PubMed]
24. Xiang, W.; Ming, C. Implementing extended producer responsibility: Vehicle remanufacturing in China. *J. Clean. Prod.* **2011**, *19*, 680–686. [CrossRef]
25. Ifang, S.; Gallus, M.; Liedtke, S.; Kurtenbach, R.; Wiesen, P.; Kleffmann, J. Standardization methods for testing photo-catalytic air remediation materials: Problems and solution. *Atmos. Environ.* **2014**, *91*, 154–161. [CrossRef]
26. Chen, J.; Poon, C.-S. Photocatalytic construction and building materials: From fundamentals to applications. *Build. Environ.* **2009**, *44*, 1899–1906. [CrossRef]
27. Seo, D.; Yun, T.S. NOx removal rate of photocatalytic cementitious materials with TiO_2 in wet condition. *Build. Environ.* **2017**, *112*, 233–240. [CrossRef]
28. Song, Y.W.; Kim, M.Y.; Chung, M.H.; Yang, Y.K.; Park, J.C. NO_x-Reduction Performance Test for TiO_2 Paint. *Molecules* **2020**, *25*, 4087. [CrossRef]

29. Gallus, M.; Akylas, V.; Barmpas, F.; Beeldens, A.; Boonen, E.; Boréave, A.; Cazaunau, M.; Chen, H.; Daële, V.; Doussin, J. Photocatalytic de-pollution in the Leopold II tunnel in Brussels: NO_x abatement results. *Build. Environ.* **2015**, *84*, 125–133. [CrossRef]
30. Ballari, M.; Hunger, M.; Hüsken, G.; Brouwers, H. NO_x photocatalytic degradation employing concrete pavement containing titanium dioxide. *Appl. Catal. B Environ.* **2010**, *95*, 245–254. [CrossRef]
31. Folli, A.; Strøm, M.; Madsen, T.P.; Henriksen, T.; Lang, J.; Emenius, J.; Klevebrant, T.; Nilsson, Å. Field study of air purifying paving elements containing TiO_2. *Atmos. Environ.* **2015**, *107*, 44–51. [CrossRef]
32. Boonen, E.; Beeldens, A. Photocatalytic roads: From lab tests to real scale applications. *Eur. Transp. Res. Rev.* **2013**, *5*, 79–89. [CrossRef]
33. Boonen, E.; Beeldens, A. Recent photocatalytic applications for air purification in Belgium. *Coatings* **2014**, *4*, 553–573. [CrossRef]
34. Cassar, L. Photocatalysis of cementitious materials: Clean buildings and clean air. *MRS Bull.* **2004**, *29*, 328–331. [CrossRef]
35. Guerrini, G.L. Photocatalytic performances in a city tunnel in Rome: NO_x monitoring results. *Constr. Build. Mater.* **2012**, *27*, 165–175. [CrossRef]
36. Yu, Q.; Hendrix, Y.; Lorencik, S.; Brouwers, H. Field study of NOx degradation by a mineral-based air purifying paint. *Build. Environ.* **2018**, *142*, 70–82. [CrossRef]
37. Cordero, J.; Hingorani, R.; Jiménez-Relinque, E.; Grande, M.; Borge, R.; Narros, A.; Castellote, M. NO_x removal efficiency of urban photocatalytic pavements at pilot scale. *Sci. Total Environ.* **2020**, *719*, 137459. [CrossRef]

 materials

Article

The Influence of Dry Hydrated Limes on the Fresh and Hardened Properties of Architectural Injection Grout

Andreja Padovnik * and Violeta Bokan-Bosiljkov

Faculty of Civil and Geodetic Engineering, University of Ljubljana, Jamova 2, SI-1000 Ljubljana, Slovenia; violeta.bokan-bosiljkov@fgg.uni-lj.si
* Correspondence: andreja.padovnik@fgg.uni-lj.si

Abstract: Dry hydrated lime is an air binder often used in architectural injection grouts. This study compared the influences of three commercially available dry hydrated limes on the injection grouts' workability and mechanical properties. The main differences between the limes were in their chemical and mineralogical composition and Blaine specific surface area. The grouts were composed of dry hydrated lime, finely ground limestone filler, water, and super plasticiser. Subsequent results obtained revealed that the Blaine specific surface area is not directly related to the fresh grout properties. Grain size distribution and shape of lime particles and their aggregates in the water suspension are key parameters influencing the following fresh grout properties: fluidity, injectability, the mixture's stability, and water retention capacity. However, the lime injection grouts' mechanical strengths were higher in relation to an increase in the content of portlandite and the Blaine specific surface area of the dry hydrate.

Keywords: architectural injection grout; dry hydrated lime; particle density; specific surface area; workability; porosity; mechanical properties

Citation: Padovnik, A.; Bokan-Bosiljkov, V. The Influence of Dry Hydrated Limes on the Fresh and Hardened Properties of Architectural Injection Grout. *Materials* **2021**, *14*, 5585. https://doi.org/10.3390/ma14195585

Academic Editor: Yeonung Jeong

Received: 18 August 2021
Accepted: 23 September 2021
Published: 26 September 2021

Publisher's Note: MDPI stays neutral with regard to jurisdictional claims in published maps and institutional affiliations.

1. Introduction

Hydrated lime was one of the prevailing binders for renders (external wall mortar layers), plasters (internal wall mortar layers) and masonry mortars up to the 20th century, when cement-based materials took the dominant role in the building sector. Unfortunately, cement-based materials were also applied to repair historic buildings where hydrated lime composites were used to bond the masonry units and protect masonry walls. Due to incompatibility with the historic masonry fabrics and additional unfavourable characteristics of the Portland cement binder—such as salt formation—the historic masonry buildings suffered new extensive damage. Over the last decades, the hydrated lime binder that provides similar composition and properties as the original historical architectural fabrics has become widely used to repair and restore historic lime plasters and renders. Where consolidation or re-attachment of such architectural surfaces is needed, architectural injection grouts prepared using the hydrated lime binder are often used to ensure compatibility between the new and historical materials and components [1]. A comprehensive state of the art regarding the composition of architectural injection grouts used in restoration practise between 1950 and 2015 is given in [2].

The reproducibility of the injection grouts' properties is better when dry hydrated lime is applied as the binder. Moreover, the dry hydrate enables an easy application of different chemical admixtures in the grout mixtures—such as superplasticisers that reduce the water content of the grout considerably and thus increase its mechanical properties [3,4]. Additionally, a range of the lime-based grouts' proportions are easily prescribed [5].

Historically, hydrated lime was produced by burning limestone in a lime kiln to obtain calcium oxide or quicklime, which was subsequently slaked with an excess of water in an exothermic reaction to form calcium hydroxide. Slaked lime or lime putty obtained in the process was stored in pits for at least three months before use. In the 20th century, industrial production fulfilled the need for ready-to-use hydrated lime in larger quantities, promoting hydrated lime as a dry powder [6]. In today's industrial process, the production of dry hydrate is based on quicklime hydration with a controlled excess of water. Technically, the terms hydration and slaking are synonymous. That said, slaking involves a higher water amount and produces a wet hydrate in the form of lime putty, while hydration yields the dry powdered hydrate [7]. The dry hydrate is produced by mixing one part by weight of quicklime with about 0.5 to 0.75 parts of water. This value is significantly above the theoretical amount of water (0.245) required for complete hydration. A higher amount of water is necessary due to water evaporating during the hydration process [8].

The shape and particle size distribution of calcium hydroxide depend on the slaking process parameters. Several studies highlight the importance of the water/lime ratio, temperature and agitation rate on the final quality of lime putty and dry hydrate [7,9–15]. Whitman and Davis [11] studied the influence of different hydration processes on the properties of dry hydrated lime. Their study showed that a high-grade hydrate that contains many fine particles is produced when the rate of hydration is rapid compared to the particles' growth rate. Excess water, reasonably high temperatures, and agitation all favour rapid hydration and a fine product. Another study [12] showed that a more reactive hydrate with a higher resistance to segregation in suspension is produced when the quicklime with finer particle size distribution is used. Dry hydration of quicklime should occur at around 100 °C to obtain a finer product [7,11,12].

The purity of quicklime is another parameter that influences calcium hydroxide quality. It depends on the quality of limestone and the manufacturing process. The main impurities consist of silica, alumina, iron and magnesium (in high calcium lime) [16]. Magnesium oxide (MgO) is disadvantageous in high calcium lime as it affects the reactivity of the quicklime [17]. Additionally, the four impurities listed influence limestone hydraulicity calculated using the cementation index (CI), developed by Eckel [18]. According to the CI value, the lime-based binders could be classified into five categories: pure (< 0.15), subhydraulic (0.15–0.30), feebly hydraulic (0.3–0.5), moderately hydraulic (0.5–0.7), and eminently hydraulic (0.7–1.1) [19]. Impurities can also influence the optical properties of hydrated lime [9,16].

Many studies have compared the physical, chemical, and mechanical properties of lime powder and lime putty based mortars. Recent studies [6,20] show that mortars with dry hydrated lime have a higher carbonation rate and higher compressive and flexural strength values compared to the lime putty mortars subjected to ageing for more than 90 days. On the contrary, older studies promote the use of lime putty based mortars. Rodriguez-Navarro et al. [10] reported that the lime putty prepared with dry hydrate did not achieve traditional slaked lime putty's properties, such as high workability, sand-carrying capacity, and good setting, strength development, and durability. The dry hydrated lime putty exhibited oriented aggregation of nanoparticles, which is irreversible and resulted in a significant decrease in specific surface area and, consequently, lower workability and slower reactivity. Elert et al. [7] also recommended using mortars prepared with the aged lime putty, which exhibit higher porosity and water-retention capacity. The porosity and water absorption of the dry hydrated lime mortars are greater than those of the lime putty mortars [6,21]. The plasticity of lime putties prepared using seven dry hydrated calcium limes was studied by Klein et al. [22]. The higher plasticity value is related to the higher specific surface area of dry hydrates due to the interaction between the liquid phase and the calcium hydroxide particles. A higher specific surface area is the result of the finer particle size distribution in hydrated lime. On the other hand, Paiva et al. [23] found that lime mortars prepared using dry hydrated lime and left to mature for a week, isolated from atmospheric CO_2, present a thickening behaviour due to the agglomerates' gradual breakdown and the increase in the surface area of particles exposed to the binding of water. The amount of free water was reduced, and the mortars became denser in their fresh state; they showed higher strengths at the age of 90 days and achieved a lower capillary absorption coefficient than the mortars that were not subjected to a maturation process. However, the authors concluded that increasing the rotation time and speed during the mixing process could achieve the same effect.

The use of dry hydrated lime and lime putty as a binder in architectural injection grouts is addressed mainly in combination with pozzolanic additives and/or chemical admixtures [2,24–26]. Relevant experimental studies [2,24–29] show that various key parameters can influence the properties of injection grout in its fresh and hardened state. These include grout composition, water-to-binder ratio, the reactivity of pozzolanic material used, chemical admixture type and content, incorporation of additives such as fibres and hollow glass bubbles, and the mixing and curing process. However, at the age of 90 days, the compressive strength of injection grouts based on dry hydrated lime [2,25] or lime putty [26], with or without pozzolanic constituent and/or superplasticiser, is in the range of 2.10 to 3.13 MPa.

The present study compares and evaluates the fresh and hardened properties of architectural injection grouts with the same composition. The grouts were prepared using three commercially available dry hydrated limes (powders) produced in Slovenia and Switzerland.

2. Materials and Methods

2.1. Materials and Composition

Three commercially available dry hydrated lime types were used to prepare the injection grouts. The limes are classified according to the EN 459-1:2015 standard [30]. The limes of classes CL 70-S (IAK, Kresnice, Slovenia) and CL 90-S (IGM, Zagorje, Slovenia) were produced in Slovenia (denotations SI-CL70 and SI-CL90) and the third, of class CL 90-S (KFN, Netstal, Switzerland), was produced in Switzerland (denotation CH-CL90). A finely ground limestone supplied from Slovenia (CALCIT, Stahovica, Slovenia) was used as a filler. The chemical compositions of the limes and limestone filler, determined by the X-ray fluorescence analysis (Bruker S8 TIGER, Anhovo, Slovenia) according to the EN 196-2:2013 standard [31], are shown in Table 1. Table 2 gives contents of crystalline phases in the three limes and filler, determined by the X-ray powder diffraction (XPert Pro X-ray diffractometer; measurement parameters: Cu-Kα radiation λ = 1.54 Å, exploration range from 20° and 70° 2θ, (University of Ljubljana, Ljubljana, Slovenia). The quantitative phase analysis of the samples was completed using the Rietveld method.

Table 1. Chemical composition of the dry hydrated limes and limestone filler.

Sample	CaO (%)	MgO (%)	Al_2O_3 (%)	Fe_2O_3 (%)	SO_3 (%)	SiO_2 (%)	I.L. (%)
SI-CL70 lime	71.25	2.09	0.60	0.19	0.06	0.79	25.69
SI-CL90 lime	71.01	3.05	0.58	0.20	0.14	2.14	23.38
CH-CL90 lime	74.90	0.40	0.02	0.01	0.02	0.05	25.00
Limestone filler	55.38	0.76	0.15	0.01	0.01	<0.01	44.02

Table 2. Contents of crystalline phases in the powders, obtained by the Rietveld method.

Sample	Portlandite ($Ca(OH)_2$)	Calcite ($CaCO_3$)	Periclase (MgO)	Quartz (SiO_2)	Lime (CaO)	Magnesite $MgCO_4$	Larnite (Ca_2SiO_4)	Dolomite ($CaMg(CO_3)_2$)
SI-CL70 lime	95.8	2.9	0.2			0.3	0.8	
SI-CL90 lime	92.5	1.2	2.3	0.2	0.1	0.4	3.5	
CH-CL90 lime	97.0	3.0						
Limestone filler		95.3						4.7

The results in Tables 1 and 2 show that the CH-CL90 lime contains the highest portlandite content and a higher purity compared to the two limes from Slovenia. The highest content of impurities can be found in the SI-CL90 lime from Slovenia, which also contains the highest MgO content that can negatively influence the slaking process of the quicklime and thus the quality of the hydrated lime [17]. All these hydrated lime products can be characterised as high-calcium lime ($Ca(OH)_2 \geq 90\%$) with traces of $CaCO_3$ ($\leq 6\%$ by mass) [32].

The limestone filler is a very pure calcite powder, composed of 95.3% calcite and 4.7% dolomite (Table 2).

The specific surface area determined by the Blaine method [33] may be one of the hydrated limes' most important physical properties [8]. The pycnometer method was applied to determine particle density [33]. Table 3 shows the Blaine specific surface area and density values for the studied dry hydrated limes and limestone filler. The SI-CL70, SI-CL90, and CH-CL90 limes have different fineness values; they are equal to 9623 cm^2/g, 8767 cm^2/g, and 16198 cm^2/g, respectively. It is evident that the Slovenian limes SI-CL70 and SI-CL90 are much coarser than the CH-CL90 lime from Switzerland, which indicates that Slovenian limes possess lower reactivity. The particle density of a particular lime is well correlated ($R^2 = 0.997$) to its specific surface area. The North America National Lime Association uses hydrated lime particle density to classify it as a high-calcium lime (densities between 2.3 and 2.4 g/cm^3) [34]. According to this criterium, the CH-CL90 lime is the only one that can be classified as a high-calcium hydrate (particle density of 2.34 g/cm^3). The particle densities of hydrated limes used in studies on lime-based mortars and grouts range from approximately 2.2 g/cm^3 [35,36] to 2.47 g/cm^3 [37]. The hydrated lime particles are characterised by their irregular block-like shape, with each particle being a porous cluster of small grains [38]. It can be concluded that a lower density of hydrated lime results from higher porosity of the cluster, where smaller particles present a reduced internal porosity.

Table 3. Particle density and specific surface area of hydrated limes and limestone filler.

Sample	Particle Density (g/cm^3)	Blaine Fineness (cm^2/g)
SI-CL70 lime	2.237	9623
SI-CL90 lime	2.217	8767
CH-CL90 lime	2.343	16,198
Limestone filler	2.764	3194

The limestone filler particles had a density of 2.76 g/cm^3 and water absorption of 0.4%; their maximum size was 100 μm, with 10%, 20%, 50%, and 90% of particles smaller or equal to 3 μm, 9 μm, 15 μm, and 40 μm, respectively.

The composition of grout mixtures in this study is based on the 1:3 (lime: filler) volume ratio composition from the previous study [2]. The volume ratios of the components in [2] were converted to mass ratios to provide identical compositions of the grouts tested. Each grout mixture was composed of 290 g lime, 1030 g filler, and 540 g water (water/binder ratio of 1.86). The grout's adequate consistency and workability were assured using a polycarboxylate ether-based (PCE) super plasticiser. The super plasticiser content was equal to 0.5% of the total mass of solids (lime + filler).

The grout mixtures were prepared with a simple kitchen mixer to simulate the preparation of grout mixtures on a construction site. The mixer was a small hand-held electric whisk with a power of 300 W. The lime and filler were mixed first. This was followed by 70% of the water being added and mixed for 2 min at a low speed (540 rpm). In the last 15 s of the low-speed mixing, the PCE-SP and 30% of the water were added. Finally, each grout was mixed for 3 min at high speed (1200 rpm).

2.2. Methods

Test methods used to evaluate the properties of the grouts in the fresh and hardened state are predominantly the same as described in [28]. Thus, in continuation, a short list of the tests with essential information is given for procedures already described in [28], and a more detailed description is available for the rest of the tests. At least three repetitions of each test were carried out per grout mixture.

The fresh properties of the injection grouts were evaluated as follows:

The mini slump-flow test [39] was used to determine the flow behaviour of the grout mixtures.

According to the modified ASTM C940 standard procedure [40], the bleeding test was carried out. The modification applied reduced the grout volume from 800 $\pm$ 10 mL to 80 $\pm$ 1 mL.

The wet density of the grout was measured according to the modified EN 1015–6:1998 standard procedure [41]. The volume of the sample was reduced from 1000 mL to 10 mL [42].

The water-retaining ability of the grout was evaluated by the prEN 1015–8:1999 standard procedure [43].

The drying shrinkage test with mortar cups was used to determine the reduction in grout volume after hardening [42].

An injectability test with a syringe was used to determine the ability of the grout to fill a capillary network of dry granular materials under pressure [42]. First, 20 mL of the grout was poured into a vertically held syringe that was partially filled with 20 mL of granular material. Subsequently, the pressure was applied to the grout with a plunger. The crushed lime mortar was used as a granular material, with a grain size between 2 and 4 mm, which simulates an approximately 0.3–0.6 mm large crack or void width. After 10 min, the water absorption coefficient of the mortar was 11 kg/(m$^2\sqrt{}$min). The injectability of the grout is classified as the following: easy (E)—if the grout flows through the granular material and out of the syringe tip when pressure is applied; feasible (F)—if the grout flows through the granular material and reaches the tip but does not flow through it; and difficult (DL)—if the grout stops in the granular material before reaching the tip [42]. The penetration distance, measured from the top of the granular material to the level of the grout, is recorded as L in millimetres.

The hardened properties of the gout samples were evaluated at the age of 90 days. The grouts were cast in cylindrical moulds, with a diameter and height equal to 50 mm, and demoulded after 48 h. Curing was executed under controlled ambient conditions (RH 60 $\pm$ 10% and 19 $\pm$ 1 °C) until the test day.

The grouts' dry density and water absorption by capillarity were determined using the EN 1015-10:1999 [44] and EN 1015-18:2004 [45] standard procedures, respectively. The total and capillary porosities were measured in accordance with the Appendix A of Swiss standard SIA 262/ 1:2008 [46]. The compressive test was carried out in accordance with the EN 1015–11:1999/A1:2006 standard [47]. The splitting tensile test followed the ASTM C496/C496 M-1 standard [48]. The compressive and splitting tensile strengths were determined on four specimens per composition. Tests were performed by a Roel Amsler HA 100 servo-hydraulic testing machine (Zwick GmbH & Co. KG, Ulm, Germany), complemented by a load cell with the capacity adjusted to the compressive (25 kN) and splitting tensile (5 kN) strength of the tested specimens.

3. Results and Discussion

3.1. Fresh State Properties

The average values and the corresponding standard deviations of wet density, mini-slump flow, and bleeding after 3 h, and water retention capacity of the SI-70, SI-90, and CH-90 grouts are listed in Table 4. The fresh state properties of the SI-70 and CH-90 grouts are approximately the same. The SI-90 grout, on the other hand, has a higher mini-slump-flow value and increased bleeding compared to the SI-70 and CH-90 grouts. As all constituents in the studied mixtures were of the same mass, the hydrated lime particle density (Table 3) and entrapped air content—as a result of mixing and casting procedures—determined the actual volumes of the lime particles, water, and limestone filler in the grout's unit volume. This is reflected in the fresh grouts' densities. The mini-slump-flow value, which is a measure of flowability and consistency of the fresh grout, is often related to the paste's yield stress τ_0. "Paste" is a generic term for the mixture of binder, water and filler particles that are smaller than 0.1 mm; it can also contain a chemical admixture. The highest mini slump-flow value of 300 mm was measured for the SI-90 mixture, prepared with the coarsest lime. The two finer limes had a 15% lower slump-flow value compared to that of the SI-90 mixture.

Table 4. Wet density (g/cm^3); mini-slump flow (mm); bleeding after 3 h (%); and water retention capacity (%) of the fresh SI and CH mixtures.

Mixture	Wet Density (g/cm^3)	Mini Slump Flow (mm)	Bleeding after 3 h (%)	Water Retention Capacity (%)
SI-70	1.74 ± 0.02	259 ± 16	1.0 ± 0.3	83 ± 1
SI-90	1.74 ± 0.00	300 ± 15	1.5 ± 0.3	82 ± 2
CH-90	1.76 ± 0.02	254 ± 10	0.9 ± 0.4	85 ± 2

Ince et al. [49] observed a loss of mortars' fluidity due to the finer particle size distribution of the hydrated lime. By maintaining the standard Vicat consistency of lime putties, Klein et al. [22] showed that the particle size distribution of dry hydrated lime affected the plasticity of the lime putty, which increased with decreasing lime particles' size. Plasticity is a rheological property that relates to lime putty workability. Both approaches confirmed the vital influence of dry hydrated lime fineness on the fluidity and workability of the fresh lime mortars and grouts. It can be concluded that dry hydrated limes with higher total specific surface areas of particles and aggregates require higher mixing water content to produce a water suspension (hydrated lime putty) with the same consistency. This is related to the water physically adsorbed onto the surface of solids in suspension. When using the mini-slump-flow test method to evaluate the flowability and consistency of lime grout, the relation between water content and relative flow area (R = (slump–flow value/100)^2 − 1) is often considered as a parameter that measures the sensitivity of the grout flowability to increasing water content [39]. The retained water, which provides sufficient particle dispersal for flow to commence, is comprised of the water adsorbed onto solid particles and that which is required to fill the voids in the powder system. A water content higher than the retained water content is needed for a slump–flow value of more than 100 mm. At the constant water content in the grout mixtures, lime particles with a higher surface area physically bind more water, and thus less water is available for the grout to flow. The oriented aggregation of dry hydrated lime particles in the fresh grout could be the reason for the surface area reduction and the change in a grout's workability [10]. These observations explain the similar flowability of the SI-70 and CH-90 grouts, which could result from the finest CH-CL90 lime particles' oriented aggregation in the grout suspension.

Figure 1 shows the highest volume of bleed water accumulating on a particular fresh grout surface at prescribed intervals. After 3 h of testing, the grout mixture SI-90 exhibited a higher average bleeding (1.5%) than mixtures SI-70 and CH-90, where the average bleeding was 1.0 and 0.9%, respectively (Table 4). However, the final average bleeding—measured after 5 h, when the two successive measurements showed no further bleeding—was increased to 1.6% for the SI-90 grout and 1.3% for the CH-90 grout. On the other hand, the mixture SI-70 showed no further changes in bleed water up to the 5th hour of testing (Figure 1). The results of the bleeding test indicate a faster segregation and a subsequently reduced stability of the coarser dry hydrate and a much slower bleeding of the finest dry hydrate (Figure 1). However, the final bleeding appears not to be directly related to the measured Blaine specific surface area of the dry hydrated lime. Again, a possible oriented aggregation of lime particles in the water suspension [10] can explain the observed bleeding behaviour of the CH-90 grout. The smaller $Ca(OH)_2$ particles and their aggregates provide a larger surface area for wetting and bonding with the filler particles, which improves the stability of the mixture and, consequently, its resistance to bleeding [50]. Moreover, additional influencing parameters—such as the packing of particles, oriented $Ca(OH)_2$ aggregates, impurities, etc.—can impact the stability of the grout [7,10]. Oriented aggregates reduce the packing density of solid particles in the suspension and thus increase the retained water content. Increased volume of the retained water filling voids between the particles can be responsible for the time-dependent bleeding observed in the CH-90 grout.

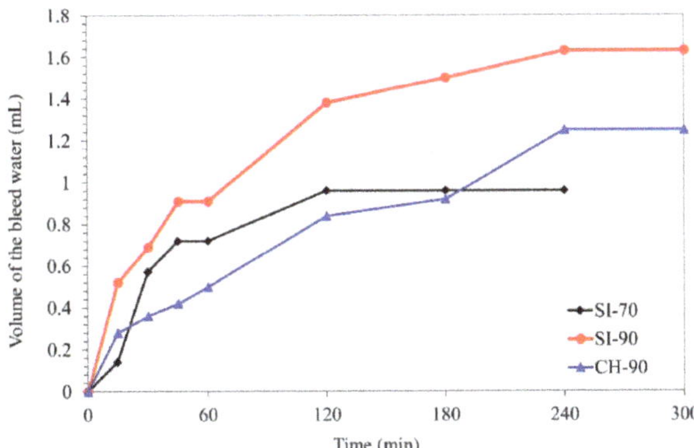

Figure 1. The bleed water volume, measured over five hours for the three grout compositions.

Although there is no significant difference in the water retention capacity of the studied grouts (Table 4), due to the test method's poor repeatability, the test results indicated a lower water retention capacity for the coarser SI-CL90 lime. The highest average water retention capacity of 85% was measured for the CH-90 grout and the lowest (82%) for the SI-90 grout. These results are in line with the results of previous studies [49,51], where it was shown that the water retention capacity increases with a decrease in the particle size of the lime. Ince et al. [50] attributed the observed behaviour to the small radii of curvature of the menisci between finer particles, which could contribute to the increased water-retaining ability of the grout. Biçer-Simsir et al. [42] pointed out that the high water retention capacity of the grout is an essential property for reducing the grout's shrinkage and achieving satisfactory values for other properties.

Table 5 presents the results of the injectability test for the dry and pre-wetted crushed lime mortar used as granular material inside the syringe. The SI-70 and SI-90 grouts' injectability was classified as easy (E) through dry and pre-wetted granular material. The situation was significantly different for the CH-90 mixture, which only penetrated 25 mm of the dry granulate and was thus classified as difficult (D_{25}). When the granulate was pre-wetted, the injectability of the CH-90 improved considerably and was classified as feasible.

Table 5. Results of injectability with a syringe.

Mixture	Crushed Lime Mortar	
	Dry	Wet
SI-70	E	E
SI-90	E	E
CH-90	D_{25}	F

The injectability of the grout is closely related to its fluidity, viscosity, and water retention ability. The mixture SI-90, with the highest fluidity and water release, achieved a level of injectability comparable to that of the SI-70 mixture, which exhibited a lower fluidity and similar water retention. The poorest injectability was observed in the CH-90 grout. Oriented $Ca(OH)_2$ aggregates that could form in the CH-CL90 grout result in higher resistance of the grout to the applied pressure and, consequently, in a loss of injectability of the CH-90 grout. This grout would thus require an addition of water to its composition to reach adequate fresh state properties.

The results of the drying shrinkage test inside mortar cups are presented in Table 6. Figure 2 shows the volume change for the mixtures SI-70, SI-90 and CH-90 after drying in the dry or pre-wetted mortar cups. The SI-70 and SI-90 grouts formed an approximately 0.2 mm thick separation ring (a continuous circular gap between the cup and the grout in Figure 2) between the grout and mortar cup in the dry and prewetted cups. A much thicker separation ring of 0.4 mm formed in the CH-90 grout, again in dry and pre-wetted mortar cups. The SI-90 mixture also formed cracks in the grout—small cracks with a maximum width of 0.1 mm close to the cup's wall (dry mortar cups) and extensive cracks of 0.2 mm width (pre-wetted mortar cups). The mixtures SI-70 and CH-90 did not form cracks in the grout when dry mortar cups were used. On the other hand, extensive cracks of 0.3 mm were formed in the CH-90 grout's pre-wetted cups.

Table 6. Drying shrinkage in dry and pre-wetted mortar cups.

Mixture	Dry Mortar Cup		Pre-Wetted Mortar Cup	
	Separation Size (mm)	Crack Size (mm)	Separation Size (mm)	Crack Size (mm)
SI-70	0.2	0	0.2	0.1
SI-90	0.2	0.1	0.2	0.2
CH-90	0.4	0	0.4	0.3

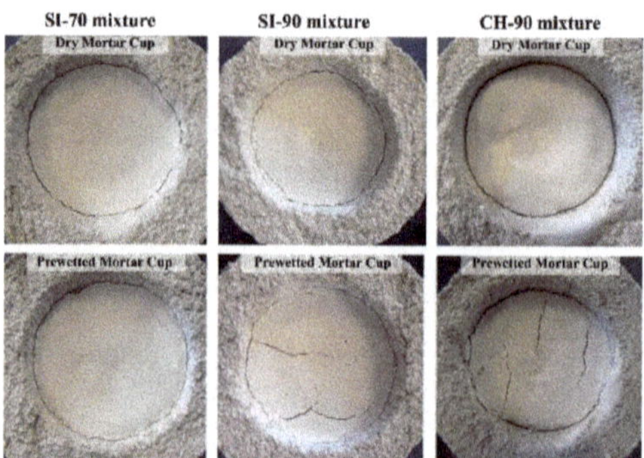

Figure 2. Grout mixtures SI and CH after drying in mortar cups.

The mortar cup shrinkage test simulates the absorption effect of porous mortar and demonstrates the grout's ability to retain moisture inside its structure when exposed to dry for an extended period (24 h or more). The obtained test results indicate that the SI-70 grout possesses the best resistance to the suction of porous mortar and air drying and thus the best volume stability. The lowest volume stability was obtained for the CH-90 grout with relatively high final bleeding and the lowest injectability/penetrability, resulting in a lower bond strength between the grout and the mortar cup. The SI-90 grout's volume stability was only slightly better than that of the CH-90 grout.

Carbonation shrinkage is another mechanism that influences the hydrated lime grout shrinkage. As the mechanism is basically the same as in the case of drying shrinkage, it is often considered part of the drying shrinkage. Swenson and Sereda [52] suggested that carbonation shrinkage occurs due to a gradient of moisture content within the $CaCO_3$ passivation layer around noncarbonated portlandite; it is the highest when the relative humidity of the air is approximately 50%. They also indicated that the lime binder's rate and degree of portlandite carbonation and carbonation shrinkage increase with the lime fineness [52]. These findings explain the highest shrinkage deformations and cracking observed for the CH-90 lime grout.

From the fresh properties' point of view, the SI-70 grout showed the best overall performance appropriate for the practical application.

3.2. Properties in Hardened State

Table 7 reports the results of dry density, total and capillary porosity, and water absorption coefficient after 24 h and 10 min for the SI-70, SI-90, and CH-90 hardened grouts at the age of 90 days. The average dry density of the grouts ranged from 1.45 to 1.51 g/cm^3. Higher values measured for the SI-70 and CH-90 grouts (1.50 g/cm^3 and 1.51 g/cm^3) can be predominantly related to the carbonation effect due to a higher portlandite ($Ca(OH)_2$) content (Table 2). They can also be linked to the faster reactivity of smaller lime particles that form more calcite ($CaCO_3$) in 90 days compared to the coarser SI-CL90 lime with the lowest portlandite content, as explained in [6,52].

Table 7. Density of hardened state (g/cm^3); total and capillary porosity (%); and water absorption coefficient after 24 h (W_{24}) and 10 min (W_{10}) of the SI and CH mixtures.

Mixture	Dry Density (g/cm^3)	Total Porosity (%)	Capillary Porosity (%)	Content of Air Pores (%)	W24 (kg/(m$^2\sqrt{}$min))	W10 (kg/(m$^2\sqrt{}$min))
SI-70	1.51 ± 0.01	43 ± 0	37 ± 0	6	0.43 ± 0.02	1.01
SI-90	1.45 ± 0.01	40 ± 1	38 ± 1	2	0.46 ± 0.01	2.59
CH-90	1.50 ± 0.02	44 ± 0	38 ± 0	6	0.43 ± 0.03	3.34

The total porosity results show that the denser SI-70 and CH-90 mixtures have a higher total porosity (43–44%) compared to the SI-90 mixture which exhibits the lowest dry density—predominantly due to a higher entrapped air content (6%) inside the specimens. However, all three mixtures showed approximately the same capillary porosity (37–38%), much smaller than the initial volume of water in the grouts (about 52%). The water absorption of the limestone filler (0.4%) resulted in only 4 g of water absorbed by the filler for all three compositions. Therefore, it did not have an important influence on the capillary porosity formed due to evaporation of the excess of kneading water from residual spaces previously occupied [53]. Due to capillary forces, the evaporation of water results in the shrinkage of lime grout and thus in about 3 to 5% lower initial porosity of the dry noncarbonated grout compared to the volume percentage of kneading water [54]. Moreover, the mercury intrusion porosimetry of fully carbonated lime showed an open porosity that is about 10 to 12% lower than that of the noncarbonated grout [36]. These findings are in line with results in Table 7, where capillary porosity of about 38% is 14% lower than the volume percentage of kneading water in the lime grouts (52%).

From Table 7 and Figure 3, it can be seen that all three grouts absorbed about the same water content after 24 h, which is consistent with their capillary porosities. However, there is an important difference in the initial capillary sorptivity of the grouts, determined after 10 min; the mixture CH-90 absorbed water much faster (10.6 kg/m^2) than the SI-90 mixture (8.2 kg/m^2). Considerably slower initial sorptivity of 3.2 kg/m^2 was obtained for the SI-70 grout. As the capillary sorptivity force (as a pressure difference) increases when the pore diameter drops [55], we can conclude that the most refined capillary pore system was formed in the CH-90 grout and the coarsest in the SI-70 grout.

Figure 4 shows the compressive and splitting tensile strengths of the lime grouts at 90 days. The failure modes for compressive and splitting tensile strengths are shown in Figure 5. The average compressive and tensile strengths for the SI-70 and SI-90 grouts are in the range expected for materials with a pure hydrated lime binder—approximately 2.8 MPa and 0.16–0.34 MPa, respectively. The compressive strengths of hydrated lime-based mortars and injection grouts range from 0.2 to 4.5 MPa, with tensile strengths of between 0.07 and 1.5 MPa [2,6,21,51,56,57]. On the other hand, the CH-90 grout reached unexpectedly high compressive and tensile strengths equal to 8.1 and 0.76 MPa, respectively. These values are about 3 times higher than those seen in grouts prepared with Slovenian limes. The CH-90 grout compressive strength lies within the interval of 1.0 to 14.0 MPa, which represents the compressive strengths reported for natural hydraulic lime or lime-based mortars with pozzolanic additives [2,25,26,51,56,58,59]. Moreover, the combination of hydrated lime

CH-CL90 and PCE superplasticiser leads to compressive and tensile strengths close to the upper limits of the intervals 3–8 MPa and 0.3–1.2 MPa, respectively. These intervals were reported by Ferragni et al. [60] for hydraulic lime architectural injection grouts.

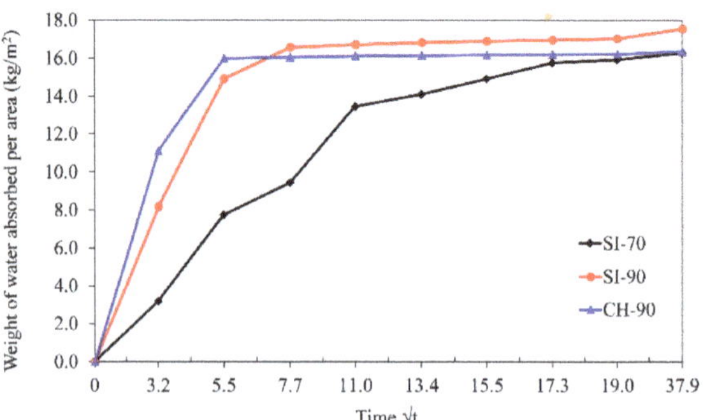

Figure 3. Capillary water absorption curves of the three injection grouts.

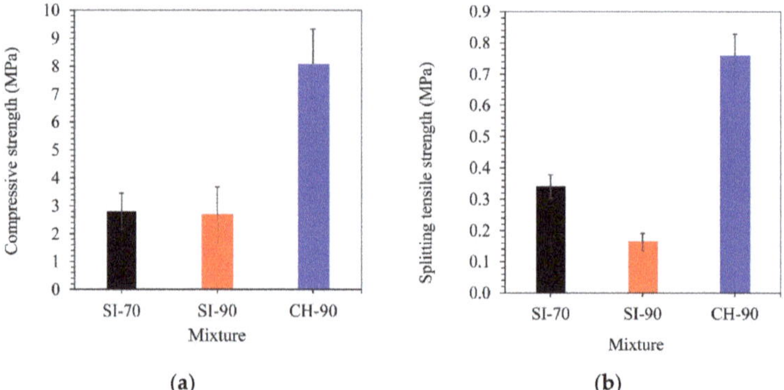

Figure 4. (**a**) Compressive and (**b**) splitting tensile strength for SI and CH mixtures.

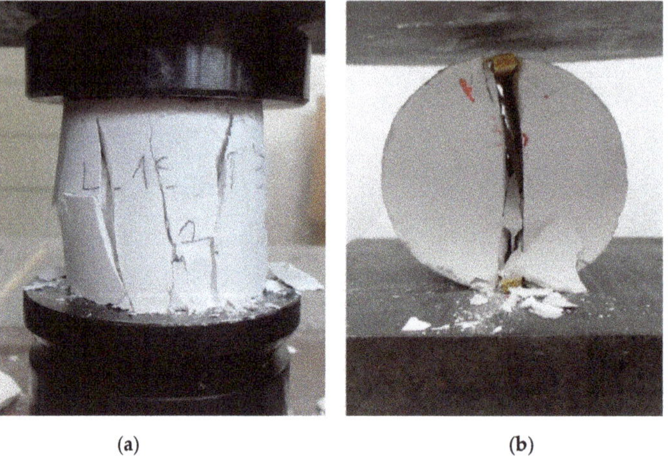

Figure 5. Examples of failure modes: (**a**) Compressive strength and (**b**) splitting tensile strength for SI and CH mixtures.

Dry hydrated lime properties that govern the grout strengths include impurities (Table 1), portlandite content (Table 2), and specific surface area of the lime particles. The smaller particles are more reactive compared to coarser particles (Table 3). The CH-CL90 lime is very pure, with the highest portlandite content and the highest specific surface area. Therefore, its carbonation rate and final volume of $CaCO_3$ in the grout should be the highest among studied limes [52]. This conclusion is also in line with nanoscience findings; application of so-called "nano-lime" showed that reducing the size of calcium hydroxide particles leads to higher suspension stability. Moreover, smaller particles have a higher reactivity to form a calcite crystal structure, which improves the cohesion and mechanical strength of the wall paintings' substrate layers [61]. Incorporating the PCE super plasticiser into the grout composition, combined with high-speed mixing (1200 rpm), disintegrates the lime and limestone filler particles agglomerates in the suspension, thus increasing the surface available for carbonation in the hardened grout. Simultaneously, a more homogeneous and compact hardened grout with much finer capillary pores is formed in the case of the CH-90 grout (Table 7). Fernandez et al. [62] and Silva et al. [63] showed that PCE superplasticiser in the lime mortar results in a strong reduction in macropores (in the diameter range of 1–10 μm) and a more compact, homogeneous and continuous mortar matrix. Compared to the lime mortar with the same water-to-binder ratio and without the PCE super plasticiser, compressive and flexural strengths increase considerably, especially at the mortar's age of 6 months and one year. However, the porosity reduction can obstruct an adequate CO_2 flow through lime-based mortars and grouts. Therefore, an open homogeneous microstructure supports a better and efficient carbonation rate and promotes a well-developed carbonate morphology [64,65]. Padovnik et al. [2] showed that the PCE superplasticiser was very effective at lowering the amount of water in hydrated lime based grout, which increased its mechanical strengths. According to Van Balen [66], one of the mechanisms that control the carbonation rate can be the dissolution of portlandite at the water-adsorbed surface; therefore, the carbonation rate depends on the hydrated lime specific surface area. Paiva et al. [23] stated that intense, rapid mixing might break up the hydrated lime particle aggregates, causing a reduction in the capillary water absorption and a mechanical strength increase.

In addition to the beneficial properties of the CH-CL90 lime in relation to strength properties of the architectural grout (as discussed earlier), there must be other influencing parameters responsible for such high compressive and tensile strengths of the CH-90 grout. The load-bearing capacity of the crystal lattice, which makes up the solid mass of the hardened grout, must be significantly higher for the CH-90 grout than the SI-70 and SI-90 grouts. It appears that the CH-CL90 lime (Table 3) can bind limestone filler particles more efficiently compared to the Slovenian limes and can form a composite solid mass with highly increased strength. The absence or reduced content of pre-existing cracks due to shrinkage, combined with the refined capillary porosity, could present another parameter responsible for the high strengths observed. In [67], authors showed that pre-existing cracks and their inclination angles to nearby entrapped air pores influence the loadbearing behaviour of the mortar.

4. Conclusions

This study addresses the influence of three dry hydrated limes on architectural injection grouts' fresh and hardened properties. The main differences between the hydrated limes could be found in their chemical and mineralogical compositions, particle density, and specific surface area values, which indicate that the industrial production and limestone composition were not the same for the studied limes. The grouts' composition was identical regarding mass ratios between lime, water, limestone filler and PCE super plasticiser. The PCE super plasticiser, combined with high-speed grout mixing (1200 rpm), disintegrates the lime and limestone filler particle agglomerates in the suspension and thus influences the fresh grout properties and the increased surface available for carbonation in the hardened grout. Additionally, oriented $Ca(OH)_2$ aggregates can gradually form in the fresh grout and change the fresh grout properties. The main conclusions of this study can be summarised as follows.

The mixtures' consistency, stability, water-retaining ability, and injectability were influenced by the specific surface area of the lime binder in the grout, governed by its particle and agglomerate size, distribution, and shape. Although the highest Bleine specific surface area was measured for the dry hydrated CH-CL90 lime, the CH-90 grout's fresh properties were approximately the same compared to the SI-70 grout prepared with lower Blaine specific surface area dry hydrate. The only exception was injectability, which was the poorest for the CH-90 grout. Possible oriented aggregation of the CH-90 lime particles in water suspension can explain the observed behaviour. The SI-70 grout showed the most balanced fresh properties among the three grouts, most probably due to the appropriate combination of lime particles specific surface area and solid particles (lime and limestone filler particles) packing property.

The shrinkage test in mortar cups revealed an important influence of carbonation shrinkage on the CH-90 grout. The rate and degree of portlandite carbonation were the highest among the studied grouts, which resulted in extensive carbonation shrinkage. As a consequence, poor volume stability of the CH-90 grout was observed in the study.

The three grouts showed the same capillary porosity, which is consistent with the same water content of the fresh mixtures and the same water content absorbed in the hardened state. On the other hand, faster carbonation and a higher volume of $CaCO_3$ in the CH-CL90 lime did not reflect in the CH-90 grout open porosity. There was an important difference in the initial capillary sorptivity of the grouts (determined after 10 min) that indicated the formation of the most refined capillary pore system in the CH-90 grout, while the coarsest capillary pore system appeared in the SI-70 grout.

The CH-CL90 lime resulted in an unexpectedly high compressive and tensile strength of the CH-90 grout at 90 days, 8.1 and 0.76 MPa, respectively. These strengths are about 3-times the strengths measured for the SI-70 and SI-90 grouts. They are close to the upper limit of the intervals reported by Ferragni et al. [37] for hydraulic lime architectural injection grouts. The main parameters that increase the lime grout strengths are the highest portlandite content and specific surface area of the lime particles and more homogeneous and compact hardened grout with the finer capillary network that still provides efficient carbonation. Additionally, the much higher particle density of the CH-CL90 lime and its ability to bind limestone filler particles efficiently can form a composite solid mass with highly increased load-bearing ability. Further work is required to evaluate the microstructure of limes and grout mixtures and provide answers to the remaining questions.

The use of dry hydrated limes is increasingly recommended, as they have a lower mass volume and facilitate a range of mix proportions and reproducibility of the properties in a fresh and hardened state. The application of the PCE super plasticiser can significantly reduce the kneading water in the dry hydrated lime-based mixtures, which, in turn, compensates for the deficiencies compared to the hydrated lime putty mixtures. Moreover, in the case of architectural injection grouts, the quantities of dry hydrated lime are low enough to use the best possible product available on the international market. In Slovenia, for example, products available in other European countries should be reviewed. The Swiss product used in this study showed that commercially available dry hydrated limes could possess highly diverse properties and provide high enough strengths for different applications, especially in historical structures' repair and strengthening.

Author Contributions: Conceptualisation, A.P. and V.B.-B.; methodology, A.P. and V.B.-B.; validation, V.B.-B.; formal analysis, A.P. and V.B.-B.; investigation, A.P.; resources, V.B.-B.; writing—original draft preparation, A.P.; writing—review and editing, V.B.-B.; visualisation, A.P.; supervision, V.B.-B.; project administration, A.P.; and funding acquisition, A.P. and V.B.-B. All authors have read and agreed to the published version of the manuscript.

Funding: This research was funded by the project C3330-17-529030 "Raziskovalci-2.0-UL-FGG-529030". Ministry of Education, Science and Sport of the Republic of Slovenia has approved the project. The investment is co-financed by the Republic of Slovenia and the European Union under the

European Regional Development Fund. The authors acknowledge also the financial support from the Slovenian Research Agency through the research core funding No. P2-0185.

Institutional Review Board Statement: Not applicable.

Informed Consent Statement: Not applicable.

Data Availability Statement: Data sharing is not applicable to this article.

Conflicts of Interest: The authors declare no conflict of interest.

References

1. *Venice Charter. International Charter for the Conservation and Restoration of Monuments and Sites;* Committee for Drafting the International Charter for the Conservation and Restoration of Monuments: Venice, Italy, 1964.
2. Padovnik, A.; Piqué, F.; Jornet, A.; Bokan-Bosiljkov, V. Injection grouts for the re-attachment of architectural surfaces with historic value—Measures to improve the properties of hydrated lime grouts in Slovenia. *Int. J. Archit. Herit. Conserv. Anal. Restor.* **2016**, *10*, 993–1007. [CrossRef]
3. Padovnik, A. Consolidation of Detached Plaster Layers of Mural Paintings with Non-Structural Grouting. Ph.D. Thesis, Faculty of Civil and Geodetic Engineering, University of Ljubljana, Ljubljana, Slovenia, 26 May 2016.
4. González-Sánchez, J.F.; Taşcı, B.; Fernández, J.M.; Navarro-Blasco, Í.; Alvarez, J.I. Combination of polymeric superplasticizers, water repellents and pozzolanic agents to improve air lime-based grouts for historic masonry repair. *Polymers* **2020**, *12*, 887. [CrossRef] [PubMed]
5. Zacharopoulou, G. Interpreting chemistry and technology of lime binders and implementing it in the conservation field. *Conserv. Património* **2009**, *10*, 41–53. [CrossRef]
6. Aggelakopoulou, E.; Bakolas, A.; Moropoulou, A. Lime putty versus hydrated lime powder: Physicochemical and mechanical characteristics of lime based mortars. *Constr. Build. Mater.* **2019**, *225*, 633–641. [CrossRef]
7. Elert, K.; Rodriguez-Navarro, C.; Pardo, E.S.; Hansen, E.; Cazalla, O. Lime mortars for the conservation of historic buildings. *Stud. Conserv.* **2002**, *47*, 62–75. [CrossRef]
8. Boynton, R.S. *Chemistry and Technology of Lime and Limestone;* John Wiley & Sons: Hoboken, NJ, USA, 1966.
9. Kemperl, J.; Maček, J. Precipitation of calcium carbonate from hydrated lime of variable reactivity, granulation and optical properties. *Int. J. Miner. Process.* **2009**, *93*, 84–88. [CrossRef]
10. Rodriguez-Navarro, C.; Ruiz-Agudo, E.; Ortega-Huertas, M.; Hansen, E. Nanostructure and irreversible colloidal behavior of $Ca(OH)_2$: Implications in cultural heritage conservation. *Langmuir* **2005**, *21*, 10948–10957. [CrossRef]
11. Whitman, W.G.; Davis, G.H.B. The hydration of lime. *Ind. Eng. Chem.* **1926**, *18*, 118–120. [CrossRef]
12. Adams, F.W. Effect of particle size on the hydration of lime. *Ind. Eng. Chem.* **1927**, *19*, 589–591. [CrossRef]
13. Navrátilová, E.; Tihlaříková, E.; Neděla, V.; Rovnaníková, P.; Pavlík, J. Effect of the preparation of lime putties on their properties. *Sci. Rep.* **2017**, *7*, 17260. [CrossRef]
14. Ruiz-Agudo, E.; Rodriguez-Navarro, C. Microstructure and rheology of lime putty. *Langmuir* **2010**, *26*, 3868–3877. [CrossRef] [PubMed]
15. Ghiasi, M.; Abdollahy, M.; Khalesi, M. Removal of iron from milk of lime to produce pure precipitated calcium carbonate. *Sep. Sci. Technol.* **2020**, *55*, 1425–1435. [CrossRef]
16. Schweigert, K. The Effects of Impurities on Lime Quality. Available online: http://www.penta.net/wp-content/uploads/2018/10/Effects-of-Impurities-on-Lime-Quality.pdf (accessed on 10 September 2021).
17. Kang, S.-H.; Lee, S.-O.; Hong, S.-G.; Kwon, Y.-H. Historical and scientific investigations into the use of hydraulic lime in korea and preventive conservation of historic masonry structures. *Sustainability* **2019**, *11*, 5169. [CrossRef]
18. Eckel, E.C. *Cements, Limes and Plasters: Their Materials, Manufacture, and Properties*, 1st ed.; John Wiley & Sons: New York, NY, USA, 1905.
19. Holmes, S.; Wingate, M. *Building with Lime: A Practical Introduction*, 2nd ed.; Practical Action Publishing: Rugby, UK, 2002.
20. Mendoza, E.; Alonso-Guzman, E.; Ruvalcaba-Sil, J.; Sánchez Calvillo, A.; Martinez Molina, W.; García, H.; Bedolla-Arroyo, J.; Becerra-Santacruz, H.; Borrego, J. Compressive strength and ultrasonic pulse velocity of mortars and pastes, elaborated with slaked lime and high purity hydrated lime, for restoration works in México. *Key Eng. Mater.* **2020**, *862*, 51–55. [CrossRef]
21. Faria, P.; Henriques, F.; Rato, V. Comparative evaluation of lime mortars for architectural conservation. *J. Cult. Herit.* **2008**, *9*, 338–346. [CrossRef]
22. Klein, D.; Haas, S.; Schmidt, S.-O.; Middendorf, B. Morphological and chemical influence of calcium hydroxide on the plasticity of lime based mortars. In *Historic Mortars;* Válek, J., Hughes, J.J., Groot, C.J.W.P., Eds.; Springer: Dordrecht, The Netherlands, 2012; pp. 319–328.
23. Paiva, H.; Velosa, A.; Veiga, R.; Ferreira, V.M. Effect of maturation time on the fresh and hardened properties of an air lime mortar. *Cem. Concr. Res.* **2010**, *40*, 447–451. [CrossRef]
24. Azeiteiro, L.C.; Velosa, A.; Paiva, H.; Mantas, P.Q.; Ferreira, V.M.; Veiga, R. Development of grouts for consolidation of old renders. *Constr. Build. Mater.* **2014**, *50*, 352–360. [CrossRef]

25. Pachta, V.; Papayianni, I.; Spyriliotis, T. Assessment of laboratory and field testing methods in lime-based grouts for the consolidation of architectural surfaces. *Int. J. Archit. Herit.* **2020**, *14*, 1098–1105. [CrossRef]
26. Pasian, C.; Piqué, F.; Jornet, A.; Cather, S. A 'Sandwich' specimen preparation and testing procedure for the evaluation of non-structural injection grouts for the re-adhesion of historic plasters. *Int. J. Archit. Herit.* **2021**, *15*, 455–466. [CrossRef]
27. Luso, E.; Lourenço, P.B. Experimental laboratory design of lime based grouts for masonry consolidation. *Int. J. Archit. Herit.* **2017**, *11*, 1143–1152. [CrossRef]
28. Padovnik, A.; Bokan-Bosiljkov, V. Effect of ultralight filler on the properties of hydrated lime injection grout for the consolidation of detached historic decorative plasters. *Materials* **2020**, *13*, 3360. [CrossRef]
29. Pachta, V. The role of glass additives in the properties of lime-based grouts. *Heritage* **2021**, *4*, 906–916. [CrossRef]
30. UNI EN 459-1. *Building Lime—Part 1: Definitions, Specifications and Conformity Criteria*; CEN: Brussels, Belgium, 2010.
31. EN 196-2. *Method of Testing Cement—Part 2: Chemical Analysis of Cement*; CEN: Brussels, Belgium, 2013.
32. Cizer, Ö.; Van Balen, K.; Elsen, J.; Van Gemert, D. Real-time investigation of reaction rate and mineral phase modifications of lime carbonation. *Constr. Build. Mater.* **2012**, *35*, 741–751. [CrossRef]
33. EN 196-6. *Methods of Testing Cement—Part 6: Determination of Fineness*; CEN: Brussels, Belgium, 2018.
34. Fact Sheet, Properties of Typical Commercial Lime Products. Available online: https://www.lime.org/documents/lime_basics/lime-physical-chemical.pdf (accessed on 25 June 2021).
35. Papayianni, I.; Stefanidou, M. Strength-porosity relationships in lime-pozzolan mortars. *Constr. Build. Mater.* **2006**, *20*, 700–705. [CrossRef]
36. Arandigoyen, M.; Bicer-Simsir, B.; Alvarez, J.I.; Lange, D.A. Variation of microstructure with carbonation in lime and blended pastes. *Appl. Surf. Sci.* **2006**, *252*, 7562–7571. [CrossRef]
37. Pachta, V.; Goulas, D. Fresh and hardened state properties of fiber reinforced lime-based grouts. *Constr. Build. Mater.* **2020**, *261*, 119818. [CrossRef]
38. Beruto, D.T.; Botter, R. Liquid-like H2O adsorption layers to catalyse the $Ca(OH)_2/CO_2$ solid–gas reaction and to form a non-protective solid product layer at 20 °C. *J. Eur. Ceram. Soc.* **2000**, *20*, 497–503. [CrossRef]
39. Domone, P.; Hsi-Wen, C. Testing of binders for high performance concrete. *Cem. Concr. Res.* **1997**, *27*, 1141–1147. [CrossRef]
40. ASTM C940-16. *Standard Test Method for Expansion and Bleeding of Freshly Mixed Grouts for Preplaced-Aggregate Concrete in the Laboratory*; ASTM International: Harrisburg, PA, USA, 2016.
41. EN 1015-6. *Methods of Test for Mortar for Masonry—Part 6: Determination of Bulk Density of Fresh Mortar*; CEN: Brussels, Belgium, 1999.
42. Biçer-Şimşir, B.; Rainer, H.L. *Evaluation of Lime-Based Hydraulic Injection Grouts for the Conservation of Architectural Surfaces: A Manual of Laboratory and Field Test Methods*; Getty Conservation Institute: Los Angeles, CA, USA, 2013; ISBN 978-1-937433-15-4.
43. PrEN 1015-8. *Methods of Test for Mortar for Masonry—Part 8: Determination of Water Retentivity of Fresh Mortar*; CEN: Brussels, Belgium, 1999.
44. EN 1015-10. *Methods of Test for Mortar for Masonry—Part 10: Determination of Dry Bulk Density of Hardened Mortar*; CEN: Brussels, Belgium, 1999.
45. SIST EN 1015-18. *Methods of Test for Mortar for Masonry—Part 18: Determination of Water Absorption Coefficient Due to Capillary Action of Hardened Mortar*; CEN: Brussels, Belgium, 2004.
46. SIA 262/1. *Construction En Béton-Spécifications Complémentaires, Appendix A*; SIA: Zurich, Switzerland, 2003.
47. EN 1015-11. *Methods of Test for Mortar for Masonry—Part 11: Determination of Flexural and Compressive Strength of Hardened Mortar*; CEN: Brussels, Belgium, 1999.
48. ASTM C496/C496M-1. *Standard Test Method for Splitting Tensile Strength of Cylindrical Concrete Specimens*; ASTM International: Harrisburg, PA, USA, 2004.
49. Ince, C.; Ozturk, Y.; Carter, M.A.; Wilson, M.A. The influence of supplementary cementing materials on water retaining characteristics of hydrated lime and cement mortars in masonry construction. *Mater. Struct.* **2014**, *47*, 493–501. [CrossRef]
50. Ince, C.; Carter, M.A.; Wilson, M.A.; Collier, N.C.; El-Turki, A.; Ball, R.J.; Allen, G.C. Factors affecting the water retaining characteristics of lime and cement mortars in the freshly-mixed state. *Mater. Struct.* **2011**, *44*, 509–516. [CrossRef]
51. Jornet, A.; Mosca, C.; Cavallo, G.; Corredig, G. Comparison between traditional, lime based, and industrial, dry mortars. In *Historic Mortars*; Válek, J., Hughes, J.J., Groot, C.J.W.P., Eds.; Springer: Dordrecht, The Netherlands, 2012; pp. 227–237.
52. Swenson, E.G.; Sereda, P.J. Mechanism of the carbonatation shrinkage of lime and hydrated cement. *J. Appl. Chem.* **1968**, *18*, 111–117. [CrossRef]
53. Sahu, S.; Badger, S.; Thaulow, N.; Lee, R.J. Determination of water–cement ratio of hardened concrete by scanning electron microscopy. *Scanning Electron Microsc. Cem. Concr.* **2004**, *26*, 987–992. [CrossRef]
54. Arandigoyen, M.; Bernal, J.L.P.; López, M.A.B.; Alvarez, J.I. Lime-pastes with different kneading water: Pore structure and capillary porosity. *Appl. Surf. Sci.* **2005**, *252*, 1449–1459. [CrossRef]
55. Martys, N.S.; Ferraris, C.F. Capillary transport in mortars and concrete. *Cem. Concr. Res.* **1997**, *27*, 747–760. [CrossRef]
56. Pavía, S.; Aly, M. Influence of aggregate and supplementary cementitious materials on the properties of hydrated lime (CL90s) mortars. *Mater. Constr.* **2016**, *66*, e104. [CrossRef]
57. Veiga, R. Air lime mortars: What else do we need to know to apply them in conservation and rehabilitation interventions? A review. *Constr. Buld. Mater.* **2017**, *157*, 132–140. [CrossRef]

58. Papayianni, I.; Pachta, V. Experimental study on the performance of lime-based grouts used in consolidating historic masonries. *Mater. Struct.* **2015**, *48*, 2111–2121. [CrossRef]
59. Vavričuk, A.; Bokan-Bosiljkov, V.; Kramar, S. The influence of metakaolin on the properties of natural hydraulic lime-based grouts for historic masonry repair. *Constr. Build. Mater.* **2018**, *172*, 706–716. [CrossRef]
60. Ferragni, D.; Forti, M.; Malliet, J.; Mora, P.; Teutonico, J.M.; Torraca, G. Injection Grouting of mural paintings and mosaics. *Stud. Conserv.* **1984**, *29*, 110–116. [CrossRef]
61. Baglioni, P.; Chelazzi, D.; Giorgi, R. *Nanotechnologies in the Conservation of Cultural Heritage, A Compendium of Materials and Techniques*; Springer Science and Business Media: New York, NY, USA, 2015; pp. 1–144. ISBN 978-94-017-9302-5.
62. Fernández, J.M.; Duran, A.; Navarro-Blasco, I.; Lanas, J.; Sirera, R.; Alvarez, J.I. Influence of nanosilica and a polycarboxylate ether superplasticizer on the performance of lime mortars. *Cem. Concr. Res.* **2013**, *43*, 12–24. [CrossRef]
63. Silva, B.; Ferreira Pinto, A.P.; Gomes, A.; Candeias, A. Fresh and hardened state behaviour of aerial lime mortars with superplasticizer. *Constr. Build. Mater.* **2019**, *225*, 1127–1139. [CrossRef]
64. Lanas, J.; Alvarez, J.I. Masonry repair lime-based mortars: Factors affecting the mechanical behavior. *Cem. Concr. Res.* **2003**, *33*, 1867–1876. [CrossRef]
65. Zacharopoulou, G. The Effect of wet slaked lime putty and putty prepared from dry hydrate on the strength of lime mortars. In *2nd Conference on Historic Mortars–HMC 2010 and RILEM TC 203-RHM Final Workshop*; Válek, J., Groot, C., Hughes, J.J., Eds.; RILEM Publications SARL Publication: Prague, Czech Republic, 2010; pp. 1283–1291.
66. Van Balen, K. Carbonation reaction of lime, kinetics at ambient temperature. *Cem. Concr. Res.* **2005**, *35*, 647–657. [CrossRef]
67. Zhou, S.; Zhu, H.; Yan, Z.; Ju, J.W.; Zhang, L. A Micromechanical study of the breakage mechanism of microcapsules in concrete using PFC2D. *Constr. Build. Mater.* **2016**, *115*, 452–463. [CrossRef]

Article

Service Life Evaluation for RC Sewer Structure Repaired with Bacteria Mixed Coating: Through Probabilistic and Deterministic Method

Hyun-Sub Yoon [1], Keun-Hyeok Yang [1], Kwang-Myong Lee [2] and Seung-Jun Kwon [3,*]

1 Department of Architectural Engineering, Kyonggi University, Suwon 16227, Korea; lonsohs@naver.com (H.-S.Y.); yangkh@kgu.ac.kr (K.-H.Y.)
2 Department of Civil, Architectural and Environmental System Engineering, Sungkyunkwan University, Suwon 16419, Korea; leekm79@skku.edu
3 Department of Civil and Environmental Engineering, Hannam University, Daejeon 34430, Korea
* Correspondence: jjuni98@hannam.ac.kr

Abstract: Since a concrete structure exposed to a sulfate environment is subject to surface ion ingress that yields cracking due to concrete swelling, its service life evaluation with an engineering modeling is very important. In this paper, cementitious repair materials containing bacteria, *Rhodobacter capsulatus,* and porous spores for immobilization were developed, and the service life of RC (Reinforced Concrete) structures with a developed bacteria-coating was evaluated through deterministic and probabilistic methods. Design parameters such protective coating thickness, diffusion coefficient, surface roughness, and exterior sulfate ion concentration were considered, and the service life was evaluated with the changing mean and coefficient of variation (COV) of each factor. From service life evaluation, more conservative results were evaluated with the probabilistic method than the deterministic method, and as a result of the analysis, coating thickness and surface roughness were derived as key design parameters for ensuring service life. In an environment exposed to an exterior sulfate concentration of 200 ppm, using the deterministic method, the service life was 17.3 years without repair, 19.7 years with normal repair mortar, and 29.6 years with the application of bacteria-coating. Additionally, when the probabilistic method is applied in the same environment, the service life was changed to 9.2–16.0 years, 10.5–18.2 years, and 15.4–27.4 years, respectively, depending on the variation of design parameters. The developed bacteria-coating technique showed a 1.47–1.50 times higher service life than the application of normal repair mortar, and the effect was much improved when it had a low COV of around 0.1.

Keywords: *Rhodobacter capsulatus*; sewage concrete; surface coating; sulfate ion; service life; cover depth

Citation: Yoon, H.-S.; Yang, K.-H.; Lee, K.-M.; Kwon, S.-J. Service Life Evaluation for RC Sewer Structure Repaired with Bacteria Mixed Coating: Through Probabilistic and Deterministic Method. *Materials* **2021**, *14*, 5424. https://doi.org/10.3390/ma14185424

Academic Editor: Angelo Marcello Tarantino

Received: 24 August 2021
Accepted: 15 September 2021
Published: 19 September 2021

1. Introduction

RC (Reinforced Concrete) sewage structures are a life-line system for the public life of users and require periodic repair and maintenance due to its deterioration. In general, RC structures are commonly used for sewage drainage systems, but the ingress of sulfate ions results in concrete surface layer deterioration, which yields reinforcement and leakage in the joints, thus incurring a large amount of cost to address these problems [1,2]. In the case of sulfate ingress, unlike the case of chloride attack and carbonation, cracking and disintegration from concrete expansion on the surface are major problems.

As for the bio-chemical deterioration of concrete, with the identification of Thiobacillus thiooxidans (the sulfate reducing bacteria), concrete surface erosion and the deterioration of internal hydrates have been reported [3]. Unlike the water supply pipes, the municipal wastewater flowing into the sewer has varying volume and velocity of inflow, and discharge in the pipe, and therefore, slime and sludge are usually deposited at the bottom of the duct. Anaerobic bacteria present in the depositions generate a large amount of hydrogen sulfide gas by reducing sulfate ions (SO_4^{2-}) to hydrogen sulfide (H_2S) in the process of decomposing and consuming the deposited organic matter required as nutrients for growth [4]. This is due to the process of anaerobic respiration, in which anaerobic bacteria oxidize organic matter by using oxygen bonded with sulfur (SO_4^{2-}) instead of pure oxygen (O_2) for the purpose of protein synthesis and the energy acquisition necessary for growth. In this process, hydrogen sulfide is generated, which has long been considered as the main deterioration factor of sewage facilities [5,6]. Several studies have started for the evaluation of concrete degradation due to sulfuric acid and hydrogen sulfide in sewage structures. These techniques can be used for indirect and quantitative methods to determine the propagation and growth level of sulfate-reducing bacteria. However, when using the procedure for determining the breeding and growth level of the bacteria, it is difficult to quantitatively evaluate the sulfate reducing bacteria due to various bacteria existing in the sewage environment; therefore, different technologies have been proposed for evaluating the concrete erosion progress by detecting changes in humidity and temperature in RC sewage structures considering the growth and reproduction of sulfate reducing bacteria [7,8]. In addition, techniques for predicting the depth of corrosion through monitoring the changing pH of surface concrete due to sulfuric acid and hydrogen sulfide gases formed by the growth of these bacteria have been proposed [9].

Figure 1 shows the metabolic process of generation of hydrogen sulfide due to the growth of anaerobic bacteria, and Figure 2 shows the deterioration mechanism of concrete caused by the effects of hydrogen sulfide and sulfuric acid.

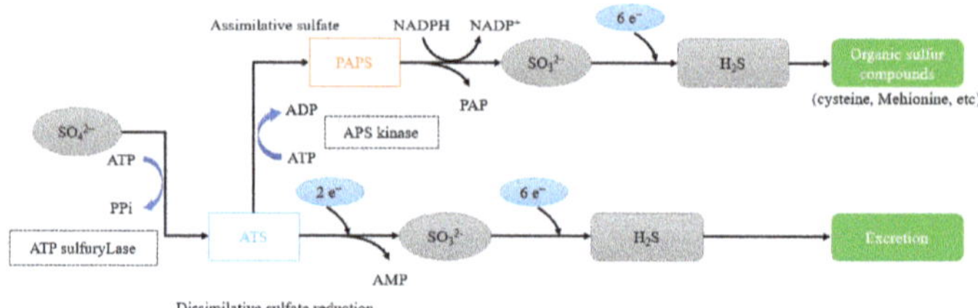

Figure 1. Metabolic reaction process of sulfate reducing bacteria.

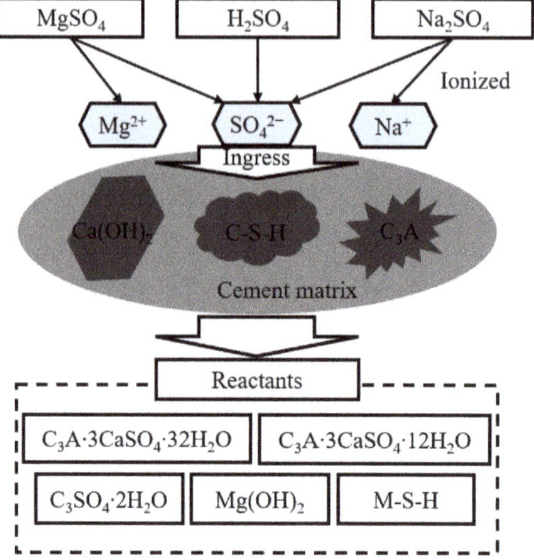

Figure 2. Deterioration mechanism of cement matrix exposed to sulfate ion ingress.

In recent years, there has been active research on the development of repair techniques for protecting sewage structures and controlling deterioration factors, and most of the developed techniques are focused on crack healing and durability improvement of cementitious composites using bacteria [10–12]. These studies have shown the mechanisms in which calcium ions (Ca^{+}) in cement mixtures decompose urea into ammonium and calcium carbonate ($CaCO_3$) is formed through bacterial urea hydrolysis and the biomineralization of organic acid oxidation, thereby healing internal voids and microcracks. On the other hand, research on the use of bacteria with the function of inhibiting the penetration and diffusion of deterioration factors using a viscous slime formed on the cell surface have drawn attention. Studies have been conducted to form a surface protective coating on the concrete structures exposed to deterioration environments by culturing *Rhodobacter capsulatus*, a slime-forming bacterium [12–14]. A previous study showed that the coating repairing with bacteria demonstrated a superior sulfate resistance performance and showed an improvement in durability and mechanical performance such as reducing the diffusion of sulfate ions and maintaining strength at the exposure condition to sulfuric acid, compared to conventional concrete repair mortars [15]. When these repair materials are used in the RC sewage structure, the service life can increase, which reduces the maintenance cost, thus enabling the establishment of an effective maintenance plan. For a maintenance plan, an assessment of the RC structure through NDT (Non-Destructive Technique) such as 3D scanning and visual inspection is effective. As for cracks or surface disintegration in concrete due to chemical deterioration, an evaluation of the crack width and depth is necessary; however, the NDTs still have the limitation of application [16,17]. For the evaluation of cracks and large voids inside concrete, GPR (Ground Penetration Radar) is reported to be effective as well [18].

For the maintenance of the RC sewage structure, it is necessary to determine a limit state for service life, and a deterministic method is generally used. This method aims to secure the required performance within the intended service life considering the environmental factor and the durability reduction factor. In the specifications and guidelines of many countries, five deteriorations such as chloride attack, carbonation, sulfate ingress, freeze–thaw, and alkali aggregate reaction are considered as durability design items for RC structures [19–21].

For chloride attack and carbonation, a number of models for service life prediction have been proposed, which are based on a diffusion and convection mechanism. In the case of service life under sulfate attack, the deterioration depth equation [1,22] that takes into account the diffusion coefficient of sulfate ions and the mixing characteristics, service life evaluation considering the decrease in strength through acceleration test [23,24], and service life evaluation for multilayer diffusion considering coating diffusion have been studied [25]. The variation of chemical compositions and pH under sulfate ingress has been studied with lab-scaled, test, and corrosion depth, and weight loss has been related with duration period [9,26].

Many studies on the probabilistic service life evaluation have also been carried out in order to consider engineering uncertainties for quantitative materials, design, and construction. The probabilistic technique means a design method in which the probability of exceeding the critical condition during the intended service life is considered to be lower than the preset target durability probability [27–30]. For this purpose, the probability distribution of each design parameter should be determined, and the target durability probability (failure probability) according to the target service life should be defined by the design engineer. Since the 1990s, for chloride attack and carbonation, a number of design techniques have been proposed with deterministic and stochastic methods, and studies on analysis considering spatial variability reflecting time-dependent characteristics have been conducted [31,32]. However, there have been limitations in studies on the service life change in the case of repaired structures exposed to sulfate ingress such as RC sewage structures. This is because there may be a complicated deterioration process such as freeze–thaw, and cracks due to the expansion of the concrete and the consequent penetration of deterioration factors, and also sulfate ion ingress and reactions with internal hydrates (calcium hydroxide and gypsum) occur continuously. There have been several studies on sulfate-service life with deterministic methods, but there are very few studies on sulfate-service life considering probability theory and design parameters with variation.

Actually, durability design has developed from deterministic and probabilistic engineering uncertainties such as construction skill variation, design errors, and material quality variation. Regarding sulfate ion ingress, service life evaluation through a probabilistic approach is very limited since the representative evaluation method is not suggested in specifications and design code; therefore, neither the effect of the design parameter on service life nor target durability failure is not clear. The variation of design parameters also affects service life, and the allowable variation of design parameters and safety factors can be reasonably derived through a comparison of the results from the deterministic and probabilistic approach. In this study, the changing service life was analyzed through deterministic and probabilistic methods for concrete repaired with normal repair material and bacteria coating for RC sewage structures. To this end, four major design parameters such as the exterior sulfate ion concentration, diffusion coefficient, cover (coating) depth, and surface roughness, were defined. Furthermore, the effect of each parameter on the service life was evaluated and the results from the two methods were compared considering the design parameter variation. The developed bacteria coating was evaluated to be effective to extend service life for its engineering advantages and the service life variations with changing characteristics of design parameters were quantitatively analyzed in the study.

2. Service Life Evaluation of RC Sewage Structure with Bacteria-Coating Material with Deterministic and Probabilistic Methods

2.1. Governing Equation of Service Life under Sulfate Exposure

Many studies have been performed for relating a damage from sulfate ingress with service life evaluation considering the changes in the chemical composition, pH, weight and strength loss, and resistance to freezing and thawing actions [1,9,26]; however, a quantitative equation with design parameters is necessary for service life evaluation in a deterministic and probabilistic manner. Among the techniques with design parameters such as the deterioration depth method, the multi-layer diffusion method, and the strength reduction method, the deterioration velocity model, linear deterioration depth with sulfate ion diffusion, proposed by American Concrete Institute (ACI) is most widely used [1,21].

This model assumes the deterioration velocity (R) as a linear function of time considering the penetration of sulfate ions into concrete, the reaction between sulfate and aluminum hydrates, and surface expansion. This is expressed as Equation (1) below.

$$R = \frac{E \cdot B^2 \cdot c_0 \cdot C_E \cdot D_i}{\alpha \cdot \gamma_f (1 - \nu)} \quad (1)$$

where E is the modulus of elasticity of concrete (MPa), B is the linear coefficient of deformation by 1 mol of sulfate ions reacting in a unit volume (=1.8 × 10^{-6} m^3/mol), c_0 is the concentration of sulfate ion (mol/m^3), D_i is the sulfate ion diffusion coefficient (m^2/s), α is the surface roughness, γ_f is the concrete failure energy (=10 J/m^2), ν is the concrete Poisson's ratio, and C_E is the sulfate ion concentration reacting with ettringite (mol/m^3).

For applying this equation to the probabilistic method, design parameters with probabilistic variation are required. With respect to Equation (1), a probabilistic service life evaluation equation as expressed in Equation (2) can be set with probabilistic design parameters. In the equation, the probability of exceeding the limit state condition during the target service life is defined and the service life is calculated based on this relationship.

$$p_f \left[\frac{E \cdot B^2 \cdot c_0(\mu, \sigma) \cdot C_E \cdot D_i(\mu, \sigma)}{\alpha(\mu, \sigma) \cdot \gamma_f (1 - \nu)} (t) > C_d(\mu, \sigma) \right] > p_d \quad (2)$$

where $p_f(t)$ is the durability failure probability for the deterioration depth that increases with time, $C_d(\mu, \sigma)$ is the probability distribution for the cover depth, and p_d is the target durability probability assumed during durability design. In Equation (2), the exterior sulfate ion concentration (c_0), diffusion coefficient (D_i), surface roughness (α), and cover depth (c_d) were defined as random variables for design parameters.

In the case of chloride attack and carbonation, the service life limit state conditions and target durability failure probability are determined, respectively [21,30,33], but in the case of sulfate attack, there are currently no clearly set limit conditions. Therefore, as in the case of carbonation, in which the increasing carbonation depth is compared with design cover depth, the time that the deterioration depth reaches the cover depth is assumed as the limit state condition. Figure 3 shows a general overview of the probabilistic service life evaluation method, and Table 1 outlines the engineering uncertainties that require probabilistic analysis. In Figure 3, the concept of evaluating the respective deteriorating factors and design resistance is referred to as the service period concept (two lines with $S(t)$ and $R(t)$), and the concept considered as one probability distribution is referred to as the lifetime concept (single line with $f[R, T]$), and each concept is presented in Equations (3) and (4), respectively [28].

Table 1. Engineering uncertainties in durability design.

Uncertainty Type	Limitation
Physical	Inherent random nature of a basic variables - Concrete cover depth - Concentration of exterior sulfate ions - Quality of concrete (diffusion coefficient of sulfate ion) - Local condition (cracks and joints)
Statistical	Assumption for probability density functions - Limited sample size
Model	Governing mechanism for deterioration by sulfate - Simplified equation of deterioration by sulfate without considering sulfuric reaction - Assumption of material properties (sulfurictable and reactant material) - Assumption of non-correlated variables
Decision	Definition of durability failure criteria - The period that carbonation depth exceeds the cover depth

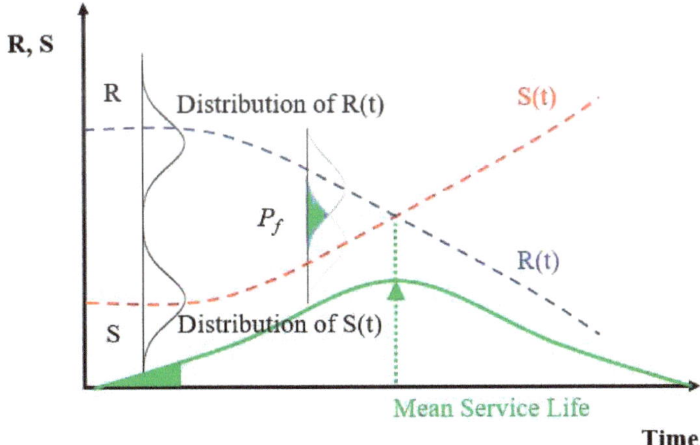

Figure 3. Durability design of probabilistic model with service life and lifetime concept.

$$P_{L,T} = P[R(t) - S(t) < 0]_T < P_{target} \quad (3)$$

$$L = f[R, S] \quad (4)$$

here, $R(t)$ and $S(t)$ represent the resistance function and deterioration function with time, and P_{target} denotes the allowable value of the durability probability maintained during the target service life [25]. In Equation (2), since there is no prior study clearly indicating this value, P_{target} was assumed to be 10%, which is the same level as the case of chloride attack, and the sum of the period when the durability failure probability of the repair material reaches 10% and the durability failure probability reaches 10% at the cover depth was defined as the service life. In addition, during the probabilistic design, the MCS (Monte Carlo Simulation) was performed, and the schematic diagram of the analysis process is presented in Figure 4.

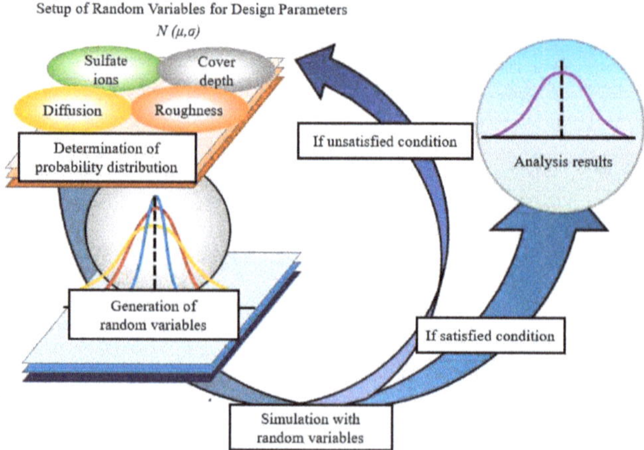

Figure 4. Analysis step of MCS with random variables of design parameters.

For the deterministic design method, the period in which the deterioration depth exceeds the sum of the repair material and the cover depth was considered, and the environmental factor (γ_p) 1.1 and durability reduction factor (ϕ_k) 0.92 proposed in the specifications were adopted as shown in Equation (5) [20,33].

$$\gamma_p Z_p \leq \phi_k Z_{\lim} \quad (5)$$

where Z_p is a predicted value of the chemical erosion depth and is derived from the deterioration velocity formula of Equation (1). $Z_{\lim}$ is the design cover depth considering the construction variation.

2.2. *Bio-Coating Materials Mixed with Bacteria*

2.2.1. Preparation of Repair Coating Material with Bacteria

A bacteria-coating material was developed using the culture medium in which the *Rhodobacter capsulatus* strain selected in the previous study [15] was cultured at a concentration of 10^9 cells/mL. The mixing proportion details of the bacteria-coating material are presented in Table 2, and the detailed mixing proportions of the normal repair mortar in which the bacteria-immobilizing material is not mixed are also shown. Table 3 shows the mixing proportions of the concrete used in the existing RC sewage structure. The chemical properties of ordinary Portland cement (OPC), fly ash (FA), and ground granulated blast furnace slag (GGBFS) are shown in Table 4, and the physical properties of each material are summarized in Table 5. For the aggregates used for the preparation of bacteria-coating material, silica sand with grain diameters of 0.05–0.17, 0.17–0.25, and 0.25–0.70 mm was mixed in the same ratio for use. Figure 5 shows an overview of the improvement of the resistance performance to sulfate deterioration in the RC sewage structure with the application of the developed bacteria-coating material.

Table 2. Mixing proportions of repair mortar and bacteria-coating material.

Specimens	*W/B* (%)	*S/B*	Binder Mixing Ratio (%)			Replacement Ratio of Bacteria Carrier by the Volumetric Fraction of Sand (%)
			OPC	FA	GGBFS	
Repair mortar	35	2	35	20	45	-
Bacteria-coating material						35

Table 3. Mixing proportion of sewage concrete.

W/C (%)	*S/a* (%)	Unit Weight (kg/m³)			
		Water	Cement	Sand	Gravel
45	47	180	400	786	903

Table 4. Chemical properties of OPC, FA, and GGBFS.

Materials	Composition (wt.%)									
	SiO_2	Al_2O_3	Fe_2O_3	CaO	MgO	SO_3	TiO_2	Na_2O	K_2O	LOI
OPC	21.7	5.3	3.1	62.4	1.6	1.7	-	-	-	0.8
GGBFS	33.5	15.2	0.5	45.6	2.6	2.5	0.9	0.2	0.4	0.7
FA	53.3	27.9	7.8	6.8	1.1	0.8	-	0.6	0.8	0.9

Table 5. Physical properties of the used cementitious materials.

Materials	Density (g/cm³)	Fineness (cm²/g)
OPC	3.15	3260
GGBFS	2.94	4355
FA	2.23	3720

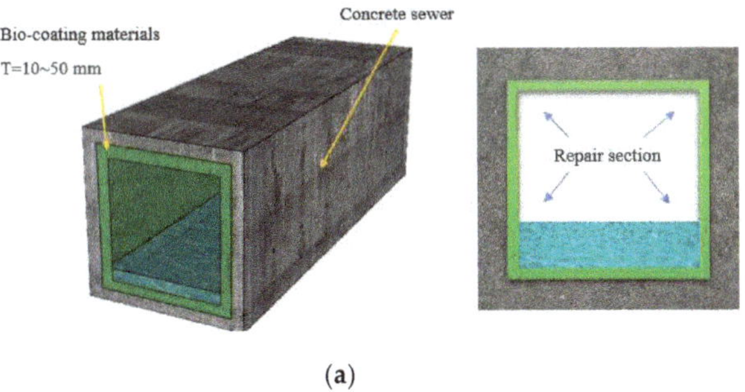

(a)

Figure 5. *Cont.*

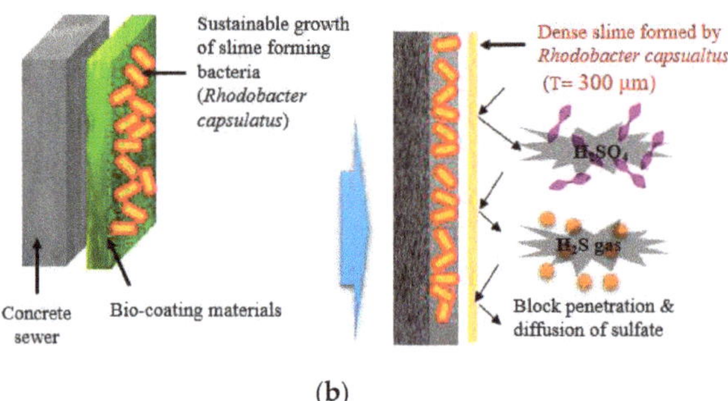

(**b**)

Figure 5. Schematic design for repair with bacteria and RC sewage structure: (**a**) RC culvert with bacteria coating repair; (**b**) Metabolism of resistance to sulfate attack by bacteria-formed slime.

2.2.2. Diffusion Coefficient of Sulfate Ion through Natural Diffusion Cell Test

In the deterioration depth evaluation, the diffusion coefficient is the governing factor, and it changes with the mixing proportions. In the previous study, the sulfate ion diffusion coefficient was evaluated using the ratio between the sulfate and chloride ion diffusion coefficients [24], but in this study, 10 diffusion cells were prepared for defining the coefficient of variation (COV). By performing the natural diffusion cell test [34], the sulfate ion diffusion coefficient of bacteria mixed coating, concrete, and normal repair material were evaluated. The equation for derivation is presented in Equation (6), and the related experimental photos are shown in Figure 6.

$$D_i = \frac{V\Delta Q}{A\Delta T} \times \frac{L}{c_1 - c_2} \tag{6}$$

where V is the volume of the diffusion cell (m^3),ΔQ is the increase in sulfate ion concentration in the cell containing distilled water (kg/m^3), A is the area of the exposure of the slice specimen (m^2), ΔT is the elapsed time after the test starts (sec), L is the slice specimen thickness (m), c_1 is the concentration of sulfate ion solution (kg/m^3), and c_2 is the average concentration of the cell containing distilled water (kg/m^3).

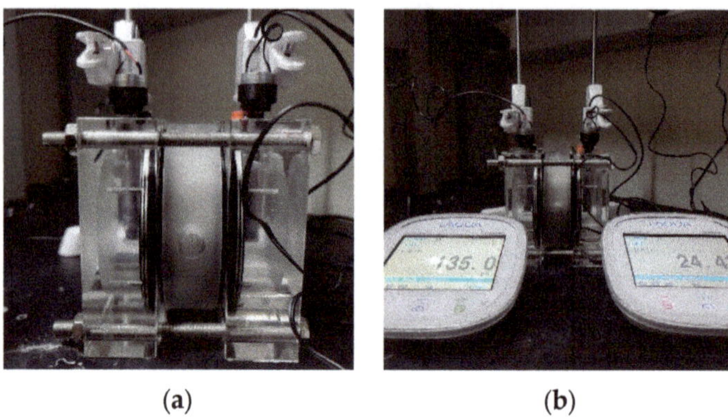

(**a**) (**b**)

Figure 6. Photo for sulfate ion diffusion test: (**a**) setting image of test specimens; (**b**) setting image of ion meter for recoding the chaining sulfate ion concentration.

3. Changing Service Life with Probabilistic Characteristics of Design Parameters

3.1. *Service Life Evaluation with the Deterministic Method*

As described above, the deterioration depth relation shown in Equation (1) is linearly proportional to the sulfate ion diffusion coefficient, and the period when the deterioration depth reaches the cover depth is calculated with an increasing period. The variables for deterministic service life evaluation are shown in Table 6. The service life of the structure was determined as the period when the deterioration depth reached the cover depth in the case without coating, and as the period when the deterioration depth reached the sum of the coating thickness and cover depth in the case with a coated structure. In addition, the cover depth of the concrete base material was set to 30 mm, and the coating thickness of the normal repair mortar and bacteria-coating material were set to 5.0 mm, which were conventionally applied. As a result of the analysis, the deterioration depth according to the exterior environment conditions is shown in Table 7, and the service life evaluation results are shown in Figure 7 and Table 8. The bacteria coating showed a much lower diffusion coefficient of 8.3–8.4% compared to normal repair mortar and sewage concrete; therefore, a considerably long service life was evaluated, even with a small coating thickness. The exterior sulfate concentration varies with the distance from the commercial facilities' sewage outlet and 120–200 ppm of concentration level was reported [35]. In this analysis, two levels of concentration were considered, one for normal condition (120 ppm) and the other is severe condition (200 ppm), respectively.

Table 6. Analysis conditions on service life evaluation (Deterministic method).

Variables	Sewage Concrete	Repair Mortar	Bacteria-Coating Material
c_0	200	200	200
D_i	2.12×10^{-12}	2.09×10^{-12}	0.17×10^{-12}
E	25,700	21,500	21,500
ν	0.17	0.27	0.27
α	1.5	1.5	1.5
γ_f	10	10	10
Binder weight	400	300	300
C_E	207	196	462
Cover depth (mm)	30	5	5

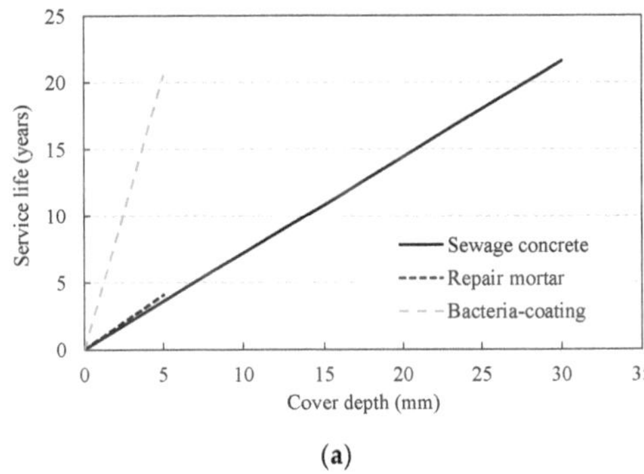

(**a**)

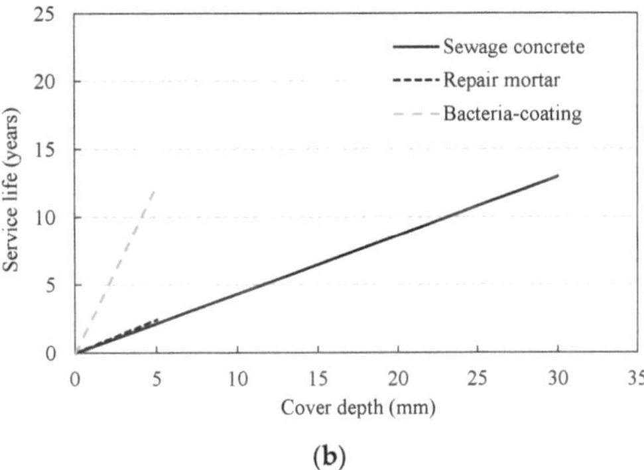

(**b**)

Figure 7. Service life evaluation through deterministic approach: (**a**) sulfate concentration = 120 ppm (normal condition); (**b**) sulfate concentration = 200 ppm (severe condition).

Table 7. Service life through deterministic technique.

Exterior Condition (Sulfate Concentration)	Deterioration Velocity (mm/Year)		
	Sewage Concrete	Repair Mortar	Bacteria-Coating Material
120 ppm	1.16	1.03	0.20
200 ppm	1.94	1.71	0.34

3.2. Service Life Evaluation Using Probabilistic Method

3.2.1. Preparation of Bacteria-Coating Material

Among the design parameters, four random variables were considered, which include cover depth, sulfate ion diffusion coefficient, surface roughness, and exterior sulfate concentration. The surface roughness and exterior sulfate concentration were assumed to be 0.1, which is a reasonable level in engineering aspect. As previously explained, the sulfate ion concentration level in RC sewage structure was 120–200 ppm [35]; however, the more severe condition of 280 ppm was added considering rapidly increasing, domestic and industrial sewage pollution. As for the thickness of the repair material and sewage RC concrete, and the diffusion coefficient, their COVs were obtained from testing 10 times, and all had variations between 0.10 and 0.14. Figure 8 shows the probability characteristics of each design parameter.

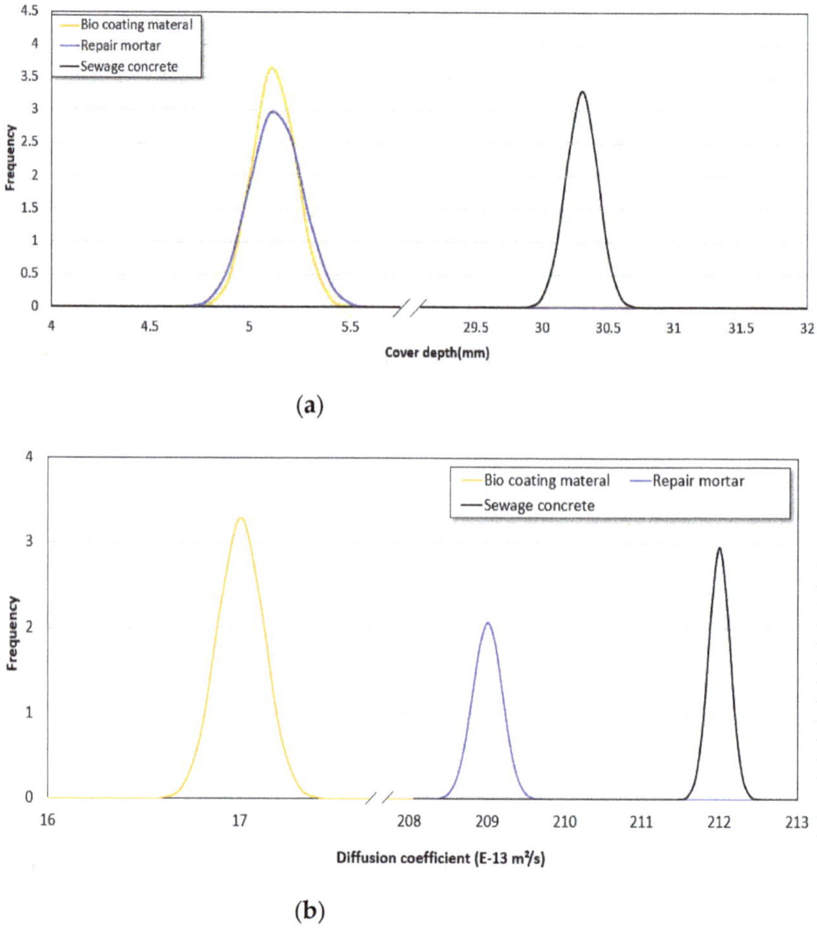

Figure 8. Probability distributions of design: (**a**) probabilistic distribution of thickness; (**b**) probabilistic distributions of diffusion coefficient.

3.2.2. Service Life Change of Bacteria-Coating Material

(1) Overview of service life evaluation of repair materials

In order to evaluate the durability failure probability and service life of the developed bacteria coating, the effects of the design parameters on the service life were analyzed. For the design parameters of the bacteria coating, a normal distribution and a 0.1 COV were used. An MSC was performed with the design parameter values in Table 6 as the mean values. The changes in service life were analyzed considering the changes in the mean and variation.

(2) Changes in failure probability and service life according to varying mean value

As a result of analyzing the changes in failure probability while increasing the external sulfate ion concentration at three levels from 120, 200, to 280 ppm, the service life showed a rapid decrease to 19.5, 11.2, and 8.2 years, respectively. In terms of the coating thickness, it was analyzed as the thickness increased from 5.0 mm, which is the typical coating thickness, to 15 mm. The service life was evaluated to be 11.2, 22.2, and 33.9 years, respectively, indicating more a significant impact compared to the other parameters. The surface roughness showed a wide range of values (1–10) in previous studies [4,36], and the failure probability was analyzed with a changing surface roughness in the range of 1.5–3.5. The failure probability decreased with an increasing surface roughness, and the service life accordingly increased to 11.2, 18.8, and 26.1 years, respectively. The diffusion coefficient is linearly proportional to the deterioration depth and, since it varies widely depending on the mixing proportion, it is an important design parameter. The failure probability and service life were derived by changing the sulfate ion diffusion coefficients derived through the natural diffusion cell test to 0.5, 1.0, and 2.0 times, and the service life showed a rapid decrease to 22.1, 11.2, and 5.4 years, respectively.

Figure 9 shows the changes in durability failure probability for each service life evaluation, and the service life corresponding to the target durability probability of 10% is plotted in Figure 10, which shows service life from the probabilistic method.

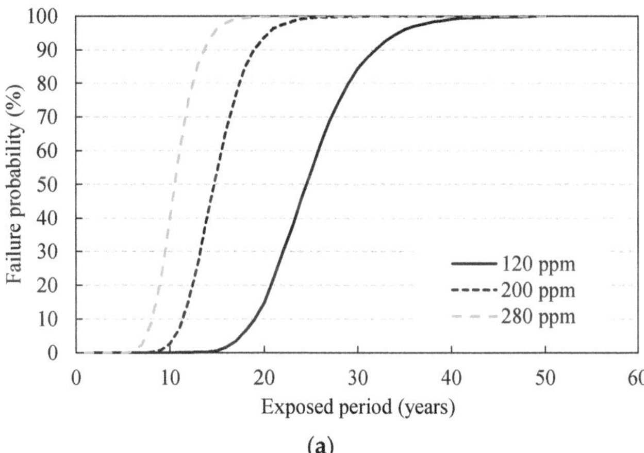

Figure 9. *Cont.*

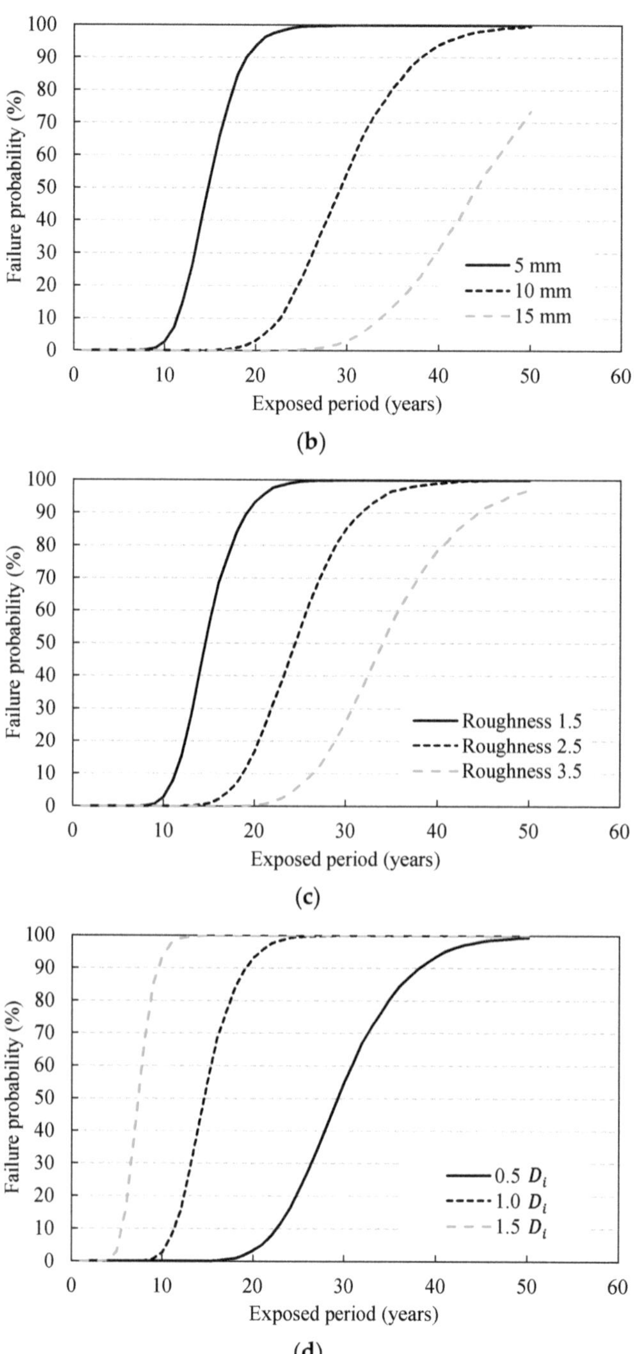

Figure 9. Changes in failure probability with design variables: (**a**) exterior sulfate ions; (**b**) coating thickness; (**c**) surface roughness; (**d**) diffusion coefficient.

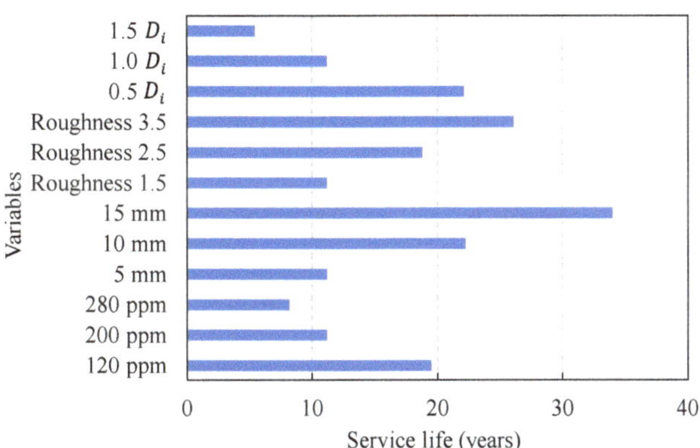

Figure 10. Changes in service life with design variables.

Based on the design constants for actual construction, if the change ratio with reference to the baseline value (average) of each design parameter can be illustrated, the effect of each parameter on the service life can be quantitatively evaluated. The changes in service life gradient with respect to the baseline conditions (exterior sulfate concentration, 200 ppm; coating thickness, 5.0 mm; surface roughness, 1.5; and D_i, 1.0) are depicted in Figure 11. The exterior sulfate concentration and diffusion coefficient show almost the same effect, and surface roughness and coating thickness have a relatively large effect. That is, surface roughness has a close relationship with the strength between the aggregate and the base concrete, and cover depth is also a primary mechanism to block the direct deteriorating factors from the outside, and therefore, the careful determination of the two design parameters is required for securing performance during service life.

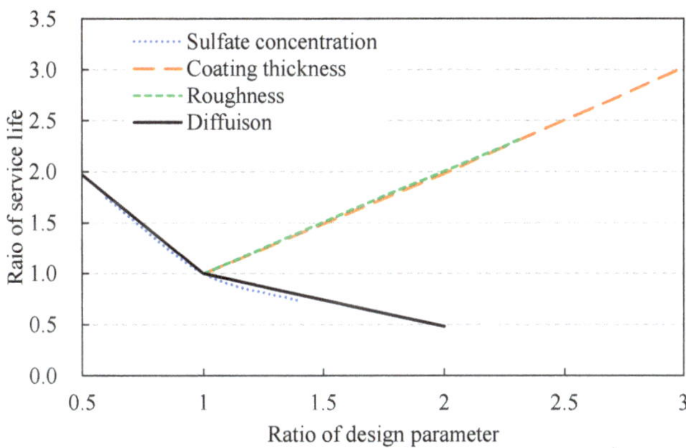

Figure 11. Relationship between design parameters and service life.

(3) Failure probability and changes in service life according to COV

In this section, the variation in the service life is evaluated while increasing the COV of each design parameter from 0.1 to 0.3. Even with the change in COV, the period with a failure probability at 50% is evaluated to be the same result from the deterministic method under the assumption of normal distribution. In other words, in the domain before the failure probability at 50%, the larger COV causes the shorter service life, and therefore, reducing their COV, considering the construction and material quality, is instrumental. Figure 12 shows the changes in the failure probability of each design parameter with changing COV. When the COV of one variable increased, the COV of the other design parameters was all fixed to 0.1 for evaluation. The changes in service life derived according to the variation of design parameters are shown in Figure 13. When the COV increases from 0.1 to 0.3, the service life of the coating decreased by 2.5 to 2.6 years, amounting to a 13–22% reduction in service life. Therefore, reducing the variation in quality during construction is highly important.

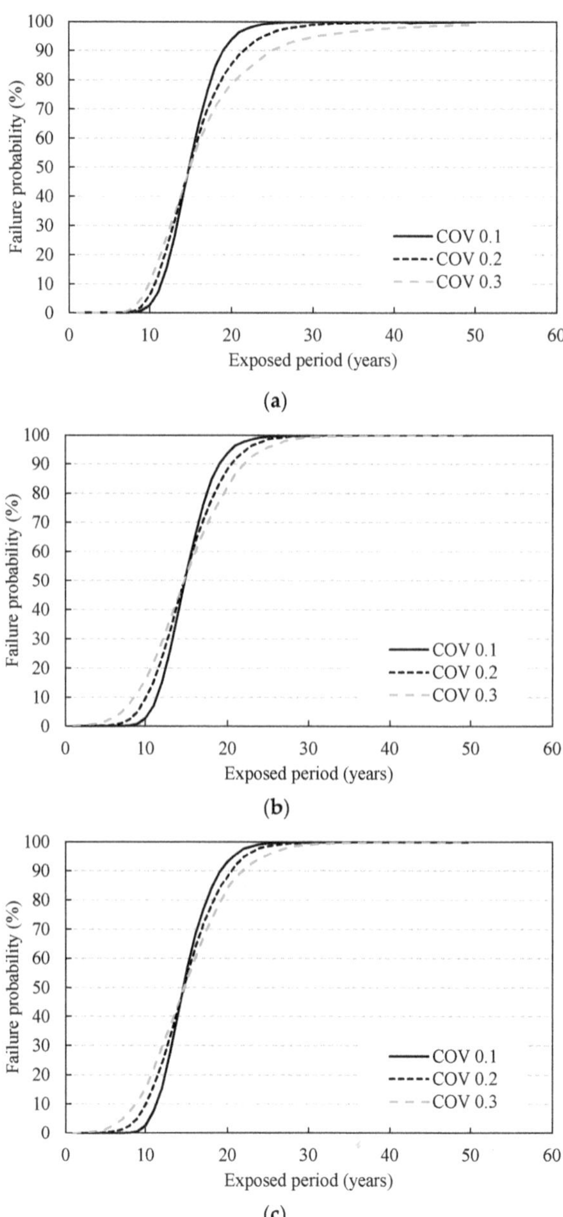

Figure 12. *Cont.*

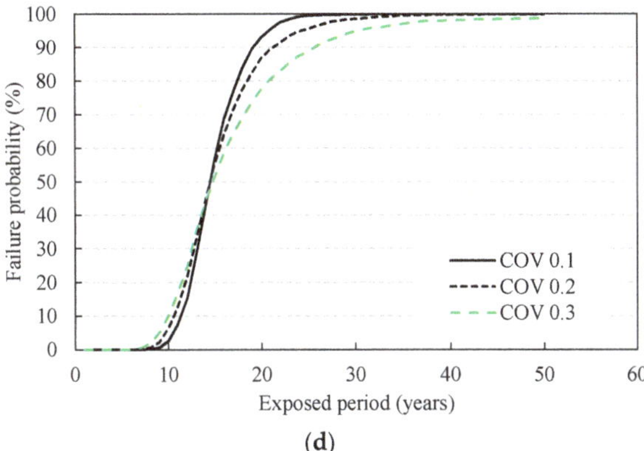

Figure 12. Changes in failure probability with COV variation: (**a**) exterior sulfate ions; (**b**) coating thickness; (**c**) surface roughness; (**d**) diffusion coefficient.

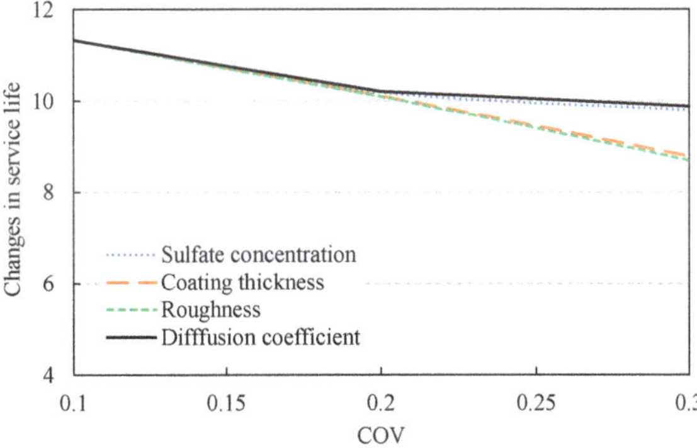

Figure 13. Decreasing service life with increasing COV.

4. Service Life Changes in RC Sewage Structure with Different Repair Materials

In this section, the service life was evaluated with two exterior sulfate concentrations, a normal sulfate ion level (120 ppm) and a slightly higher level (200 mm) referred to the previous field assessment [35]. As previously explained, the service life was defined to be the period when the deterioration depth exceeds the sum of the cover depth and coating thickness, and it was evaluated with the following three cases: concrete only without repair material, protection with normal repair mortar, and protection with a bacteria coating. In addition, the service life of the structure was evaluated by assigning each design parameter a variation of 0.1, 0.2, and 0.3.

Figure 14 shows the changes in the service life of the RC sewage structure with the exterior sulfate concentration. In the deterministic method, the service life was evaluated as 21.6 years in the case without repair under the environmental exposure to 120 ppm, it was 25.7 years when a normal repair mortar was applied, and 42.1 years when a bacteria-coating material was applied. When the probabilistic method was adopted, the service life was slightly reduced. When the COV increased from 0.1 to 0.3, the service life showed a sharp decrease from 26.5 to 15.0 years for normal concrete, from 30.6 to 17.2 years when the normal repair mortar was used, and from 45.1 to 25.3 years when bacteria-coating was applied (Figure 15). The reduction rate was evaluated to be between 56.1 and 56.6%, and in the case of 200 ppm, the reduction rate was from 56.2 to 57.7%, indicating a similar level. In the case of a sulfate concentration of 200 ppm, with the deterministic method, the service life was evaluated at 12.9 years without repair; this increased to 15.4 years when normal repair mortar was applied, and further increased to 25.3 years when a bacteria coating was used. When the probabilistic method was applied in the same environment, the service life was in the range of 9.2–16.0, 10.5–18.2, and 15.4–27.4 years, respectively, depending on the variation, showing a significant difference depending on the type of repair material.

The smaller the coating thickness of the normal RC sewage structure is, the less clogging there will be in the flow. Considering the coating thickness of 5 mm, the bacteria coating demonstrated a 1.47–1.50 times longer service life than normal repair mortar, and the service life was evaluated to be longer as the COV decreased. In the case of the developed bacteria mixed coating, it was shown that the quality can be maintained steadily under the influence of continuous sulfide ions and humidity. The service life of an RC sewage structure repaired with a bacteria-coating material under sulfate ion ingress can be extended considerably due to a low diffusion coefficient and low COV.

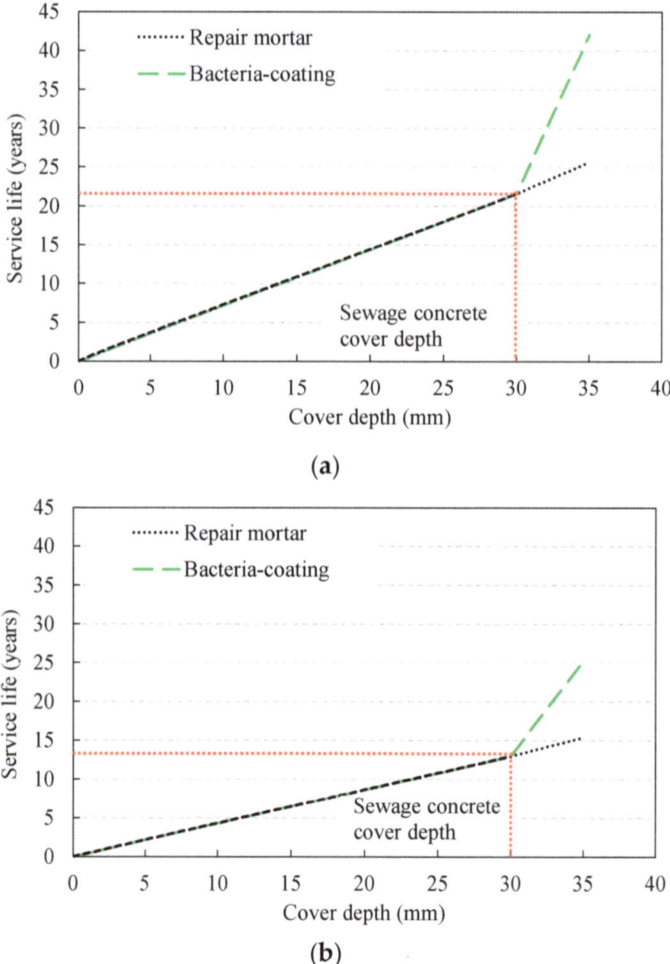

Figure 14. Service life variation according to deterministic method: (**a**) sulfate concentration = 120 ppm; (**b**) sulfate concentration = 200 ppm.

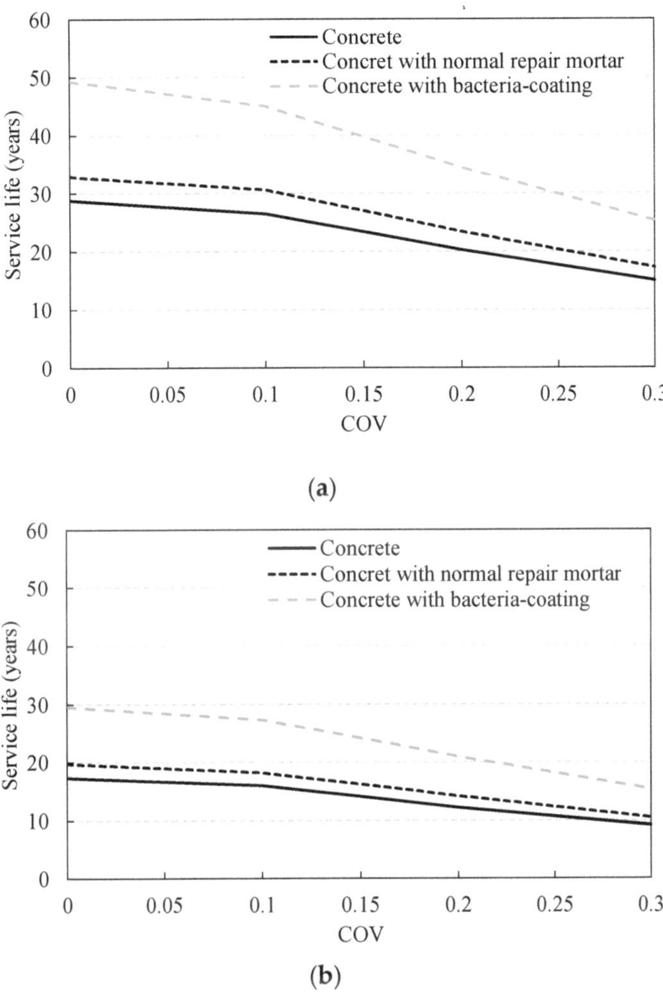

Figure 15. Service life variation with increasing COV according to probabilistic method: (**a**) sulfate concentration = 120 ppm; (**b**) sulfate concentration = 200 ppm.

Considering the conventional unit repair cost and intended service life (60 years), the repair cost can be reduced due to extended service life and decreasing the number of repair events. The unit repair cost [37] and total repair cost under 120 and 200 ppm of sulfate ingress are shown in Table 8 and Figure 16. As shown, the repair events were reduced from nine to one time and the repair cost was reduced from 441.9 to 65.8 USD/m^2 under 120 ppm. For the 200 ppm case, they were reduced from 19 to 3 times and 932.9 to 98.7 USD/m^2, respectively.

Table 8. Unit repair cost for each technique (thickness = 5 mm).

	Repair Mortar	**Bacteria-Coating Material**
Repair cost (USD/m^2)	49.1	32.9

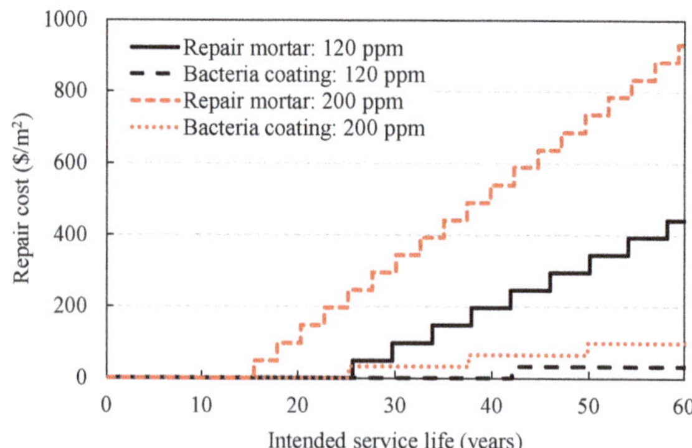

Figure 16. Variation in repair cost due to increase in service life with normal repair mortar and bacteria-coating material.

5. Conclusions

In this study, a repair coating technique with bacteria was developed and the service life of an RC sewage structure under sulfate ion ingress was evaluated with the following two different methods: the deterministic and probabilistic method. The conclusions from this study are summarized as follows:

- An FA-GGBFS-based repair material with *Rhodobacter capsulatus* was developed, which consumes sulfate ions as nutrients. For applying probabilistic service life evaluation, probability distribution characteristics were derived through diffusion cell tests and cover depth measurements. The derived diffusion coefficient of the developed repair material with *Rhodobacter capsulatus* showed a very low value in the level of 8.3–8.4% compared to that of the base concrete and normal repair material, and the COV was around 10%, indicating stable quality assurance.
- Five major design parameters (surface coating, coating thickness, diffusion coefficient, exterior sulfate ion concentration, and surface roughness) were defined as random variables for analyzing the service life with a probabilistic approach. The effect of the mean value of surface roughness and cover depth was evaluated to be dominant; therefore, it is important to pay special attention to the surface treatment and coating thickness during repair construction. When the service life was evaluated while increasing the COV of each design parameter from 0.1 to 0.3, the service life was reduced by 1.5 to 2.6 years, even with the application of the appropriate coating thickness, which yields a decrease in service life by more than 15%.
- When the service life of the RC sewage structure was evaluated considering under 120 ppm of exterior sulfate ingress, it was evaluated to be 21.6 years without repair, but this increased to 25.7 years with a normal repair mortar, and further increased to 42.1 years enhanced with a bacteria-coating material. When the probabilistic method was applied, it was reduced due to the low target durability probability. With the increasing COV of the design parameter, it was greatly reduced from 26.5 to 15.0 years for base concrete, from 30.6 to 17.2 years for normal repair mortar, and from 45.1 to 25.3 years with the developed bacteria-coating application. The service life reduction rates were evaluated to be around 56.1–56.6% (120 ppm of sulfate ion), and 56.2–57.7% (200 ppm of sulfate ion). Even when the applied coating thickness of the sewage structure is thin (5 mm), the bacteria-coating shows a 1.47–1.50 times longer service life than that with the normal repair mortar.
- Utilizing the unit repair cost and applying the developed bacteria repair technique reduced the total repair cost from 441.9 to 65.8 USD/m^2 (120 ppm) and from 932.9 to 98.7 USD/m^2 (200 ppm), respectively. It is because of the low diffusion coefficient of the bacteria repair material and low unit repair price.

- The developed repair material with *Rhodobacter capsulatus* consumes sulfide ions as nutrients; therefore, it can assure a continuous maintenance of quality level; however, further research is still required on the rapid increase in flow velocity due to flooding, damage to collisions with floats, temperature management and quality control during coating composite manufacturing, and temperature effects during construction.

Author Contributions: Conceptualization, K.-H.Y. and K.-M.L.; methodology, S.-J.K.; formal analysis, H.-S.Y., K.-H.Y. and S.-J.K.; investigation, H.-S.Y.; writing—original draft preparation, H.-S.Y.; writing—review and editing, K.-H.Y., S.-J.K. and K.-M.L.; visualization, H.-S.Y.; supervision, S.-J.K. All authors have read and agreed to the published version of the manuscript.

Funding: This research was supported by a grant (21SCIP-C158976-02) from Construction Technology Research Program funded by Ministry of Land, Infrastructure, and Transport of Korean government.

Institutional Review Board Statement: Not applicable.

Informed Consent Statement: Not applicable.

Data Availability Statement: The data presented in this study are available on request from the corresponding author.

Conflicts of Interest: The authors declare no conflict of interest.

References

1. Lee, H.J.; Cho, M.S.; Lee, J.S.; Kim, D.G. Prediction model of life span degradation under sulfate attack regarding diffusion rate by amount of sulfate ions in seawater. *Int. J. Mater. Mech. Manuf.* **2013**, *1*, 251–255. [CrossRef]
2. Khan, H.A.; Castel, A.; Khan, M.S.H. Corrosion investigation of fly ash based geopolymer mortar in natural sewer environment and sulphuric acid solution. *Corros. Sci.* **2020**, *168*, 108586. [CrossRef]
3. Parker, C. The Corrosion of the concrete I. The isolation of a species of bacterium associated with the corrosion of concrete exposed to atmospheres containing hydrogen sulfide. *Aust. J. Exp. Biol. Med. Sci.* **1945**, *23*, 81–90. [CrossRef]
4. De Belie, N.; Monteny, J.; Beeldens, A.; Vincke, E.; Van Gemert, D.; Verstraete, W. Experimental research and prediction of the effect of chemical and biogenic sulfuric acid on different types of commercially produced concrete sewer pipes. *Cem. Concr. Res.* **2004**, *34*, 2223–2236. [CrossRef]
5. Barton, L.L. *Sulfate-Reducing Bacteria*; Springer: Boston, MA, USA, 1995.
6. Takahashi, H.; Kopriva, S.; Giordano, M.; Saito, K.; Hell, R. Sulfur assimilation in photosynthetic organisms: Molecular functions and regulations of transporters and assimilatory enzymes. *Annu. Rev. Plant Biol.* **2011**, *62*, 157–184. [CrossRef] [PubMed]
7. Thiyagarajan, K.; Kodagoda, S.; Ranasinghe, R.; Vitanage, D.; Iori, G. Robust sensing suite for measuring temporal dynamics of surface temperature in sewers. *Sci. Rep.* **2018**, *8*, 16020. [CrossRef] [PubMed]
8. Thiyagarajan, K.; Kodagoda, S.; Ranasinghe, R.; Vitanage, D.; Iori, G. Robust sensor suite combined with predictive analytics enabled anomaly detection model for smart monitoring of concrete sewer pipe surface moisture conditions. *IEEE Sens. J.* **2020**, *20*, 8232–8243. [CrossRef]
9. Valix, M.; Zamri, D.; Mineyama, H.; Cheung, W.H.; Shi, J.; Bustamante, H. Microbiologically induced corrosion of concrete and protective coatings in gravity sewer. *Chin. J. Chem. Eng.* **2012**, *20*, 433–438. [CrossRef]
10. Kang, T.W. *Development of Repair Material and Development/Repair Method Development Utilizing Amine Derivatives and Ionic Reactions to Repair Concrete Structures in Sewage Systems*; Technical Report; Sammyung ENC: Busan, Korea, 2019.
11. Oh, S.R.; Choi, Y.W.; Lee, K.M. Effect of cement powder based self-healing solid capsule on the quality of mortar. *Const. Build. Mater.* **2019**, *214*, 574–580. [CrossRef]
12. Yoo, K.S.; Jang, S.Y.; Lee, K.M. Recovery of chloride penetration resistance of cement-based composites due to self-healing of cracks. *Materials* **2021**, *14*, 2501. [CrossRef]
13. Yang, K.H.; Yoon, H.S.; Lee, S.S. Feasibility tests toward the development of protective biological coating mortars. *Constr. Build. Mater.* **2018**, *181*, 1–11. [CrossRef]
14. Sotiriadis, K.; Mácová, P.; Mazur, A.S.; Viani, A.; Peter, M.T.; Tsivilis, S. Long-term thaumasite sulfate attack on portland-limestone cement concrete: A multi-technique analytical approach for assessing phase assemblage. *Cem. Concr. Res.* **2020**, *130*, 105995. [CrossRef]
15. Yoon, H.S.; Yang, K.H.; Lee, S.S. Evaluation of sulfate resistance of protective biological coating mortars. *Constr. Build. Mater.* **2021**, *270*, 121381. [CrossRef]
16. Gholizadeh, S. A review of non-destructive testing methods of composite materials. *Procedia Struct. Integr.* **2016**, *1*, 50–57. [CrossRef]
17. Moosavi, R.; Grunwald, M.; Redmer, B. Crack detection in reinforced concrete. *NDT E Int.* **2020**, *109*, 102190. [CrossRef]
18. Rasol, M.A.; Pérez-Garcia, V.; Fernades, F.M.; Pais, J.C.; Solla, M.; Santos, C. NTD assessment of rigid pavement damages with ground penetrating radar: Laboratory and field tests. *Int. J. Pavement Eng.* **2020**, *2020*, 1778692.

19. Hilsdof, H.K.; Kroff, J. *Performance Criteria for Concrete Durability*; Taylor & Francis: Milton Park, UK, 1995.
20. *Standard Specification for Concrete Structures–Design, JSCE Guidelines for Concrete*; JSCE (Japan Society of Civil Engineering): Tokyo, Japan, 2007.
21. *Eurocode 1: Basis of Design and Actions on Structures*; EN-1991; CEN (European Committee for Standardization): Brussels, Belgium, 2000.
22. Atkinson, A.; Hearne, J.A. Mechanistic model for the durability of concrete barriers exposed to sulphate-bearing groundwaters. *Mater. Res. Soc. Symp. Proc.* **1989**, *176*, 149–156. [CrossRef]
23. Zhang, M.; Yang, L.M.; Guo, J.J.; Liu, W.L.; Chen, H.L. Mechanical properties and service life prediction of modified concrete attacked by sulfate corrosion. *Adv. Civ. Eng.* **2018**, *2018*, 8907363. [CrossRef]
24. Qin, S.; Zou, D.; Liu, T.; Jivkov, A. A chemo-transport-damage model for concrete under external sulfate attack. *Cem. Concr. Res.* **2020**, *132*, 106048. [CrossRef]
25. Yang, K.H.; Lim, H.S.; Kwon, S.J. Effective bio-slime coating technique for concrete surfaces under sulfate attack. *Materials* **2020**, *13*, 1512. [CrossRef] [PubMed]
26. Wells, W.; Melchers, R.E. An observation-based model for corrosion of concrete sewers under aggressive conditions. *Cem. Concr. Res.* **2014**, *61*, 1–10. [CrossRef]
27. Gjorv, O.E. Steel corrosion in concrete structures exposed to norwegian marine environment. *Concr. Int.* **1994**, *16*, 35–39.
28. DuraCrete. *Probabilistic Performance Based Durability Design of Concrete Structures*; European Brite-Euram Programme: Gouda, The Netherlands, 2000.
29. Bentz, E.C. Probabilistic modeling of service life for structures subjected to chlorides. *ACI Mat. J.* **2003**, *100*, 391–397.
30. Kwon, S.J.; Na, U.J.; Park, S.S.; Jung, S.H. Service life prediction of concrete wharves with early-aged crack: Probabilistic approach for chloride diffusion. *Struct. Saf.* **2009**, *31*, 75–83. [CrossRef]
31. Stewart, M.G.; Mullard, J.A. Spatial time-dependent reliability analysis of corrosion damage and the timing of first repair for RC Structures. *Eng. Struct.* **2007**, *29*, 1457–1464. [CrossRef]
32. Na, U.J.; Kwon, S.J.; Chaudhuri, S.R.; Shinozuka, M. Stochastic model for service life prediction of RC structures exposed to carbonation using random field simulation. *KSCE J. Civ. Eng.* **2012**, *16*, 133–143. [CrossRef]
33. *Model Code for Service Life Design*; FIB Bulletin 34; FIB (International Federation for Structural Concrete): Lausanne, Switzerland, 2006.
34. Tritthart, J. Chloride binding in cement, II. The influence of hydroxide concentration in the pore solution of hardened cement paste on chloride binding. *Cem. Concr. Res.* **1989**, *19*, 683–691. [CrossRef]
35. Yoon, H.S.; Yang, K.H. Evaluation of effectiveness of concrete coated with bacterial glycocalix under simulated sewage environments. *J. Rec. Const. Resour.* **2020**, *8*, 97–104. (In Korean)
36. Barnat-Hunek, D.; Widomski, M.S.; Szafraniec, M.; Łagód, G. Impact of different binders on the roughness, adhesion strength, and other properties of mortars with expanded cork. *Materials* **2018**, *11*, 364. [CrossRef] [PubMed]
37. Dongyang Economic Research Institute. *Quantity-Per-Unit Costing of Bacteria Based Deteriorated Concrete Section Repair Method*; Technical Report; Dongyang Economic Research Institute: Seoul, Korea, 2020.

 materials

MDPI

Article

The Effect of Incorporating Industrials Wastewater on Durability and Long-Term Strength of Concrete

Ehsan Nasseralshariati [1], Danial Mohammadzadeh [2,3,4], Nader Karballaeezadeh [4,5], Amir Mosavi [6,7,*], Uwe Reuter [6] and Murat Saatcioglu [1]

1 Department of Civil Engineering, University of Ottawa, Ottawa, ON K1N 6N5, Canada; Enase063@uottawa.ca (E.N.); Murat.Saatcioglu@uOttawa.ca (M.S.)
2 Department of Civil Engineering, Ferdowsi University of Mashhad, Mashhad 9177948974, Iran; Danial.mohammadzadehshadmehri@mail.um.ac.ir
3 Department of Civil Engineering, Mashhad Branch, Islamic Azad University, Mashhad 9187147578, Iran
4 Department of Civil Engineering, Faculty of Montazeri, Khorasan Razavi Branch, Technical and Vocational University (TVU), Mashhad 9176994594, Iran; N.karballaeezadeh@shahroodut.ac.ir
5 Faculty of Civil Engineering, Shahrood University of Technology, Shahrood 3619995161, Iran
6 Faculty of Civil Engineering, Technische Universität Dresden, 01069 Dresden, Germany; uwe.reuter@tu-dresden.de
7 John von Neumann Faculty of Informatics, Obuda University, 1034 Budapest, Hungary
* Correspondence: amir.mosavi@uni-obuda.hu

Citation: Nasseralshariati, E.; Mohammadzadeh, D.; Karballaeezadeh, N.; Mosavi, A.; Reuter, U.; Saatcioglu, M. The Effect of Incorporating Industrials Wastewater on Durability and Long-Term Strength of Concrete. *Materials* **2021**, *14*, 4088. https://doi.org/10.3390/ma14154088

Academic Editor: Yeonung Jeong

Received: 27 June 2021
Accepted: 14 July 2021
Published: 22 July 2021

Abstract: Concrete, as one of the essential construction materials, is responsible for a vast amount of emissions. Using recycled materials and gray water can considerably contribute to the sustainability aspect of concrete production. Thus, finding a proper replacement for fresh water in the production of concrete is significant. The usage of industrial wastewater instead of water in concrete is considered in this paper. In this study, 450 concrete samples are produced with different amounts of wastewater. The mechanical parameters, such as slump, compressive strength, water absorption, tensile strength, electrical resistivity, rapid freezing, half-cell potential and appearance, are investigated, and a specific concentration and impurities of wastewater that cause a 10% compressive strength reduction were found. The results showed that the usage of industrial wastewater does not significantly change the main characteristics of concrete. Although increasing the concentration of wastewater can decrease the durability and strength features of concrete nonlinearly, the negative effects on durability tests are more conspicuous, as utilizing concentrated wastewaters disrupt the formation of appropriate air voids, pore connectivity and pore-size distribution in the concrete.

Keywords: sustainable concrete; wastewater; industrial waste management; sustainable development; sustainable construction materials; circular economy; recycling; materials design; construction materials; materials properties

1. Introduction

In the modern era, concrete is one of the most used materials in the construction industry. In fact, the only other substance that humans use more than concrete is water, which indicates the importance of concrete and the water used for it [1–4]. Since the first time concrete was utilized as a building material, fresh water was used to cure the cement [5]. The performance of concrete that is made of wastewater has also been investigated; however, further research is essential for examining whether using wastewater is financially feasible and could meet construction standards [6]. There is a research gap in the life cycle assessment, environmental, functional, physical and economic aspects of using wastewater; filling this gap could lead to a revolutionary movement in the construction industry [7]. Bearing in mind the amount of water required for construction projects, if potable water could be substituted with recycled water, it would reduce costs but it would also prevent the wastage of an enormous amount of drinkable water resources [8]. Rivers and fountains that are not contaminated by domestic wastewaters and do not have a salty taste are appropriate for concrete mix designs [9]. Researchers also have indicated that the lake water, which contains less silt, organic materials and impurities, has insignificant adverse effects on concrete features; however, other comprehensive studies are needed on other potential replacements [10]. In industrial and urban areas with limited drinkable resources, and according to fast enhancement in the industry, the demand for water storage is being felt more and more [11]. According to the majority of scientists, the best way to make construction materials is to use the residue of materials, and one of the most prominent construction materials is concrete, of which approximately 5 million cubic meters is used per year globally [11]. This significant value can be seen as an excellent opportunity to use wastewater in concrete, containing 28% of the water cycle [12]. It is undeniable that one of the most usable basic materials in industrial towns is water, which becomes wastewater after use, and is highly harmful to human health and the environment. Concerning the potable water crisis, especially in the Middle East and Africa, finding other water resources as a suitable replacement rather than drinkable water for producing and curing concrete has drawn significant attention, leading to a search for solutions that not only economize cost and energy but also present novel methods. As a result, burying harmful materials and better productivity are obtainable, and less detrimental influences on the environment are expected. According to the United Nations (UN) world water development report, a series of global actions have been taken over five years, costing over 25 billion dollars in order to have healthy infrastructures for water and wastewater [12]. It is worth mentioning that the amounts of produced industrial wastewater and sludge in the United States of America are 119 billion gallons and 17 million tons per year, respectively. These statistics for Europe are 123 billion gallons and 18.9 million tons, respectively [12]. Therefore, according to the huge volume of industrial wastewater and its harmful impacts on the environment, the current study is urgently required.

In this research, help is provided to find the level of wastewater refinement to be used in concrete production. This can help to keep the wastewater infrastructures well maintained due to the massive amount of caustic materials in industrial wastewater. It defines what amount of impurity in a sample can cause a less than 10% loss of compressive strength, compared with a control sample; this is a crucial factor because it can help with the approval of the use of different types of wastewater as appropriate replacements for drinkable water. Al-Ghusain et al. [13] reported on primary, secondary and tertiary treated wastewater, which was taken from the local wastewater plant. The water they utilized did not change the slump; however, the setting time was increased by lowering the water quality. They explained that impurities in the water of a concrete mix design impose different effects on setting time and strength, and also create some stains on the concrete's external surface. Not all impurities harm concrete and some reactions can be neutral or even suitable for concrete. Shekarchi [14] used biologically treated wastewater in concrete mixing and curing. Physical and mechanical tests were performed on mortar and concrete cube specimens. Some durability tests of concrete were also evaluated. When the mixing and curing of concrete was carried out in primary and secondary water, the compressive strength increased up to 17% more than in the concrete mixed and cured in tap water, for up to 180 days. After 180 days, concrete that was mixed and cured in primary treated water showed a small reduction, and when secondary treated water was used for mixing or curing in concrete, compressive strengths were decreased from 9% to 18%. The water absorption of the concrete mixed in tap water and that mixed in treated wastewater was identical. Curing in secondary wastewater increased the water absorption of the specimen. These results showed the feasibility of biologically treated water in the concrete production industry. Asadollahfardi et al. [15] studied using concrete wash water to produce concrete. Their results indicated that concrete wash water is suitable for producing fresh concrete. This research is based on the compressive strength, flexural strength, abrasion resistance, chlorine resistance and carbonation resistance of treated wastewater concrete (10%, 25%, 50% and 100% replacement with tap water) and compares the results with control concrete. This research shows the feasibility of using treated wastewater in concrete to reduce the consumption of fresh water in the concrete industry, as well as solving the problem of disposing of industrial wastewater. Asadollahfardi et al. [16] used treated domestic wastewater instead of drinking water to produce and cure concrete samples. Their results indicated that the compressive strength of the samples made with treated domestic wastewater at the age of 28 days was 93–96% of the compressive strength of the control samples made with drinking water. The use of treated domestic wastewater also did not have much effect (less than a 4% decrease in resistance) on the tensile strength of the concrete samples; however, a final setting time of the cement was delayed by 15 min was observed. Domenico et al. [17] conducted research on the structural behavior of RC beams containing EAF slag as recycled aggregate. The authors stipulated that EAF slag has a remarkably higher specific weight (evaluated macroscopically with the pycnometer test method), which provides roughly the upper limit of the slag. This is in principle, because of the high content of metallic iron, iron and manganese oxides (which have a density higher than 5000 kg/m^3), which compose the slag. Their results indicated that the existence of steel slags results in more shear potential than the traditional RC beams, crack widths are smaller and the basic ductility is augmented. Bahraman et al. [18] carried out research on the feasibility of using both wash water from a ready mixed concrete plant and synthetic wastewater. According to the visual stability index and slump flow time results, they reported that the utilization of either wash water from the ready-mixed concrete plant or synthetic wastewater impose destructive impacts on the workability of concrete in comparison with the control sample. Likewise, it was indicated that outcomes of the J-ring and column segregation index of individual self-compacting concrete escalated in comparison with the control sample. Besides, although the 28 days' compressive strength of all specimens, using wash water or synthetic wastewater instead of tap water, was reduced, the concrete that contained synthetic wastewater (1000 mg/L total dissolved solid) had 13.335% more compressive strength. Taherlou et al. [19] studied the practicability of using a variety of percentages of simultaneous municipal solid waste incineration bottom ash and treated industrial wastewater in self-compacting concrete. It was illustrated that the workability of different self-compacting concrete mixtures, including different percentages of municipal solid waste, can reach a satisfactory level within the European guidelines (ASTM C1585) by utilizing the rate of the superplasticizer. In addition, the compressive strength increased more by using solid waste and the treated industrial wastewater compared to self-compacting concrete samples using tap water. The SEM images showed fewer pores and cracks while utilizing the treated industrial wastewater in self-compacting concerts. Ali Raza et al. [20] assessed the mechanical and durability behavior of recycled aggregate concrete made with different kinds of wastewater, including that from domestic sewerage, a fertilizer factory, a textile factory, a sugar factory and a service station. It was observed that, by utilizing the wastewater taken from the textile factory, the compressive strength and split tensile strength were 19% and 16% higher compared to concrete produced with drinkable water. Moreover, the specimen made with domestic sewerage for the mixing had 13.88% improvement, which was the highest water absorption of all utilized wastewaters.

Undeniably, concrete production is the main reason for a considerable amount of energy consumption as well as CO_2 production. Therefore, it is vital to substitute new promising ways in which components are replaced by other materials, yet rapt attention should be paid to recycled keys [21]. Clearly, there is still a research gap in the functional and economical aspects of using wastewater. Due to development in all industrial sectors, which has flourished and enlarged industrial towns as well as increased the human population, coupled with the demand to hamper expenditure in different government budget sectors, attention should be focused on the reuse of resources if possible. In previous studies, the feasibility of using wastewater, including treated domestic wastewater and concrete wash water, as well as primary, secondary and tertiary treated wastewater, was evaluated. Nevertheless, the precise effect of a variety of industrial wastewater types with different concentrations was not specified, especially concerning its behavior in terms of diluting and concentrating. Another important reason to research industrial wastewater is that the result can be expanded to other kinds of wastewater because it has the highest level of impurity and chemical parameters, so any solution for this type of wastewater could be applied to other weaker wastewaters. Thus, detailed research with different durability and strength test ranges, from primary industrial wastewater, a variety of treated wastewater concentrations and a standard control sample, was required. In brief, a few unique results of this research are as follows: (i) presenting an optimum level of industrial refinement for using it in concrete; (ii) providing a vivid understanding of the linearity or non-linearity behavior of specimens and their performances by diluting or concentrating industrial wastewater; (iii) defining the impurity level of a sample that can cause a 10% compressive strength reduction in comparison with a control sample; (iv) the effect of impurity and industrial parameters on concrete specimens, including ITZ region, pore connectivity, air void parameters, pore structure and size-distribution, air content and so forth.

In the present research, different concentrations of industrial wastewater were used for producing concrete specimens. Subsequently, the durability and strength of concrete specimens within 365 days were assessed and compared with the control specimen, which was produced with drinkable water. Eventually, a statistical analysis is presented to augment the level of strength prediction in concrete based on the obtained results.

2. Materials and Methods

2.1. Method of Examination

The wastewaters were gathered from Toos industrial town, Mashhad, Iran and within a maximum of three hours, they were analyzed in the laboratory. The analyses were performed on industrial primary wastewater, treated wastewater, diluted treated wastewater, and concentrated treated wastewaters. The control specimen was produced with drinkable water from Mashhad City, which is a standard water. Altogether, 430 specimens were created, pouring concrete ten times and fourteen skilled operators participated in producing them, which took two hours in total. The number of completed tests on specimens were as follows: slump 10 samples, compressive strength 240 samples, electrical resistivity 20 samples, water absorption of thirty minutes 10 samples, mass water absorption 10 samples, capillary water absorption 30 samples, tensile strength 40 samples, rapid freezing and thawing 40 samples and half-cell 30 samples. All of the tables, results, and tests were carried out exclusively for this research, and no archive data were used. The Technical and Vocational University (TVU), Mashhad, Iran, provided the researchers with testing facilities. The used standards are shown in Table 1.

Table 1. The types of testing and the corresponding Standards.

Type of Testing	Method of Testing
Chemical and physical properties of treated wastewater	APHA [22]
Standard Specification for Mixing Water Used in the Production of Hydraulic Cement Concrete	ASTM C1602M-18 [23]
Standard specification for Portland cement	ASTM-C150 [24]
Standard test method for density of hydraulic cement	ASTM C188-15 [25]
Standard test method for sieve analysis of fine and coarse aggregates	ASTM-C136 [26]
Standard specifications of concrete aggregates	ASTM-C33 [27]
Standard test methods for the time of setting of hydraulic cement by Vicat needle	ASTM-C191 [28]
The slump of hydraulic-cement concrete	ASTM C143 [29]
Testing hardened concrete. Compressive strength of test specimens	BS EN 12390-3 [30]
Standard test method for splitting tensile strength of cylindrical concrete	ASTM-C496 [31]
Absorption of concrete water	BS1881-122 [32]
Florida method of test For Concrete resistivity as an electrical indicator of its permeability	FM-5-578 [33]
Water absorption rate by hydraulic-cement concretes	ASTM-C1585 [34]
Concrete resistance against thawing and rapid freezing	ASTM-666 [35]
The standard method for the test of half-cell potentials of uncoated reinforcing steel	ASTM- C876-15 [36]
Standard test method for density, absorption, and voids in hardened concrete	ASTM C642-13 [37]
Standard test method for air content of freshly mixed concrete by the pressure method	ASTM C231/M17a [38]

The methodology consisted of several stages of operation and processing. Figure 1 represents the methodology strategy and functional stages in detail. Sampling, conditional stages, and experimental tests are the foundation of the methodology described in Figure 1.

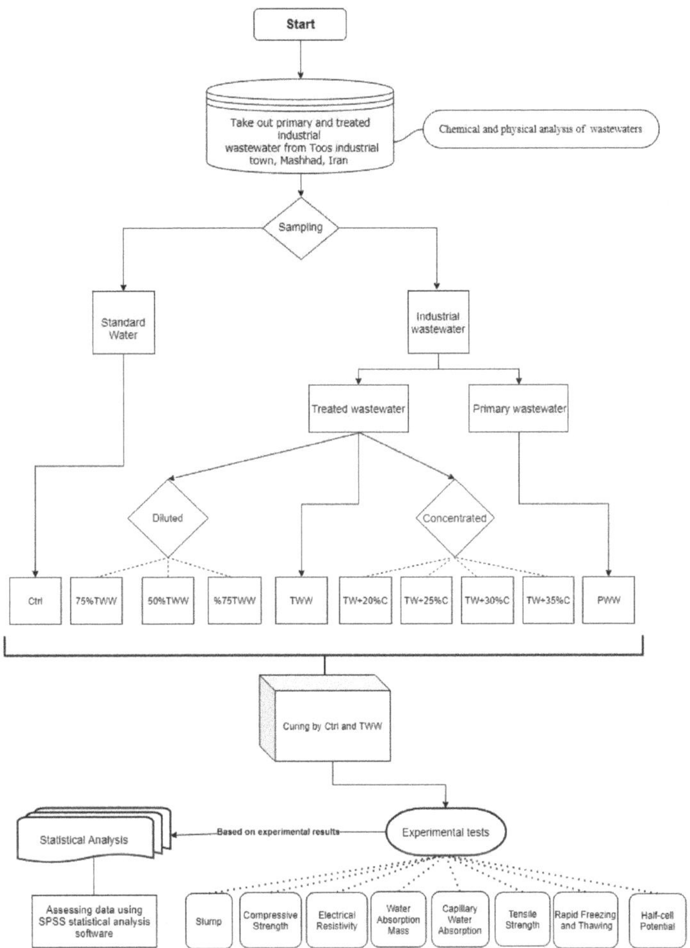

Figure 1. Flowchart of methodology.

In this research, ten different groups of specimens were produced with different wastewater concentrations. All groups of specimens had the same mix design, and no additive was used in order to figure out the exact effect of wastewater concentration on concrete durability and strength features. In this study, one of the targets was to find the optimum concentration of treated wastewater that may cause a less than 10% compressive strength loss compared with the control sample (made with drinkable standard water). Technically, 10% compressive strength reduction could still be counted as an acceptable substitution for the water in a concrete mix design [39]. The main control sample was produced with the potable water of Mashhad city. The used industrial wastewaters were categorized into four groups including treated wastewater, diluted treated wastewater, concentrated treated wastewater, and primary wastewater. All groups of concrete specimens were produced in a similar situation and were cured in drinkable water or treated wastewater according to the test standards and intended purposes. The parameters, such as concrete density, temperature, moisture, cement type and aggregates characteristics, were used in the same condition for all specimens.

2.2. *Wastewaters*

For producing concrete with wastewaters, the amount of distilled water was considered, based on the quality of the control specimen, and all other extra substances were subtracted. The majority of the time, there is an allowable limit for the water of mix design; within those restrictions, the impurity can be harmless and acceptable. Nevertheless, there is no limitation for organic materials in concrete and it is assumed that only wastewater impurities are the reasons for negative effects on the water in concrete mix design.

2.2.1. Treated Wastewater (TWW)

Treated wastewater is also known as output wastewater and goes through three steps of refinement including filters, aeration and chlorination. Treated wastewater was used as the main replacement for drinkable water and for producing distilled and concentrated specimens as well. TWW was used for curing the specimens if they were intended to be cured by wastewater separately. The characteristics of TWW are presented in Table 2.

Table 2. Chemical and physical characteristics of treated wastewater and primary wastewater.

No.	Parameter	Unit	Treated Wastewater	Primary Wastewater	Mashhad Potable Water (Ctrl)
1	pH	-	7.92	7.68	7.2
2	TDS	mg/L	1870	2541	412
3	SALT	mg/L	2.4	2.51	40
4	EC	mg/L	3950	4120	193
5	COD	mg/L	150	3215	0
6	BOD	mg/L	114	1240	3
7	TSS	mg/L	25	451	121
8	Ammonium	mg/L	2	3	0.4
9	Detergent	mg/L	1.25	3.1	-
10	Color	-	Light brown	Black	Transparent
11	Temperature	°C	17	17–19	25
12	Sulfate	mg/L	80	145	50
13	Chloride	mg/L	1230	740	94
14	Chromium	mg/L	0.9	1.89	0.1
15	Cadmium	mg/L	0.7	2.95	-
16	Lead	mg/L	2.85	2.85	0.02
17	Turbidity	Nephelometric Turbidity Unit	10	800	2

2.2.2. Diluted Treated Wastewater (%TW)

Diluted specimens were produced by TWW plus mixing with distilled water. They contained 75% wastewater (75%TW), 50% wastewater (50%TW), and 25% wastewater (25%TW), respectively, and the remainder was distilled water. These water percentages of the mix designs were selected in order to investigate the existence of linear or non-linear relationships in strength and durability features by diluting treated wastewater as the water of the mix design. Based on the laboratory results, the number of parameters was reduced correctly by dilution percentages. In order to obtain the number of parameters in diluted specimens, the characteristics of treated wastewater (Table 2) should be reduced by dilution percentages.

2.2.3. Concentrated Treated Wastewater (TW + %C)

Concentrated specimens were produced from TWW by evaporation; concentrating percentages were 20% (TW + 20%C), 25% (TW + 25%C), 30% (TW + 30%C), and 35% (TW + 35%C), respectively. Based on laboratory results, the parameters of the thickened specimens were increased by the concentration percentages. So, concentrated specimens had the same parameters as the treated wastewater (Table 2) but their characteristics were 20%, 25%, 30% and 35% more than the characteristics of the treated wastewater, respectively. According to the intended concentration, the amount of surplus treated wastewater was added and after time at a precise warming temperature, the intended concentration was achieved. Reaching the intended concentration by way of evaporation is almost acceptable, but sufficient accuracy for important parameters such as COD, BOD, Sulfate, Chromium, Cadmium and Salt was considered and double-checked.

2.2.4. Primary Wastewater

The initial discharge of industrial wastewater is primary wastewater, which is from a collection of several polluting industries such as pharmaceutical, food, ironmaking and chemical. It contains many organic materials and caustic heavy metals such as Cadmium and Chromium because it does not go through any refinement process and, technically, this is the TWW before the refinement procedure. PWW contains a huge amount of organic materials, microorganisms and heavy metals, which are mostly harmful and caustic for both the environment and concrete. Table 2 shows the characteristics of primary wastewater (PWW). In Table 2, TDS, EC, COD, BOD and TSS stand for Total Dissolved Solids, Electrical Conductivity, Chemical Oxygen Demand, Biochemical Oxygen Demand and Total Suspended Solids, respectively.

2.3. Concrete Preparation

For producing the control sample and curing all groups with normal water, the potable water of Mashhad, Iran, was used. In order to conduct this research, concrete cubic samples ($100 \times 100 \times 100$ mm), including 400 kg cement per cubic meter, were made and tested; for each test, the required standard, specification and introduction were fully considered. The Portland cement type II, produced by Mashhad cement factory, Mashhad (Iran), was chosen and its quality was tested according to the ASTM-C150. The strength class of the Portland cement was 42.5R, and the desired strength of the concrete was 35 Mpa, which was achieved based on breaking cylinder specimens and regarding the average result. Moreover, the concrete grade was M35, which was commensurate with the achieved results. The desired workability based on NS 8500 EN 165 was S3. The quality of the required material in the production of concrete, the chemical and physical analysis of water and wastewaters, and the sieve analysis of the aggregates were experimentally assessed. The resistance of degradation in large size coarse aggregate to abrasion and impact was carried out using a Los Angeles machine. In addition, the soundness of the aggregates was evaluated using the sodium sulfate or magnesium sulfate and sulfate content of the aggregates. In this research, the following parameters were also checked: the hydraulic cement autoclave expansion, the amount of water essential for an ordinary consistency of hydraulic cement, and the setting time of hydraulic cement using a Vicat needle. The ASTM C33 standard was adopted for checking the coarse and sand sizes using a sieve assessment and the precise percentage passing through sieve number 200 was determined. Table 3 shows the chemical and physical properties of the cement. Tor educe the effect of other parameters on the concrete, except for wastewater, a good-quality, continuous, less flawed aggregate was used [40]. The ASTM-C33 [27] standard was adopted to test the characteristics of the aggregates. It should be noted that the aggregates used in this study were kept in SSD condition. The mix design for all groups of specimens was the same and the concrete mixture is presented in Table 4. The concrete specimens were molded in metal molds and cured based on ASTM-C31 [41]. A separate set of specimens was cured by potable water, and a separate set of specimens was cured by treated wastewater. In addition, all requirements were considered in terms of curing and the storage of test specimens before rupture based on ASTM-C31 [41]. The concrete mixture is presented in Table 4. For curing purposes, the temperature in the laboratory was 25 centigrade while the relative humidity varied from %30 to %45 throughout the time of curing and testing.

For both curing methods, curing in water and wastewater was conducted at the same time. All specimens were dried by oven prior to testing for water absorption. %clearpage

Table 3. Chemical and physical properties of cement.

Chemical & Physical Measurands	Units	Test Method	ISIRI 389	EN 197-1: 2011	Sample Analysis
SiO_2	%	ASTM C114:2011b	>20.00	-	21.77
Al_2O_3	%	ASTM C114:2011b	<6.00	-	5
Fe_2O_3	%	ASTM C114:2011b	<6.00	-	4.3
CaO	%	ASTM C114:2011b	-	-	63.13
mgO	%	ASTM C114:2011b	<5.00	<5.00	1.78
L.I.O	%	EN 196-2:2013	<3.00	<5.00	1.38
SO_3	%	EN 196-2:2013	<3.00	<3.5	2.22
IR	%	EN 196-2:2013	-	<5.00	0.63
Na_2O	%	EN 196-2:2013	-	-	0.32
K_2O	%	EN 196-2:2013	-	-	0.83
CI	%	EN 196-2:2013	-	<0.10	0.010
Free CaO	%	EN 196-2:2013	-	-	1.10
Cao/SIO_2	-		-	>2.0	2.90
$C_3S + C_2S$	%		-	>66.667	73.48
Fineness	cm^2/gr		>2800	-	3000
Le Chatelier Expansion	mm	EN 196-3:2005	-	<10.00	0.9
Initial Setting Time	min	EN 196-3:2005	>45	>75	116
Final Setting Time	min	EN 196-3:2005	<360	-	175
3 days Com. Strength	MPa	EN 196-3:2005	-	-	16.8
7 days Com. Strength	MPa	EN 196-3:2005	-	-	23.2
28 days Com. Strength	MPa	EN 196-3:2005	-	>32.5, <52.5	45.3

Table 4. The detail of concrete mixture designs.

Sample	Free Water Mass	Wastewater Mass	Cement Mass	Sand Mass
Control (Ctrl)	168 kg	-	400 kg	974 kg
Treated wastewater (TWW)	-	168 kg	400 kg	974 kg
Concentrated treated wastewater (TW + %C)	-	168 kg	400 kg	974 kg
Diluted treated wastewater (%TW)		168 kg	400 kg	974 kg

To reach the optimum mix design, the ACI method of concrete mix design was used based on the water–cement ratio of 0.42 and, for the mechanical mixing of the cement, ASTM-C305 [42] was adopted. The good-quality and washed aggregates were selected after several initial samples according to the details in Table 5. In Appendix A, Tables A1 and A2 present other details of the aggregates used in this study for constructing concrete.

Table 5. Detail of mix design of concrete samples.

Parameter	Control (Ctrl)	Treated Wastewater (TWW), Concentrated Treated Wastewater (TW + %C), Diluted Treated Wastewater (%TW)	Primary Wastewater (PWW)
Free water mass	168 kg	-	-
Wastewater mass	-	168 kg	168 kg
Cement mass	400 kg	400 kg	400 kg
Sand mass	974 kg	974 kg	974 kg
Fine gravel mass	185 kg	185 kg	185 kg
Coarse gravel mass	576 kg	576 kg	576 kg
Stone powder	74	74	74

3. Results and Discussion

3.1. Slump

The slump shall be consistent with the placement and consolidation methods, equipment, and site conditions and shall be identified by the contractor and concrete supplier prior to construction. According to the achieved results, TWW had less workability than the control sample. Diluted specimens reacted like TWW, which shows that the existence of the wastewater can affect the workability even at low percentages. The concentrated specimens followed the same method of treated TWW, but TW + 25%C had a reduction and stayed in the next specimens too. The TWW had 13.3% lower workability than the control specimen and by increasing the concentration of treated wastewater TW + 25%C by 25%, the workability declined 20% more than the control. It clearly showed that wastewater has a subtractive effect on workability and it is dependent on wastewater concentration. So, it is highly recommended, for projects with a high required workability, that the additives should be considered to increase the slump, especially when more concentrated wastewaters are used as the water of the mix design. No linear relationship was observed in any specimen when their concentration was increased or decreased. The concentrated specimens were more viscous and greasy, which is one of the reasons why concentrated specimens had less workability; it was obvious in the PWW sample, which had the highest impurities. Figure 2 represents the slump test results.

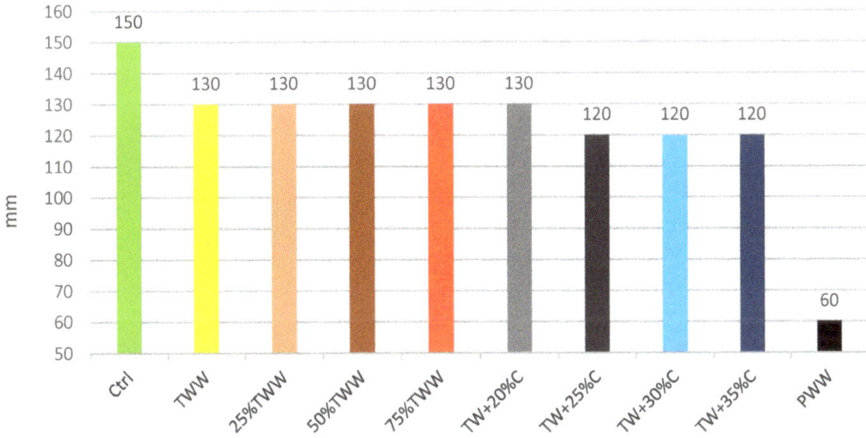

Figure 2. Slump test results.

3.2. Compressive Strength

The compressive strength results at different ages and days are shown for specimens cured by drinkable water (Figure 3) and cured by treated wastewater (Figure 4). The compressive strength was obtained by testing cubic specimens according to the BS EN 12390-3. The cubes were tested in a 3000 kN testing machine at a rate of 2.5 kN/s. The control sample had the highest strength at all ages, substantiating that the best result can be achieved by using drinkable water. TWW had lower strength than the control, but its reduction was insignificant. So, it demonstrated that treated industrial wastewater is applicable for use in concrete. The compressive strength in wastewater specimens was better when the concentrations of water for producing and curing were the same. It vividly showed the homogeneity and similarity features between the curing situation and water of the mix design. For instance, at the age of 7, 28, 90 and 365 days, when TWW and 75%TW were cured by treated wastewater, they had 0.54%, 1.65% 1.06%, 1.55%, 2.86%, 0%, 3.6% and 1.06% more compressive strength than when cured by standard water, respectively. Besides, 25%TW, in which its mixed design water was roughly similar to drinkable water, had 1.62%, 1.1%, 1.6% and 1.02% more compressive strength when it was cured by standard water at different ages. The concentrated specimens at low ages had a better performance when they were cured by treated wastewater but at late ages, better results were shown for those cured by drinkable water. The positive effect of curing with treated wastewater for those specimens produced by treated wastewater disappeared by increasing the specimens' concentration and was changed adversely. For instance, TW + 35%C cured by treated wastewater had 3%, 3.1%, 2.4% and 2.6% less compressive strength when it was cured by treated wastewater than when it was cured by drinkable water. PWW produced by primary wastewater had the highest impurity, and corroborated this result, having 2.6%, 5.3%, 4.5% and 4.3% less compressive strength when it was cured by treated wastewater. Neither in the diluted specimens nor the concentrated specimens was a linear relationship was observed and a non-linear relationship was dominant; however, the concentration of specimens was increased and decreased in order.

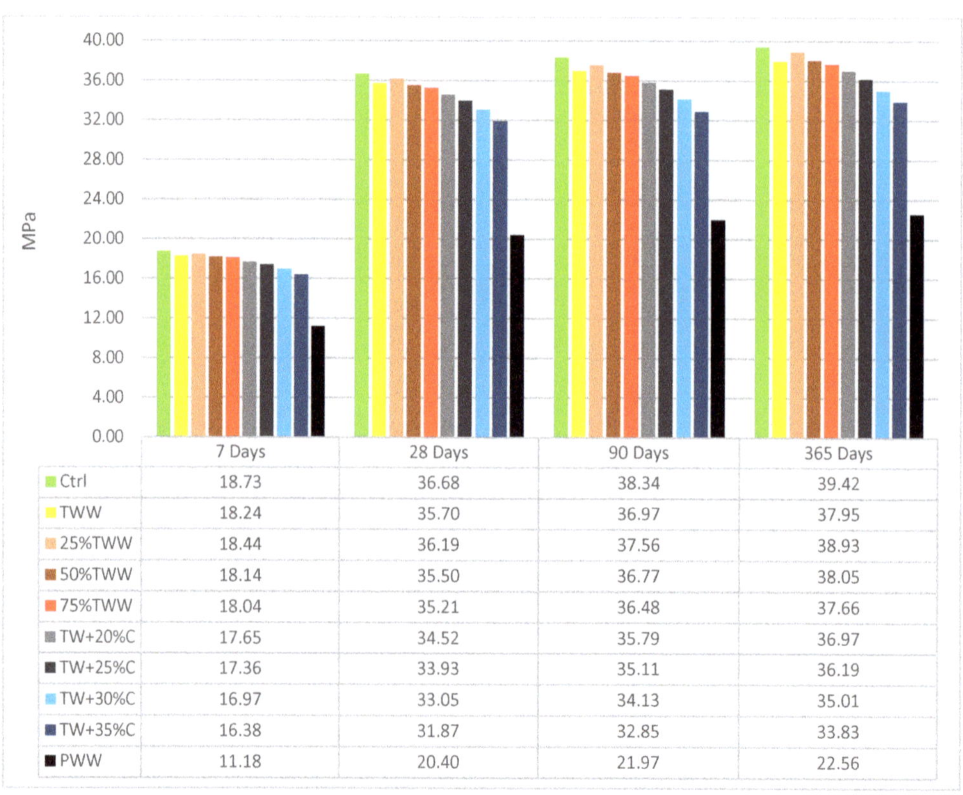

	7 Days	28 Days	90 Days	365 Days
Ctrl	18.73	36.68	38.34	39.42
TWW	18.24	35.70	36.97	37.95
25%TWW	18.44	36.19	37.56	38.93
50%TWW	18.14	35.50	36.77	38.05
75%TWW	18.04	35.21	36.48	37.66
TW+20%C	17.65	34.52	35.79	36.97
TW+25%C	17.36	33.93	35.11	36.19
TW+30%C	16.97	33.05	34.13	35.01
TW+35%C	16.38	31.87	32.85	33.83
PWW	11.18	20.40	21.97	22.56

Figure 3. Compressive strength cured by drinkable water.

By aging, concentrated specimens had less compressive strength growth than the control sample and, by increasing the specimens' concentration, the reduction increased. One of the most important intentions of this study was to find the impurity and concentration of wastewater which causes a 10% reduction in the compressive strength of concrete in comparison to the control after 28 days. Based on Figures 3 and 4, TW + 30%C, cured by drinkable water and treated wastewater at the age of 28 days, had 9.9% and 10.7% less compressive strength than the control, respectively. This clarified the largest amount of impurity in industrial treated wastewater which can be still acceptable for use in concrete mix design [36]. The chemical and physical characteristics of TW + 30%C were as follows: BOD: 150 mg/L, COD: 200 mg/L; Total Dissolved Solid: 1924 mg/L; Total Suspended

Solids: 33 mg/L; Sulfate: 98 mg/L. It exhibited the optimum concentration of wastewater for refining to balance mechanical, durability and physical characteristics.

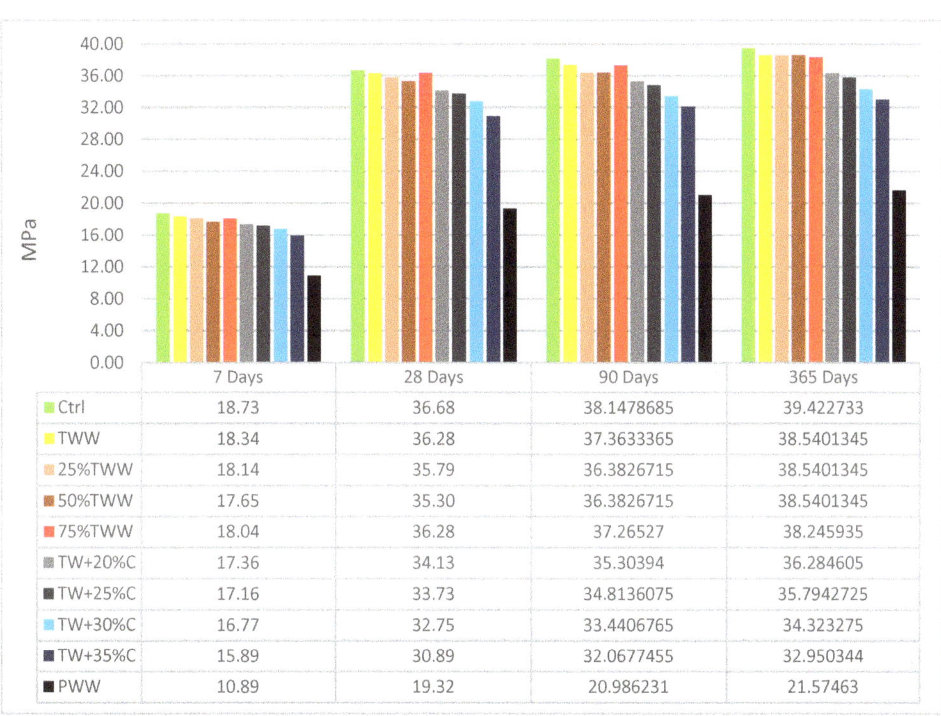

	7 Days	28 Days	90 Days	365 Days
Ctrl	18.73	36.68	38.1478685	39.422733
TWW	18.34	36.28	37.3633365	38.5401345
25%TWW	18.14	35.79	36.3826715	38.5401345
50%TWW	17.65	35.30	36.3826715	38.5401345
75%TWW	18.04	36.28	37.26527	38.245935
TW+20%C	17.36	34.13	35.30394	36.284605
TW+25%C	17.16	33.73	34.8136075	35.7942725
TW+30%C	16.77	32.75	33.4406765	34.323275
TW+35%C	15.89	30.89	32.0677455	32.950344
PWW	10.89	19.32	20.986231	21.57463

Figure 4. Compressive strength cured by treated wastewater.

3.3. Electrical Resistivity

The level of permeability of concrete has a direct effect on the electrical resistivity of specimens. This test indicates specimens' permeability and specifies existing voids and cracks in the concrete structure, which have a significant effect on concrete durability [43]. Figures 5 and 6 present the electrical resistivity of specimens at the age of 7, 28, 90, 180 and 365 days.

The control sample had better resistance when it was cured by drinkable water and a reduction was observed when cured by treated wastewater. By increasing the concentration, electrical homogeneity decreased. The diluted specimens' behavior was inclined towards the TWW results and not the control, even when insignificant amounts of treated wastewater were involved. For example, 25%TW, which is 75% drinkable water, followed the TWW's resistance and not that of the control, cured by either drinkable water or treated wastewater. It showed that whenever wastewater parameters are involved in the specimens, they could exceedingly influence the concrete's structure and create voids and porosity in specimens. So, diluting the concentration has an insignificant effect on electrical resistivity enhancement. TWW and specimens with concentrations close to that of TWW had better resistivity when they were cured by treated wastewater at lower ages; however, with aging the positive effect declined even on them, as if being cured by treated wastewater in the long term has caustic effects on concrete structure and causes more avenues of penetration. Nevertheless, in specimens with a lower concentration at an early age, a resistance growth was observed which again supported the positive effect of homogeneity features, as well as the negative effect of being cured by treated wastewater in the long term.

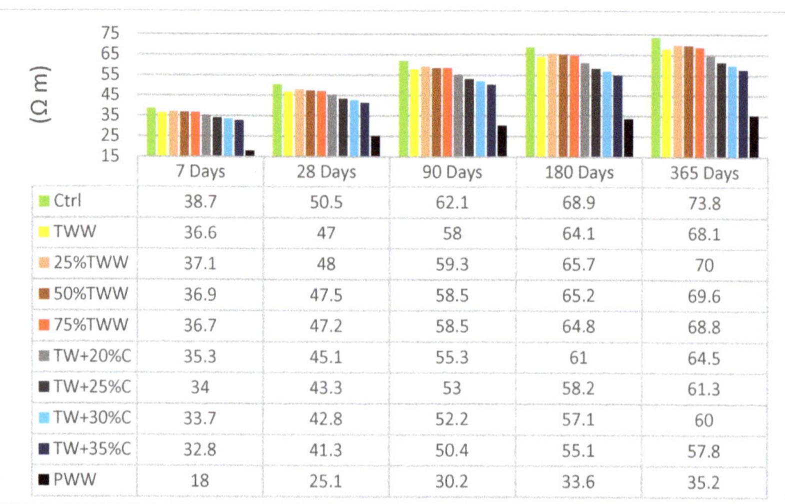

	7 Days	28 Days	90 Days	180 Days	365 Days
Ctrl	38.7	50.5	62.1	68.9	73.8
TWW	36.6	47	58	64.1	68.1
25%TWW	37.1	48	59.3	65.7	70
50%TWW	36.9	47.5	58.5	65.2	69.6
75%TWW	36.7	47.2	58.5	64.8	68.8
TW+20%C	35.3	45.1	55.3	61	64.5
TW+25%C	34	43.3	53	58.2	61.3
TW+30%C	33.7	42.8	52.2	57.1	60
TW+35%C	32.8	41.3	50.4	55.1	57.8
PWW	18	25.1	30.2	33.6	35.2

Figure 5. The results of concrete electrical resistivity tests cured by standard water.

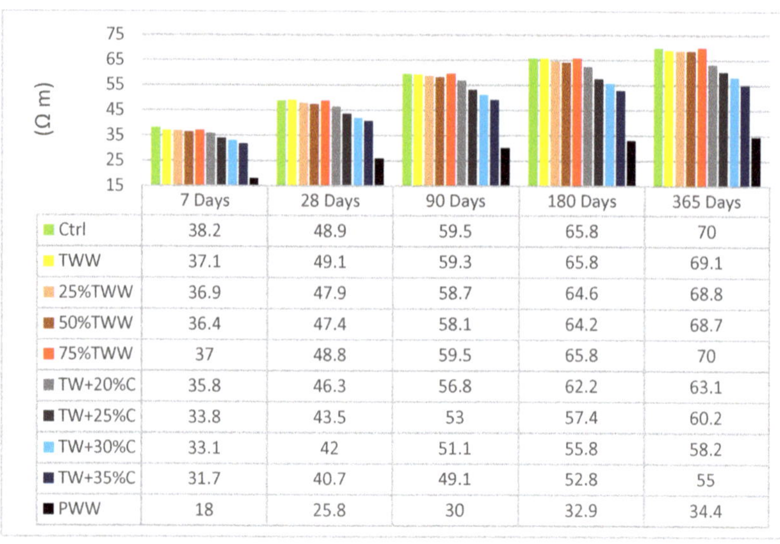

	7 Days	28 Days	90 Days	180 Days	365 Days
Ctrl	38.2	48.9	59.5	65.8	70
TWW	37.1	49.1	59.3	65.8	69.1
25%TWW	36.9	47.9	58.7	64.6	68.8
50%TWW	36.4	47.4	58.1	64.2	68.7
75%TWW	37	48.8	59.5	65.8	70
TW+20%C	35.8	46.3	56.8	62.2	63.1
TW+25%C	33.8	43.5	53	57.4	60.2
TW+30%C	33.1	42	51.1	55.8	58.2
TW+35%C	31.7	40.7	49.1	52.8	55
PWW	18	25.8	30	32.9	34.4

Figure 6. The results of concrete electrical resistivity tests cured by TWW.

At the age of 28 days, TW + 30%C, cured by drinkable and treated wastewater, had 15% and 14% less electrical resistivity than the control sample, respectively. This indicated that using wastewater in concrete has more negative effects on the concrete's durability features than on its strength, because it had 10% reductions in the compressive strength test but a 15% reduction in electrical resistivity. Therefore, it is recommended that treated wastewater should not be used for projects with high contact with caustic material or marine projects.

3.4. Water Absorption Mass

For this test, specimens are dried using an industrial oven. In the drying process, heating and cooling at a special set of temperatures in both the oven and desiccator over a determined period are used. At every step, the materials are weighed. The workflow of a water absorption test is based on drying specimens in a laboratory or industrial oven according to BS1881-122 [30]. The time and temperature for using the oven and the cooling time are set for the test. Specimens are weighed both before and after the cooling. The test is followed by immersing the materials in the water, at 23 °C, for one day. Materials can be kept in the water either for the whole day or until equilibrium. Table 6 indicates the results of mass water absorption, which has a significant relationship with concrete permeability. The less porous the structure of concrete is and the less cracks it has, the less possibility exists for the movement of harmful parameters into the structure of the concrete; consequently, concrete corrosion is less expected. Hence, based on BS1881-122, the allowable water absorption is restricted to between 2% and 5%. In this test, except for PWW, all specimens stood in the allowable limitation after 72 h; however, TW + 35%C stood at the edge of rejection. This test showed not only that using wastewater increases water absorption, but also that the rate of age to age water absorption growth is more than that of the control, which is improper. For instance, control samples from 1 h to 72 h had 49.5% water absorption growth but TWW and TW + 30%C had 53.3% and 63.4%, respectively. The TW + 30%C had 39.5% more mass water absorption than the control, which shows the exactness of 30 min water absorption results. Table 6 shows the mass water absorption.

Table 6. Mass water absorption.

	1 H (%)	3 H (%)	24 H (%)	72 H (%)
Ctrl	2.10	2.62	2.93	3.14
TWW	2.42	3.05	3.54	3.71
25%TW	2.30	2.78	3.05	3.2
50%TW	2.30	2.85	3.25	3.45
75%TW	2.34	2.94	3.38	3.52
TW + 20%C	2.55	3.21	3.60	3.88
TW + 25%C	2.64	3.39	3.85	4.18
TW + 30%C	2.68	3.48	4.00	4.38
TW + 35%C	2.90	3.83	4.40	4.92
PWW	8.6	11.04	12.45	13.60

3.5. Capillary Water Absorption

The capillary test evaluates the process of non-saturated concrete water absorption by capillary suction while it is in touch with water. Table 7 shows the results of capillary water absorption at 3, 6, 24 and 72 h. Basically, the more moisture concrete contains, the less capillary water absorption will be measured. The capillary water absorption increased by using wastewater; even 25%TW, which contained 75 percent distilled water, had 11.19% more capillary water absorption than the control at 72 h. It showed that treated wastewater, even at low concentrations, influences capillarity absorption and subsequently reduces concrete durability. Using wastewater causes bigger and looser capillary pipes, which are connected to each other and intensify the concrete corrosion. The more and larger capillary pores a concrete has, the more deleterious substances will pass into it superficial and interior layers. For instance, after 72 h, TWW samples had 25.87% more capillary water absorption growth than the control; this growth for TW + 30%C was 95.30%.

Table 7. The results of capillary water absorption ($\frac{gr}{mm^2}$ or mm).

Sample	3 H (%)	6 H (%)	24 H (%)	72 H (%)
CTRL	1.35	1.72	2.00	2.86
TWW	1.66	2.14	2.54	3.70
25%TW	1.46	2.05	2.28	3.18
50%TW	1.53	2.12	2.40	3.38
75%TW	1.57	2.08	2.40	3.44
TW + 20%C	1.80	2.35	2.84	4.22
TW + 25%C	1.88	2.52	3.18	4.86
TW + 30%C	1.94	2.62	3.40	5.22
TW + 35%C	2.05	2.84	3.70	5.65
PWW	8.90	12.88	16.85	29.32

Figure 7 indicates the rate of growth during the test period. The wastewater specimens had more capillary water absorption and growth rates than the control sample. For example, from 24 to 72 h, TWW and TW + 30%C had 4% and 10% more growth than the control. So, based on Tables 8–10, it is highly recommended that wastewater at high concentrations should not be used as the water of concrete mix design when it is going to be used in caustic environments because of the high possibility of corrosion.

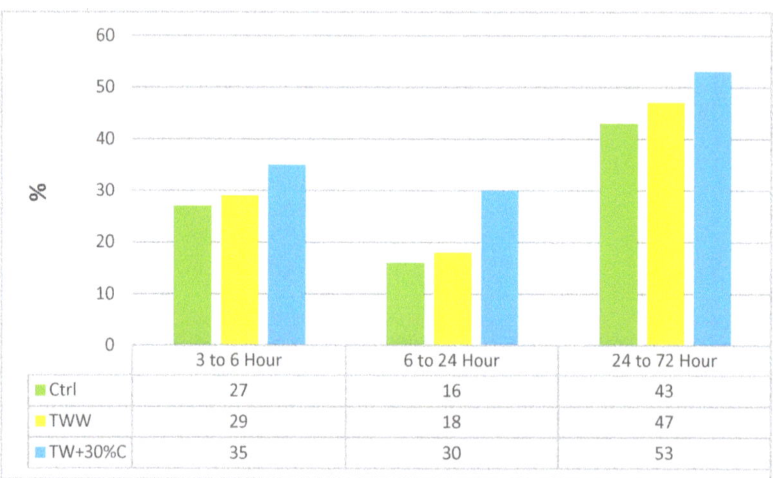

Figure 7. The capillary water absorption growth hour to hour (%).

Table 8. The results of resistance of concrete to rapid freezing and thawing.

Samples	28-Day Compressive Strength (MPa)	50 Cycles (MPa)	100 Cycles (MPa)	150 Cycles (MPa)	200 Cycles (MPa)
Ctrl	36.68	35.70	34.62	33.24	31.58
TWW	35.70	34.13	33.05	31.09	28.73
25%tw	36.19	35.30	34.23	32.75	31.09
50%tw	35.50	34.42	33.34	31.97	29.62
75%tw	35.21	34.13	33.15	31.28	29.13
TW + 20%C	34.52	33.54	32.26	30.01	27.46
TW + 25%C	33.93	32.36	30.99	28.54	25.99
TW + 30%C	33.05	31.38	29.81	27.46	24.81
TW + 35%C	31.87	29.91	28.34	25.99	23.44
PWW	20.40	18.34	16.87	15.20	13.34

Table 9. The start age of armature corrosion.

Sample	Age of Corrosion Start (Day)
Ctrl	32
TWW	27
25%TW	30
50%TW	28
75%TW	28
TW + 20%C	24
TW + 25%C	24
TW + 30%C	24
TW + 35%C	22
PWW	8

Table 10. Statistical characteristics of variables.

Variables	Mean	Maximum	Minimum	Kurtosis	Skewness	Variance	Std. Deviation
TDS (mg/L)	1879.2500	2431.00	1014.00	−1.249	−0.522	272,887.933	522.38677
EC (mg/L)	3935.5000	3959.00	3880.00	−0.461	−1.251	1064.000	32.61901
TSS (mg/L)	25.6250	38.00	10.00	−1.607	−0.306	108.517	10.41713
Detergent (mg/L)	1.2587	1.80	0.70	−1.494	−0.206	0.152	0.39002
Sulfate (mg/L)	77.0000	101.00	36.00	−1.169	−0.680	596.267	24.41857
Chromium (mg/L)	0.8400	0.99	0.50	0.553	−1.242	0.026	0.16199
Cadmium (mg/L)	0.7463	0.90	0.50	0.386	−0.854	0.015	0.12099
Compressive Strength (kg/cm^2)	351.2500	370.00	315.00	−0.221	−0.784	281.133	16.76703
Electrical Resistivity (Ω m)	45.4938	49.10	40.70	−1.340	−0.439	7.898	2.81033
Tensile Strength (kg/cm^2)	25.7875	28.40	21.60	−1.625	−0.346	6.276	2.50516
Rapid Freezing and Thawing (kg/cm^2)	350.3750	370.00	322.00	0.076	−0.842	245.411	15.66559
Water Absorption in 30 min (%)	1.9137	2.30	1.72	−0.829	0.826	0.050	0.22328
Water Absorption Mass 72 h (%)	3.9050	4.92	3.20	−0.031	0.708	0.317	0.56346
Capillary Water Absorption 72 h (%)	4.2063	5.65	3.18	−1.504	0.491	0.874	0.93496

3.6. Tensile Strength

The tensile strength of concrete is a prominent property when it is to be utilized for making prestressed concrete structures, roads and runways. Figures 8 and 9 illustrate the results of tensile strength testing on days 7 and 28 on concrete cured by drinkable water and treated wastewater, respectively. This test illustrated that the behavior of specimens in the tensile strength test is approximately similar to that in the compressive strength test but the situation is worse in concentrated specimens. For example, the TW + 30%C sample cured by drinkable water and treated wastewater on day 28 had almost 10% less compressive strength than the control and it had 19% less tensile strength. This indicated that Interfacial Transition Zone (ITZ) area is weaker in wastewater specimens and the capacity for water absorption is more in this area. Some wastewater parameters, such as sludge, have spongy features and they reduce the water available for hydration reactions, while the water–cement ratio needs to be more in the ITZ region [39]. In addition, some other greasy wastewater, such as oils, cover the aggregates' surface and hamper the proper connection between cement and aggregates in the ITZ region [39]. That is why the tensile strength is more affected by increasing the concentration compared to the compressive strength. Although the specimens' concentration was increased and decreased in order, no linear relationship was observed between diluted or concentrated specimens.

Not only did the tensile strength decline by increasing the concentration, but the rate of tensile strength growth was also lower than in the control sample. For instance, within days 7 to 28, the control sample cured by drinkable water and treated wastewater had 85.3% and 84.4% growth, but TW + 30%C had 78% and 77% tensile strength growth and TW + 35%C had 77% and 75% growth, respectively.

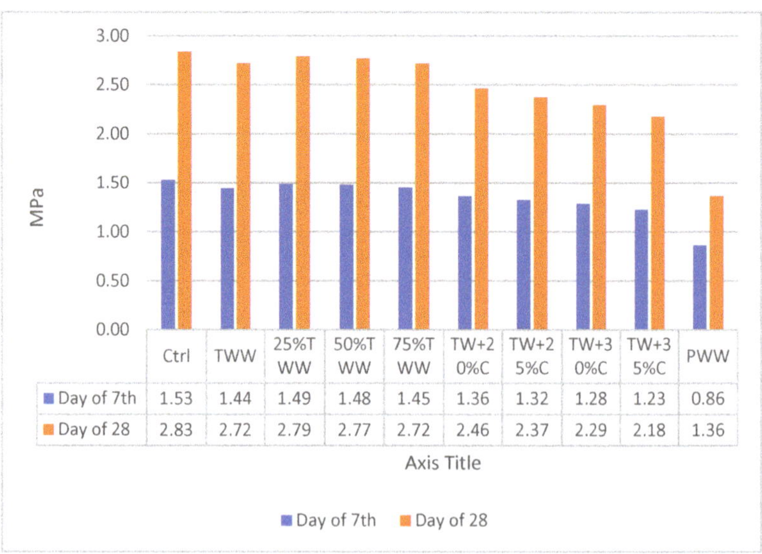

Figure 8. The results of tensile strength cured by drinkable water.

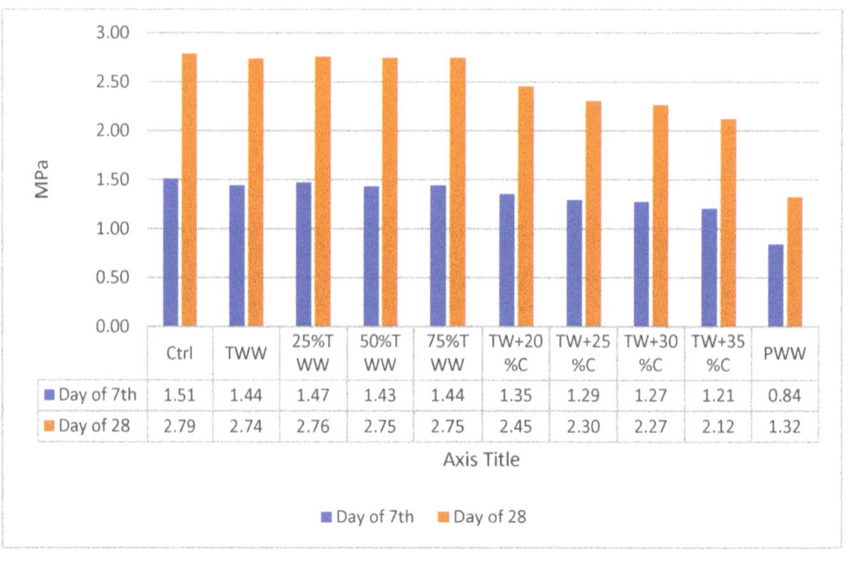

Figure 9. The results of tensile strength cured by treated wastewater.

3.7. *Rapid Freezing and Thawing*

Table 8 indicates the compressive strength results of specimens at the age of 28 days, and the strength reduction of each cycle due to the rapid freezing and thawing test. Change in compressive strength: a decline of more than 10% is the sign of failure. The volume expansion is the first reason for cracking in concrete. This expansion is caused by frozen water inside the concrete. Another reason for cracking is thermal stress. Thermal stresses appear because of repeated freeze–thaw cycles. By increasing the number of fast freeze–thaw cycles, the value of the mechanical property declines. Basically, existing air bubbles in the concrete are effective at enhancing its resistance to disintegration when exposed to cycles of freezing and thawing in a censoriously saturated state, and at decreasing the scaling that involves the application of chemicals for ice removal [44]. The tiny air voids work as empty chambers in the paste for the freezing and moving water to enter, hence mitigating the pressure in the pores and intercepting damage to the concrete. However, it is very unlikely that air-void clustering can happen, causing a loss of compressive strength. Likewise, the pore connectivity, as well as pore-size distribution, are known as the main factors which remarkably affect the freezing and thawing resistance [45].

The recommended bubble distribution is less than 0.2 mm, but by using wastewater instead of drinkable water, the pores become harmful for the concrete [46]. The better pore connectivity leads to a better performance of the concrete regarding the freezing and thawing test, so the higher the connectivity in the concrete, the higher the resistance of the specimen [47]. The impurities of wastewaters disrupt the formation of appropriate air voids, pore structure and pore-size distribution in the concrete. Consequently, low connectivity of pores and larger harmful pores (pores greater than 0.064 μm) are developed. By using wastewater with high concentrations, the free spacing and pore structure are not formed properly and somehow get clogged by TDS and other wastewater impurities, including lead, cadmium and detergent [48]. Therefore, those tiny air voids do not act like chambers in wastewater specimens in comparison with the control sample produced by standard drinkable water. Technically, the air bubbles in the concrete provide protection from the strain that originated from the freezing of water in the capillary gaps in the concrete specimens and thus minimize damage to the hardened paste.

Results indicated that the reduction rate in the control and TWW samples was almost the same until 100 cycles, but then TWW demonstrated different behavior and declined more than the control sample. For instance, until 100 cycles, in comparison to day 28 compressive strength, the control and TWW had an almost 3% strength reduction, whereas in 150 and 200 cycles, the control sample had 4% and 5% reductions while TWW had 6% and 7.6% compressive strength reduction, respectively. It was observed that the strength of the specimens was negatively affected by the lack of proper pore connectivity and air voids in the specimens produced by wastewater, especially when the smooth exterior layer of the specimens was gone due to initial cycles of the test, and the inner pore structures were the main effective factor. This illustrated the destructive effect of harmful pores and tiny air voids on the concrete structure caused by wastewater impurities. Interestingly, 75%TW also had the same reaction as TWW and the rate of reduction rose after cycle 100. No significant difference was observed in 25%TW; its concentration was fairly close to that of the control specimen and it supported the negative impact of larger pores in the concrete structure even after using diluted wastewater. Compared to the control sample, concentrated specimens had more compressive strength reduction in all cycles, and by increasing the concentration and cycles, the rate of reduction increased. Clearly, PWW, which contained the highest impurity, had the lowest strength due to having harmful pores. It should be noted that the exterior layer of PWW was already honeycombing with vulnerability, yet cracks and flaws intensified the volume expansion and failure in this specimen.

Figure 10 shows the compressive strength reduction rate due to rapid freezing and thawing, which has a direct relationship with poor void parameters and porosity connectivity in the concrete structure. In fact, by having more sulfate, TDS, BOD and COD in the utilized wastewaters, the availability of well-developed voids decreased, which increased the pressure in the concrete. The existing oil and impurities in the wastewater led to losing the connectivity of the pores, and then freezing and thawing resistance declined. Again, these reduction rate differences were more conspicuous after 100 cycles when the pressure reached a peak. For example, by concentrating the TWW up to 35%, the comprehensive strength after 200 cycles of compression with the control and TWW dropped by 26% and 19%, respectively.

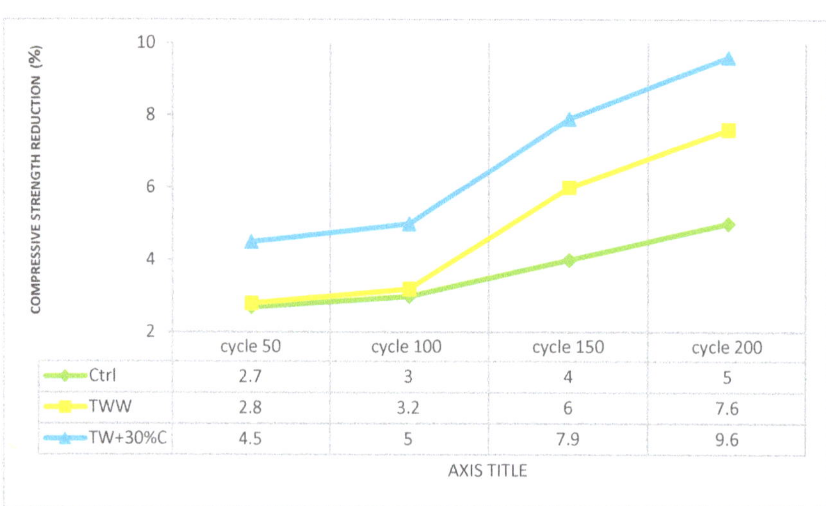

	cycle 50	cycle 100	cycle 150	cycle 200
Ctrl	2.7	3	4	5
TWW	2.8	3.2	6	7.6
TW+30%C	4.5	5	7.9	9.6

Figure 10. The compressive strength reduction in each rapid freezing and thawing cycle.

Using wastewater causes larger pores and more voids in the concrete's structure and these specimens can contain water and, subsequently, more frozen water inside of the concrete and more corrosion are expected in the rapid freezing and thawing test. TW + 30%C at 50, 100, 150 and 200 cycles had 12%, 13.9%, 17.4% and 21.4% less compressive strength than the control. The rate of compressive strength reduction was higher in the specimens with higher concentrations. Therefore, the improper pore structures and clogged air voids caused by the total dissolved and all impurities in the more concentrated wastewaters can be interpreted as the primary factors of strength reduction in the concrete specimens produced by wastewater. Based on the achieved results, void parameters, air voids and size-distribution have more prominent effects on the resistance than the air-void spacing. Therefore, based on the obtained results, air void structure in air-entrained concretes utilizing a Protected Paste Volume (PPV) parameter is recommended as it protects the paste area with air voids in the total paste area.

3.8. Half-Cell Potential

Table 9 illustrates the half-cell potential, which is influenced by chloride ions and the internal alkaline environment of concrete. Basically, increasing the wastewater concentration caused a larger reduction in the reinforcement corrosion potential. In other words, by aging the specimens and increasing the concentration of used wastewater as the water of the mixture design, the possibility of corrosion increased [39]. For instance, the corrosion in TW + 30%C and PWW started after 24 and 8 days; this is clearly because of the wastewater parameters. Half-cell potential decreased when chloride content and sulfate content increased in the wastewaters' concentration. In fact, those specimens with higher compressive strength demonstrated better half-cell potential. Likewise, the extent of corrosion escalated with the reduction of the half-cell potential. According to the voltage, corrosion was conspicuous when the half-cell potential was lower than −450 mV in dry conditions. In addition, using more concentrated wastewater not only decreased the level of half-cell potential but it also negatively affected the level of corrosion and the age at which corrosion starts. For instance, in PWW, which had the highest chloride and sulfate in comparison to other specimens, corrosion occurred almost four times sooner than in the control or even in TWW. Clearly, the level of corrosion sped up with the decline of the half-cell potential.

3.9. Statistical Analysis

In scientific research, researchers seek to present results as practically and, of course, as easily as possible. One of the most important ways that other researchers can make practical use of research is to provide mathematical models for use in future experiments and research [49]. In this study, after completing the laboratory phase, the authors collected laboratory data to examine the data's relationships. Data were analyzed using SPSS statistical analysis software. In this statistical analysis, some input parameters were used, including TDS, TSS, EC, Detergent, Sulfate, Chromium and Cadmium, to predict output parameters including Compressive Strength, Electrical Resistivity, Tensile Strength, Rapid Freezing and Thawing, Water Absorption in 30 min, Water Absorption Mass 72 h, and Capillary Water Absorption 72 h. The statistical indicators of all parameters are shown in Table 10. Standard division was reported using Equation (1).

$$\sigma = \sqrt{\frac{\Sigma(x_i - \mu)^2}{N}} \quad (1)$$

where σ, N, x_i and μ are the population standard deviation, the size of the population, each value from the population and the population mean, respectively.

In the next step, the normality of the data should be checked. The normality of data is generally determined by examining the Skewness and Kurtosis coefficients [50–52]. Achieving a situation where data distribution is perfectly normal is very rare, so in scientific texts, data is normal when the coefficients of Kurtosis and Skewness are in the range of −2 to 2 [53]. According to the coefficients of Kurtosis and Skewness in Table 11, all variables have a normal distribution. Once the normality of the data is determined, it is time to determine the correlation coefficient between the variables. For normal data, the Pearson correlation test is used. Table 11 presents the results of the Pearson correlation test.

Table 11. Pearson correlation coefficients between inputs and outputs in this study.

		Inputs						
		TDS	**EC**	**TSS**	**Detergent**	**Sulfate**	**Chromium**	**Cadmium**
Outputs	Compressive Strength	−0.793	−0.525	−0.834	−0.85	−0.747	−0.678	−0.831
	Electrical Resistivity	−0.81	−0.536	−0.858	−0.856	−0.77	−0.688	−0.817
	Tensile Strength	−0.898	−0.661	−0.933	−0.934	−0.868	−0.777	−0.874
	Rapid Freezing and Thawing	−0.853	−0.658	−0.879	−0.891	−0.819	−0.796	−0.889
	Water Absorption in 30 min	0.842	0.602	0.888	0.896	0.807	0.728	0.832
	Water Absorption Mass 72 Hour	0.896	0.697	0.927	0.942	0.869	0.831	0.902
	Capillary Water Absorption	0.906	0.673	0.942	0.942	0.876	0.807	0.886

Once the correlation coefficients have been determined, it is time to determine the estimation models for research outputs. Multivariate linear regression is used for this purpose. The general form of multivariate linear regression is as follows [54–56]:

$$Y = a_0 + a_1X_1 + a_2X_2 + a_3X_3 + \ldots, \quad (2)$$

where Y is the dependent variable (output of model), X_i is the independent variable (inputs of the model), and ai are the regression coefficients of the model. The models obtained from the analysis performed with SPSS software are presented in Table 12. To measure the accuracy of the models, the two criteria R^2, standard error and standard deviation were used.

The equations presented in Table 12 are very accurate, but they include an important point. These equations were constructed based on the laboratory data of this study. If the number of laboratory data is increased, the coefficients of the models will change slightly. Therefore, the authors recommend that other researchers, before applying these equations, first calibrate the models for their project or research conditions and then use them.

Table 12. Prediction models for determining characteristics of concrete.

Equations	R^2	Std. Error	Std. Deviation
Compressive Strength = 2504.102 − 0.017TDS − 4.693TSS − 24.11Detergent + 2.869Sulfate + 93.896Chromium − 134.054Cadmium − 0.551EC	0.964	4.37321	61.84
Electrical Resistivity = 467.794 − 1.426TSS + 7.087Detergent + 0.522Sulfate + 15.7Chromium − 17.614Cadmium − 0.110EC	0.960	0.76933	10.88
Tensile Strength = 280.918 + 0.001TDS − 0.066EC − 0.758TSS + 2.536Detergent + 0.197Sulfate + 19.169Chromium − 17.064Cadmium	0.994	0.25981	3.68
Rapid Freezing and Thawing = 3634.478 − 0.021TDS − 0.847EC − 7.396TSS − 1.118Detergent + 4.532Sulfate − 24.176Chromium − 63.520Cadmium	0.999	0.0001	0.0015
Water Absorption in 30 min = 0.0006092TDS + 0.011EC + 0.112TSS − 0.19Detergent − 0.051Sulfate − 0.739Chromium + 0.945Cadmium − 39.212	0.998	0.0002	0.0029
Water Absorption Mass 72 hours = 0.02EC + 0.178TSS + 1.037Detergent − 0.104Sulfate + 0.515Chromium + 1.866Cadmium − 74.792	0.997	0.0005	0.0071
Capillary Water Absorption 72 HOUR = 0.001TDS + 0.032EC + 0.389TSS − 1.269Detergent − 0.161Sulfate − 1.902Chromium + 2.821Cadmium − 117.984	0.999	0.0001	0.0015

In sum, the achieved results were commensurate with other tests with data integrity, and no significant contradiction was observed. The negative impact of the wastewater was conspicuous in durability results and curing with wastewater is not recommended. This research indicated that using industrial wastewater could decrease the quality of produced concrete according to its concentration but, based on the results, a proper understanding of the use of different concentrations was presented, which helps to reach the economic level of refinement in industrial towns for using their wastewater in concrete.

4. Conclusions

In this paper, ten groups of concrete specimens with different industrial wastewater concentrations were produced and cured by drinkable water and treated wastewater separately according to tests and standards. Using wastewater as the water for mix design reduces the strength and durability, but industrial TWW can be a good and acceptable replacement for the water in mix design, having insignificant strength and durability reduction effects on concrete in all tests. By concentrating the treated wastewater properties by up to 30% in the TW + 30%C specimen, the compressive strength declined by almost 10% after 28 days, which showed the optimum level of concentration for an industrial town's refinement. However, concentration had more adverse effects on durability tests, which showed that using wastewater causes more negative impacts on the durability than on the strength features of concrete. PWW did not have acceptable behavior in any tests, and it was rejected. Although in concentrated and diluted specimens the percentage of wastewater was increased or decreased in order, no linear relationship in the strength and durability features of the tests was observed. Control specimens showed better strength and durability when they were cured by drinkable water, which proved homogeneity and similarity features; nonetheless, wastewater specimens showed better strength and durability when they were cured by wastewater only at lower ages, whereas the good effect disappeared at late ages; as if being in touch with treated wastewater for curing deteriorates the specimens' strength and causes corrosion. All of the specimens had less growth in terms of strength and durability when they were cured by treated wastewater in comparison to being cured by drinkable water. Using wastewater reduced the electrical resistivity and increased the water absorption of samples, yet diluting the treated wastewater could not correct the negative effects on concrete durability. Diluted specimens' results were closer to those of TWW, not of the control and, by increasing the concentration, negative effects became more conspicuous. Using wastewater in the concrete had a negative impact on the rapid freezing and thawing results. It is concluded that the improper pore-size distribution, lack of pore connectivity and air voids caused by total dissolved and impurities of the wastewaters can be counted as the main reasons of strength reduction in the specimens produced by wastewater. Despite dilution, it was not an adequate solution for resolving the strength reduction of specimens in the freezing and thawing test, and even low concentrations of wastewater disrupt the formation of appropriate air voids and pore structure void parameters in the wastewater specimens in comparison with the control. By concentrating industrial wastewater, not only were the freezing–thawing properties negatively affected, but also the strength reduction rate increased, especially after 100 cycles. In the half-cell potential test, using wastewater insignificantly damaged the reinforcement, but specimens with more concentrated wastewater started to corrode faster. The specimens' appearances had insignificant differences, except for PWW, which had more discontinuity and a distinguishable lack of hydration; however, by increasing the concentration, uneven and small cracks on the exterior layers were observed. Using wastewater increased water absorption and decreased workability. Therefore, it is highly recommended that it not be used in projects with caustic materials, exposed surfaces, or when a high slump is needed.

Author Contributions: Conceptualization, E.N. and D.M.; methodology, E.N. and D.M.; software, N.K.; validation, N.K. and A.M.; formal analysis, D.M. and N.K.; investigation, E.N.; resources, E.N. and D.M.; data curation, E.N. and D.M.; writing—original draft preparation, E.N. and D.M.; writing—review and editing, E.N., U.R. and A.M.; visualization, E.N., D.M.; supervision, U.R., M.S.; project administration, A.M.; funding acquisition, A.M. All authors have read and agreed to the published version of the manuscript.

Funding: This research received no funding.

Institutional Review Board Statement: Not applicable.

Informed Consent Statement: Not applicable.

Data Availability Statement: Not applicable.

Acknowledgments: Mosavi would like to acknowledge his support from the project GINOP-2.2.1-18-2018-00015.

Conflicts of Interest: The authors declare no conflict of interest.

Appendix A

This Appendix presents the granular details of the materials used in this study for producing concrete.

Table A1. Gradation test results.

Sieve		Percentage Passing					
No.	Size (mm)	Coarse Gravel	Fine Grave	Sand	Fine Sand	Stone Powder	Final Composition
1 1/2 inch	37.5	100	100	100	100	100	100
1 inch	25	100	100	100	100	100	100
3/4 inch	19	93	100	100	100	100	98
1/2 inch	12.5	28	99	100	100	100	78
3/8 inch	9.5	1	78	100	100	100	67
No. 4	4.75	0.2	3	92	98	100	55
No. 8	2.38	0.2	0.6	71	95	100	43
No. 16	1.19	0.1	0.5	42	91	100	27
No. 30	0.6	0.1	0.5	23	88	98	17
No. 50	0.3	0.1	0.5	9.3	69	87	9
No. 100	0.15	0.1	0.5	3.1	20	70	5
No. 200	0.075	0.1	0.5	1.7	16	57	3

Table A2. Aggregates' parameters.

Parameter	Sand	Fine Gravel	Coarse Gravel	Stone Powder
Specific density (kg/m^3)	1730	1582	1576	2600
SSD density (kg/L)	2.580	2.70	2.71	2.70
Water absorption (%)	2.49	0.910	0.670	0.00
Sand equivalent	82	-	-	-
Passed from #200 (%)	1.70	0.5	0.1	56.60
Fineness modulus	3.60	6.16	7.05	0.45

References

1. Sharbatdar, M.K.; Abbasi, M.; Fakharian, P. Improving the Properties of Self-compacted Concrete with Using Combined Silica Fume and Metakaolin. *Period. Polytech. Civ. Eng.* **2020**. [CrossRef]
2. Naderpour, H.; Noormohammadi, E.; Fakharian, P. Prediction of Punching Shear Capacity of RC Slabs using Support Vector Machine. *Concr. Res.* **2017**, *10*, 95–107.
3. Gandomi, A.; Alavi, A.H.; Shadmehri, D.M.; Sahab, M.; Gandomi, A.; Alavi, A.H.; Shadmehri, D.M.; Sahab, M. An empirical model for shear capacity of RC deep beams using genetic-simulated annealing. *Arch. Civ. Mech. Eng.* **2013**, *13*, 354–369. [CrossRef]
4. Gandomi, A.H.; Mohammadzadeh, D.S.; Ordóñez, J.L.P.; Alavi, A.H. Linear genetic programming for shear strength prediction of reinforced concrete beams without stirrups. *Appl. Soft Comput.* **2014**, *19*, 112–120. [CrossRef]
5. Guest, J.; Skerlos, S.J.; Barnard, J.L.; Beck, M.B.; Daigger, G.T.; Hilger, H.; Jackson, S.J.; Karvazy, K.; Kelly, L.; Macpherson, L.; et al. *A New Planning and Design Paradigm to Achieve Sustainable Resource Recovery from Wastewater*; American Chemical Society (ACS): Washington, DC, USA, 2009; Volume 43, pp. 6126–6130.
6. González, O.; Bayarri, B.; Aceña, J.; Pérez, S.; Barceló, D. Treatment Technologies for Wastewater Reuse: Fate of Contaminants of Emerging Concern. In *The Handbook of Environmental Chemistry*; Springer Science and Business Media LLC: Berlin, Germany, 2015; Volume 45, pp. 5–37.
7. Gerlach, E.; Franceys, R. Regulating Water Services for All in Developing Economies. *World Dev.* **2010**, *38*, 1229–1240. [CrossRef]
8. Asanoa, T.; Cotruvob, J.A. Groundwater recharge with reclaimed municipal wastewater: Health and regulatory considerations. *J. Water Res.* **2004**, *38*, 1941–1951. [CrossRef]
9. Naderpour, H.; Rafiean, A.H.; Fakharian, P. Compressive strength prediction of environmentally friendly concrete using artificial neural networks. *J. Build. Eng.* **2018**, *16*, 213–219. [CrossRef]
10. Molinos-Senante, M.; Hernández-Sancho, F.; Sala-Garrido, R. Cost–benefit analysis of water-reuse projects for environ-mental purposes: A case study for Spanish wastewater treatment plants. *J. Environ. Manag.* **2011**, *92*, 3091–3097. [CrossRef]
11. Hanjra, M.A.; Blackwell, J.; Carr, G.; Zhang, F.; Jackson, T.M. Wastewater irrigation and environmental health: Implications for water governance and public policy. *Int. J. Hyg. Environ. Health* **2012**, *215*, 255–269. [CrossRef] [PubMed]
12. The United Nation Website. 2017 UN World Water Development Report, Wastewater: The Untapped Resource (UNESCO WWAP). ISBN 978-92-3-100201-4. Available online: https://www.unwater.org/publications/world-water-development-report-2017/ (accessed on 22 March 2017).
13. Al-Ghusain, I.; Terro, M. Mechanical properties of concrete made with treated wastewater at ambient and elevated temperatures. *Kuwait J. Sci. Eng.* **2003**, *30*, 214–244.
14. Shekarchi, M.; Yazdian, M.; Mehrdadi, N. Use of biologically treated domestic wastewater in concrete. *Kuwait J. Sci. Eng.* **2009**, *39*, 97–111.
15. Asadollahfardi, G.; Asadi, M.; Jafari, H.; Moradi, A.; Asadollahfardi, R. Experimental and statistical studies of using wash water from ready-mix concrete trucks and a batching plant in the production of fresh concrete. *Constr. Build. Mater.* **2015**, *98*, 305–314. [CrossRef]
16. Asadollahfardi, G.; Delnavaz, M.; Rashnoiee, V.; Ghonabadi, N. Use of treated domestic wastewater before chlorination to produce and cure concrete. *Constr. Build. Mater.* **2016**, *105*, 253–261. [CrossRef]
17. De Domenico, D.; Faleschini, F.; Pellegrino, C.; Ricciardi, G. Structural behavior of RC beams containing EAF slag as recycled aggregate: Numerical versus experimental results. *Constr. Build. Mater.* **2018**, *171*, 321–337. [CrossRef]
18. Bahraman, M.; Asadollahfardi, G.; Salehi, A.M.; Yahyaei, B. Feasibility study of using wash water from ready mixed concrete plant and synthetic wastewater based on tap water with different total dissolved solid to produce self-compacting concrete. *J. Build. Eng.* **2021**, *41*, 102781. [CrossRef]
19. Taherlou, A.; Asadollahfardi, G.; Salehi, A.M.; Katebi, A. Sustainable use of municipal solid waste incinerator bottom ash and the treated industrial wastewater in self-compacting concrete. *Constr. Build. Mater.* **2021**, *297*, 123814. [CrossRef]
20. Raza, A.; Rafique, U.; Haq, F.U. Mechanical and durability behavior of recycled aggregate concrete made with different kinds of wastewater. *J. Build. Eng.* **2021**, *34*, 101950. [CrossRef]
21. Małek, M.; Łasica, W.; Kadela, M.; Kluczyński, J.; Dudek, D. Physical and Mechanical Properties of Polypropylene Fibre-Reinforced Cement–Glass Composite. *Materials* **2021**, *14*, 637. [CrossRef] [PubMed]

22. APHA. *Standard Methods for the Examination of Water and Wastewater*, 22nd ed.; American Public Health Association; American Water Works Association; Water Environment Federation: Washington, DC, USA, 2012.
23. ASTM C1602M-18. *Standard Specification for Mixing Water Used in the Production of Hydraulic Cement Concrete*; ASTM International: West Conshohocken, PA, USA, 2018. [CrossRef]
24. ASTM C150-04. *Standard Specification for Portland Cement*; ASTM International: West Conshohocken, PA, USA, 2004. [CrossRef]
25. ASTM C188-15. *Standard Test Method for Density of Hydraulic Cement*; ASTM International: West Conshohocken, PA, USA, 2015. [CrossRef]
26. ASTM C136M-19. *ASTM C136/C136M-19, Standard Test Method for Sieve Analysis of Fine and Coarse Aggregates*; ASTM International: West Conshohocken, PA, USA, 2019. [CrossRef]
27. ASTM C33/C33M-18. *Standard Specification for Concrete Aggregates*; ASTM C191-19, Standard Test Methods for Time of Setting of Hydraulic Cement by Vicat Needle; ASTM International: West Conshohocken, PA, USA, 2019. [CrossRef]
28. ASTM C191-19. *Standard Test Methods for Time of Setting of Hydraulic Cement by Vicat Needle*; ASTM International: West Conshohocken, PA, USA, 2019. [CrossRef]
29. ASTM C143/C143M-15a. *Standard Test Method for Slump of Hydraulic-Cement Concrete*; ASTM International: West Conshohocken, PA, USA, 2015. [CrossRef]
30. BS EN 12390-3. *Testing Hardened Concrete Cement and Concrete Technology, Compressive Strength, Test Specimens, Concretes, Compression Testing, Failure (Mechanical), Mechanical Testing 2019*; British Standards Institution: London, UK, 2019.
31. ASTM C496/C496M-17. *Standard Test Method for Splitting Tensile Strength of Cylindrical Concrete Specimens*; ASTM International: West Conshohocken, PA, USA, 2017. [CrossRef]
32. BS 1881-122:2011. *Testing Concrete. Method for Determination of Water Absorption 2011*, 11th ed.; British Standards Institution: London, UK, 2011.
33. FM 5-578. *Florida Method of Test for Concrete Resistivity as an Electrical Indicator of Its Permeability Florida a Method of Test for Concrete Resistivity as an Electrical Indicator of Its Permeability Designation*; Florida Department of Transportation, Ed.; Elsevier: New York, NY, USA, 2004.
34. ASTM C1585-13. *Standard Test Method for Measurement Rate of Absorption of Water by Hydraulic-Cement Concretes*; ASTM International: West Conshohocken, PA, USA, 2013. [CrossRef]
35. ASTM C666/C666M-15. *Standard Test Method for Resistance of Concrete to Rapid Freezing and Thawing*; ASTM International: West Conshohocken, PA, USA, 2015. [CrossRef]
36. ASTM C876-15. *Standard Test Method for Corrosion Potentials of Uncoated Reinforcing Steel in Concrete*; ASTM International: West Conshohocken, PA, USA, 2015. [CrossRef]
37. ASTM C642-13. *Standard Test Method for Density, Absorption, and Voids in Hardened Concrete*; ASTM International: West Conshohocken, PA, USA, 2013. [CrossRef]
38. ASTM C231/C231M-17a. *Standard Test Method for Air Content of Freshly Mixed Concrete by the Pressure Method*; ASTM International: West Conshohocken, PA, USA, 2017. [CrossRef]
39. Meena, K.; Luhar, S. Effect of wastewater on properties of concrete. *J. Build. Eng.* **2019**, *21*, 106–112. [CrossRef]
40. Xiao, J. Recycled Aggregate Concrete. In *Environmental and Human Impact of Buildings*; Springer Science and Business Media LLC: Berlin, Germany, 2017; pp. 65–98.
41. ASTM C31/C31M-19a. *Standard Practice for Making and Curing Test Specimens in the Field*; ASTM International: West Conshohocken, PA, USA, 2019. [CrossRef]
42. ASTM C305-14. *Standard Practice for Mechanical Mixing of Hydraulic Cement Pastes and Mortars of Plastic Consistency*; ASTM International: West Conshohocken, PA, USA, 2014. [CrossRef]
43. Medeiros-Junior, R.A.; Lima, M.G. Electrical resistivity of unsaturated concrete using different types of cement. *Constr. Build. Mater.* **2016**, *107*, 11–16. [CrossRef]
44. Wang, L.; Jin, M.; Wu, Y.; Zhou, Y.; Tang, S. Hydration, shrinkage, pore structure and fractal dimension of silica fume modified low heat Portland cement-based materials. *Constr. Build. Mater.* **2021**, *272*, 121952. [CrossRef]
45. Wang, Y.; Cao, Y.; Zhang, P.; Ma, Y.; Zhao, T.; Wang, H.; Zhang, Z. Water absorption and chloride diffusivity of concrete under the coupling effect of uniaxial compressive load and freeze–thaw cycles. *Constr. Build. Mater.* **2019**, *209*, 566–576. [CrossRef]
46. Yang, Z.; Weiss, W.J.; Olek, J. Water Transport in Concrete Damaged by Tensile Loading and Freeze–Thaw Cycling. *J. Mater. Civ. Eng.* **2006**, *18*, 424–434. [CrossRef]
47. Grubeša, I.N.; Marković, B.; Vračević, M.; Tunkiewicz, M.; Szenti, I.; Kukovecz, Á. Pore Structure as a Response to the Freeze/Thaw Resistance of Mortars. *Materials* **2019**, *12*, 3196. [CrossRef]
48. Zhang, W.; Pi, Y.; Kong, W.; Zhang, Y.; Wu, P.; Zeng, W.; Yang, F. Influence of damage degree on the degradation of concrete under freezing-thawing cycles. *Constr. Build. Mater.* **2020**, *260*, 119903. [CrossRef]
49. Mohammadzadeh, D.S.; Seyed-Farzan, K.; Mosavi, A.; Nasseralshariati, E.; Tah, J.H.M. Prediction of Compression Index of Fine-Grained Soils Using a Gene Expression Programming Model. *Infrastructures* **2019**, *4*, 26. [CrossRef]
50. Cembrowski, G.S.; Westgard, J.O.; Conover, W.J.; Toren, E.C. Statistical Analysis of Method Comparison Data: Testing Normality. *Am. J. Clin. Pathol.* **1979**, *72*, 21–26. [CrossRef] [PubMed]
51. Ghasemi, A.; Zahediasl, S. Normality Tests for Statistical Analysis: A Guide for Non-Statisticians. *Int. J. Endocrinol. Metab.* **2012**, *10*, 486–489. [CrossRef]

52. Thode, H.C. *Testing for Normality*; CRC Press: Boca Raton, FL, USA, 2002; Volume 164.
53. Nabipour, N.; Karballaeezadeh, N.; Dineva, A.; Mosavi, A.; Mohammadzadeh, D.S.; Shamshirband, S. Comparative Analysis of Machine Learning Models for Prediction of Remaining Service Life of Flexible Pavement. *Mathematics* **2019**, *7*, 1198. [CrossRef]
54. Khademi, F.; Akbari, M.; Jamal, S.M.; Nikoo, M. Multiple linear regression, artificial neural network, and fuzzy logic prediction of 28 days compressive strength of concrete. *Front. Struct. Civ. Eng.* **2017**, *11*, 90–99. [CrossRef]
55. Chakraborty, A.; Goswami, D. Prediction of slope stability using multiple linear regression (MLR) and artificial neural network (ANN). *Arab. J. Geosci.* **2017**, *10*, 385. [CrossRef]
56. Falade, F.; Iqbal, T. Compressive strength Prediction recycled aggregate incorporated concrete using Adaptive Neuro-Fuzzy System and Multiple Linear Regression. *Int. J. Civ. Environ. Agric. Eng.* **2019**, *1*, 19–24. [CrossRef]

 materials

MDPI

Article

Effect of Cement Types and Superabsorbent Polymers on the Properties of Sustainable Ultra-High-Performance Paste

Mei-Yu Xuan [1], Yi-Sheng Wang [2], Xiao-Yong Wang [1,2,*], Han-Seung Lee [3] and Seung-Jun Kwon [4]

1 Department of Architectural Engineering, Kangwon National University, Chuncheon-si 24341, Korea; xuanmeiyu@kangwon.ac.kr
2 Department of Integrated Energy and Infra System, Kangwon National University, Chuncheon-si 24341, Korea; wangyisheng@kangwon.ac.kr
3 Department of Architectural Engineering, Hanyang University, Ansan-si 15588, Korea; ercleehs@hanyang.ac.kr
4 Department of Civil and Environmental Engineering, Hannam University, Daejeon-si 34430, Korea; jjuni98@hannam.ac.kr
* Correspondence: wxbrave@kangwon.ac.kr; Tel.: +82-033-250-6229

Abstract: This study focuses on the effects of superabsorbent polymers (SAP) and belite-rich Portland cement (BPC) on the compressive strength, autogenous shrinkage (AS), and micro- and macroscopic performance of sustainable, ultra-high-performance paste (SUHPP). Several experimental studies were conducted, including compressive strength, AS, isothermal calorimetry, X-ray diffraction (XRD), thermogravimetric analysis (TGA), attenuated total reflectance (ATR)–Fourier-transform infrared spectroscopy (FTIR), ultra-sonic pulse velocity (UPV), and electrical resistivity. The following conclusions can be made based on the experimental results: (1) a small amount of SAP has a strength promotion effect during the first 3 days, while BPC can significantly improve the strength over the following 28 days. (2) SAP slows down the internal relative humidity reduction and effectively reduces the development of AS. BPC specimens show a lower AS than other specimens. The AS shows a linear relationship with the internal relative humidity. (3) Specimens with SAP possess higher cumulative hydration heat than control specimens. The slow hydration rate in the BPC effectively reduces the exothermic heat. (4) With the increase in SAP, the calcium hydroxide (CH) and combined water content increases, and SAP thus improves the effect on cement hydration. The contents of CH and combined water in BPC specimens are lower than those in the ordinary Portland cement (OPC) specimen. (5) All samples display rapid hydration of the cement in the first 3 days, with a high rate of UPV development. Strength is an exponential function of UPVs. (6) The electrical resistivity is reduced due to the increase in porosity caused by the release of water from SAP. From 3 to 28 days, BPC specimens show a greater increment in electrical resistivity than other specimens.

Keywords: superabsorbent polymer; belite-rich Portland cement; sustainable ultra-high-performance paste; internal curing; autogenous shrinkage

Citation: Xuan, M.-Y.; Wang, Y.-S.; Wang, X.-Y.; Lee, H.-S.; Kwon, S.-J. Effect of Cement Types and Superabsorbent Polymers on the Properties of Sustainable Ultra-High-Performance Paste. *Materials* **2021**, *14*, 1497. https://doi.org/10.3390/ma14061497

Academic Editor: Yeonung Jeong

Received: 24 February 2021
Accepted: 16 March 2021
Published: 18 March 2021

1. Introduction

Ultra-high-performance concrete (UHPC) has a high cementitious material content, low water–cement ratio (w/c), dense microstructure, and very low porosity [1]. Therefore, it has superior mechanical properties and durability performance. To achieve a high-strength and dense structure in UHPC, the mechanical properties and rheology can be improved by adding microfillers [2]. Usually, silica fume is added as a typical reactive powder for UHPC, but this causes autogenous shrinkage (AS) to be high [3]. In addition, limestone fillers and blast-furnace slag are also used to increase UHPC sustainability. The addition of limestone and slag can increase cement's degree of hydration, improve mechanical strength, and lower CO_2 emissions [3–5]. In addition, as a result of the extremely low w/c of UHPC, the final hydration value can be less than 50% [6]. AS and autogenous cracking limit the application of UHPC in practical engineering [7].

To effectively promote cement hydration and reduce AS, the internal relative humidity of UHPC can be increased by providing it with additional water [8]. However, the extremely low porosity (very low permeability) of UHPC makes it impossible for water from outside to enter [6]. UHPC was found to be more suitable than the internal curing method for maintenance [9]. Internal curing is the process of mixing and dispersing materials with high water absorption capacity into concrete and gradually releasing water to the surrounding area during the cement's hydration process [10]. Common internal curing methods are lightweight aggregates (LWA) [11] and SAP [9]. The water absorption ability of SAP is much higher than LWA; consequently, the internal curing efficiency of SAP is higher [12,13]. Moreover, SAP is more conducive to controlling the distribution, shape, and size of defects and pores [14]. For the internal curing of UHPC, SAP can effectively maintain internal relative humidity and reduce AS [15–17].

SAP materials have hydrophilic networks that can absorb large amounts of water (other solutions) without being dissolved [18]. Further, SAP absorbs water and swells to form a hydrogel, providing additional water during the concrete hardening process [19]. Liu et al. [20] indicated that SAP, as an internal curing agent, could delay the appearance of cracks and reduce the early AS of UHPC. In composite cement systems with fly ash or slag, the water absorbed/released by SAP has an effective shrinkage reduction [21]. Justs et al. [6] explored the internal curing effect of SAP on UHPC, which reduced shrinkage by around 75% when 2% SAP was added. When SAP is used as an internal curing agent, it can promote hydration to improve the microstructure, but the formation of pores during the release of water also increases the porosity [22] and affects the mechanical properties [7]. Song et al. [23] indicated that SAP mitigated the internal relative humidity drop and was effective in reducing AS, but the strength of the samples was also reduced. In mortars with a w/c of 0.55, the addition of SAP reduces the early strength, but the effect gradually decreases at a later stage [24]. From previous studies, it can be concluded that SAP as an internal curing agent has excellent performance in reducing AS, but with a corresponding reduction in strength.

Although many studies have been conducted to examine SAP internal curing of UHPC, previous studies show some weak points. (1) The binder used in UHPC in previous studies mainly consists of cement and silica fume. The study of sustainable, ultra-high-performance paste (SUHPP) with other SCMs and fillers, such as limestone and slag, was insufficient. Moreover, the internal curing of SUHPP was rarely studied. (2) The cement used in UHPC in previous studies mainly consisted of type I Portland cement. Other types of cement, such as BPC, were seldom used. Compared with OPC, the CO_2 emission of BPC is around 10% lower [25]. Moreover, BPC can reduce hydration heat, which is helpful for reducing thermal cracking. Hence, BPC is a sustainable material compared with OPC. In addition, compared with OPC, the rate of hydration of BPC is much slower, which may be helpful for reducing AS in UHPC [26]. In addition, UHPC has a high binder content and experiences a high temperature rise. The utilization of BPC can lower the temperature rise of UHPC. Furthermore, the hydration of C_2S in BPC can enhance the late-age strength of UHPC. Due to these advantages of BPC, concrete factories were eager to know whether BPC was suitable for producing UHPC. (3) Previous studies mainly focus on the AS and strength of SAP-blended UHPC. Studies on other aspects have also been insufficient, i.e., studies regarding internal relative humidity and temperature of hardening specimens, strength monitoring with ultrasonic pulse velocity (UPV), and electrical resistivity development.

Therefore, this study focuses on the effect of adding SAP and replacing BPC in SUHPP with silica fume, limestone, and slag. Several experimental studies were conducted, including compressive strength, AS coupled with internal relative humidity and temperature, isothermal calorimetry, X-ray diffraction (XRD), thermogravimetric analysis (TGA), attenuated total reflectance (ATR)–Fourier-transform infrared spectroscopy (FTIR), UPV, and electrical resistivity.

The innovation points of this study can be summarized as follows: First, silica fume, limestone, and slag were used to produce SUHPP. The internal curing effect of SAP on

SUHPP was investigated. Second, we clarified the effects of the type of cement on the differences in the performance of SUHPP. Finally, a systematical experimental investigation into hydration, AS, strength, and durability was performed.

The aims of the research questions in our research are (1) to find feasible methods for reducing the AS of SUHPP and to clarify the mechanism of AS in SUHPP; (2) to conduct detailed and various experimental studies on SUHPP containing SAP and BPC and to explore the relations among the various results; and (3) to discuss the strong points and weak points of paste containing SAP or BPC and to determine the expected practical applications of the tested specimens.

2. Materials and Experimental Methods

2.1. Material and Ratio Design

In this study, SUHPP was prepared with a fixed water–binder ratio of 0.2. The chemical composition of OPC, BPC, silica fume, limestone, and slag used in mixtures is listed in Table 1. Based on the chemical composition listed in Table 1 and calculated by the Bogue formula, the mineralogical composition of the clinker is shown in Table 2. The particle size distribution of each component is shown in Figure 1. According to Figure 1, the average particle sizes of OPC, BPC, limestone, and slag were 16.4, 14.5, 5.21, and 12.7 μm, respectively. In addition, the water absorption of SAP was measured using the "tea bag" method [27]. In water, the swelling capacity of SAP was equal to 295.5 g/g_{SAP} after 10 min and 291.5 g/g_{SAP} after 12 h. The swelling capacity of SAP in cement paste was much lower than when in water [12]. In cement paste, the swelling capacity of SAP was approximately 10.4 g/g_{SAP}. A scanning electron microscope (SEM) (S-4800, Hitachi, Tokyo, Japan) image of the SAP is shown in Figure 2. In general, SAP particles have an irregular shape.

Table 1. The chemical composition of the materials.

Oxides	Ordinary Portland Cement (%)	Belite-Rich Portland Cement (%)	Silica Fume (%)	Limestone Filler (%)	Blast-Furnace Slag (%)
CaO	64.7	63.27	0.57	59.19	38.29
SiO_2	21.1	26.65	94.52	–	36.12
Al_2O_3	5.07	2.36	0.73	0.21	14.82
Fe_2O_3	3.14	2.85	0.22	–	0.47
MgO	0.89	0.96	0.49	0.45	6.49
Na_2O	0.19	–	–	–	0.06
TiO_2	0.22	–	–	–	0.62
SO_3	1.61	2.12	0.26	–	1.61
Loss onignition	2.32	0.89	1.59	39.21	1.16

Table 2. Mineralogical composition of the clinker.

Mineralogical Composition	OPC (%)	BPC (%)
$3CaO{\cdot}SiO_2$ (C_3S)	59.89	29.06
$2CaO{\cdot}SiO_2$ (C_2S)	15.31	54.48
$3CaO{\cdot}Al_2O_3$ (C_3A)	8.12	1.43
$4CaO{\cdot}AlO_3{\cdot}Fe_2O_3$ (C_4AF)	9.56	8.67

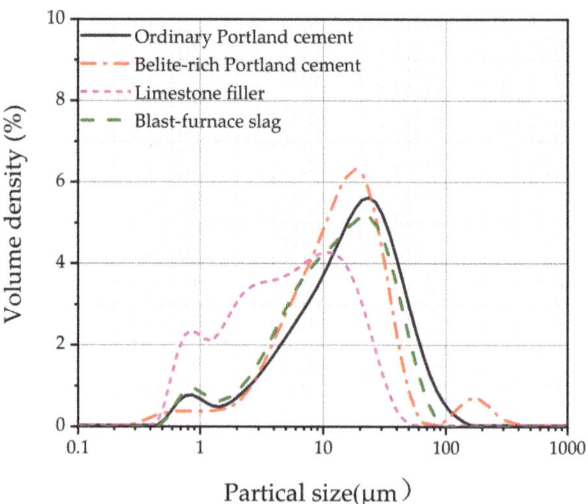

Figure 1. Particle size distribution of each component.

Figure 2. Scanning electron microscope (SEM) images of dried superabsorbent polymer (SAP) particles (magnification scale is 100).

Regarding the quaternary mixture, the weight ratio of silica fume, limestone, and slag was 1:2:2. As shown in Table 3, the effect of SAP content was studied. We added 0.25% and 0.5% (% by mass of binder) SAP content, respectively. In addition, the effect of the cement type was studied using OPC–0SAP and BPC–0SAP. In order to obtain proper workability, 1.2% (% by mass of binder) by weight of superplasticizer (SP) was added.

Table 3. Mixture design of sustainable ultra-high-performance paste (SUHPP).

Number	Binders (%)				SAP (%)	Water (%)	Additional Water (%)	SP (%)
	Cement (OPC/BPC)	Silica Fume	Limestone	Slag				
OPC–0SAP	50	10	20	20	0	20	0	1.2
BPC–0SAP	50	10	20	20	0	20	0	1.2
OPC–0.25SAP	50	10	20	20	0.25	20	2.5	1.2
OPC–0.5SAP	50	10	20	20	0.5	20	5	1.2

2.2. Mixture Preparation Method

Figure 3 shows the specific steps of the mixing procedure of SUHPP. Step 1: dry mix all the binding materials for 30 s. Step 2: add 50% of water, SP, and/or additional water, mixing for 60 s. Step 3: add the dry SAP, mixing for 60 s. Considering that the presoaked SAP turns into a hydrogel, it is difficult to disperse in the mixture [12]. Therefore, it is reasonable to add the dry SAP directly [12]. Step 4: add the rest of the water, SP, and/or additional water; mix slowly for 120 s; and then mix quickly for 180 s.

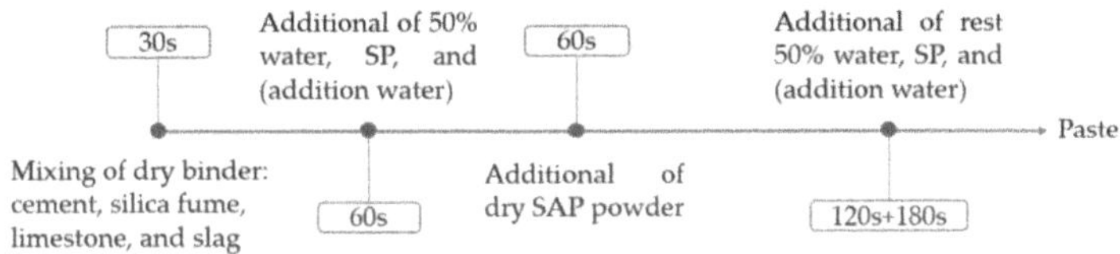

Figure 3. Mixing procedure for composite mixture paste.

2.3. Experimental Methods

Table 4 lists the experimental methods used in this study and the range of tests performed. First, we measured the compressive strength of the sample based on ASTM C39 [28]. Second, the AS coupled with internal relative humidity and temperature measurements were performed using a bellows self-shrinkage tester (Instrument Creation Era, Beijing, China) (based on American Standard ASTM C1698-09) [29]. The isothermal calorimeter TAM Air (TA Instruments, New Castle, CO, USA) was used to measure the heat flow and cumulative hydration heat of the samples (mixing out-of-bottle). After 28 days, XRD analysis was conducted using PANalytical X'pert pro MPD diffractometers (Panalytical, Almelo, The Netherlands). Scanning measurements were performed in the 2θ range from 5° to 75° in steps of 0.02° under CuKα radiation (λ = 1.5404 Å) [30]. At 28 days, a thermal analysis system (SDT Q600, TA Instruments, Santa Clara, CA, USA) was used for TGA. We used temperatures ranging from 20 to 1050 °C at a rate of 10°/min. At 28 days, the samples were scanned using a frontier spectrometer (PerkinElmer, Waltham, MA, USA). The resolution was 0.4 cm^{-1}, and each scan ranged from 2000 to 500 cm^{-1} (7–8). Moreover, at 1, 3, 7, and 28 days, the ultra-sonic pulse velocity and electrical resistivity evolution of the samples were recorded using a nondestructive digital indicator tester (Pundit Lab, Proceq Company, Schwerzenbach, Switzerland) and a four-point Wenner probe surface testing device (Proceq Company, Schwerzenbach, Switzerland).

Table 4. Experimental methods and test ranges.

NO.	Experimental Method	Total Number of Test Samples	Standard Deviation	Sample Size	Test Time
1	Compressive strength	36	Within ±0.05% of the indicated load	50 × 50 × 50 mm	3, 7, and 28 days
2	Autogenous shrinkage coupled with relative humidity and temperature	4	0.001 μm/m; ±0.5% RH; ±0.1 °C	Ø29 × 430 mm	1–7 days
3	Isothermal calorimetry	4	±20 μW	5 g paste	72 h
4	X-ray diffraction	4	λ = 1.5406 Å 2θ = 0.013°	Powder	28 days
5	Thermogravimetric analysis	4	0.1 μg	Powder	28 days
6	Attenuated total reflectance–Fourier-transform infrared spectroscopy	4	±0.01 cm^{-1}	Powder	28 days
7	Ultrasonic pulse velocity	36	0.5 μs	50 × 50 × 50 mm	1, 3, 7, and 28 days
8	Electrical resistivity	12	±0.2 to ±2 kΩ cm	Ø100 × 200 mm	1, 3, 7, and 28 days

3. Experimental Results

3.1. Compressive Strength

Figure 4 shows the effects of the amount of SAP added and the type of cement on the compressive strength development of the mixture samples at 3, 7, and 28 days. The addition of SAP to a mixture has both pros and cons. First, SAP becomes a hydrogel by absorbing water and swelling in the mixing process. In the process of resolution, larger pores are formed, resulting in lower strength [6]. Secondly, SAP releases the absorbed water to the surrounding area in the reaction process (w/c decreases), which plays the role of internal curing, increases hydration, improves the microstructure (pore densification), and promotes the development of strength [17,23]. In general, the effect of SAP on the development of strength depends on the competing effects of SAP on the increase and decrease in compressive strength. When BPC is used instead of OPC, the structure of belite (e.g., irregular structure, few cavities, and low activity) leads to a slow strength increase at an early age and a faster strength increase at a later age [31].

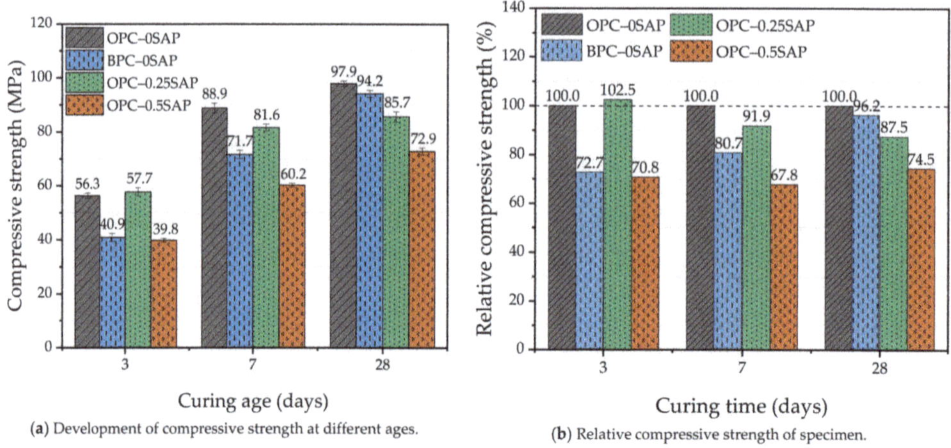

Figure 4. The effect of the amount of SAP added at different ages (3, 7, and 28 days) and the type of cement on the compressive strength. (**a**) Development of compressive strength at different ages; (**b**) relative compressive strength of specimen.

At 3 days of curing age, the compressive strength of OPC–0.25SAP was slightly higher than that of the control group due to the improved degree of hydration from the internal curing of water. The compressive strength of OPC–0.5SAP samples was 29.2% lower than that of the control group due to the increase in porosity caused by SAP [16]. Compared with higher content, a lower amount of SAP was observed for the early-age strength.

At 7 and 28 days of curing age, the compressive strength of the samples with SAP was lower than that of the control group. At 7 days, the compressive strengths of OPC–0.25SAP and OPC–0.5SAP were 8.1% and 32.2% lower than that of the control group, respectively. At 28 days, the compressive strengths of OPC–0.25SAP and OPC–0.5SAP were 12.5% and 25.5% lower than that of the control group, respectively. This indicates that the internal curing of SAP at a later age improves hydration, promotes the pozzolanic reaction, and effectively increases the compressive strength [7,16]. However, the SAP internal maintenance effect does not fully compensate for the strength loss caused by the voids created by the SAP water release [7]. The higher the amount added, the more obvious the reduction phenomenon.

In contrast, the effect of the replacement cement type on strength ranged between 0.25% and 0.5% of SAP. At 3 and 7 days of hydration, the strength of BPC–0SAP was only 72.7%–80.7% of OPC–0SAP. This was attributed to the slow hydration rate of C_2S in BPC–0SAP, resulting in a slow increase in strength in the early stages [32]. Although the early compressive strength developed slowly due to the higher C_2S and lower C_3A in the BPC cement, the later (28 days) strength was significantly improved [33]. The compressive strength of BPC–0SAP samples at 28 days could reach 96.2% of the control group. For strength, the BPC–0SAP samples had a higher late-stage strength than the samples doped with SAP.

3.2. Autogenous Shrinkage (AS) Coupled with Internal Relative Humidity and Temperature

As shown in Figure 5, the development of AS can be divided into the following three stages: rapid growth period at initial ages, stable period at middle ages, and continuous growth period at later ages. There are five reasons for the change in AS: (i) cement hydration consumes water and the internal relative humidity decreases, resulting in a rapid increase in AS (shrinkage factor) [12]; (ii) the densification of the internal pore structure, resulting in increased AS (shrinkage factor) [33]; (iii) the pozzolanic reaction of silica fume and slag to produce dense C-S-H, as well as to promote AS development (i.e., shrinkage factor) [34]; (iv) the production of AFm and internal temperature changes leading to expansion (expansion factor) [35]; (V) limestone stabilizes the produced ettringite and causes expansion (expansion factor) [36,37]. The final macroresponse depends on changes in the dominant expansion and contraction factors.

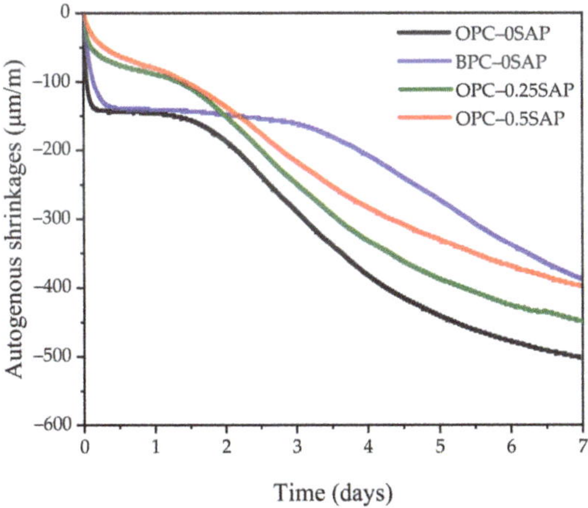

Figure 5. Autogenous shrinkage (AS) of sustainable ultra-high-performance paste (SUHPP) hydrated for 7 days.

At the initial ages, the AS contraction of all samples was significantly greater, while the AS of samples with SAP was significantly lower than that of the control group. This was due to the release of water from SAP during the reaction process, slowing down the decrease in internal relative humidity and effectively reducing AS development [12]. This decrease became more pronounced as the SAP content increased. It can be observed from Figure 6 that the internal relative humidity of samples containing SAP was significantly higher than that of the control at the beginning of the reaction.

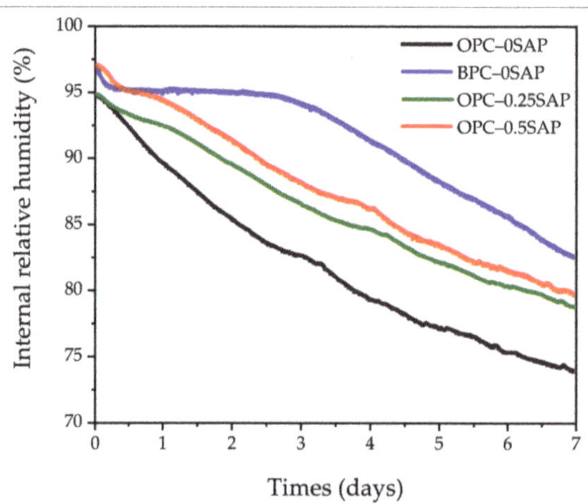

Figure 6. Variation in internal relative humidity with the age of curing.

In the stable period of middle ages, it was clearly observed that the OPC–0SAP and BPC–0SAP samples were stable and lasted longer than the samples with SAP. The macroperformance was more stable because the factors causing expansion and contraction offset each other.

During the continuous growth period of the late ages, the AS continued to increase as the reaction proceeded. The rate of AS development was significantly accelerated in the BPC–0SAP sample. Consistent with the strength trend, it showed a faster development trend at a later age. The AS of the BPC–0 samples at 7 days was lower than that of the samples doped with SAP. From the experimental results, it can be seen that the BPC can effectively reduce AS.

At 7 days of age, the AS values of OPC–0SAP, OPC–0.25SAP, and OPC–0.5SAP were -504.1, −447.6, and −399.2 μm/m, respectively. Compared with the control group, the reduction ratios of AS for OPC–0.25SAP and OPC–0.5SAP were 11.2% and 20.8%, respectively. Moreover, as shown in Figure 4, compared with the control group, the reduction ratios of strength for OPC–0.25SAP and OPC–0.5SAP were 8.1% and 32.2%, respectively. For the case of OPC–0.25SAP, the reduction in the AS sample was greater than the strength, while, for the case of OPC–0.5SAP, the reduction in the AS sample OPC–0.5SAP was lower than the strength. It should be noted that strength and AS are measured by different units, and they have different levels of importance. It is not wise to judge the benefit based on the comparisons between the strength loss ratio and AS reduction ratio.

At the age of 7 days, the AS values of OPC–0SAP and BPC–0SAP were −504.1 and −388.5 μm/m, respectively. Compared with the control group, the reduction ratio of AS for BPC–0SAP was 22.9%. Moreover, as shown in Figure 4, compared with the control group, the reduction ratio of strength for BPC–0SAP was 19.2%. The case of BPC–0SAP demonstrated that the reduction in AS was more obvious than for the strength.

Overall, the addition of SAP effectively reduced AS. The AS in the samples doped with BPC developed slowly and increased rapidly after the stable period. This trend is consistent with the internal relative humidity variation shown in Figure 6. The internal relative humidity of all samples gradually decreased when the curing age increased. This was due to the hydration of the cement and the decrease in water in the capillaries. However, the samples with SAP significantly delayed the rate of internal relative humidity reduction. In addition, the BPC-doped samples also retarded the reduction in internal relative humidity due to the slow reaction rate of C_2S [38]. The low reaction rate of the BPC–0SAP sample at the early stage could be verified from the internal temperature shown in Figure 7. The internal temperature of the BPC–0SAP sample was lower compared to the other samples.

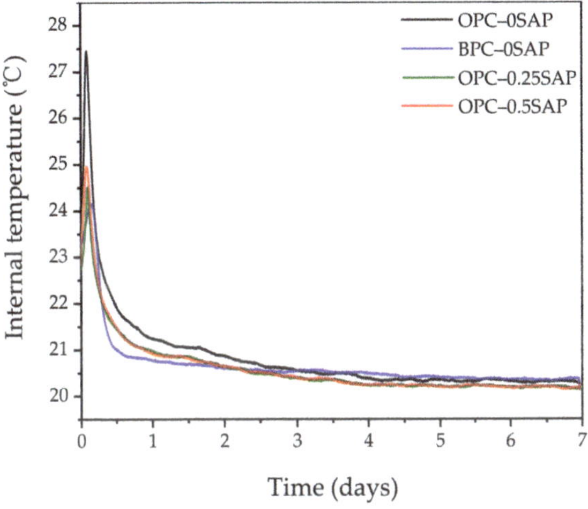

Figure 7. Variation in internal temperature with the age of curing.

As shown in Figure 7, after the age of 1 day, the internal temperature was almost constant. We plotted the AS as a function of internal relative humidity from days 1 to 7. As shown in Figure 8, AS showed a linear relationship with the internal relative humidity. The coefficients of determination between AS and internal relative humidity were higher than 98%; hence, the reduction in internal relative humidity is the main reason for AS in SUHPP.

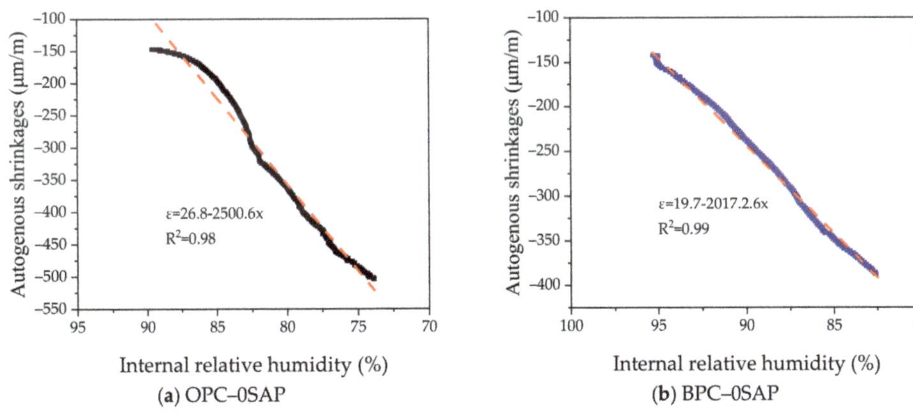

Figure 8. *Cont.*

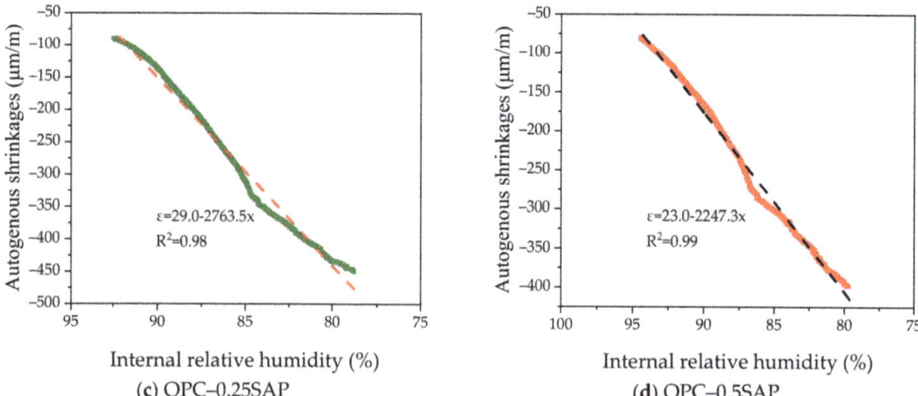

Figure 8. Correlation between AS and the internal relative humidity of specimen.

3.3. *Isothermal Calorimetry*

Figures 9 and 10 show the rate of hydration heat and cumulative hydration heat change in the mixture. We found that hours 1–7 represented the dormant period, which was due to the increase in the effective w/c of samples OPC–0.25SAP and OPC–0.5SAP [39,40]. Yang et al. [41] indicated that when w/c increased, the transition from dormancy to the acceleration period was prolonged, which was consistent with the results obtained in this experiment.

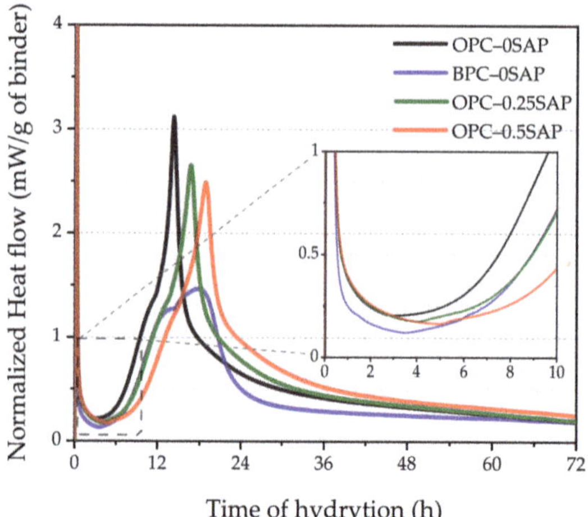

Figure 9. Exothermic rate of hydration of the mixed paste at 20 °C.

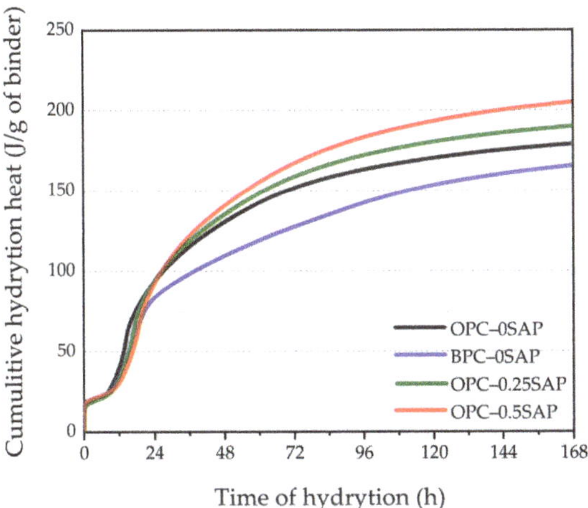

Figure 10. Cumulative heat change of the mixed paste during 72 h of mixing.

The acceleration period occurred from hours 7 to 19. The peak of the main peak decreased significantly when SAP increased and became blunter. Moreover, the appearance of the main peak was delayed by 2.5–4.5 h due to the rapid resolution of SAP, which increased the effective w/c and reduced the initial reduction concentration in the pore solution, resulting in delayed cement hydration [41]. The experimental results of Justs et al. [40] indicated that within 30 h of the initial reaction, a reduced w/c shortened the dormancy period, and the main peak appeared earlier and more sharply. In addition, the BPC–0SAP sample showed a shoulder peak whose peak was higher than the peak of the main peak. The content of C_3S in BPC was lower than OPC, resulting in a lower main peak in the BPC–0SAP sample than in the OPC–0SAP. In addition, limestone reacts with alumina in slag to form calcium alumina, thus forming a shoulder peak [42]. Shoulder peaks were also present in the other three groups of samples, but the main peaks and shoulder peaks were close to each other and did not show obvious shoulder peaks [34]. The heat flow of all samples stabilized after 36 h of hydration reaction.

As shown in Figure 10, the cumulative heat of hydration increases with the addition of SAP. The cumulative heat of hydration of both samples OPC–0.25SAP and OPC–0.5SAP was higher than that of the control after 24 h. The water entrained in SAP increased the w/c, and the higher w/c increased the degree of hydration of the sample, resulting in a higher cumulative hydration heat [43]. In contrast, the cumulative hydration heat of the BPC–0SAP samples was much lower compared to the control group. This was due to the fact that the early C_2S hydration reaction was slower than C_3S, releasing less hydration heat [33]. The cumulative hydration heat was released at 72 h and reduced by 15.9% compared to the control group. Further, it had much lower cumulative hydration heat than the samples mixed with SAP. The trend of cumulative hydration heat generally agreed with that of the internal temperature, as shown in Figure 7.

3.4. X-ray Diffraction (XRD) and Thermogravimetric Analysis (TGA)

Figure 11 shows the XRD spectra of the mixed cements at 28 days of hydration. It can be observed from the plots that the peaks of $Ca(OH)_2$ in OPC–0.25SAP and OPC–0.5SAP were higher than OPC–0SAP. This was due to the release of water from SAP, which promoted the hydration reaction and produced more hydration product [16]. Moreover, compared with OPC, BPC had lower reactivity, and the content of calcium hydroxide (CH) in BPC-0SAP was lower than that in OPC–0SAP. The peak of C_2S in BPC–0SAP was higher than other specimens. In addition, the formation of hemicarbonaluminate (Hc) was due to the reaction between the alumina phase and limestone powder.

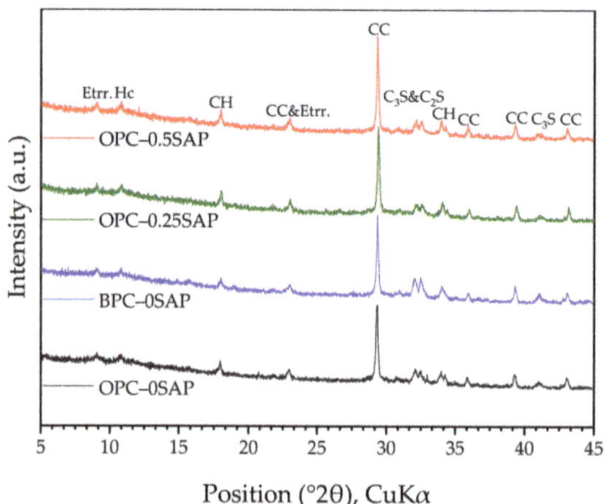

Figure 11. X-ray diffraction (XRD) spectra of the mixed cement paste at 28 days. Ettr.: ettringite; Hc: hemicarboaluminate; CH: calcium hydroxide; CC: calcite.

In addition to the XRD spectral scan, the amount of hydration product in the samples was analyzed using thermal analysis techniques. Figures 12 and 13 show the TGA and derivative thermogravimetric analysis (DTG) curves of the samples for 28 days. Weight loss can be observed in the temperature ranges of 400–450 °C and 650–750 °C, corresponding to dehydroxylation of CH and decarbonation of $CaCO_3$, respectively [38,44]. The amount of CH was calculated as follows [45]:

$$CH = \frac{w_{400} - w_{450}}{w_{550}} \times \frac{74}{18} \times 100\% \tag{1}$$

where w_{400} and w_{450} are the masses of samples at the temperatures of 400 and 450 °C, respectively. The item 74/18 indicates the ratio of CH molar weight to water.

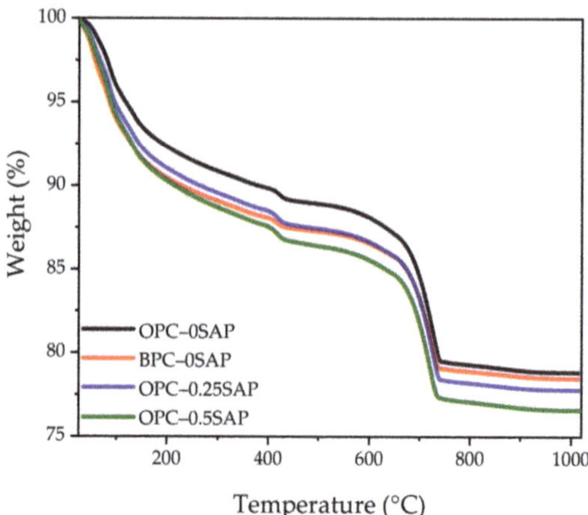

Figure 12. Thermogravimetric analysis (TGA) curves of all samples.

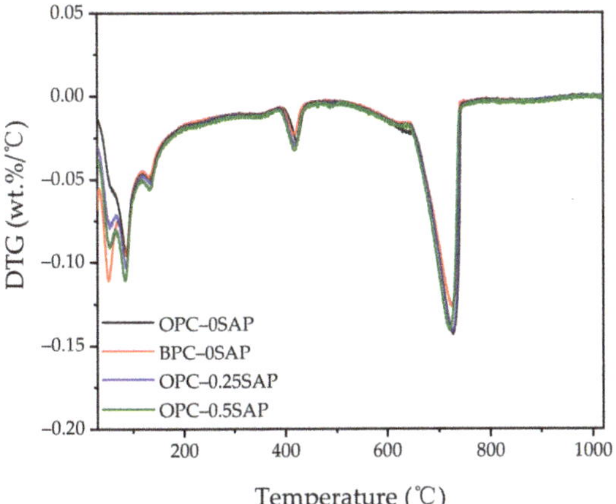

Figure 13. Derivative thermogravimetric analysis (DTG) curves of all samples.

The amount of combined water was calculated as follows [45]:

$$\text{water} = \frac{w_{105} - w_{550}}{w_{550}} \times 100\% \tag{2}$$

where w_{105} is the mass of samples at the temperature of 105 °C.

Based on the above equations, the masses of CH and combined water (W) can be determined and are shown in Table 5.

Table 5. The mass of calcium hydroxide (CH) and combined water (W).

Samples	CH (g/g)	Combined Water (g/g)
OPC–0SAP	3.44%	7.90%
BPC–0SAP	2.99%	7.62%
OPC–0.25SAP	3.95%	8.38%
OPC–0.5SAP	4.32%	8.94%

Regarding the CH content, our experimental results showed that $CH_{OPC\text{-}0.5SAP}$ > $CH_{OPC\text{-}0.25SAP}$ > $CH_{OPC\text{–}0SAP}$ > $CH_{BPC\text{–}0SAP}$. The increase in CH content with increasing SAP content indicates that the addition of SAP increased hydration. Moreover, the samples with BPC had a lower CH content compared to the control group because the reactivity of the BPC was lower than that of the OPC. The TGA analysis results showed agreement with the XRD results.

Regarding the combined water content, our experimental results show that $W_{OPC\text{-}0.5SAP}$ > $W_{OPC\text{-}0.25SAP}$ > $W_{OPC\text{–}0SAP}$ > $W_{BPC\text{–}0SAP}$. The trend of combined water was similar to that of CH. The combined water content was much less than 25%, which was the theoretical maximum of the combined water for the full hydration of 1 g cement. This was due to the low water/binder ratio and dilution effect of mineral mixtures, such as silica fume, slag, and limestone. A low water/binder ratio restricts cement-based materials from hydrating, and the addition of the mineral admixture can lower the combined water content.

3.5. Attenuated Total Reflectance (ATR)–Fourier-Transform Infrared Spectroscopy (FTIR)

Figure 14 shows the ATR–FTIR spectra of the mixed samples in the range of 500–2000 cm^{-1}. The 1420, 873 (out-of-plane vibration), and 709 cm^{-1} (in-plane bending) spectra in the scanned spectra correspond to $CO_3{}^{2-}$ (limestone) [46–49]. The 1118 and 954 cm^{-1} spectra represent the asymmetric stretching vibration of $SO_4{}^{2-}$ and the stretching vibration of Si-O at C-S-H, respectively [50,51]. The absorption peaks of the spectrum corresponding to the samples containing SAP at 954 cm^{-1} were significantly greater than those of the other two groups. The relative absorption peaks more obviously increased when the SAP content increased, which was attributed to the reaction of the C_3S clinker at 800–970 cm^{-1} and the C-S-H formed by silica polymerization [46,52].

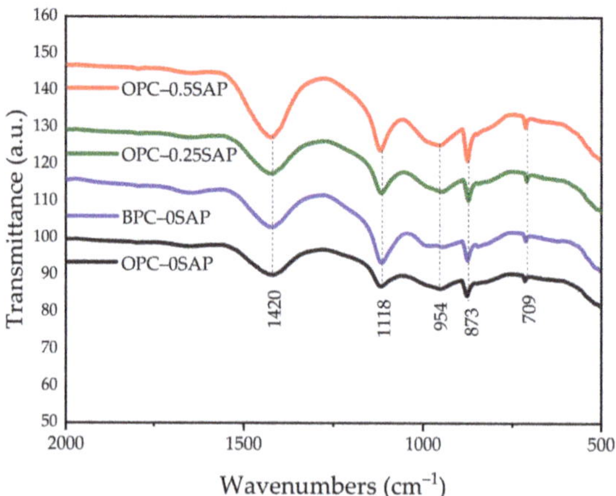

Figure 14. ATR–FTIR spectra of all samples for 28 days.

3.6. Ultrasonic Pulse Velocity (UPV)

Figure 15 shows the UPV mixture trend. The evolution of UPV was not only related to the density and elasticity of the material but also to the water content [53]. Higher SAP content makes higher effective w/c, and lower UPV [54,55]. After the dormant period, cement hydration produced a large amount of hydration product, wherein the degree of hydration increased, and the mixture had a lower porosity and a higher solid volume fraction, leading to an increase in UPV [55,56].

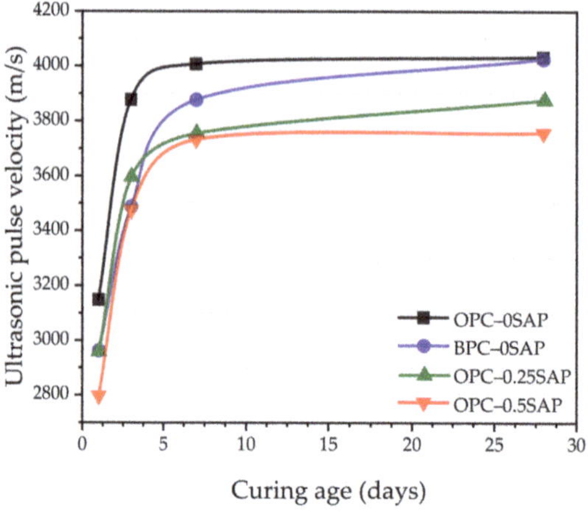

Figure 15. Variation in ultrasonic pulse velocity (UPV) of the mixture with the curing age.

The UPV for all samples increased when curing age increased. There was fast growth at the early age and slow growth at the later age. The UPV of the OPC–0SAP sample was higher than that of the OPC–0.25SAP and OPC–0.5SAP samples. The difference gradually became larger as the SAP content increased. This was due to the fact that the higher the SAP addition and w/c, the lower the UPV. The BPC–0SAP sample grew at a lower rate than the other three sample groups (OPC–0SAP, OPC–0.25SAP, and OPC–0.5SAP) at the early age, but the growth rate significantly increased at the later age due to the high content of C_2S contained in the BPC, which resulted in a slow initial reaction. C_2S reacted in the late stages of hydration, increasing the degree of hydration and helping to increase the UPV at the later age [57,58]. This was in line with the trend of strength. The compressive strength of the samples at 3, 7, and 28 days regressed against the UPV, as shown in Figure 16. The strength was found to have a high exponential correlation with UPV. UPV can be a non-destructive test method for evaluating the strength development of SUHPP.

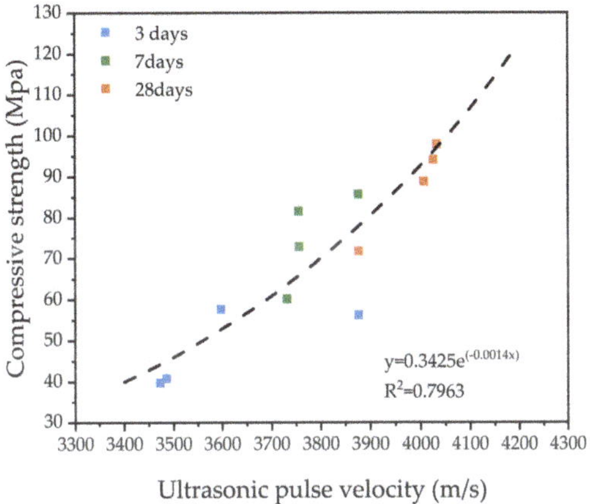

Figure 16. Correlation between UPV and compressive strength corresponding to 3, 7, and 28 days of hydration.

3.7. Electrical Resistivity

In this study, durability was evaluated based on electrical resistivity tests. For cement-based materials, as electrical resistivity increases, chloride diffusivity decreases and the service life of structures increases. Moreover, the increase in the electrical resistivity can lower the corrosion current, which is helpful for extending the service life.

Figure 17 shows the evolution of the samples' electrical resistivity with the curing age. Electrical resistivity was mainly influenced by porosity and pore structure (microstructural characteristics) [59]. SAP was added to the mixture to act as an internal curing agent and to promote hydration. However, SAP forms large pores during the resolution process, resulting in increased porosity. The negative effect of porosity on electrical resistivity far exceeds the promoting effect of internal curing [51]. The C_2S of BPC was hydrated at a later age and gradually refined the pore structure, resulting in a gradual increase in electrical resistivity [33].

The electrical resistivity of the samples OPC–0.25SAP, OPC–0.5SAP, and BPC–0SAP was lower than that of the control samples. The higher the admixture of SAP, the lower the electrical resistivity. This is attributed to the increased w/c of the samples with SAP and the increased porosity due to SAP resolution. In addition, the electrical resistivity of the BPC–0SAP sample was significantly lower than that of the other samples. However, a gradual increase in the growth rate was clearly observed with the increase in the curing age. This was due to the late C_2S hydration, which gradually improved the microstructure. Wang et al. [60] also showed that cements containing higher C_2S have a much denser pore structure at a later age, which can improve the properties of the mixture. A gradual acceleration of the resistivity growth rate of BPC–0SAP can be clearly observed in later stages. The densification of the pore structure also promoted the strength development (shown in Figure 4).

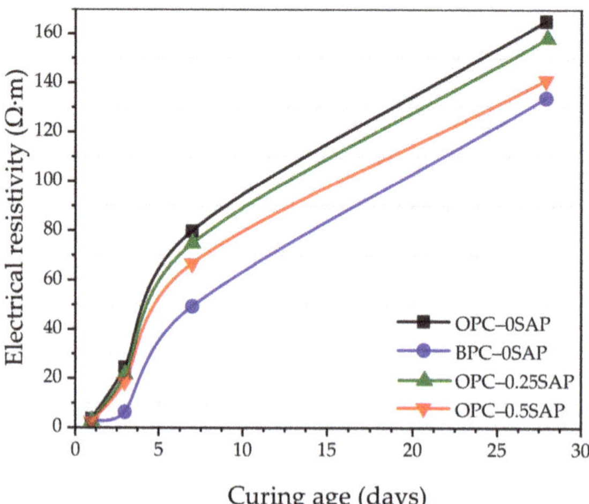

Figure 17. Changes in electrical resistivity of samples at different curing ages.

4. Discussion

The discussion items consist of (1) the comparison of SAP with classic fibers, (2) comparison of BPC with OPC, (3) comparison of the AS reduction mechanisms of SAP and BPC, (4) the expected practical application of SAP and BPC, and (5) future studies.

First, regarding AS reduction, SAP shows superior performance to classic polyester fibers. The mechanisms of AS reduction for SAP and classic polyester fibers are different. SAP can replenish water through internal curing, and classic polyester fibers can serve as internal restraints of AS [26]. Moreover, SAP affects the technological processes of concrete mixing. The mixing procedure of SAP-blended concrete is slightly more complex than that of traditional concrete, because the presoaked SAP turns into a hydrogel and is difficult to disperse in the mixture [12]. As shown in the mixing procedure (Figure 3), Step 2 involves adding 50% of water, SP, and/or additional water and mixing for 60 s, and Step 3 involves adding dry SAP powder into the mixer. In addition, compared with classic polyester fibers, the cost of SAP is much higher. However, considering the superior performance of SAP regarding AS reduction, SAP is a promising material for AS reduction in SUHPP.

Second, BPC specimens present lower AS than OPC specimens. Furthermore, BPC specimens have a lower hydration heat, a lower CH and combined water content, less electrical resistivity, and a similar late-age strength and UPV to OPC specimens. Moreover, because BPC concrete has a higher late-age strength than OPC concrete, BPC concrete has a stronger abrasion resistance [61]. In addition, compared with OPC, the cost of BPC is slightly higher. However, considering the advantages of BPC, such as low AS, low hydration heat, low CO_2 emission, and high late-age strength, we believe that BPC is a promising material for the production of SUHPP.

Third, the mechanisms of AS reduction in SAP and BPC are different. SAP can be used for internal curing, and BPC can control the rate of hydration [26]. The AS reduction level of SAP can be designed through SAP content, absorption ability, and particle size [26]. The AS reduction level of BPC mainly depends on the reactivity of BPC. Moreover, the addition of SAP impairs late-age strength, while BPC concrete has a similar strength to OPC concrete.

Fourth, based on the experimental results of this study, SAP can be used as a component of high-strength concrete which has a low water/binder ratio [37]. SAP can lower the AS of hardening high-strength concrete and mitigate cracks due to AS. Moreover, BPC can be used as a binder of high-strength concrete and mass concrete because BPC can lower the AS of high-strength concrete and hydration heat [61]. BPC is helpful for the mitigation of early-age AS cracks and thermal cracks.

Fifth, this study focuses only on the basic properties of SUHPP. In future studies, more investigations should be carried out to evaluate other highly important material properties—for example, tensile strength, modulus of elasticity, and fracture energy.

5. Conclusions

This study presents a systematical experimental investigation of the effects of SAP and cement type on SUHPP performance. Four group specimens were produced: OPC–0SAP, OPC–0.25SAP, OPC–0.5SAP, and BPC–0SAP. The SAP contents of OPC–0SAP, OPC–0.25SAP, and OPC–0.5SAP were 0%, 0.25%, and 0.5%, respectively. The cement types used in OPC–0SAP and BPC–0SAP specimens were ordinary Portland cement and belite-rich Portland cement, respectively. Experimental items consisted of compressive strength, AS, isothermal calorimetry, XRD, TGA, ATR–FTIR, UPV, and electrical resistivity. The following conclusions were obtained:

(i) At the early age of 3 days, the strength of OPC–0.25SAP samples was 2.5% higher than that of the control specimen. At the later age of 28 days, the compressive strengths of OPC–0.25SAP and OPC–0.5SAP were 12.5% and 25.5% lower than that of the control specimen, OPC–0SAP. This was due to the water release cavities created by SAP. In contrast, the strength of BPC–0SAP developed slowly at an early age, while the strength was similar to that of the control specimen, OPC–0SAP, at the later age of 28 days.

(ii) For the case of OPC–0.25SAP and BPC–0SAP, compared to the control group, the reduction in the AS sample was greater than the strength. However, for the OPC–0.5SAP sample, AS reduction was lower than the strength. Moreover, AS showed a linear relationship with the internal relative humidity. The coefficients of determination between AS and the internal relative humidity were higher than 98%. The reduction in internal relative humidity was the main reason for AS in the SUHPP.

(iii) In terms of hydration heat, the additional water entrained by SAP increased the effective w/c and reduced the initial concentration of the pore solution, resulting in a delayed and blunter main peak of 2.5–4.5 h. The higher w/c increased the degree of hydration, and the cumulative hydration heat at 72 h was significantly higher than that of the control group. The mixture doped with BPC released 15.9% lower cumulative hydration heat at 72 h compared to the control group due to the slow rate of C_2S development responses.

(iv) TGA showed the following sequence of CH content: $CH_{OPC\text{-}0.5SAP} > CH_{OPC\text{-}0.25SAP} > CH_{OPC\text{–}0SAP} > CH_{BPC\text{–}0SAP}$. When the SAP content increased, the CH content also increased, indicating that the addition of SAP increased hydration. The samples with BPC had a lower CH content compared to the control group. The analysis results of CH from TGA showed agreement with those of the XRD. Moreover, TGA analysis showed that the trend of combined water was similar to CH. The combined water content was much lower than the theoretical maximum of combined water for the full hydration of 1 g cement. In addition, the ATR–FTIR results showed that the Si-O stretching vibration of C-S-H was enhanced when the SAP content increased.

(v) In terms of UPV, the samples showed faster growth in 3 days because they were able to reach 86.3%–96.1%. For all specimens, there was an exponential correlation between compressive strength and UPV. UPV is a non-destructive test method for evaluating the strength development of SUHPP.

(vi) In terms of electrical resistivity, the OPC–0.25SAP and OPC–0.5SAP samples showed lower electrical resistivity than the control group due to the increased porosity from SAP. The BPC–0SAP sample showed lower resistivity in the early stage. However, from 3 to 28 days, the electrical resistivity of BPC–0SAP showed a higher increment than other specimens.

In summary, as the content of SAP increases, the AS, strength, UPV, and electrical resistivity decrease, and the hydration heat, CH content, and combined water content increase. Moreover, compared with OPC specimens, BPC specimens show lower AS, electrical resistivity, hydration heat, CH, and combined water content and similar late-age strength and UPV. BPC was suitable for producing SUHPP.

Author Contributions: Conceptualization, M.-Y.X. and X.-Y.W.; methodology, M.-Y.X. and X.-Y.W.; validation, M.-Y.X., Y.-S.W., H.-S.L. and S.-J.K.; writing—original draft preparation, M.-Y.X.; writing—review and editing, M.-Y.X., X.-Y.W., Y.-S.W. and H.-S.L.; visualization, M.-Y.X. and S.-J.K.; supervision, X.-Y.W. All authors have read and agreed to the published version of the manuscript.

Funding: This research was supported by the Basic Science Research Program through the National Research Foundation of Korea (NRF), funded by the Ministry of Science, ICT and Future Planning (no. 2015R1A5A1037548), and an NRF grant (no. NRF-2020R1A2C4002093).

Institutional Review Board Statement: Not applicable.

Informed Consent Statement: Not applicable.

Data Availability Statement: The data presented in this study are available from the corresponding author upon a reasonable request.

Conflicts of Interest: The authors declare no conflict of interest.

References

1. Liu, J.; Khayat, K.H.; Shi, C. Effect of superabsorbent polymer characteristics on rheology of ultra-high performance concrete. *Cem. Concr. Compos.* **2020**, *112*, 103636. [CrossRef]
2. Heikal, M.; El Aleem, S.A.; Morsi, W. Characteristics of blended cements containing nano-silica. *HBRC J.* **2013**, *9*, 243–255. [CrossRef]
3. Ghafari, E.; Ghahari, S.A.; Costa, H.; Júlio, E.; Portugal, A.; Durães, L. Effect of supplementary cementitious materials on autogenous shrinkage of ultra-high performance concrete. *Constr. Build. Mater.* **2016**, *127*, 43–48. [CrossRef]
4. Yazıcı, H.; Yiğiter, H.; Karabulut, A.Ş.; Baradan, B. Utilization of fly ash and ground granulated blast furnace slag as an alternative silica source in reactive powder concrete. *Fuel* **2008**, *87*, 2401–2407. [CrossRef]
5. Wang, C.; Yang, C.; Liu, F.; Wan, C.; Pu, X. Preparation of Ultra-High Performance Concrete with common technology and materials. *Cem. Concr. Compos.* **2012**, *34*, 538–544. [CrossRef]
6. Justs, J.; Wyrzykowski, M.; Bajare, D.; Lura, P. Internal curing by superabsorbent polymers in ultra-high performance concrete. *Cem. Concr. Res.* **2015**, *76*, 82–90. [CrossRef]
7. Kang, S.-H.; Hong, S.-G.; Moon, J. The effect of superabsorbent polymer on various scale of pore structure in ultra-high per-formance concrete. *Constr. Build. Mater.* **2018**, *172*, 29–40. [CrossRef]
8. Hisseine, O.A.; Soliman, N.A.; Tolnai, B.; Tagnit-Hamou, A. Nano-engineered ultra-high performance concrete for controlled autogenous shrinkage using nanocellulose. *Cem. Concr. Res.* **2020**, *137*, 106217. [CrossRef]
9. Li, L.; Dabarera, A.G.; Dao, V. Time-zero and deformational characteristics of high performance concrete with and without superabsorbent polymers at early ages. *Constr. Build. Mater.* **2020**, *264*, 120262. [CrossRef]
10. Tu, W.; Zhu, Y.; Fang, G.; Wang, X.; Zhang, M. Internal curing of alkali-activated fly ash-slag pastes using superabsorbent polymer. *Cem. Concr. Res.* **2019**, *116*, 179–190. [CrossRef]
11. Bentur, A.; Igarashi, S.-I.; Kovler, K. Prevention of autogenous shrinkage in high-strength concrete by internal curing using wet lightweight aggregates. *Cem. Concr. Res.* **2001**, *31*, 1587–1591. [CrossRef]
12. De Meyst, L.; Kheir, J.; Filho, J.R.T.; Van Tittelboom, K.; De Belie, N. The Use of superabsorbent polymers in high performance concrete to mitigate autogenous shrinkage in a large-scale demonstrator. *Sustain. J. Rec.* **2020**, *12*, 4741. [CrossRef]
13. Jensen, O.M.; Hansen, P.F. Water-entrained cement-based materials: I. Principles and theoretical background. *Cem. Concr. Res.* **2001**, *31*, 647–654. [CrossRef]
14. Piérard, J.; Pollet, V.; Cauberg, N. In Mitigating autogenous shrinkage in HPC by internal curing using superabsorbent polymers. In Proceedings of the International RILEM Conference on Volume Changes of Hardening Concrete: Testing and Mitigation, Lyngby, Denmark, 20–23 August 2006; pp. 97–106.
15. Kang, S.-H.; Hong, S.-G.; Moon, J. Shrinkage characteristics of heat-treated ultra-high performance concrete and its mitigation using superabsorbent polymer based internal curing method. *Cem. Concr. Compos.* **2018**, *89*, 130–138. [CrossRef]
16. Liu, J.; Farzadnia, N.; Shi, C.; Ma, X. Shrinkage and strength development of UHSC incorporating a hybrid system of SAP and SRA. *Cem. Concr. Compos.* **2019**, *97*, 175–189. [CrossRef]
17. Craeye, B.; Geirnaert, M.; De Schutter, G. Super absorbing polymers as an internal curing agent for mitigation of early-age cracking of high-performance concrete bridge decks. *Constr. Build. Mater.* **2011**, *25*, 1–13. [CrossRef]
18. Zohourian, M.M.; Kabiri, K. Superabsorbent polymer materials: A review. *Iran. Polym. J.* **2008**, *17*, 451-477.

19. Stefan, L.; Boulay, C.; Torrenti, J.-M.; Bissonnette, B.; Benboudjema, F. Influential factors in volume change measurements for cementitious materials at early ages and in isothermal conditions. *Cem. Concr. Compos.* **2018**, *85*, 105–121. [CrossRef]
20. Liu, J.; Shi, C.; Ma, X.; Khayat, K.H.; Zhang, J.; Wang, D. An overview on the effect of internal curing on shrinkage of high performance cement-based materials. *Constr. Build. Mater.* **2017**, *146*, 702–712. [CrossRef]
21. Snoeck, D.; Jensen, O.; De Belie, N. The influence of superabsorbent polymers on the autogenous shrinkage properties of cement pastes with supplementary cementitious materials. *Cem. Concr. Res.* **2015**, *74*, 59–67. [CrossRef]
22. Mönnig, S. Superabsorbing Additions in Concrete: Applications, Modelling and Comparison of Different Internal Water Sources. Master's Thesis, Universität Stuttgart, Stuttgart, Germany, 2009.
23. Song, C.; Choi, Y.C.; Choi, S. Effect of internal curing by superabsorbent polymers—Internal relative humidity and autogenous shrinkage of alkali-activated slag mortars. *Constr. Build. Mater.* **2016**, *123*, 198–206. [CrossRef]
24. Beushausen, H.; Gillmer, M.; Alexander, M. The influence of superabsorbent polymers on strength and durability properties of blended cement mortars. *Cem. Concr. Compos.* **2014**, *52*, 73–80. [CrossRef]
25. Naqi, A.; Jang, J.G. Recent Progress in green cement technology utilizing low-carbon emission fuels and raw materials: A review. *Sustainability* **2019**, *11*, 537. [CrossRef]
26. Yang, L.; Shi, C.; Wu, Z. Mitigation techniques for autogenous shrinkage of ultra-high-performance concrete—A review. *Compos. Part B Eng.* **2019**, *178*, 107456. [CrossRef]
27. Kang, S.-H.; Hong, S.-G.; Moon, J. Absorption kinetics of superabsorbent polymers (SAP) in various cement-based solutions. *Cem. Concr. Res.* **2017**, *97*, 73–83. [CrossRef]
28. ASTM International. *Standard Test Method for Compressive Strength of Hydraulic Cement Mortars (Using Portions of Prisms Broken in Flexure)*; ASTM International: West Conshohocken, PA, USA, 2014.
29. ASTM International. *Standard Test Method for Autogenous Strain of Cement Paste and Mortar*; ASTM International: West Conshohocken, PA, USA, 2014.
30. Lin, R.-S.; Park, K.-B.; Wang, X.-Y.; Zhang, G.-Y. Increasing the early strength of high-volume Hwangtoh–Cement systems using bassanite. *J. Build. Eng.* **2020**, *30*, 101317. [CrossRef]
31. Kacimi, L.; Simon-Masseron, A.; Salem, S.; Ghomari, A.; Derriche, Z. Synthesis of belite cement clinker of high hydraulic re-activity. *Cem. Concr. Res.* **2009**, *39*, 559–565. [CrossRef]
32. Koumpouri, D.; Angelopoulos, G. Effect of boron waste and boric acid addition on the production of low energy belite cement. *Cem. Concr. Compos.* **2016**, *68*, 1–8. [CrossRef]
33. Wang, L.; Jin, M.; Wu, Y.; Zhou, Y.; Tang, S. Hydration, shrinkage, pore structure and fractal dimension of silica fume modified low heat Portland cement-based materials. *Constr. Build. Mater.* **2021**, *272*, 121952. [CrossRef]
34. Zhang, G.-Z.; Cho, H.-K.; Wang, X.-Y. Effect of nano-silica on the autogenous shrinkage, strength, and hydration heat of ultra-high strength concrete. *Appl. Sci.* **2020**, *10*, 5202. [CrossRef]
35. Darquennes, A.; Staquet, S.; Delplancke-Ogletree, M.-P.; Espion, B. Effect of autogenous deformation on the cracking risk of slag cement concretes. *Cem. Concr. Compos.* **2011**, *33*, 368–379. [CrossRef]
36. Duran-Herrera, A.; De-León-Esquivel, J.; Bentz, D.; Valdez-Tamez, P. Self-compacting concretes using fly ash and fine limestone powder: Shrinkage and surface electrical resistivity of equivalent mortars. *Constr. Build. Mater.* **2019**, *199*, 50–62. [CrossRef]
37. Xuan, M.-Y.; Han, Y.; Wang, X.-Y. The Hydration, mechanical, autogenous shrinkage, durability, and sustainability properties of cement–limestone–slag ternary composites. *Sustain. J. Rec.* **2021**, *13*, 1881. [CrossRef]
38. Jang, J.; Lee, H. Microstructural densification and CO_2 uptake promoted by the carbonation curing of belite-rich Portland cement. *Cem. Concr. Res.* **2016**, *82*, 50–57. [CrossRef]
39. Langan, B.W.; Weng, K.; Ward, M.A. Effect of silica fume and fly ash on heat of hydration of Portland cement. *Cem. Concr. Res.* **2002**, *32*, 1045–1051. [CrossRef]
40. Justs, J.; Wyrzykowski, M.; Winnefeld, F.; Bajare, D.; Lura, P. Influence of superabsorbent polymers on hydration of cement pastes with low water-to-binder ratio. *J. Therm. Anal. Calorim.* **2014**, *115*, 425–432. [CrossRef]
41. Yang, J.; Wang, F.; Liu, Z.; Liu, Y.; Hu, S. Early-state water migration characteristics of superabsorbent polymers in cement pastes. *Cem. Concr. Res.* **2019**, *118*, 25–37. [CrossRef]
42. Scrivener, K.L.; Juilland, P.; Monteiro, P.J. Advances in understanding hydration of Portland cement. *Cem. Concr. Res.* **2015**, *78*, 38–56. [CrossRef]
43. Dai, J.; Wang, Q.; Lou, X.; Bao, X.; Zhang, B.; Wang, J.; Zhang, X. Solution calorimetry to assess effects of water-cement ratio and low temperature on hydration heat of cement. *Constr. Build. Mater.* **2021**, *269*, 121222. [CrossRef]
44. Lin, R.-S.; Lee, H.-S.; Han, Y.; Wang, X.-Y. Experimental studies on hydration–strength–durability of limestone-cement-calcined Hwangtoh clay ternary composite. *Constr. Build. Mater.* **2021**, *269*, 121290. [CrossRef]
45. De Weerdt, K.; Ben Haha, M.; Le Saout, G.; Kjellsen, K.; Justnes, H.; Lothenbach, B. Hydration mechanisms of ternary Portland cements containing limestone powder and fly ash. *Cem. Concr. Res.* **2011**, *41*, 279–291. [CrossRef]
46. Ylmén, R.; Jäglid, U.; Steenari, B.-M.; Panas, I. Early hydration and setting of Portland cement monitored by IR, SEM and Vicat techniques. *Cem. Concr. Res.* **2009**, *39*, 433–439. [CrossRef]
47. Trezza, M.; Lavat, A. Analysis of the system 3CaO Al2O3–CaSO4· 2H2O–CaCO3–H2O by FT-IR spectroscopy. *Cem. Concr. Res.* **2001**, *31*, 869–872. [CrossRef]

48. Hughes, T.L.; Methven, C.M.; Jones, T.G.; Pelham, S.E.; Fletcher, P.; Hall, C. Determining cement composition by Fourier transform infrared spectroscopy. *Adv. Cem. Based Mater.* **1995**, *2*, 91–104. [CrossRef]
49. Lin, R.-S.; Wang, X.-Y.; Lee, H.-S.; Cho, H.-K. Hydration and microstructure of cement pastes with calcined hwangtoh clay. *Materials* **2019**, *12*, 458. [CrossRef]
50. Mollah, M.; Yu, W.; Schennach, R.; Cocke, D.L. A Fourier transform infrared spectroscopic investigation of the early hydration of Portland cement and the influence of sodium lignosulfonate. *Cem. Concr. Res.* **2000**, *30*, 267–273. [CrossRef]
51. Wehbe, Y.; Ghahremaninezhad, A. Combined effect of shrinkage reducing admixtures (SRA) and superabsorbent polymers (SAP) on the autogenous shrinkage, hydration and properties of cementitious materials. *Constr. Build. Mater.* **2017**, *138*, 151–162. [CrossRef]
52. Lin, R.-S.; Han, Y.; Wang, X.-Y. Macro–meso–micro experimental studies of calcined clay limestone cement (LC3) paste sub-jected to elevated temperature. *Cem. Concr. Compos.* **2021**, *116*, 103871. [CrossRef]
53. Shariq, M.; Prasad, J.; Masood, A. Studies in ultrasonic pulse velocity of concrete containing GGBFS. *Constr. Build. Mater.* **2013**, *40*, 944–950. [CrossRef]
54. Van Breugel, K.; Fraaij, A. *Experimental Study on Ultrasonic Pulse Velocity Evaluation of the Microstructure of Cementitious Material at Early Age*; Delft University of Technology: Delft, The Netherlands, 2001.
55. Ye, G.; Lura, P.; Van Breugel, K.; Fraaij, A. Study on the development of the microstructure in cement-based materials by means of numerical simulation and ultrasonic pulse velocity measurement. *Cem. Concr. Compos.* **2004**, *26*, 491–497. [CrossRef]
56. Huang, H.; Gao, X.; Wang, H.; Ye, H. Influence of rice husk ash on strength and permeability of ultra-high performance concrete. *Constr. Build. Mater.* **2017**, *149*, 621–628. [CrossRef]
57. Wang, L.; Dong, Y.; Zhou, S.; Chen, E.; Tang, S. Energy saving benefit, mechanical performance, volume stabilities, hydration properties and products of low heat cement-based materials. *Energy Build.* **2018**, *170*, 157–169. [CrossRef]
58. Chen, T.; Gao, X.; Qin, L. Mathematical modeling of accelerated carbonation curing of Portland cement paste at early age. *Cem. Concr. Res.* **2019**, *120*, 187–197. [CrossRef]
59. Neithalath, N.; Weiss, J.; Olek, J. Characterizing Enhanced Porosity Concrete using electrical impedance to predict acoustic and hydraulic performance. *Cem. Concr. Res.* **2006**, *36*, 2074–2085. [CrossRef]
60. Wang, L.; Yang, H.; Zhou, S.; Chen, E.; Tang, S. Mechanical properties, long-term hydration heat, shinkage behavior and crack resistance of dam concrete designed with low heat Portland (LHP) cement and fly ash. *Constr. Build. Mater.* **2018**, *187*, 1073–1091. [CrossRef]
61. Wang, L.; Yang, H.; Dong, Y.; Chen, E.; Tang, S. Environmental evaluation, hydration, pore structure, volume deformation and abrasion resistance of low heat Portland (LHP) cement-based materials. *J. Clean. Prod.* **2018**, *203*, 540–558. [CrossRef]

Article

Use of Coal Bottom Ash and CaO-$CaCl_2$-Activated GGBFS Binder in the Manufacturing of Artificial Fine Aggregates through Cold-Bonded Pelletization

Dongho Jeon [1], Woo Sung Yum [1], Haemin Song [1], Seyoon Yoon [2], Younghoon Bae [3,*] and Jae Eun Oh [1,*]

1 School of Urban and Environmental Engineering, Ulsan National Institute of Science and Technology (UNIST), UNIST-gil 50, Ulju-gun, Ulsan 44919, Korea; jeondongho@unist.ac.kr (D.J.); wsyum@kict.re.kr (W.S.Y.); haemin93@unist.ac.kr (H.S.)

2 Department of Civil Engineering, Kyonggi University, 154-42, Gwanggyosan-ro, Yeongtong-gu, Suwon-Si, Gyeonggi-do 16227, Korea; yoonseyoon@kyonggi.ac.kr

3 Advanced Railroad Civil Engineering Division, Korea Railroad Research Institute, 176, Cheoldobangmulgwan-ro, Uiwnag-si, Gyeonggi-do 16105, Korea

* Correspondence: yhbae@krri.re.kr (Y.B.); ohjaeeun@unist.ac.kr (J.E.O.)

Received: 6 November 2020; Accepted: 7 December 2020; Published: 8 December 2020

Abstract: This study investigated the use of coal bottom ash (bottom ash) and CaO-$CaCl_2$-activated ground granulated blast furnace slag (GGBFS) binder in the manufacturing of artificial fine aggregates using cold-bonded pelletization. Mixture samples were prepared with varying added contents of bottom ash of varying added contents of bottom ash relative to the weight of the cementless binder (= GGBFS + quicklime (CaO) + calcium chloride ($CaCl_2$)). In the system, the added bottom ash was not simply an inert filler but was dissolved at an early stage. As the ionic concentrations of Ca and Si increased due to dissolved bottom ash, calcium silicate hydrate (C-S-H) formed both earlier and at higher levels, which increased the strength of the earlier stages. However, the added bottom ash did not affect the total quantities of main reaction products, C-S-H and hydrocalumite, in later phases (e.g., 28 days), but simply accelerated the binder reaction until it had occurred for 14 days. After considering both the mechanical strength and the pelletizing formability of all the mixtures, the proportion with 40 relative weight of bottom ash was selected for the manufacturing of pilot samples of aggregates. The produced fine aggregates had a water absorption rate of 9.83% and demonstrated a much smaller amount of heavy metal leaching than the raw bottom ash.

Keywords: fine aggregate; bottom ash; GGBFS; cold-bonded pelletization; heavy metal leaching

---

1. Introduction

Concrete is an artificial material that is widely used for building infrastructure in civil engineering [1] and it is estimated that around 30 billion tons of concrete are produced worldwide every year [2]. However, concrete production requires large quantities of natural fine and coarse aggregates, as about 70% of the total volume of concrete is composed these of aggregates. Therefore, natural aggregates, such as river and sea sand, are constantly being excavated, which destroys the natural environment and depletes natural resources. As interest in environmental conservation grows, many studies have been conducted with the goal of replacing natural concrete aggregates with either recycled concrete aggregates [3] or waste materials [4,5].

Coal bottom ash (bottom ash) is an incombustible residue waste collected from coal-fired power plants [6]. In the Republic of Korea (South Korea), coal fly ash (fly ash), which is also an industrial by-product of coal-fired power plants, has been sold as either a mineral admixture for concrete or as a raw material for Portland cement manufacturing for USD 20–25 per ton, while bottom ash is mostly disposed of in landfills or in ash ponds at a disposal cost of USD 10–20 per ton. Additionally, as bottom ash contains toxic heavy metals, its disposal can cause serious environmental pollution [7].

Previous studies have suggested that bottom ash can be used as a mineral admixture for concrete after (1) grinding to increase its pozzolanic reactivity [8], (2) sieving to increase its fineness [9], (3) the addition of a superplasticizer to reduce its water absorption [10], or (4) putting it in a cement kiln as a raw material [11]. However, it is difficult to use raw bottom ash directly as a mineral admixture for concrete because its use generally decreases the workability of fresh concrete due to the angular grain texture [12] and bottom ash has a lower pozzolanic reactivity than fly ash due to its larger particle size [12].

Conversely, the granulation of bottom ash can be a successful way to increase its utilization rate because it can then be used as an artificial aggregate in concrete production due to its good volumetric stability in hardened concrete [13]. Among the various granulating techniques, the cold-bonded pelletization is a decent candidate as it can shape powdered materials into spherical pellets. By simply spraying water on the powdered materials in a rotating pelletizing disk, powder particles are agglomerated by gravity and friction over time and become spherical pellets [14,15]. Therefore, previous studies have attempted to manufacture artificial aggregates using waste powders such as fly ash, bottom ash, or iron ore fines through the cold-bonded pelletization technique [16–19].

Additionally, rotary kilns are generally used to manufacture artificial aggregates after the granulation process. However, rotary kilns generally use a sintering process at temperatures near to 1200 °C to form aggregates, which requires the consumption of a great deal of energy [15,20]. It is therefore necessary to develop new methods of producing artificial aggregates that work at relatively low temperatures (e.g., room temperature).

Therefore, this study investigated the use of bottom ash and cementless binder in the manufacturing of artificial fine aggregates using cold-bonded pelletization. This study used the CaO-$CaCl_2$-activated ground granulated blast furnace slag (GGBFS) as a cementless binder because it is more economical and environmentally friendly than Portland cement and geopolymer [21,22]. Specifically, the use of geopolymer binder is inappropriate for cold-bonded pelletization because spraying high pH alkaline solutions such as NaOH or water glass on powder materials in pelletizing disks would cause serious safety issues for workers. To this end, mixture samples were prepared using a CaO-$CaCl_2$-GGBFS binder with varying additions of bottom ash for strength testing and material analyses, including powder X-ray diffraction (XRD), thermogravimetry (TG), inductively coupled plasma-optical emission spectrometry (ICP-OES), and ion chromatography (IC). After evaluating the mechanical strength and pelletizing formability of the mixtures, one proportion was selected for use in the production of pilot samples of artificial fine aggregates using a disk pelletizer and room temperature curing; water absorption and leaching tests for heavy metals were then conducted for the produced aggregates.

2. Materials and Methods

2.1. *Raw Materials*

Bottom ash was collected at the Ha-dong thermal power plant in South Korea. Commercial GGBFS (Chunghae, Korea) was used. Analytical grade CaO and $CaCl_2$ (extra pure, 98%) were used as activators.

The oxide and elemental compositions of the raw materials, GGBFS and bottom ash, were determined using an X-ray fluorescence (XRF) spectrometer (S8 Tiger; Bruker, Billerica, MA, USA) and their losses on ignition (LOI) were measured with a thermal analyzer (SDT Q600; TA Instruments, New Castle, DE, USA). Table 1 shows the elemental and oxide compositions of the GGBFS and bottom ash.

Table 1. Elemental and oxide compositions of ground granulated blast furnace slag (GGBFS) and bottom ash by X-ray fluorescence (XRF).

GGBFS				Bottom Ash			
Element (atomic%)		Oxide (wt %)		Element (atomic%)		Oxide (wt %)	
Ca	59.5	CaO	45.1	Si	50.2	SiO_2	60.4
Si	23.1	SiO_2	33.6	Al	15.4	Al_2O_3	18.4
Al	9.7	Al_2O_3	13.3	Fe	13.3	Fe_2O_3	7.4
Mg	2.5	MgO	3.2	Ca	11.0	CaO	6.8
S	1.4	SO_3	2.1	K	2.0	MgO	1.6
Ti	1.0	TiO_2	0.8	Ti	1.7	Na_2O	1.4
Mn	0.8	MnO	0.5	Na	1.5	TiO_2	1.2
Fe	0.7	K_2O	0.5	Mg	1.5	K_2O	1.1
K	0.7	Fe_2O_3	0.4	Sr	0.8	MoO_3	0.4
Na	0.3	Na_2O	0.3	Mo	0.8	P_2O_5	0.3
Sr	0.1	SrO	0.1	Ba	0.7	BaO	0.3
Ba	0.1	BaO	0.1	Nb	0.5	SrO	0.3
Zr	0.09	ZrO_2	0.1	P	0.3	Tb_4O_7	0.1
V	0.03	V_2O_5	0.0	Mn	0.2	MnO	0.08
Y	0.02	P_2O_5	0.0	Cl	0.06	SO_3	0.06
P	0.01	Y_2O_3	0.0	S	0.03	Cl	0.03
Nb	0.00	Cr_2O_3	0.0	Cu	0.03	V_2O_5	0.03
LOI (wt %)		0.41%		LOI (wt %)		0.68%	

Powder X-ray diffraction (XRD) patterns for the GGBFS and bottom ash were collected using a high-power X-ray diffractometer (D/MAX 2500V/PC; Rigaku, Tokyo, Japan) with a Cu-Kα radiation (λ = 1.5418 Å) for a 2θ scanning range of 8°–60° in 2θ degrees. The obtained XRD patterns were analyzed with the X'pert High Score program [23] using the International Center for Diffraction Data (ICDD) PDF-2 database [24] and the Inorganic Crystal Structure Database (ICSD) [25].

Figure 1 shows the measured XRD patterns and identified phases of the raw GGBFS and bottom ash. In the mineralogical composition, the GGBFS contained only akermanite ($Ca_2MgSi_2O_7$, ICSD PDF-2 no. 01-079-2425) while the bottom ash included several crystalline minerals such as quartz (SiO_2, ICSD PDF-2 no. 01-087-2096), calcium aluminum silicate ($Al_{1.77}Ca_{0.88}O_8Si_{2.23}$, ICSD PDF-2 no. 00-052-1344), mullite ($3Al_2O_32SiO_2$, ICSD PDF-2 no. 01-079-1455), diopside ($Ca_1Mg_1O_6Si_2$, ICDD PDF-2 no. 98-015-9054), and magnetite (Fe_3O_4, ICDD PDF-2 no. 98-015-8743). Amorphous humps were observed in both materials and are marked as shaded areas in Figure 1. The GGBFS mostly consisted of amorphous phase while the bottom ash contained a much smaller portion.

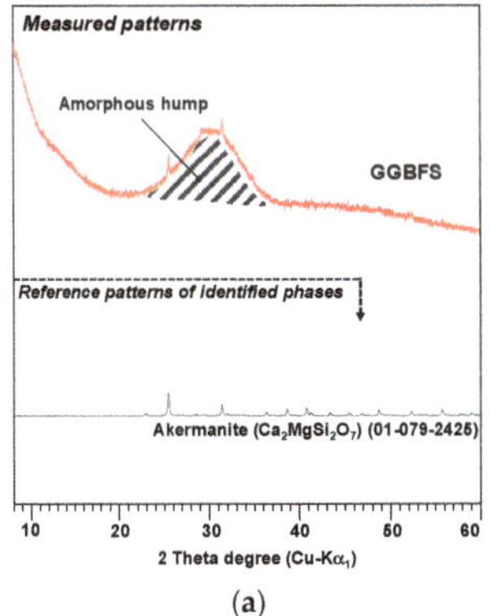

(**a**)

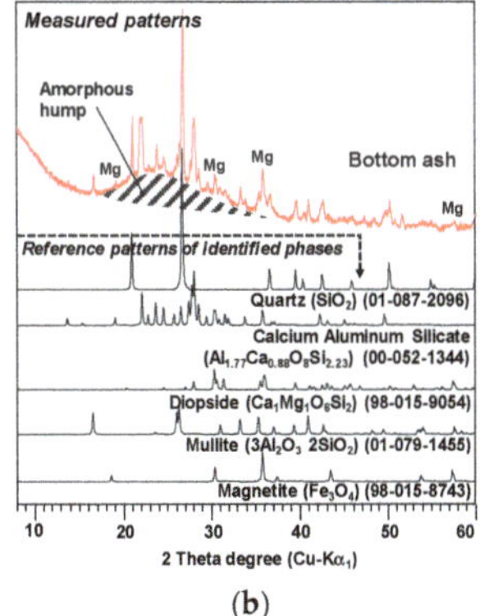

(**b**)

Figure 1. XRD patterns and identified phases of the (**a**) raw GGBFS and (**b**) raw bottom ash. Mg: magnetite.

The particle size distributions of the GGBFS and bottom ash were examined with a laser scattering particle size analyzer (HELOS; Sympatec, Clausthal-Zellerfeld, Germany) (Figure 2). It is worth noting that the mean particle size of the bottom ash (276.31 μm) was approximately 10 times larger than that of the GGBFS (27.00 μm).

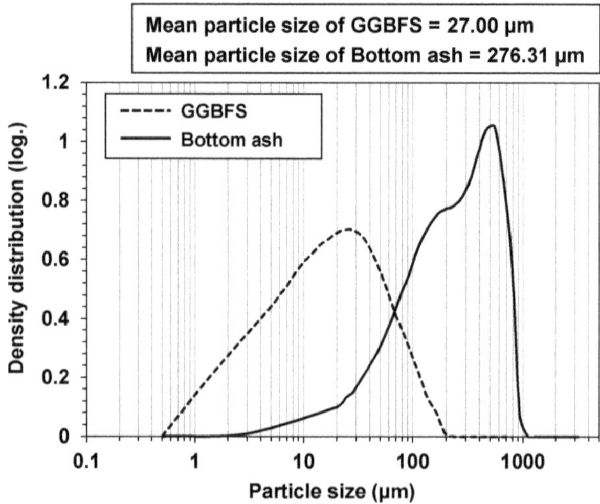

Figure 2. Particle size distributions of raw GGFBS and raw bottom ash with their mean particle sizes.

The GGBFS and bottom ash were studied using a field emission scanning electron microscope (SEM) (Quanta200; FEI, Eindhoven, The Netherlands) in the secondary electron (SE) mode. The powdered raw materials were placed on double-faced carbon tape with a platinum (Pt) coating. Figure 3 shows SEM SE images of the raw GGBFS and bottom ash.

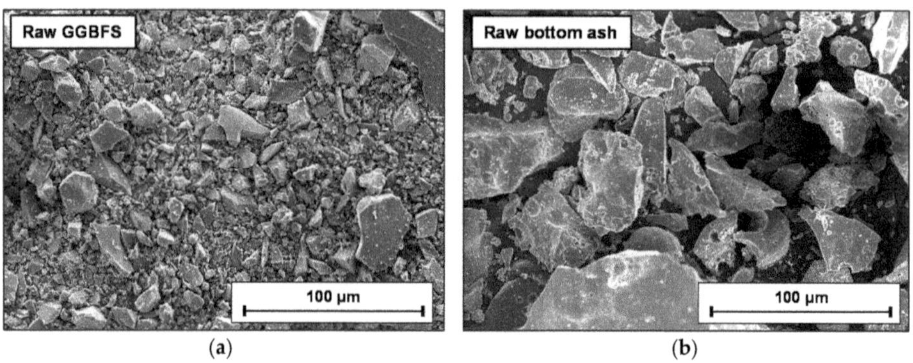

Figure 3. SEM images of (**a**) raw GGBFS and (**b**) raw bottom ash.

2.2. Sample Preparation and Tests

2.2.1. Mortar Samples

The mixture proportions are given in Table 2. The bottom ash was about 10 times larger in particle size than GGBFS and so it is likely that it acts as a fine aggregate; the mixtures will be referred to as "mortar" in this study although no sand was included in the mixture proportions.

Table 2. Mixture proportions of mortar samples in relative weight.

Label	Binder (b)				Bottom Ash (B)	Binder Fraction in Mixture (Fr) (= S/(S + B))	Water (w)
	GGBFS	CaO	$CaCl_2$	Total (S)			
B/S0.0	94	4	2	100	0	1.00	40 (w/b = 0.4)
B/S0.2					20	0.83	
B/S0.4					40	0.71	
B/S0.6					60	0.63	
B/S0.8					80	0.56	
B/S1.0					100	0.50	

Note: For inductively coupled plasma-optical emission spectrometry (ICP-OES) and ion chromatography (IC) testing, the solution samples at w/b = 2.0 were used while the samples at w/b = 0.4 were prepared for all other testing.

The mortar mixture samples were prepared by adding bottom ash in relative weight ratios of 0, 20, 40, 60, 80, and 100 to the binder weight (GGBFS + CaO + $CaCl_2$ = 100) (i.e., B/S0.0, B/S0.2, B/S0.4, B/S0.6, B/S0.8, and B/S1.0 in Table 2, respectively). As the quantity of added bottom ash increased, the relative weight of the binder to the total weight of mixture, which is the binder fraction in mixture (Fr) in Table 2 decreased from 1.0 to 0.5 (e.g., 1.0 for B/S0.0, and 0.5 for B/S1.0). Although the presence of $CaCl_2$ could result in the corrosion of the embedded steel bars in concrete, $CaCl_2$ was used in the mixture to increase the overall compressive strength of the binder used in this study [21]. The previous study reported that the use of $CaCl_2$ in the CaO-activated GGBFS binder system promoted the early dissolution of the amorphous phase of GGBFS considerably [21].

Two values of water-to-binder weight ratio (w/b) were used in the sample preparation: w/b = 2 for diluted samples for ICP-OES and IC and w/b = 0.4 for all other testing samples; the relative weight of water to the total weight of the mixture decreased as the weight of the added bottom ash increased. Thereby, the flowability of the freshly mixed mortar decreased.

The bottom ash was prepared under the surface-dry (SSD) condition before mixing. The GGBFS and CaO powders were dry-mixed with $CaCl_2$ and then mixed with varying contents of bottom ash for two minutes in a mechanical mixer. The mixtures were then mixed with de-ionized water for three minutes. The fresh mortars were cast in brass cubic molds (50 × 50 × 50 mm) and then cured for 3, 7, 14, and 28 days. All samples were stored in a humidity chamber at 23 °C with 99% relative humidity for all curing periods.

Compressive strength tests were performed on the triplicate cubic samples for each mixture after 3, 7, 14, and 28 days. After testing, the fractured specimens were collected and finely ground with an agate mortar and pestle. The powdered samples were subjected to a solvent exchange process using isopropyl alcohol to stop further hydration and carbonation for the XRD and TG tests [26].

The XRD patterns for the ground samples were collected and analyzed using the same XRD instrument and analysis program with the database used for analyzing raw materials.

Thermogravimetry (TG) (Q500; TA Instruments, Newcastle, DE, USA) was performed for the ground hardened samples with an alumina pan and nitrogen gas. The heating temperature ranged from room temperature to 1000 °C with a heating rate of 30 °C/min.

Inductively coupled plasma-optical emission spectrometry (ICP-OES) and ion chromatography (IC) were conducted to examine the dissolution behaviors of the GGBFS and bottom ash using a spectrometer (700-ES; Varian, Palo Alto, CA, USA) and a reagent-free ion chromatography system (ICS-3000; Thermo Scientific, Waltham, MA, USA), respectively. In this study, the target elements were calcium (Ca), silicon (Si), aluminum (Al), magnesium (Mg), sulfur (S), and iron (Fe) because the raw materials, GGBFS and bottom ash, were mainly composed of these elements. This was determined by XRF (Table 1).

The sample preparation for ICP-OES and IC analyses was conducted with a water-to-binder weight ratio of 2.0 (w/b = 2.0) at 23 °C for all diluted samples as mentioned earlier. Two hundred grams of water and 100 g of the binder were mixed and then the quantity of bottom ash was increased in increments of 20 g from 0 to 100 g. It is important to note that bottom ash was not included in the binder weight and the bottom ash in SSD was used. Then, the diluted mixtures were consistently agitated using a magnetic stirrer for 24 h at room temperature. After agitation, the filtrated liquid phases from the mixtures were tested for ICP-OES and IC measurement [27].

2.2.2. Aggregate Samples

Pilot samples of the artificial fine aggregates were manufactured for all mixture proportions in Table 2 through cold-bonded pelletization using a disk pelletizer. First, the binder was dry-mixed with varying quantities of bottom ash using the same mixture proportions as in Table 2, except for the fact that the bottom ash was used in a dry state because wet mixing is not necessary for pelletization.

The mixed powders were put in a disk pelletizer with a diameter of 80 cm and then pelletized by spraying water. Previous studies noted that disk inclination angle and disk revolution speed affect the efficiency of pelletization [28,29] and after several adjustments were made, the best disk inclination angle and revolution speed were found to be 34 RPM and 42°, respectively, as illustrated in Figure 4.

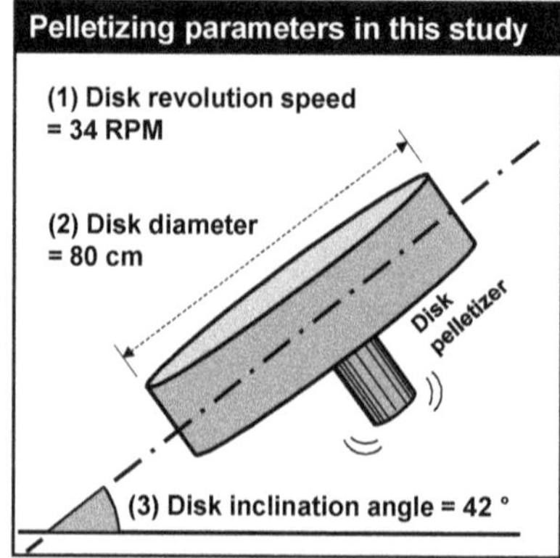

Figure 4. Pelletizing parameters for maximum pelletization efficiency in this study.

The freshly pelletized aggregates were cured in a humidity chamber for 28 days at room temperature. Finally, cured aggregates smaller than 4.75 mm were separated via sieving for use as fine aggregates.

After considering the results of the strength test on the mortar samples and the pelletizing formability after the cold-bonded pelletization, the best mixture proportion, B/S0.4, was selected from Table 2 for the production of artificial fine aggregates. The produced aggregates using B/S0.4 were cured in a humidity chamber at 23 °C with 99% relative humidity for 28 days. Aggregates were then used for a water absorption test and a leaching test for heavy metals. Although it is best to conduct the compressive strength and the instrumental analyses for the aggregate sample as well, since some mixture proportions could not be well manufactured into aggregate samples and it is difficult to set the same w/b between the aggregate samples, the identification of reaction products, which is largely influenced by w/b, was conducted only for the samples after the compressive strength test.

The water absorption was measured in the developed 28-day-cured fine aggregate sample according to the ASTM C128-15 [30]. After the fine aggregates of 500 g were submerged in water for 24 h at room temperature, the samples were slowly and equally dried until they reached the SSD condition, as determined by a cone test. The weight of the saturated samples (S) was measured. The samples were dried in an oven at 100 °C for 24 h to determine the oven dry mass (A). The value of the water absorption of the fine aggregates was then calculated following the formula from the ASTM C128-15: Absorption, % = $[(S - A)/A] \times 100$.

The leaching test was performed for both the raw bottom ash and the manufactured fine aggregates [31]. The sample preparation for the leaching test was conducted by following the toxicity characteristic leaching procedure (TCLP) for aggregates of the US Environmental Protection Agency (EPA) to detect concentrations of arsenic (As), cadmium (Cd), barium (Ba), lead (Pb), and chromium (Cr); a solution of pH = 5, which was diluted from 0.5 N acetic acid, was prepared for the extraction medium. The ratio of solution to solid for the samples was set as 20:1 and the diluted samples were agitated with a magnetic stirrer for 18 h. The liquid was tested using ICP-OES after filtrating. Additionally, although the TCLP regulation does not include copper (Cu), zinc (Zn), and nickel (Ni), the concentrations of these elements were also measured using the ICP-OES because a previous study [7] reported that bottom ash contained them in large quantities.

3. Results and Discussion

3.1. Test Results for Mortar Samples

3.1.1. Compressive Strength Test

Figure 5 shows the compressive strength testing results for the hardened mortar samples. The averaged values of the three testing results and their standard deviations are indicated in the inset table and as the error bars in Figure 5. All samples showed increasing strength as the curing time increased to 28 days.

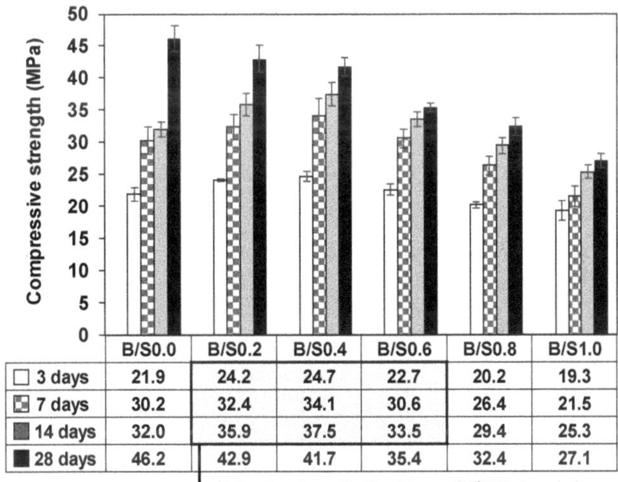

	B/S0.0	B/S0.2	B/S0.4	B/S0.6	B/S0.8	B/S1.0
3 days	21.9	24.2	24.7	22.7	20.2	19.3
7 days	30.2	32.4	34.1	30.6	26.4	21.5
14 days	32.0	35.9	37.5	33.5	29.4	25.3
28 days	46.2	42.9	41.7	35.4	32.4	27.1

Figure 5. Compressive strength testing results of the mortar mixture samples.

The influence of the added bottom ash on strength was dependent on the curing day and the quantity of additional bottom ash. As illustrated in Figure 5, in a range of 20 to 60 relative weight of bottom ash until 14 days (the gray area in the table), the added bottom ash was beneficial in increasing the mortar strength. However, all 28-day strengths were lower than that of B/S0.0 in the same range. For instance, the strength of B/S0.4 at 14 days (37.5 MPa) was ~17% greater than that of the sample without any bottom ash (B/S0.0) (32.0 MPa); however, the strength of B/S0.4 at 28 days (41.7 MPa) was ~10% lower than that of B/S0.0 (46.2 MPa). In fact, all mixture samples regardless of the bottom ash content showed lower mortar strengths at 28 days than B/S0.0 and the 28-day strength decreased as the additional bottom ash increased.

However, it is worth noting that the reduction degrees of 28-day strength were not significant for B/S0.2 and B/S0.4 as compared to those of the other samples; additionally, their strengths were relatively similar to each other despite the difference in the quantity of the bottom ash added, which was twice the ash quantity of B/S0.4 as compared to B/S0.2. As previously stated, USD 10–20 per ton can be obtained as a disposal cost for taking bottom ash from coal-fired power plants; therefore, increasing the bottom ash quantity in the mixture reduces the material cost of manufacturing artificial aggregates in this study. Given the results of strength tests and the possible material costs, the mixture B/S0.4 is likely the best candidate for manufacturing artificial aggregates.

3.1.2. XRD

Figure 6 presents the XRD patterns of the hardened mortar mixtures at 3, 7, and 28 days.

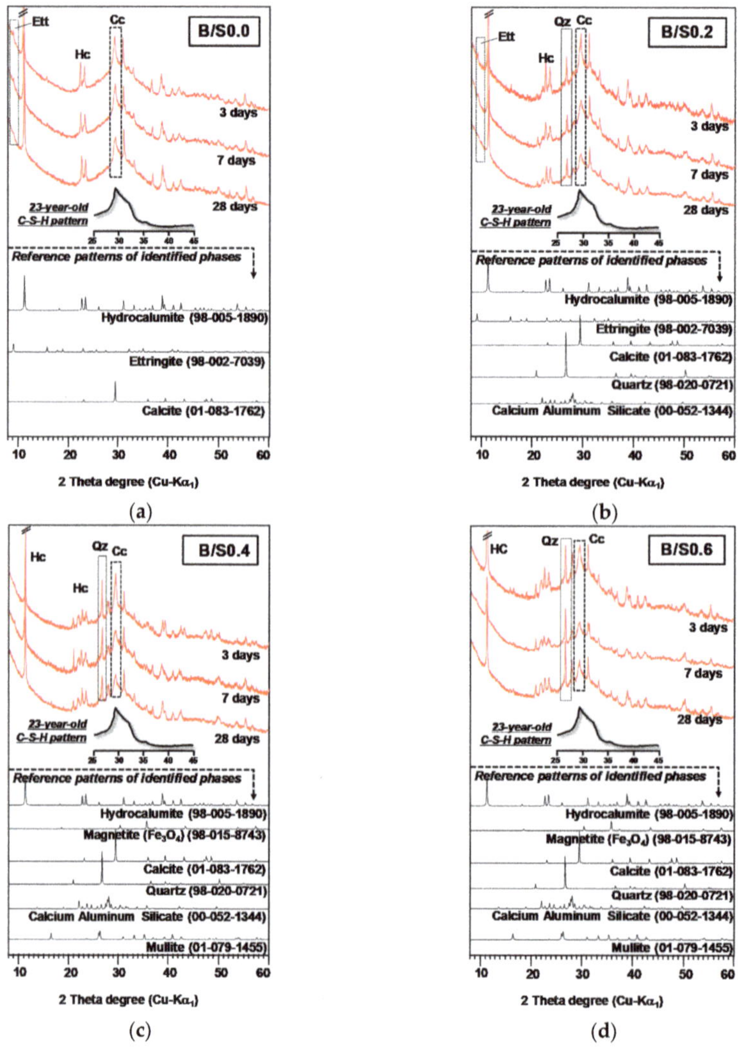

Figure 6. *Cont.*

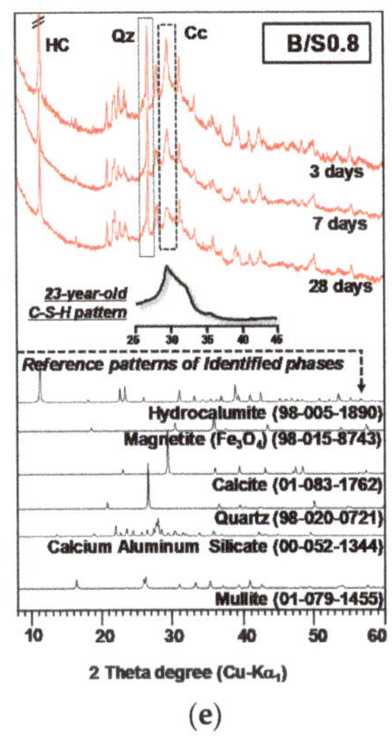

(e)

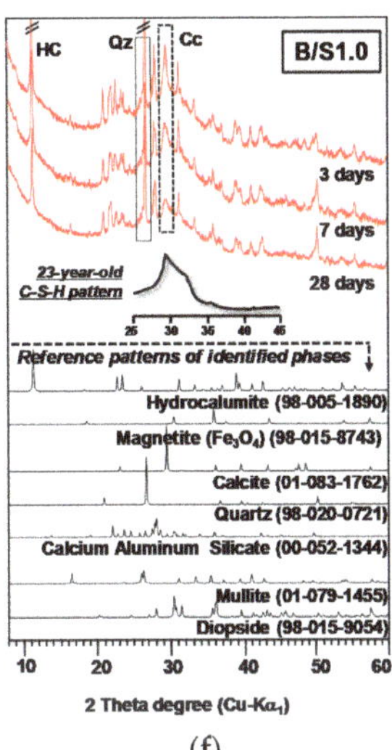

(f)

Figure 6. XRD patterns of the hardened mortar mixture samples: (**a**) B/S0.0, (**b**) B/S0.2, (**c**) B/S0.4, (**d**) B/S0.6, (**e**) B/S0.8, and (**f**) B/S1.0; Ett: ettringite, HC: hydrocalumite, Qz: quartz, and Cc: calcite. The XRD pattern of 23-year-old calcium silicate hydrate (C-S-H) was modified from a previous study [32]. The intensities of XRD patterns are modified to have the same baselines between the samples.

The sample B/S0.0 showed strong indications of hydrocalumite ($Ca_2Al(OH)_6Cl{\cdot}2H_2O$, ICDD PDF-2 no. 98-005-1890) and calcium silicate hydrate (C-S-H) as the main reaction products of the CaO-$CaCl_2$-activated GGBFS binder [21]. The formation of C-S-H is seen in the broad peak around 30° according to the XRD pattern of 23-year-old C-S-H from a previous study [32], although the strongest peak of calcite was overlapped with the C-S-H peak around 30° [33]. Additionally, very weak peaks of ettringite ($Ca_6Al_2(SO_4)_3(OH)_2{\cdot}26H_2O$, ICDD PDF-2 no.98-002-7039) were identified as a minor reaction product. Although calcite ($CaCO_3$, ICSD PDF-2 no. 01-083-1762) was also found in B/S0.0, it was likely not a reaction product of the binder but rather the result of the carbonation of CaO during curing or sample preparation.

In the samples with added bottom ash, hydrocalumite and C-S-H were also identified as the main reaction products. In terms of ettringite, as the weight fraction of bottom ash increased in the mixture, ettringite peaks decreased or disappeared.

In all the samples with added bottom ash, standard bottom ash minerals such ash quartz (SiO_2, ICDD PDF-2 no. 98-020-0721), mullite ($3Al_2O_32SiO_2$, ICSD PDF-2 no. 01-079-1455), calcium aluminum silicate ($Al_{1.77}Ca_{0.88}O_8Si_{2.23}$, ICSD PDF-2 no.00-052-1344), diopside ($Ca_1Mg_1O_6Si_2$, ICDD PDF-2 no. 98-015-9054), and magnetite (Fe_3O_4, ICDD PDF-2 no. 98-015-8743) were found, and as bottom ash content increased, the peak intensities of quartz, mullite, and calcium aluminum silicate became stronger.

3.1.3. TG

The TG and the differential thermogravimetry (DTG) curves of the hardened mortar mixtures at 28 days are presented in Figure 7. The DTG results confirmed the formation of C-S-H and hydrocalumite, however, the decomposition of calcite was not seen in the temperature range of 650–800 °C [34]. Given that the XRD patterns at 28 days in Figure 7 showed significantly reduced peaks of calcite, it was likely that the calcite amount was not high enough to be detected in TG. Ettringite is known to decompose in the temperature range of 50–200 °C which is indicated by a large DTG peak [35]. In terms of the XRD results, only a small amount of ettringite was found in B/S0.0 and B/S0.2 and no ettringite was found in the other samples. Given that the measured ettringite XRD peaks were too small, it might be difficult to identify the ettringite DTG peaks in B/S0.0 and B/S0.2 due to relatively larger DTG peaks of C-S-H and hydrocalumite. In addition, hydrocalumite was thermally decomposed in three temperature ranges in the TG: (1) 60–160 °C (dehydration), (2) 240–360 °C (dehydroxylation), and (3) above 670 °C (anion decomposition) [26,36].

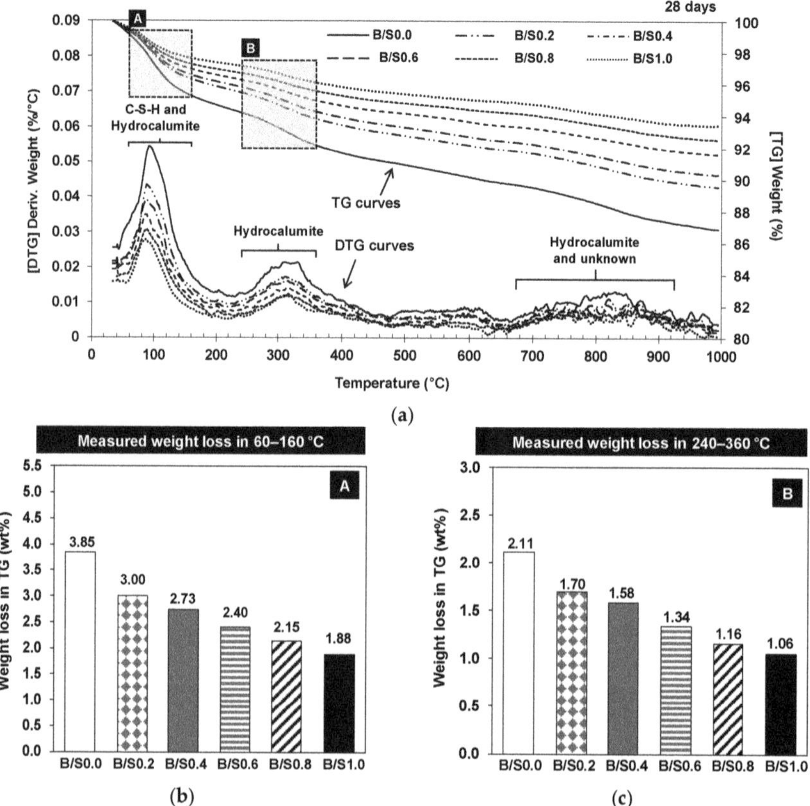

Figure 7. Thermogravimetry (TG) and differential thermogravimetry (DTG) results of the hardened samples at 28 days (**a**) and measured weight losses in 60–160 °C (box A) (**b**) and 240–360 °C (box B) (**c**).

The dehydrations of C-S-H and hydrocalumite were overlapped around 100 °C in Figure 7 [1,36–38], however, because the DTG peaks around 320 °C were only due to the dehydroxylation of hydrocalumite, the weight loss due to the dehydration of C-S-H around 100 °C can be indirectly estimated after assessing the weight loss due to the hydroxylation of hydrocalumite at 240–360 °C [26]. Thus, the influence of added bottom ash on the formation of C-S-H and hydrocalumite can be evaluated.

Figure 7b,c display as if the added bottom ash suppressed the formation of C-S-H and hydrocalumite because the measured weight losses in 60–160 °C and 240–360 °C decreased, respectively, as the added content of bottom ash increased. However, because the binder fraction in mixture (Fr in Table 2) decreased as the bottom ash content increased and C-S-H and hydrocalumite were produced primarily by the binder, the weight loss per binder weight rather than the measured weight loss must be considered in TG.

The weight loss per binder weight was calculated by dividing the measured TG weight loss by the binder weight fraction (Fr) in each mixture; Figure 8a,b provide the measured weight losses per binder weight at 60–160 °C and at 240–360 °C, respectively. The figures show that all values were similar in all samples regardless of bottom ash content. This indicates that the addition of bottom ash did not change the amount of C-S-H or the amount of hydrocalumite that was formed over 28 days.

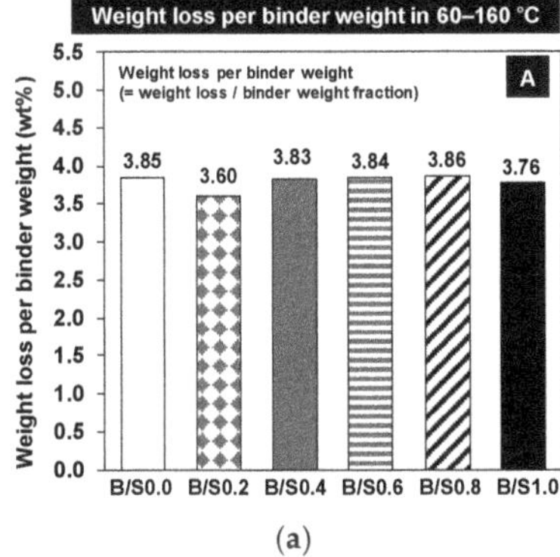

(a)

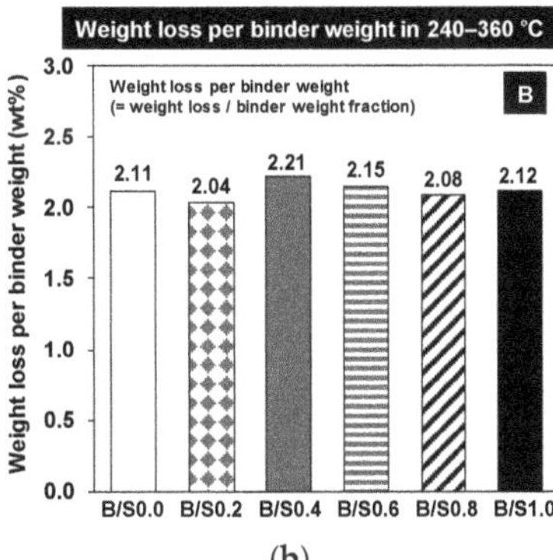

(b)

Figure 8. The calculated weight losses per binder weight for 60–160 °C (box A in Figure 7) (**a**) and 240–360 °C (**b**) (box B in Figure 7).

However, it is important to recall that the strength testing results showed that the added bottom ash improved the early strength up to 60 relative weight, although the 28-day strength was not improved. Thus, it can be inferred that the addition of bottom ash may have increased the reaction rate of C-S-H forming in the early days of the study, resulting in early strength improvement, but it did not affect the amount of C-S-H formed up to 28 days nor did it improve the 28-day strength. This is further discussed in Section 3.1.4.

3.1.4. ICP-OES and IC

Figure 9 shows the results of the ICP-OES and IC, which illustrate the ionic concentrations of calcium (Ca), silicon (Si), magnesium (Mg), aluminum (Al), sulfur (S), and iron (Fe) in the diluted mixture mortar samples with w/b = 2.0 after 24 h. In Figure 9, the Ca concentrations were significantly larger than those of any other element due to the high solubility of $CaCl_2$ [39].

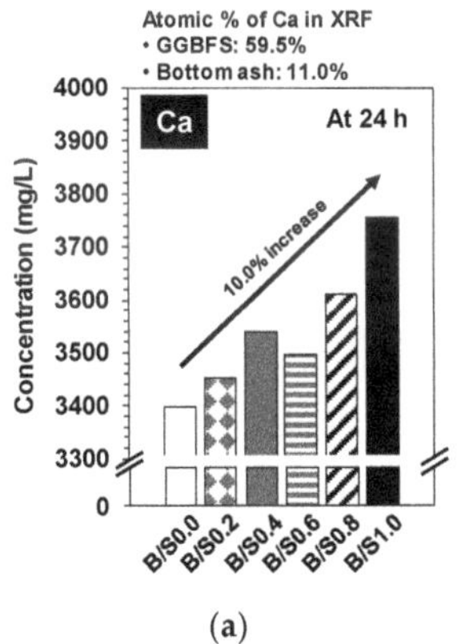

(a)

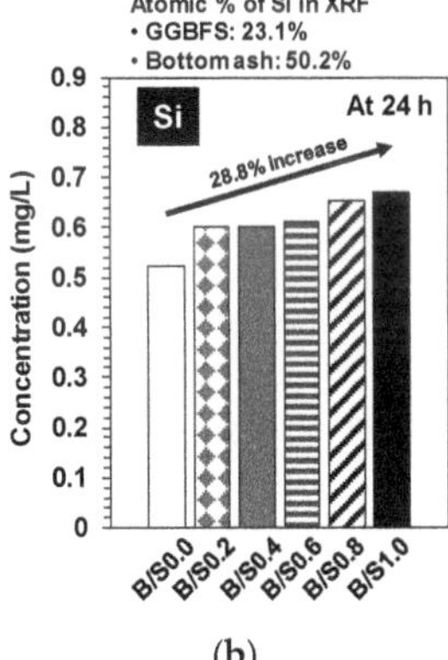

(b)

Figure 9. *Cont.*

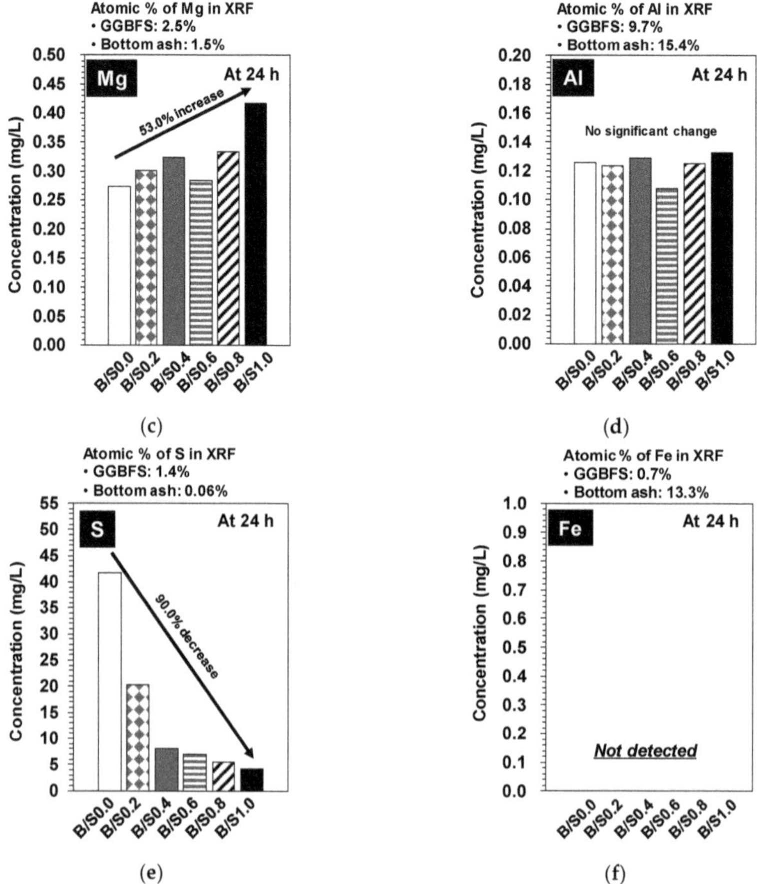

Figure 9. Concentrations of dissolved ions in filtrated liquid from the diluted samples, which were prepared with w/b = 2.0 by agitating for 24 h at 23 °C: (**a**) Ca, (**b**) Si, (**c**) Mg, (**d**) Al, (**e**) S, and (**f**) Fe.

It is worth noting that no Fe ion was detected in any of the samples regardless of the bottom ash content (Figure 9f). The GGBFS had a very small Fe content (0.06 atomic%) and the bottom ash contained considerably more (13.3 atomic%). Thus, no detection implies that Fe existed in an insoluble state in the bottom ash, and, according to the XRD result in Figure 1, most of the Fe in the bottom ash was likely in the form of magnetite (Fe_3O_4).

In this study, all elements except Ca were supplied only by GGBFS and bottom ash in the ICP-OES and IC tests. As previously mentioned, all diluted mixture samples had the same weights of binder (GGBFS + CaO + $CaCl_2$) and water with w/b = 2.0. The bottom ash was not included in the binder and was incorporated separately into the mixtures and, regardless of the bottom ash content, the weight fractions of GGBFS per water sample were the same in all mixtures. Therefore, if the bottom ash was insoluble or chemically inert, the concentration of the target elements (Ca, Si, Mg, Al, S, and Fe) should not change as the bottom ash content increases. In the results shown in Figure 9, the concentration of Al was not significantly different in all samples, which indicates that Al in the bottom ash was not dissolved in the diluted samples. However, the concentrations of Ca, Si, and Mg notably increased as the bottom ash content increased, and the increased portion of Ca, Si, and Mg were likely due to the dissolution of the additional bottom ash. This indicates that the bottom ash was not entirely inert in the binder reaction but selectively soluble depending on the type of element; that is, Ca, Si, and Mg were likely present in a dissolvable state, while Al was not. Therefore, the improved early strength of the mortar mixture samples in Figure 5 can be explained by the increased concentrations of Ca and Si at an early stage of curing because Ca and Si ions are necessary for forming more C-S-H quickly.

It is also worth noting that sulfur behaved very differently from the other elements; the sulfur concentration was substantially reduced as the bottom ash content increased. The raw GGBFS contained sulfur with 1.4 atomic%, while the raw bottom ash had very little sulfur (0.06 atomic%) in its atomic composition. Thus, unless the added bottom ash affected the dissolution behavior of sulfur in the GGBFS, the sulfur concentration should not have decreased with the addition of bottom ash. Therefore, the results suggest that the presence of the bottom ash inhibited the dissolution of sulfur in the GGBFS, although it is not clear how this inhibition occurred.

3.2. Results of Aggregate Production

In this study, the powder was not agglomerated for B/S0.0 and B/S1.0, while the other mixture proportions in Table 2 were suitable for conducting cold-bonded pelletization. In the successful mixtures, the bottom ash particles after water spraying seem to have properly acted as nucleating seeds in the pelletization, as illustrated in Figure 10, due to their rougher surfaces and larger particle sizes as compared to the GGBFS particles in Figures 2 and 3. The agglomeration of aggregate in this study also has price competitiveness compared to the existing artificial aggregate manufacturing technology as it does not require (1) a sintering process, which involves high energy consumption, and (2) additional adhesive chemicals for agglomeration (i.e., water simply acts as a binding agent, as in this study).

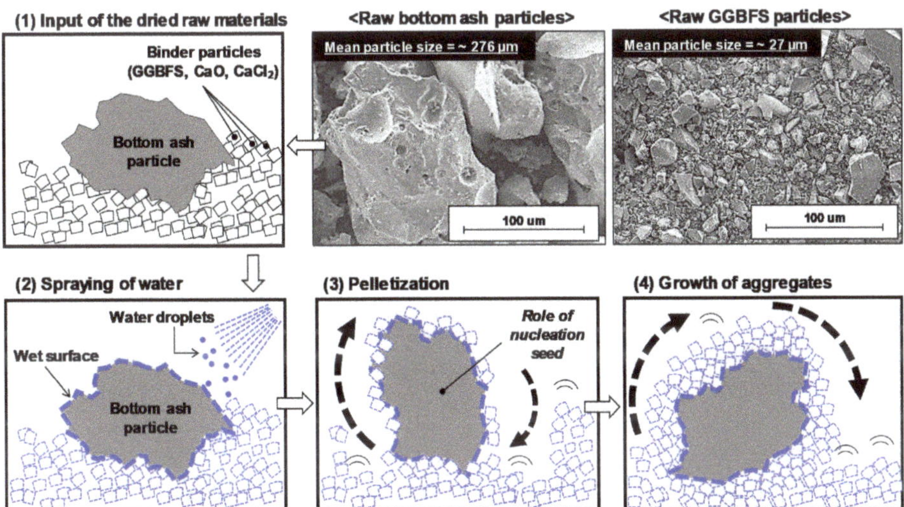

Figure 10. Schematic illustration of the cold-bonded pelletizing process using bottom ash and binder powder (GGBFS + CaO + $CaCl_2$).

Given the strength testing results of the mortar samples and their pelletizing formability, B/S0.4 was selected as the best mixture proportion in this study. The pilot samples of artificial fine aggregates were therefore produced using B/S0.4 after curing for 28 days, and were relatively spherical, as shown in Figure 11.

Figure 11. Pilot samples of artificial fine aggregates produced using B/S0.4 after 28 days of curing.

The result of the water absorption testing for the artificial fine aggregates was 9.83 wt%. According to the KS F 2527 concrete aggregate [40], natural sand should achieve an absolute dry density of 2.5 g/cm^3 or more and a water absorption of 3.0% or less. However, there is no specification regarding water absorption for artificial fine aggregate manufactured from coal bottom ash. The leaching test (TCLP) results for the fine aggregate sample and the raw bottom ash are shown in Figure 12. Heavy metals such as nickel (Ni), barium (Ba), zinc (Zn), copper (Cu), lead (Pb), chromium (Cr), cadmium (Cd), and arsenic (As) were leached from the raw bottom ash, and the results showed that the concentrations of Ba, Pb, Cr, Cd, and As were lower than the TCLP limits. Although there are no regulation limits on Ni, Cu, and Zn in TCLP, their concentrations were not negligible according to previous studies [7,41]. However, all these metals were leached far less in the produced aggregates (B/S0.4) than in the bottom ash. Specifically, no high concentration was detected for Ni, Cu, Pb, Cr, Cd, or As.

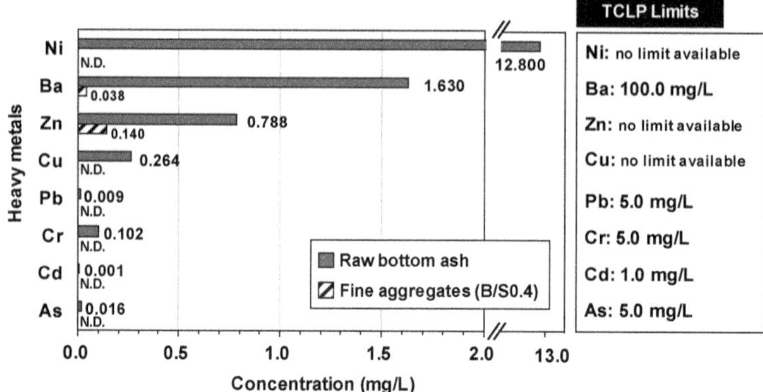

Figure 12. Leaching test results for raw bottom ash and produced fine aggregates (B/S4.0, 28 days). N.D.: not detected. No toxicity characteristic leaching procedure (TCLP) limits are available for Ni, Zn, and Cu.

4. Conclusions

This study investigated the use of coal bottom ash and CaO-$CaCl_2$-activated GGBFS binder in the manufacture of artificial fine aggregates using a cold-bonded pelletization technique.

The adding bottom ash in this study increased early compressive strength before 14 days, although it was not beneficial in improving the 28-day strength. However, when the additional bottom ash was over a certain limit (80 relative weight in this study), the early strength improvement vanished.

The results of the XRD, TG, ICP-OES, and IC tests indicate that the bottom ash was not entirely inert in the binder reaction but was selectively soluble depending on the type of element; that is, Ca, Si, and Mg were likely present in a dissolvable state while Al was not. Additionally, the presence of the bottom ash in the mixture likely suppressed the dissolution of sulfur in the GGBFS.

Although the bottom ash was dissolved to some extent, this did not affect the total amounts of main reaction products (C-S-H and hydrocalumite) formed up to 28 days. However, when a proper amount (up to 60 relative weight) of bottom ash was added, it appears to have increased the reaction rate of forming C-S-H before 14 days, as the concentrations of Ca and Si were increased at the early stage of about 24 h, resulting in increased early strength.

After considering mechanical strength and pelletizing formability, the mixture proportion B/S0.4 was selected for manufacturing artificial fine aggregates. The bottom ash particles seem to have acted as nucleating seeds during the pelletization due to their wet surfaces after water spraying. The water absorption of the artificial fine aggregates was 9.83 wt%. In addition, all target heavy metals were leached far less in the produced aggregates (B/S0.4) than in the bottom ash, and no concentration was detected for Ni, Cu, Pb, Cr, Cd, and As in the produced aggregates.

Thus, this study suggests that the addition of bottom ash to CaO-$CaCl_2$-activated GGBFS binder can result in a high-strength artificial fine aggregate for the production of cement mortar bricks and blocks, and can also be used as a concrete coarse aggregate through aggregate size selection.

Author Contributions: Conceptualization, D.J., H.S. and J.E.O.; methodology, W.S.Y., H.S. and S.Y.; investigation, D.J.; writing—original draft preparation, D.J.; writing—review and editing, J.E.O. and Y.B.; supervision, J.E.O. and Y.B. All authors have read and agreed to the published version of the manuscript.

Funding: This study was supported by the R&D Program of the Korea Railroad Research Institute, the Republic of Korea, and the Korea Agency for Infrastructure Technology Advancement (KAIA) grant funded by the Ministry of Land, Infrastructure, and Transport (20CTAP-C157281-01). This work was supported by an NRF (National Research Foundation of Korea) Grant funded by the Korean Government (NRF-2017-Global Ph.D. Fellowship Program).

Conflicts of Interest: The authors declare no conflict of interest.

References

1. Metha, K.M.; Paulo, M. *Concrete: Microstructure, Properties, and Materials*; McGraw-Hill Publishing: New York, NY, USA, 2006.
2. Wang, H.; Wang, L.; Li, L.; Cheng, B.; Zhang, Y.; Wei, Y. The Study on the Whole Stress–Strain Curves of Coral Fly Ash-Slag Alkali-Activated Concrete under Uniaxial Compression. *Materials* **2020**, *13*, 4291. [CrossRef] [PubMed]
3. Ismail, S.; Ramli, M. Engineering properties of treated recycled concrete aggregate (RCA) for structural applications. *Constr. Build. Mater.* **2013**, *44*, 464–476. [CrossRef]
4. Frigione, M. Recycling of PET bottles as fine aggregate in concrete. *Waste Manag.* **2010**, *30*, 1101–1106. [CrossRef] [PubMed]
5. Yüksel, İ.; Bilir, T.; Özkan, Ö. Durability of concrete incorporating non-ground blast furnace slag and bottom ash as fine aggregate. *Build. Environ.* **2007**, *42*, 2651–2659. [CrossRef]
6. Ghassemi, M.; Andersen, P.K.; Ghassemi, A.; Chianelli, R.R. Hazardous Waste from Fossil Fuels. In *Encyclopedia of Energy*; Cleveland, C.J., Ed.; Elsevier: New York, NY, USA, 2004; pp. 119–131. [CrossRef]
7. Hashemi, S.S.G.; Mahmud, H.B.; Ghuan, T.C.; Chin, A.B.; Kuenzel, C.; Ranjbar, N. Safe disposal of coal bottom ash by solidification and stabilization techniques. *Constr. Build. Mater.* **2019**, *197*, 705–715. [CrossRef]

8. Cheriaf, M.; Rocha, J.C.; Péra, J. Pozzolanic properties of pulverized coal combustion bottom ash. *Cem. Concr. Res.* **1999**, *29*, 1387–1391. [CrossRef]
9. Wyrzykowski, M.; Ghourchian, S.; Sinthupinyo, S.; Chitvoranund, N.; Chintana, T.; Lura, P. Internal curing of high performance mortars with bottom ash. *Cem. Concr. Compos.* **2016**, *71*, 1–9. [CrossRef]
10. Ong, S.K.; Mo, K.H.; Alengaram, U.J.; Jumaat, M.Z.; Ling, T.-C. Valorization of Wastes from Power Plant, Steel-Making and Palm Oil Industries as Partial Sand Substitute in Concrete. *Waste Biomass Valorization* **2018**, *9*, 1645–1654. [CrossRef]
11. Siddique, R. Utilization of Industrial By-products in Concrete. *Procedia Eng.* **2014**, *95*, 335–347. [CrossRef]
12. Gooi, S.; Mousa, A.A.; Kong, D. A critical review and gap analysis on the use of coal bottom ash as a substitute constituent in concrete. *J. Clean. Prod.* **2020**, *268*, 121752. [CrossRef]
13. Kim, H.; Ha, K.; Lee, H.-K. Internal-curing efficiency of cold-bonded coal bottom ash aggregate for high-strength mortar. *Constr. Build. Mater.* **2016**, *126*, 1–8. [CrossRef]
14. Gesoğlu, M.; Özturan, T.; Güneyisi, E. Shrinkage cracking of lightweight concrete made with cold-bonded fly ash aggregates. *Cem. Concr. Res.* **2004**, *34*, 1121–1130. [CrossRef]
15. Bui, L.A.-T.; Hwang, C.-L.; Chen, C.-T.; Lin, K.-L.; Hsieh, M.-Y. Manufacture and performance of cold bonded lightweight aggregate using alkaline activators for high performance concrete. *Constr. Build. Mater.* **2012**, *35*, 1056–1062. [CrossRef]
16. Gesoğlu, M.; Özturan, T.; Güneyisi, E. Effects of fly ash properties on characteristics of cold-bonded fly ash lightweight aggregates. *Constr. Build. Mater.* **2007**, *21*, 1869–1878. [CrossRef]
17. Ferone, C.; Colangelo, F.; Messina, F.; Iucolano, F.; Liguori, B.; Cioffi, R. Coal combustion wastes reuse in low energy artificial aggregates manufacturing. *Materials* **2013**, *6*, 5000–5015. [CrossRef] [PubMed]
18. Dutta, D.K.; Bordoloi, D.; Borthakur, P.C. Investigation on reduction of cement binder in cold bonded pelletization of iron ore fines. *Int. J. Min. Process.* **1997**, *49*, 97–105. [CrossRef]
19. Geetha, S.; Ramamurthy, K. Environmental friendly technology of cold-bonded bottom ash aggregate manufacture through chemical activation. *J. Clean. Prod.* **2010**, *18*, 1563–1569. [CrossRef]
20. Kockal, N.U.; Ozturan, T. Effects of lightweight fly ash aggregate properties on the behavior of lightweight concretes. *J. Hazard. Mater.* **2010**, *179*, 954–965. [CrossRef]
21. Yum, W.S.; Jeong, Y.; Yoon, S.; Jeon, D.; Jun, Y.; Oh, J.E. Effects of $CaCl_2$ on hydration and properties of lime (CaO)-activated slag/fly ash binder. *Cem. Concr. Compos.* **2017**, *84*, 111–123. [CrossRef]
22. Jeong, Y.; Yum, W.S.; Moon, J.; Oh, J.E. Utilization of precipitated $CaCO_3$ from carbon sequestration of industrially emitted CO_2 in cementless CaO-activated blast-furnace slag binder system. *J. Clean. Prod.* **2017**, *166*, 649–659. [CrossRef]
23. *Pert HighScore Plus Software*; Version 3.0 e; Malvern Panalytical: Malvern, The Netherlands, 2012.
24. Andersen, M.D.; Jakobsen, H.J.; Skibsted, J. Incorporation of aluminum in the calcium silicate hydrate (C-S-H) of hydrated Portland cements: A high-field 27Al and 29Si MAS NMR investigation. *Inorg. Chem.* **2003**, *42*, 2280–2287. [CrossRef] [PubMed]
25. Allmann, R.; Hinek, R. The introduction of structure types into the Inorganic Crystal Structure Database ICSD. *Acta Crystallogr. Sect. A Found. Crystallogr.* **2007**, *63*, 412–417. [CrossRef] [PubMed]
26. Jeon, D.; Yum, W.S.; Jeong, Y.; Oh, J.E. Properties of quicklime (CaO)-activated Class F fly ash with the use of $CaCl_2$. *Cem. Concr. Res.* **2018**, *111*, 147–156. [CrossRef]
27. Jeon, D.; Yum, W.S.; Song, H.; Sim, S.; Oh, J.E. The temperature-dependent action of sugar in the retardation and strength improvement of $Ca(OH)_2$-Na_2CO_3-activated fly ash systems through calcium complexation. *Constr. Build. Mater.* **2018**, *190*, 918–928. [CrossRef]
28. Baykal, G.; Döven, A.G. Utilization of fly ash by pelletization process; theory, application areas and research results. *Resour. Conserv. Recycl.* **2000**, *30*, 59–77. [CrossRef]
29. Bassani, M.; Diaz Garcia, J.C.; Meloni, F.; Volpatti, G.; Zampini, D. Recycled coarse aggregates from pelletized unused concrete for a more sustainable concrete production. *J. Clean. Prod.* **2019**, *219*, 424–432. [CrossRef]
30. *Standard Test Method for Density, Relative Density (Specific Gravity), and Absorption of Fine Aggregate*; ASTM International (ASTM): West Conshohocken, PA, USA, 2015.
31. *Method 1311 Toxicity Characteristic Leaching Procedure (TCLP)*; Agency, E.P. (Ed.) Environmental Protection Agency (EPA): Washington, DC, USA, 1992.
32. Taylor, H.F. *Cement Chemistry*; Thomas Telford London: London, UK, 1997; Volume 2.

33. Oh, J.E.; Monteiro, P.J.; Jun, S.S.; Choi, S.; Clark, S.M. The evolution of strength and crystalline phases for alkali-activated ground blast furnace slag and fly ash-based geopolymers. *Cem. Concr. Res.* **2010**, *40*, 189–196. [CrossRef]
34. Song, H.; Jeong, Y.; Bae, S.; Jun, Y.; Yoon, S.; Oh, J.E. A study of thermal decomposition of phases in cementitious systems using HT-XRD and TG. *Constr. Build. Mater.* **2018**, *169*, 648–661. [CrossRef]
35. Scrivener, K.; Snellings, R.; Lothenbach, B. *A Practical Guide to Microstructural Analysis of Cementitious Materials*; CRC Press: Boca Raton, FL, USA, 2018.
36. Birnin-Yauri, U.; Glasser, F. Friedel's salt, $Ca_2Al(OH)_6$ (Cl, OH) $2H_2O$: Its solid solutions and their role in chloride binding. *Cem. Concr. Res.* **1998**, *28*, 1713–1723. [CrossRef]
37. Wang, K.; Shah, S.P.; Mishulovich, A. Effects of curing temperature and NaOH addition on hydration and strength development of clinker-free CKD-fly ash binders. *Cem. Concr. Res.* **2004**, *34*, 299–309. [CrossRef]
38. Vieille, L.; Rousselot, I.; Leroux, F.; Besse, J.-P.; Taviot-Guého, C. Hydrocalumite and its polymer derivatives. 1. Reversible thermal behavior of Friedel's salt: A direct observation by means of high-temperature in situ powder X-ray diffraction. *Chem. Mater.* **2003**, *15*, 4361–4368. [CrossRef]
39. Petrucci, R.H.; Herring, F.G.; Bissonnette, C.; Madura, J.D. *General Chemistry: Principles and Modern Applications*, 11th ed.; Pearson: London, UK, 2016.
40. *Concrete Aggregate (KS F 2527)*; Korean Standards Association: Seoul, Korea, 2016.
41. Beddu, S.; Abd Manan, T.S.B.; Zainoodin, M.M.; Khan, T.; Wan Mohtar, W.H.M.; Nurika, O.; Jusoh, H.; Yavari, S.; Kamal, N.L.M.; Ghanim, A.A.; et al. Dataset on leaching properties of coal ashes from Malaysian coal power plant. *Data Brief* **2020**, *31*, 105843. [CrossRef] [PubMed]

Article

Effect of Strengthening Methods on Two-Way Slab under Low-Velocity Impact Loading

Sun-Jae Yoo, Tian-Feng Yuan, Se-Hee Hong and Young-Soo Yoon *

School of Civil, Environmental and Architectural Engineering, Korea University, 145 Anam-Ro, Seongbuk-Gu, Seoul 02841, Korea; redtoss@korea.ac.kr (S.-J.Y.); yuantianfeng@korea.ac.kr (T.-F.Y.); bestshhong@korea.ac.kr (S.-H.H.)

* Correspondence: ysyoon@korea.ac.kr; Tel.: +82-2-3290-3320

Received: 3 November 2020; Accepted: 7 December 2020; Published: 8 December 2020

Abstract: In this study, the performance of reinforced concrete slabs strengthened using four methods was investigated under impact loads transferred from the top side to bottom side. The top and bottom sides of test slabs were strengthened by no-slump high-strength, high-ductility concrete (NSHSDC), fiber-reinforced-polymer (FRP) sheet, and sprayed FRP, respectively. The test results indicated that the test specimens strengthened with FRP series showed a 4% increase in reaction force and a decrease in deflection by more than 20% compared to the non-strengthened specimens. However, the specimen enhanced by the NSHSDC jacket at both the top and bottom sides exhibited the highest reaction force and energy dissipation as well as the above measurements because it contains two types of fibers in the NSHSDC. In addition, the weight loss rate was improved by approximately 0.12% for the NSHSDC specimen, which was the lowest among the specimens when measuring the weight before and after the impact load. Therefore, a linear relationship between the top and bottom strengthening of the NSHSDC and the impact resistance was confirmed, concluding that the NSHSDC is effective for impact resistance when the top and bottom sides are strengthened. The results of the analysis of the existing research show that the NSHSDC is considered to have high impact resistance, even though it has lower resistance than the steel fiber reinforced concrete and ultra-high-performance-concrete, it can be expected to further studies on strengthening of NSHSDC.

Keywords: strengthening methods; fiber reinforced polymer; no-slump; high-strength; ductility concrete; low-velocity impact load; two-way slab

1. Introduction

Concrete structures are likely to be exposed to various impact loadings, for example, rock falls, terrorist attacks, and sudden crashes along their service life. When reinforced concrete slab is subjected to an impact load, it is expected to suffer an important local deflection, secondary damage caused by the occurrence of debris, and internal strength reduction. To prevent such damage, the demand for reinforced concrete with impact resistance has been increasing. Many researchers focus on new concrete members to resist impact loads, such as different reinforcement ratios and materials (containing fibers with various volume fractions, and fiber types) [1–4]. To improve the impact resistance, developing new construction members is also important, whereas the established concrete members are not only numerous but also less researched. Also, these few research works mainly evaluated the strengthening properties of reinforced concrete (RC) beams. The RC slab strengthened by reinforced materials (FRP and stain hardening cementitious composite) to investigate the impact test is listed in Table 1. They simply evaluated the single strengthening materials (such as just using FRP sheet, without spraying FRP) and strengthened method (without hybrid strengthening). Hence, it is necessary to evaluate the impact resistance in RC slab with a different strengthening method.

Table 1. Recent research on reinforced concrete (RC slabs strengthened by different materials).

	Researchers		Strengthening Materials	Evaluation Method
1.	Foret et al. 2008	[5]	CFRP strip (two types spacing)	Numerical analysis
2.	Bhatti et al. 2011	[6]	FRP sheet (AFRP and CFRP sheet)	Test
3.	Yoo et al. 2014	[7]	FRP sheet (AFRP and CFRP sheet)	Test and numerical analysis
4.	Radnic et al. 2015	[8]	FRP strips (two types strengthening method)	Test and numerical analysis
5.	Wang et al. 2018	[9]	Flax FRP (different strengthening method)	Test
6.	Yilmaz et al. 2018	[10]	CFRP strips (two types strengthening method)	Test and numerical analysis
7.	Elnagar et al. 2019	[11]	Strain-hardening cementitious composites	Test and numerical analysis
8.	Yang et al. 2019	[12]	FRP sheet (different strengthening method)	Numerical analysis
9.	Soltani et al. 2020	[13]	GFRP sheets (different strengthening method)	Test and numerical analysis
10.	Mugunthan et al. 2020	[14]	GFRP strips (different strengthening method)	Test
11.	Mahmoud et al. 2020	[15]	Strain-hardening cementitious composites	Test

The fiber reinforced polymer (FRP) was widely investigated in the RC members for the past few years [16–24]. Several researchers have been working toward increasing the concrete strength by attaching carbon fiber sheets (CFS) with high stiffness to the concrete surface [16–18]. CFS is not only easy to attach to concrete, but also has a high strength compared to the weight, which makes it attractive. However, the disadvantages of CFS include the requirement of an advanced surface treatment for adhesion with the reinforced concrete, as well the easy CFS peel-off that occurs after a maximum load. Furthermore, an RC member strengthened by CFS can significantly increase bending resistance property, but it does not perform effectively in compression [19].

Also, studies on the strengthening of existing structure through sprayed FRP have recently been conducted [21–24]. This method is popular because it offers the advantage of easy strengthening compared to the existing retrofitting by FRP. Sprayed FRP is easily implemented by spraying a mix of chopped fibers and resins without being affected by the directionality of the existing FRP. Banathia et al. [21] evaluated reinforced concrete beams through sprayed FRP. The test results indicated that the structure with sprayed FRP had a higher structural performance compared to the non-sprayed reinforced concrete. In addition, Han et al. [23] reinforced a structure using sprayed FRP, which produced isotropic composites; the experimental results indicated excellent structural behavior in the member applying the sprayed FRP. However, strengthening the structure through sprayed FRP may lead to construction errors relative to the design thickness and early detachment due to incomplete hardening of the reinforcing material [24].

Recently, studies on concrete structure that incorporate fibers have been actively conducted [25–33]. The fibers employed in ultra-high-performance concrete (UHPC) structure were reported to bring significant effects on the resistance and ductility of the structure after applying maximum loads. However, creating a new structure with UHPC structure is not only economically unviable due to its high cost, but also does not fully serve the purpose of reinforcing existing structure. Therefore, strengthening UHPC in existing concrete structure can be a desirable solution [33,34].

High-performance-fiber-reinforced-cement composites (HPFRCCs) exhibit high strength and are being studied and widely applied [33–37]. Among these HPFRCCs, no-slump high-strength and high-ductility concrete (NSHSDC) offers excellent adhesion with conventional concrete owing to its high viscosity, excellent load resistance, and ductility [35,36]. This is because a hybrid mix of two different fiber lengths is adopted to improve the mechanical properties during the NSHSDC manufacturing [35]. Furthermore, when mixing two types of fibers with different lengths, long fibers are considered to have an outstanding macro-crack control effect, whereas short fibers have excellent microcrack control effects [29]. In addition, NSHSDC confirmed the adhesion to the existing concrete made of high-strength concrete with a strength of 120 MPa [36].

This study evaluated the impact resistance properties for the following four strengthening methods: FRP, sprayed FRP, NSHSDC, and hybrid. Not only the bottom side, but also the top side were strengthened based on the existing research results [33,36]. The hybrid method strengthened the structure with NSHSDC on the top side and sprayed FRP on the bottom side, considering that the sprayed FRP and NSHSDC have excellent tensile and compression strengthening effects, respectively. The crack patterns and failure modes after the experiment are compared and used as reference data for the structural characteristics of reinforced concrete according to the strengthening techniques. Additionally, this study aims to contribute basic data about RC slabs strengthened by different strengthening methods to the resistant impact load.

2. Materials and Methods

2.1. Mix Proportions and Materials

The following materials were used in the normal concrete (NC) mixture, as shown in Table 2. Portland type 1 cement (specific-surface area 3492 cm^2/g, density 3.15 g/cm^3) was used, while river sand was used as the fine aggregate. In addition, crushed stones with a maximum dimension of 20 mm were used for the coarse aggregate. The details are shown in Table 2.

Table 2. Mix proportions of concrete (by cement weight ratio).

Type	w/b (%)	Unit Weight			
		Water	Cement	Fine Aggregate	Coarse Aggregate
NC	33	0.33	1.00	1.43	1.70

NC = normal strength concrete.

The NSHSDC mixture contains silica fume and silica flour for high strength, resulting in a dense cement matrix. The steel fiber (0.2 mm of diameter, 19.5 mm of length, and 7.8 g/cm^3 of density) and polyethylene (PE; 31 μm of diameter, 12 mm of length, and 0.97 g/cm^3 of density) fiber were mixed with volume fractions of 1.0% and 0.5%, respectively, and their physical properties correspond to those of existing studies [35]. Polycarboxylate high-performance reducer (liquid/brown, 1.07 g/cm^3) is adopted as superplasticizer (SP) at cement weight ratio of 3% [36] to facilitate the mixing of concrete. Table 3 presents the detailed mixture of the NSHSDC. In Table 3, w/b and SF denote the water-to-binder ratio and steel fiber, respectively.

Table 3. Mix proportions of no-slump high-strength, high-ductility concrete (NSHSDC; by cement weight ratio).

Type	w/b (%)	Unit Weight					Fiber		SP
		Water	Cement	Silica Fume	Silica Filer	Silica Sand	SF	PE	
							(%)		
NSHSDC	17.2	0.215	1.00	0.25	0.30	1.10	1.0	0.5	3.0%

NSHSDC = no-slump high-strength, high-ductility concrete; w/b = water to binder ratio; SF = steel fiber; PE = polyethylene fiber; SP = superplasticizer.

The compressive strength and elastic modulus were obtained by measuring the average compressive strain; the measurement was performed by installing three linear variable differential transformers (LVDT) on a ø100 mm × 200 mm specimen in accordance with ASTM C39 [38]. In addition, the flexural strength was measured in compliance with ASTM C1609 [39] on three specimens of square section with dimensions of 100 mm × 100 mm × 400 mm. The physical properties of NC and NSHSDC are listed in Table 4.

Table 4. Mechanical properties of concrete.

Specimens	Compressive Strength (MPa)	Flexural Strength (MPa)	Elastic Modulus (GPa)
NC	40.7	4.24	27.42
NSHSDC	123.9	18.52	41.26

2.2. Details of Test Specimens

A total of six test specimens were tested in which both top and bottom were retrofitted with externally bonded CFS, RC sprayed with carbon fiber roving, strengthened with NSHSDC, and one in which the top was strengthened with NSHSDC and the bottom was sprayed with FRP, as well as one control RC slab and one steel fiber reinforced slab. The concrete specimens are designated as follows: the first string of characters shows the type of concrete used, where NC, NSC, SC denote normal-strength concrete, normal-strength concrete strengthened with NSHSDC, and steel fiber reinforced concrete, respectively; the second string of characters indicates the attachment of FRP; test specimens retrofitted with FRP are denoted by F, and the test specimens without FRP are denoted by NF; and the last string of characters shows that if FRP was sprayed, it was marked as S, and if not, it was not marked. For example, NCS-CF-S designated the specimen for which the bottom side is sprayed with FRP after the top side is strengthened with NSHSDC. Figure 1 and Table 5 present the details of the test specimen. Figure 2 presents the designation of the test specimens.

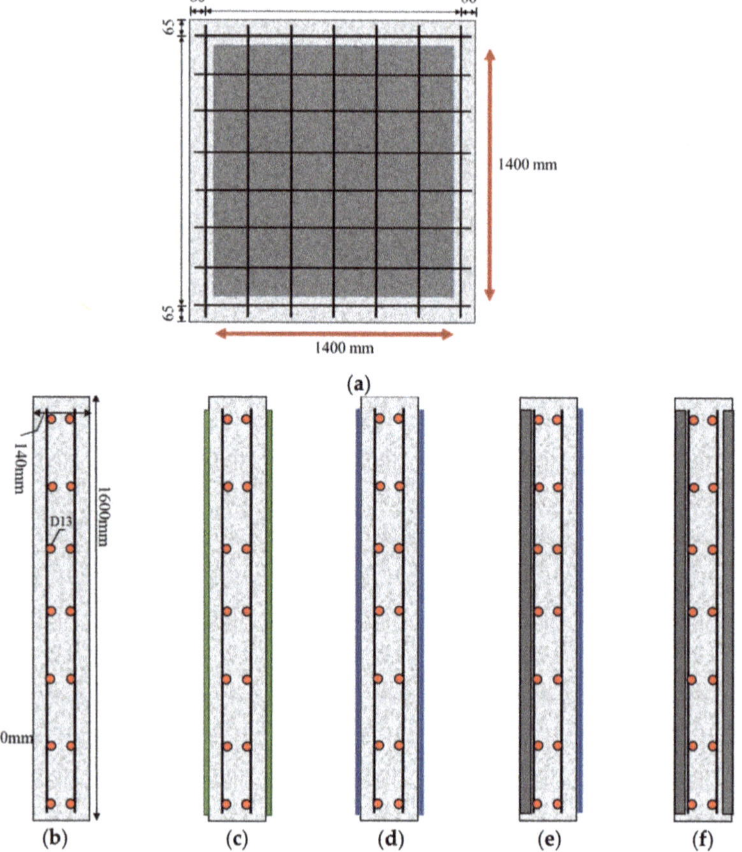

Figure 1. Dimensions and layout of test specimens: (**a**) Dimension of the test specimen; (**b**) NC-NF; (**c**) NC-F; (**d**) NC-F-S; (**e**) NSC-F-S; (**f**) NSC-NF.

Table 5. Designation and specifications of the test specimens.

Designation	Type of Strengthening Methods			Strengthening Thickness (mm)	
	NSC	Sprayed FRP	FRP Sheet	Top	Bottom
NC-NF	-	-	-	-	-
NC-F	-	-	○	0.34	0.34
NC-F-S	-	○	-	6	6
NSC-F-S	○	○	-	20	6
NSC-NF	○	-	-	20	20

NSC = no-slump high-strength, high-ductility concrete.

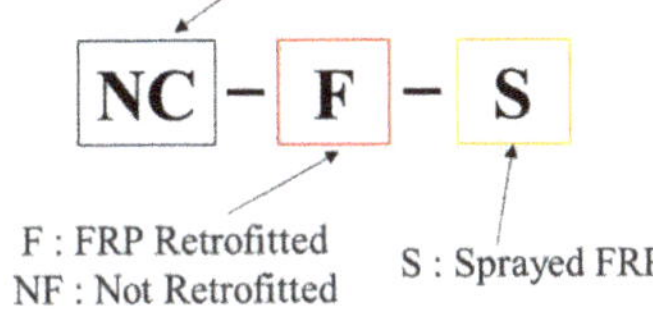

Figure 2. Designation of the test specimens.

2.3. Strengthening Methods

Figure 1a depicts the shape of the test specimens for examining the impact load. The reinforced concrete slabs present square sections with a width of 1600 mm × 1600 mm and a height of 140 mm. The tensile and compressive reinforcements are achieved by D13 steel bars with diameter of 13 mm. The rebar spacing in the horizontal plane is 240 mm in the x-direction and 210 mm in the y-direction, respectively; the corresponding rebar ratios are 0.38% and 0.43%, respectively. This satisfies the minimum reinforcement criteria specified by ACI 318-19.

CFS was attached in the x-direction of RC slab surface (ρ_{rebar} = 0.38%) and applied to the second layer equally in the traversal direction. In this study, the structural strengthening of concrete was conducted using dispensing FRP roving equipment manufactured by Magnum Venus Company (Knoxville, TN, USA) [23]. Carbon fiber roving was used as the experimental material, and vinyl ester was used as resin. The ratio of fiber and resin was set to 2:1 with reference to a previous study [24], and the curing agent was mixed at 2% of the weight of vinyl ester. The length of the carbon fiber strand was set to approximately 40 mm and the thickness was set to 6 mm considering the relationship with the FRP sheet expressed in Equation (1). The properties of the fiber and resin are listed in Table 6.

$$\frac{\sigma_{FRP}}{\sigma_{Sprayed\ FRP}} T_{FRP} = T_{Sprayed\ FRP} \quad (1)$$

where σ_{FRP} is the tensile strength of FRP Sheet, $\sigma_{Sprayed\ FRP}$ is the tensile strength of sprayed FRP, T_{FRP} is the thickness of FRP, and $T_{Sprayed\ FRP}$ is the thickness of sprayed FRP.

Table 6. Mechanical properties of carbon fiber and resin.

Series	Carbon Fiber		Resin	
	Sheet	Roving	Epoxy	Vinyl Ester
Tensile strength (MPa)	4900	4200	90	88
Elastic modulus (GPa)	230	240	30	7.3
Ultimate strain (%)	2.1	1.8	8.0	4.5
Thickness (mm)	0.167	3	-	-

To overlay the NSHSDC on the specimens, 20 mm of the concrete cover was drilled, which required careful attention because the vibration during the drilling process can deteriorate the durability of the specimens. Subsequently, the NSHSDC was poured into the top and bottom sides of the concrete slab. That is, a structural member in which NSHSDC was strengthened in the compression zone (20 mm) and tensile zone (20 mm). The test specimen obtained by the strengthening method is the same as shown in Figure 1.

2.4. Test Setup for Drop-Weight Test

Figure 3 presents the setup for the drop-weight test of the test specimens. The behavior and maximum load of the test specimens measured during impact loading may exhibit significantly different results depending on the stiffness of the test specimen as well as the load conditions, such as drop weight and impact velocity [37].

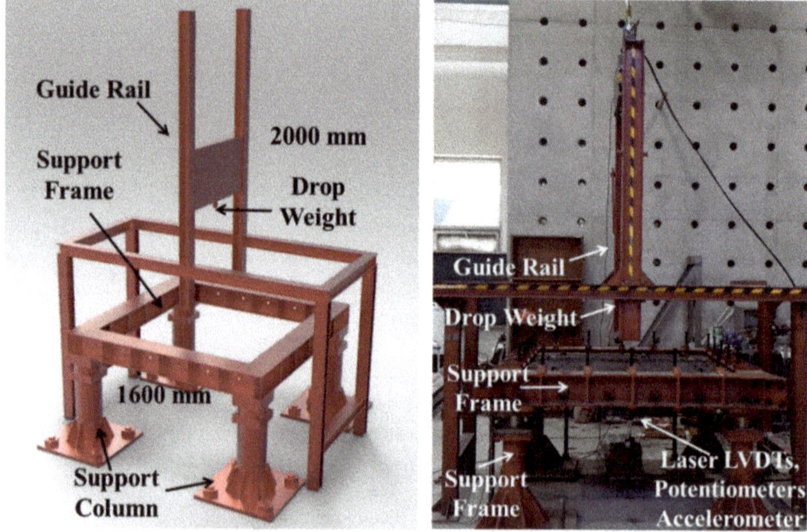

Figure 3. Test setup for drop-weight test.

As shown in Figure 3, the impact load was applied to the reinforced concrete slab by fitting the weight to the guide, pulling it up using hydraulic pressure, and letting the drop weight of 300 kg freely fall to the center of the slab. To ensure precise impact loading on the test specimens, a guide rail was used to prevent inaccuracy. In addition, a load cell with a capacity of 2000 kN was installed at the top to measure the impact force, and a load cell with a maximum capacity of 500 kN was installed in each of the four supports to measure the reaction force. The impact energy per blow applied to the test specimens was approximately 5.89 kJ at a velocity of 6.3 m/s, and all edges were fixed, with a net span of 1500 mm. The support columns were fixed with steel bars to prevent the uplift of the test specimens. All specimens were subjected to free fall at a height of 2000 mm. The criterion for the end of the experiment was based on the time when the impact or reaction force decreased rapidly, or severe cracks occurred in the specimens.

Instrumentations of the test specimen involve a laser type of LVDT installed at the bottom surface for measuring the deflection. Here, the LVDT is a laser type (Optex, Tokyo, Japan), which has the advantage of being able to record the instantaneous displacement at a rate of hundreds of thousands of data per second. An accelerometer with a limit of 5000 g (g = gravitational acceleration) and potentiometers with a maximum capacity of 100 mm were placed at the bottom of the specimen to measure the displacement. All output data were obtained through a dynamic data logger (DEWE-43, Trbovlje, Slovenia) and real-time impact data was acquired at a frequency of 200 kHz.

3. Results and Discussion

3.1. Drop-Weight Test

3.1.1. Load-Time Curves

Figure 4 and Table 7 present the measured reaction force and impact force over time. In comparison with the strengthening techniques, the impact force shows a similar strength at the first blow. This suggests that all initial stiffness values are similar regardless of the type of strengthening method used. As the blows continue, the impact force tends to decrease, which indicates a reduction in the bearing strength of the test specimen. The reaction force represents a load except for loss, such as the inertial force in the impact load, which indicates that the impact force is higher than the reaction force [40–43]. NSC-NF had a maximum reaction force of approximately 266.63 kN, which is approximately 28% to 43% higher than that of the other strengthened specimens. That is, NSC-NF had a higher strength than the test specimen strengthened with the FRP series at the first blow. This is related to the time delay in the impact energy transfer because of impact load being dominated by inertial force in the early stage, and then the impact force was dissipated, which increased the effect of the reaction force [42–44]. At this time, it is determined that the steel fiber and PE fiber in the NSC-series not only generate energy redistribution after the impact load, but also exhibit higher load resistance than the FRP series owing to time delay [45]. In NSC-F-S, the reaction force was measured to be smaller than that of NSC-NF by approximately 22%, but the reaction force was approximately 20% larger than that of NC-F and NC-F-S.

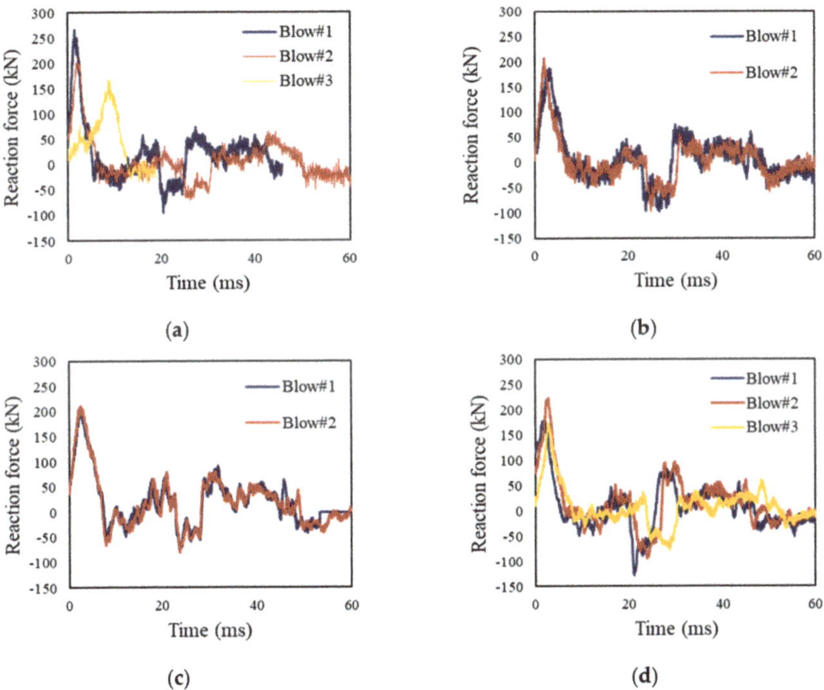

Figure 4. Time history of reaction force at each blow: (**a**) NSC-NF; (**b**) NSC-F-S; (**c**) NC-F-S; (**d**) NC-F.

Table 7. Results of impact test of two-way slabs.

Test Members	Blow No.	Impact Properties			Max. Midspan Displacement	Residual. Midspan Displacement	Max. Reaction Force
		H (mm)	E_i (kJ)	F_i (kN)	D_{max} (mm)	D_{res} (mm)	F_r (kN)
NC-NF	1	2000	5.89	410.10	23.41	7.75	179.75
NC-F	1	2000	5.89	611.97	17.31	5.92	186.36
	2			320.96	23.80	10.15	225.52
	3			282.16	19.33	19.29	176.20
NC-F-S	1	2000	5.89	637.58	17.91	6.57	187.16
	2			369.29	51.96	37.60	212.64
NSC-F-S	1	2000	5.89	587.50	18.89	6.24	206.84
	2			298.40	67.29	67.24	217.14
NSC-NF	1	2000	5.89	636.88	18.77	6.39	266.63
	2			450.13	27.65	14.07	216.97
	3			281.52	20.12	4.92	170.35

As the blow progressed, the NSC-NF tended to gradually decrease the reaction force as a number of cracks occurred, but the difference between the impact force and the reaction force gradually decreased as the time lag occurred. Accordingly, all the measured reaction forces at the second blows are similar. The reaction force of the other specimens showed a slight increment by approximately 5–20% at the second blow, which increased the reaction force as the impact force applied to the reinforced concrete specimen was transferred to a support column. As shown in Figure 4, the reaction force gradually decreases as the blows progress, and the highest reaction force in NSC-NF is measured.

3.1.2. Deflection-Time Curves

The experimental results are shown in Figure 5 and Table 8. The maximum deflection and residual deflection at the first blow were similar regardless of the strengthening method used. Among these specimens, the maximum deflection and residual deflection generated at the first blow were 17.31 mm and 5.92 mm, respectively, with the test specimen of NC-F showing the least deflection. This enhancement could be credited to the high strength of the CFS in the initial stiffness of the specimen. However, as the blows progressed, the stress concentration on the backside occurred due to accumulated energy, resulting in high maximum and residual deflections. In particular, the test specimen strengthened with sprayed-FRP series showed high maximum deflection and residual deflection at the second blow.

On the other hand, the ratio of $\Delta_{max}/\Delta_{res}$ represents the degree of recovery for the structural specimen, where the higher the ratio, the higher the return for deflection [30]. As the blow progresses, this value tends to gradually decrease. The NC-F showed the best ratio of recovery until the second blow owing to the enhancement of the initial stiffness. However, the repeated blow reduced the ratio of recovery at the third blow. In all the test specimens, except for NSC-NF, this ratio was close to 1 at the time of the final blow. Thus, the test specimen attached to the FRP series shows a high degree of recovery and deflection control in the initial blow, but it is deemed necessary to consider the sudden destruction of FRP at the repeated blow. Therefore, the residual ductility ratio, which is the ratio of the maximum deflection and residual deflection, is the lowest. In NSC-NF, the residual ductility ratio is approximately 4 and the ductility ratio is increased at the third blow, which is considered to have not yet reached the ultimate tensile strength due to the crack control effect of the steel fiber and PE fiber [45,46]. However, there was a large deflection in the test specimen, and no further experiment was conducted.

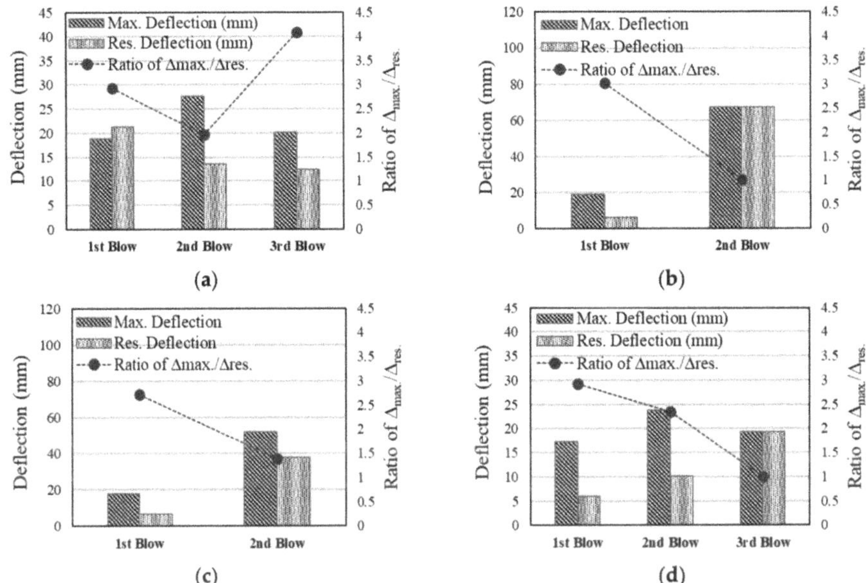

Figure 5. Blow-deflection measurement at each blow under impact loading: (**a**) NSC-NF series; (**b**) NSC-F-S series; (**c**) NC-F-S series; (**d**) NC-F series.

Table 8. Displacements of strengthened concrete slabs under impact loading.

Test Members	Blow No.	Mid-Span Displacement		
		Max. Displacement (mm)	Residual Displacement (mm)	Ratio of $\Delta_{max}/\Delta_{res}$
NC-NF	1	23.47	7.75	3.021
NC-F	1	17.31	5.92	2.924
	2	23.80	10.15	2.345
	3	19.33	19.29	1.002
NC-F-S	1	17.91	6.57	2.726
	2	51.96	37.6	1.382
NSC-F-S	1	18.89	6.24	3.027
	2	67.29	67.24	1.001
NSC-NF	1	18.77	6.39	2.937
	2	27.65	14.07	1.965
	3	20.12	4.92	4.089

3.1.3. Energy Dissipation Capacity

Figure 6 compares the energy dissipated capacity and maximum reaction force of the test specimens. In other words, the energy dissipated capacity represents the degree of absorption of the impact energy applied to the specimens. The energy dissipation capacity is used as an indicator of the degree of damage to reinforced concrete structure as the impacts drop, but it is less reliable if a large deflection occurs [33]. This was calculated from the section where the maximum deflection occurred in the load–deflection curve to the point where it fell vertically.

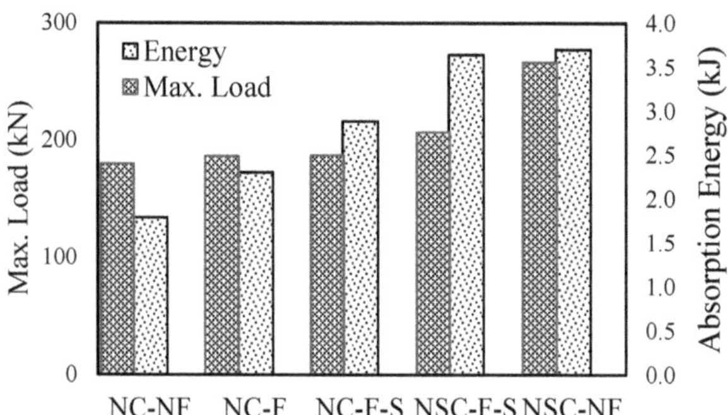

Figure 6. Energy dissipation capacity at first blow.

In this study, the energy dissipation capacity of a reinforced concrete slab was calculated based on the first blow because the deflection is excessively large at the final blow, resulting in a reliability problem in obtaining the overall energy dissipation capacity [33]. At the first blow, the energy dissipation capacities for all reinforced concrete slabs were approximately 1.79 kJ, 3.70 kJ, 3.64 kJ, 2.3 kJ, and 2.88 kJ, respectively (Figure 6). Overall, the higher the reaction force, the higher the energy dissipation ability. It appears that the energy dissipation capacity increased with the reaction force. The energy dissipation capacity of reinforced concrete slabs strengthened with NSHSDC was higher than that of slabs retrofitted with carbon fibers.

3.2. *Effect of Carbon Fiber and NSHSDC on Damage Assessment*

3.2.1. Damage Assessment Based on Support Rotation

The support rotation angle generated in the test specimen to which the impact load is applied is a criterion for evaluating the residual performance of the structure. According to UFC-3-340-02 [47], one of the commonly used design criteria, the residual performance of the structural specimen is evaluated using the rotation of the support angle. UFC-340-02 [47] defines the degree of damage to the reinforced concrete as light, moderate, or severe according to the generated angle of rotation (Table 9). Although it had three impacts, it was found that the support rotation of the final blow that occurred in the NCS-NF was the least. It appears that the NCS-NF controls the impact load owing to the high rigidity of NSHSDC. In general, the specimen strengthened with sprayed FRP had the highest value of rotation angle compared to the other specimens. The NSC-NF showed the smallest support rotation at the final blow, which is smaller than 2 degrees. It can be expected that NSC-NF is the light level of the damage assessment given in the UFC 3-340-02 criterion [47].

Table 9. Criteria of damage level according to UFC 3-340-02 [47].

Damage Level	Light	Moderate	Severe
Support rotation criteria	$0° \leq \theta \leq 2°$	$2° < \theta \leq 6°$	$6° < \theta \leq 8°$

3.2.2. Cracks and Damage

The pattern of cracks after the experiment is shown in Figure 7. Despite the first blow, the NC-NF had a crack width in the center of the bottom side. The NSC-NF showed several bending cracks, most of which were identified near the central part of the it because of the bridging effect of the steel and PE fibers [36,37]. A lot of cracks were found on the back side, and the damaged area was the smallest. Although it is not a comparison for the same blow, it is found that the micro cracks are much more frequent than the NC-NF.

Figure 7. Failure modes of tested specimens of bottom side: (**a**) NC-NF; (**b**) NC-F; (**c**) NC-F-S; (**d**) NSC-F-S; (**e**) NSC-NF.

On the other hand, the test specimen strengthened with the sprayed FRP (NC-F-S) was found to have significant damage to the back side. The tensile reinforcement was exposed, and the damaged area was found to be wider than that of the other test specimens owing to the characteristic carbon fiber roving; although the test specimen (NC-F-S) had a high resistance to the impact energy due to the high stiffness of the reinforcement at the first blow, the accumulated impact energy during the repeated blows was determined to exceed the tensile limit of the concrete [46]. In the event of an impact loading, unlike the static load condition, a large local energy was transmitted to the component, which was reinforced with sprayed FRP, and a large stress was concentrated on the backside, resulting in major destruction at the second blows [40]. Therefore, the NC-F subjected to impact loading is thought to have higher strength than that of NC-F-S. Meanwhile, hardly any cracks were identified on the NC-F due to the attachment of the carbon fiber; however, at the time of final destruction, the delamination of CFS occurred. The strength of the interface attachment between the concrete and the CFS due to the repeated blows was reduced owing to the impact load. Therefore, the least amount of damage can be expected when the top and bottom parts are strengthened with NSHSDC.

3.2.3. Weight Loss

The weight loss rate is an additional indicator of the residual performance after an impact load. The occurrence of concrete debris due to an impact load not only reduces the usability of the structure, but also causes environmental damage. Several studies are currently being conducted to reduce debris in specimens [13,48]. In this study, the residual structural performance of concrete slabs under an impact load was evaluated by comparing the occurrence rate of fragments, as shown in Figure 8. All specimens, including the control slab, did not exhibit any loss of weight on the first blow. Further experiments were not conducted for the NC-NF because they were fabricated as a comparison data to determine the strengthening effects of the test specimens. NSC-F-S and NC-CF-S showed the highest weight loss rate of approximately 5%. The NSC-NF showed a weight loss of only 1 kg, which was significantly small (approximately 0.12% of the total slab weight). This indicates that the weight loss rate of the test specimen strengthened with FRP on the bottom side is approximately 3–5%, and the lowest weight loss rate can be expected when strengthening both top and bottom sides with NSHSDC jacket. Considering the damaged areas, it is similar regardless of the test specimens. The result suggests that the brittle fracture on the top side can be fully controlled as the strengthening progresses. However, it is expected to be the least damaged in the test specimen strengthened with the NSHSDC, and the largest damaged area of failure was shown in the test specimen strengthened with the sprayed-FRP series.

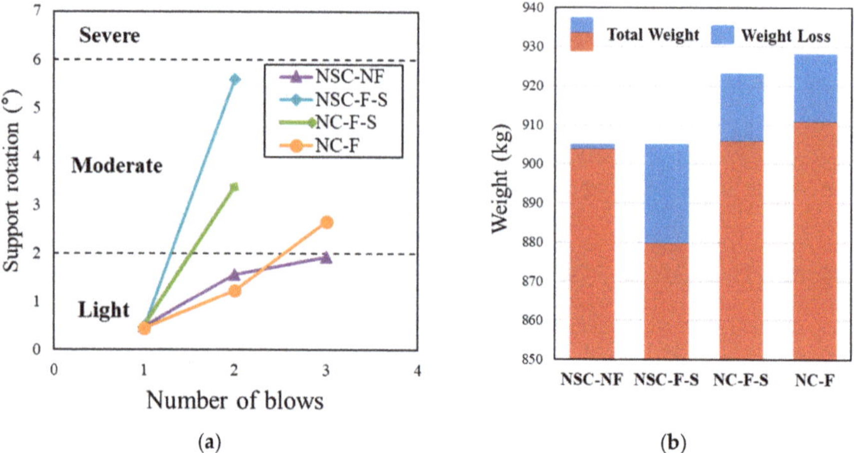

Figure 8. Support rotation and weight loss by impact loading: (**a**) Support rotation; (**b**) Weight loss.

3.3. Comparison of Previous and Experimental Test

This study evaluated the effects of reinforced concrete according to various strengthening methods. The test results showed the best impact resistance properties were shown in the NSC-NF. Furthermore, this study analyzed the two-way slabs with a similar impact energy of existing studies, as shown in Table 10. The comparison slab of the analysis consists of specimens strengthened with FRP, steel fiber reinforced concrete (SFRC), and UHPC. Bhatti et al. [6] studied the test specimens retrofitted by FRP sheets, and the size of the slab is similar to the specimen performed in this study. The reaction force was slightly smaller than that of NSC-NF, but there was a large deflection and residual deflection compared to the specimen used in this study. This suggests that the low strength of the FRP sheet is less effective than strengthening materials used in this paper. The approximate energy dissipated capacity was somewhat larger than this study, which is more exacerbated by the fact that the high rebar ratio was used, and the size of the slab was large. Trevor et al. [4] performed the drop-weight test with SFRC. The reaction force of the slab was higher than the results of this study, which is considered to have improved impact resistance due to the high rebar ratio and bridging effect of the steel fibers. Jang [49] implemented the impact test under the same conditions as the test specimens and 1% of the steel fibers were mixed with volume fractions of 1.0%. The maximum deflection was larger than that of NSHSDC in this study because the layer of NSHSDC controlled the deflection. However, the reaction force and energy dissipated ability of the SFRC is about 34% and 24% higher than NSC-NF, respectively. Although the impact resistance of NSC-NF is somewhat lower than that of the SFRC, it can be judged that the strengthening efficiency is high in that the thickness of NSHSDC is less than 30% of the thickness of SFRC. Kim [50] produced UHPC with a thickness of 25% thinner than the test specimen and the impact test was performed. The results of the experiment showed similar reaction force and deflection with NSC-NF, but the thickness of NSHSDC is only 38% of UHPC.

Table 10. Recent research about two-way RC slab under impact loadings.

No.	Researcher		Test Specimens	Impact Energy (kJ)	Reaction Force (kN)	Max. Deflection (mm)	Dissipated Energy (kJ)
1.	This study		NC-NF	5.89	179.75	23.41	1.79
2.	This study		NC-F	5.89	186.36	17.31	2.3
3.	This study		NC-F-S	5.89	187.16	17.91	2.88
4.	This study		NSC-NF	5.89	266.63	18.77	3.7
5	This study		NSC-F-S	5.89	206.84	18.99	3.64
6.	Hrynyk et al. 2014	[4]	Steel fiber (1%)	4.8	591	14	3.5
7.	Hrynyk et al. 2014	[4]	Steel fiber (2%)	4.8	621	13.5	4.2
8.	Hrynyk et al. 2014	[4]	Steel fiber (3%)	4.8	531	12.4	3.3
9.	Bhatti et al. 2011	[6]	CFRP strips	5.4	240	32	3.84
10.	Amira et al. 2019	[11]	Cementitious composites	5.93	61	2.65	2.51
11.	Mugunthan et al. 2020.	[14]	GFRP strips	4.86	-	-	-
12.	Jang 2015	[49]	Steel fiber (1%)	5.89	408.6	20.08	4.85
13.	Kim 2017	[50]	UHPC	6.13	272.72	17	4.64

4. Conclusions

This study evaluated the effects of reinforced concrete according to various reinforcement methods. Five test specimens were fabricated, including control specimens, and three strengthening methods were applied. The following results were derived from the experimental results:

(1) The NSC-NF showed excellent impact resistance and high strength. In particular, the highest strength and lowest deflection were obtained when strengthening the top and bottom sides with NSHSDC; the cracks were confirmed to spread evenly under the test specimens. This is because steel fibers and PE fibers included in the NSHSDC effectively control macro-cracking and micro-cracking and spread impact energy evenly throughout the test specimen.

(2) Comparing deflection at the first blow, the lowest maximum deflection was observed in the NC-F specimen, which was 26% lower than that of the control specimen. This is attributed to the fact that CFS increased the initial stiffness. Overall, at the first blow, the carbon fiber series showed less deflection than the NSHSDC series; however, at the final blow, the ratio of maximum deflection to residual deflection decreased.

(3) The crack of NSC-NF not only spread the crack evenly at the bottom but also improved the deflection capacity according to using NSHSDC, which was a hybrid using PE and steel fibers. However, strengthening with FRP and sprayed FRP cannot confirm the cracks due to the FRP adhesion. The NC-F was delaminated with the FRP, which showed the largest damaged area and

exposed the tensile reinforcement. The sprayed FRP may have caused brittle fracture due to the stress concentration at the tensile section.

(4) The NSC-NF showed a weight loss ratio of about 1 kg on the final blow, which is only about 0.12% of the total weight. However, the weight loss rate of the test specimen strengthened with FRP on the bottom side is approximately 2–3% of the total weight. This is due to the high ductility of the specimen as a bridging effect of steel fiber and PE fiber. Thus, the measured weight loss of the NSC-NF was smaller than that of the FRP series.

(5) Compared with the previous studies, NSC-NF showed best impact resistance in the test specimens applying the strengthening techniques, but it showed disadvantage in load and deflection SFRC concrete and UHPC. However, it can be expected to further research studies on the applicability of the NSHSDC in that it can be manufactured with the thin layer while strengthening the existing concrete.

Author Contributions: Conceptualization, S.-J.Y.; T.-F.Y. and Y.-S.Y.; methodology, S.-H.H.; experiment, S.-J.Y.; S.-H.H. and T.-F.Y.; validation, S.-J.Y. and S.-H.H. and Y.-S.Y.; investigation, Y.-S.Y.; data curation, S.-J.Y.; writing—original draft preparation, S.-J.Y.; writing—review and editing, S.-J.Y.; visualization; T.-F.Y. and S.-H.H.; supervision, T.-F.Y. and Y.-S.Y.; project administration, Y.-S.Y.; funding acquisition, Y.-S.Y. All authors have read and agreed to the published version of the manuscript.

Funding: This research was funded by a grant (19CTAP-C151911-01) from Technology Advancement Research Program (TARP) funded by Ministry of Land, Infrastructure and Transport of Korean government.

Conflicts of Interest: The authors declare no conflict of interest.

References

1. Thabet, A.; Haldane, D. Three-dimensional simulation of nonlinear response of reinforced concrete members subjected to impact loading. *Struct. J.* **2000**, *97*, 689–702.
2. Fujikake, K.; Li, B.; Soeun, S. Impact response of reinforced concrete beam and its analytical evaluation. *J. Struct. Eng.* **2009**, *97*, 938–950. [CrossRef]
3. Tomas, R.J.; Sorensen, A.D. Review of Strain Rate Effects for UHPC in Tension. *Constr. Build. Mater.* **2017**, *153*, 846–856. [CrossRef]
4. Hrynyk, T.D.; Vecchio, F.J. Behavior of Steel Fiber-Reinforced Concrete Slabs under Impact Load. *ACI Struct. J.* **2014**, *111*, 1213–1224. [CrossRef]
5. Foret, G.; Limam, O. Experimental and numerical analysis of RC two-way slabs strengthened with NSM CFRP rods. *Constr. Build. Mater.* **2008**, *20*, 2025–2030. [CrossRef]
6. Bhatti, A.Q.; Kishi, N.; Tan, K.H. Impact resistant behaviour of RC slab strengthened with FRP sheet. *Mater. Struct.* **2011**, *44*, 1855–1864. [CrossRef]
7. Yoo, D.Y.; Yoon, Y.S. Influence of steel fibers and fiber-reinforced polymers on the impact resistance of one-way concrete slabs. *J. Compos. Mater.* **2014**, *48*, 695–706. [CrossRef]
8. Radnic, J.; Matesan, D.; Grgic, N.; Baloevic, G. Impact testing of RC slabs strengthened with CFRP strips. *Compos. Struct.* **2015**, *121*, 90–103. [CrossRef]
9. Wang, W.; Chouw, N. Experimental and theoretical studies of flax FRP strengthened coconut fibre reinforced concrete slabs under impact loadings. *Constr. Build. Mater.* **2018**, *171*, 546–557. [CrossRef]
10. Yilmaz, T.; Kirac, N.; Anil, O.; Erdem, R.T.; Sezer, C. Low-velocity impact behaviour of two way RC slab strengthening with CFRP strips. *Constr. Build. Mater.* **2018**, *186*, 1046–1063. [CrossRef]
11. Elnagar, A.B.; Afefy, H.M.; Baraghith, A.T.; Mahmoud, M.H. Experimental and numerical investigations on the impact resistance of SHCC-strengthened RC slabs subjected to drop weight loading. *Constr. Build. Mater.* **2019**, *229*. [CrossRef]
12. Yang, J.Q.; Smith, S.T.; Wang, Z.; Lim, Y.Y. Numerical simulation of FRP-strengthened RC slabs anchored with FRP anchors. *Constr. Build. Mater.* **2018**, *172*, 735–750. [CrossRef]
13. Soltani, H.; Khaloo, A.; Sadraie, H. Dynamic performance enhancement of RC slabs by steel fibers vs. externally bonded GFRP sheets under impact loading. *Eng. Struct.* **2020**, *213*, 110539. [CrossRef]

14. Loganaganandan, M.; Murali, G.; Salaimanimagudam, M.P.; Haridharan, M.K.; Karthikeyan, K. Experimental Study on GFRP Strips Strengthened New Two Stage Concrete Slabs under Falling Mass Collisions. *KSCE J. Civil Eng.* **2020**, 1–10. [CrossRef]
15. Mahmoud, M.H.; Afefy, H.M.; Baraghith, A.T.; Elnagar, A.B. Impact and static behavior of strain-hardening cementitious composites-strengthened reinforced concrete slabs. *Adv. Struct. Eng.* **2020**, *23*, 1614–1628. [CrossRef]
16. Chen, J.F.; Yuan, H.; Teng, J.G. Debonding failure along a softening FRP-to-concrete interface between two adjacent cracks in concrete members. *Eng. Struct.* **2007**, *29*, 259–270. [CrossRef]
17. Dong, J.F.; Wang, Q.Y.; Guan, Z.W. Structural behaviour of RC beams externally strengthened with FRP sheets under fatigue and monotonic loading. *Eng. Struct.* **2012**, *29*, 259–270. [CrossRef]
18. Daud, R.A.; Cunningham, L.S.; Wang, Y.C. Static and fatigue behaviour of the bond interface between concrete and externally bonded CFRP in single shear. *Eng. Struct.* **2015**, *97*, 54–67. [CrossRef]
19. Mohammed, T.J.; Bakar, B.A.; Bunnori, N.M. Torsional improvement of reinforced concrete beams using ultra high-performance fiber reinforced concrete (UHPFC) jackets—Experimental study. *Constr. Build. Mater.* **2016**, *106*, 533–542. [CrossRef]
20. Attari, N.; Amziane, S.; Chemrouk, M. Flexural strengthening of concrete beams using CFRP, GFRP and hybrid FRP sheets. *Constr. Build. Mater.* **2012**, *37*, 746–757. [CrossRef]
21. Banthia, N.; Boyd, A.J. Sprayed fibre-reinforced polymers for repairs. *Can. J. Civ. Eng.* **2000**, *27*, 907–915. [CrossRef]
22. Parghi, A.; Alam, M.S. A review on the application of sprayed-FRP composites for strengthening of concrete and masonry structures in the construction sector. *Compos. Struct.* **2018**, *187*, 518–534. [CrossRef]
23. Han, S.C.; Yang, J.M.; Yoon, Y.S. Evaluation on strengthening capacities and rebound rate of structures with sprayed FRP. *J. Korea Inst. Struct. Maint. Insp.* **2008**, *12*, 193–202.
24. Chang, J.H.; Jang, K.S. Performance evaluation of structure strengthening using sprayed FRP technique. *J. Korea Inst. Struct. Maint. Insp.* **2009**, *13*, 126–136.
25. Aoude, H.; Belghiti, M.; Cook, W.D.; Mitchell, D. Response of steel fiber-reinforced concrete beams with and without Stirrups. *ACI Struct. J.* **2012**, *109*, 359–367.
26. Ong, K.C.G.; Basheerkhan, M.; Paramasivam, P. Resistance of fibre concrete slabs to low velocity projectile impact. *Cem. Concr. Compos.* **1999**, *21*, 391–401. [CrossRef]
27. Hable, K.; Paul, G. Response of ultra-high performance fiber reinforced concrete (UHPFRC) to impact and static loading. *Cem. Concr. Compos.* **2008**, *30*, 938–946. [CrossRef]
28. Yang, I.H.; Joh, C.B.; Kim, B.S. Structural behavior of ultra high performance concrete beams subjected to bending. *Eng. Struct.* **2010**, *32*, 3478–3487. [CrossRef]
29. Yoo, D.Y.; Banthia, N.; Yoon, Y.S. Impact resistance of reinforced ultra-high-performance concrete beams with different steel fibers. *ACI Struct. J.* **2017**, *114*. [CrossRef]
30. Yoo, D.Y.; Banthia, N.; Kim, S.W.; Yoon, Y.S. Response of ultra-high-performance fiber-reinforced concrete beams with continuous steel reinforcement subjected to low-velocity impact loading. *Compos. Struct.* **2015**, *126*, 233–245. [CrossRef]
31. Yoo, D.Y.; Yoon, Y.S. Structural Performance of Ultra-High-Performance Concrete Beams with Different Steel Fibers. *Eng. Struct.* **2015**, *102*, 409–423. [CrossRef]
32. Kim, D.J.; Park, S.H.; Ryu, G.S.; Koh, K.T. Comparative flexural behavior of hybrid ultra high performance fiber reinforced concrete with different macro fibers. *Constr. Build. Mater.* **2011**, *25*, 4144–4155. [CrossRef]
33. Zanuy, C.; Ulzurrun, G.S.D. Impact resisting mechanisms of shear-critical reinforced concrete beams strengthened with high-performance FRC. *Appl. Sci.* **2020**, *10*, 3154. [CrossRef]
34. Lampropoulos, A.P.; Paschalis, S.A.; Tsioulou, O.T.; Dritsos, S.E. Strengthening of reinforced concrete beams using ultra high performance fiber reinforced concrete (UHPFRC). *Eng. Struct.* **2016**, *106*, 370–384. [CrossRef]
35. Yuan, T.F.; Lee, J.Y.; Min, K.H.; Yoon, Y.S. Experimental investigation on mechanical properties of hybrid steel and polyethylene fiber reinforced no-slump high strength concrete. *Int. J. Polym. Sci.* **2019**, *2019*, 1–11. [CrossRef]
36. Yuan, T.F.; Lee, J.Y.; Min, K.H.; Yoon, Y.S. Bond strength and flexural capacity of normal concrete beams strengthened with no-slump high-strength, high-ductility concrete. *Materials* **2020**, *13*, 4218. [CrossRef]
37. Lee, N.H.; Kim, J.H.J.; Han, T.S.; Cho, Y.G.; Lee, J.H. Blast-resistant characteristics of ultra-high strength concrete and reactive powder concrete. *Constr. Build. Mater.* **2012**, *28*, 694–707. [CrossRef]

38. American Society for Testing and Materials (ASTM). ASTM C39/C39M Standard test method for compressive strength of cylindrical concrete specimens. In *Annual Book of ASTM Standards*; ASTM: West Conshohocken, PA, USA, 2014.
39. American Society for Testing and Materials (ASTM). ASTM C1609/C1609 M. Standard test method for flexural performance of fiber-reinforced concrete (using beam with third-point loading). In *Annual Book of ASTM Standards*; ASTM: West Conshohocken, PA, USA, 2012.
40. Lee, J.Y.; Yuan, T.Y.; Shin, H.O.; Yoon, Y.S. Strategic use of steel fibers and stirrups on enhancing impact resistance of ultra-high-performance fiber-reinforced concrete beams. *Cem. Concr. Compos.* **2020**, *107*. [CrossRef]
41. Isaac, P.; Darby, A.; Ibell, T.; Evernden, M. Experimental investigation into the force propagation velocity due to hard impacts on reinforced concrete members. *Int. J. Impact. Eng.* **2017**, *100*, 131–138. [CrossRef]
42. Pham, T.M.; Hao, H. Plastic hinges and inertia forces in RC beams under impact loads. *Int. J. Impact. Eng.* **2017**, *103*, 1–11. [CrossRef]
43. Huo, J.; Li, Z.; Zhao, L.; Liu, J.; Xiao, Y. Dynamic behavior of carbon fiber-reinforced polymer- strengthened reinforced concrete beams without stirrups under impact loading. *ACI Struct. J.* **2018**, *115*, 775–787. [CrossRef]
44. Saatci, S.; Vecchio, F.J. Effects of shear mechanisms on impact behavior of reinforced concrete beams. *ACI Struct. J.* **2009**, *106*, 78–86.
45. Nghiem, A.; Kang, T.H.-K. Drop-weight testing on concrete beams and ACI design equations for maximum and residual deflections under low-velocity impact. *ACI Struct. J.* **2020**, *117*, 199–210. [CrossRef]
46. Cao, Y.Y.Y.; Liu, G.; Brouwers, H.J.H.; Yu, Q. Enhancing the low-velocity impact resistance of ultra-high-performance concrete by an optimized layered-structure concept. *Compos. B Eng.* **2020**, *200*. [CrossRef]
47. United States Department of Defense. *UFC 3-340-02, Structures to Resist the Effects of Accidental Explosions*; United States Department of Defense: Washington, DC, USA, 2008.
48. Lee, J.Y.; Shin, H.O.; Min, K.H.; Yoon, Y.S. Flexural assessment of blast-damaged RC beams retrofitted with CFRP sheet and steel fiber. *Int. J. Polym. Sci.* **2018**, *2018*. [CrossRef]
49. Jang, D.S. Damage Assessment of Two-Way RC Slabs Subjected to Low-Velocity Impact Loading. Master's Thesis, University of Korea, Seoul, Korea, 5 January 2015.
50. Kim, D.H. Impact Resistance of SFRC Slabs upon Steel Fiber Content and Concrete Strength. Master's Thesis, University of Korea, Seoul, Korea, 7 January 2017.

Article

Photocatalytic Performance Evaluation of Titanium Dioxide Nanotube-Reinforced Cement Paste

Junxing Liu [1], Hyeonseok Jee [1], Myungkwan Lim [2], Joo Hyung Kim [3], Seung Jun Kwon [4], Kwang Myong Lee [5], Erfan Zal Nezhad [6] and Sungchul Bae [1,*]

1 Department of Architectural Engineering, Hanyang University, 222, Wangsimni-ro, Sungdong-gu, Seoul 04763, Korea; liujx128119@hanyang.ac.kr (J.L.); wlgustjr01@gmail.com (H.J.)
2 Department of Architecture Engineering, Songwon University, Gwangju 61756, Korea; limmk79@naver.com
3 Korea Conformity Lab, Construction Technology Research Center, Construction Division, Seoul 08503, Korea; kjhmole@kcl.re.kr
4 Department of Civil and Environmental Engineering, Hannam University, Daejeon 306-791, Korea; jjuni98@hnu.kr
5 Department of Civil, Architectural, and Environmental System Engineering, Sungkyunkwan University, Gyeonggi-Do 16419, Korea; leekm79@skku.edu
6 Department of Chemical and Biomedical Engineering, University of Texas at San Antonio (UTSA), San Antonio, TX 78249, USA; erfan.zalnezhad@utsa.edu
* Correspondence: sbae@hanyang.ac.kr

Received: 2 November 2020; Accepted: 27 November 2020; Published: 28 November 2020

Abstract: Considering the increase in research regarding environmental pollution reduction, the utilization of cementitious material, a commonly used construction material, in photocatalysts has become a desirable research field for the widespread application of photocatalytic degradation technology. Nano-reinforcement technology for cementitious materials has been extensively researched and developed. In this work, as a new and promising reinforcing agent for cementitious materials, the photocatalytic performance of titanium dioxide nanotube (TNT) was investigated. The degradation of methylene blue was used to evaluate the photocatalytic performance of the TNT-reinforced cement paste. In addition, cement paste containing micro-TiO_2 (m-TiO_2) and nano-TiO_2 (n-TiO_2) particles were used for comparison. Moreover, the effect of these TiO_2-based photocatalytic materials on the cement hydration products was monitored via X-ray diffraction (XRD) and thermogravimetric analysis (TG). The results indicated that all the TiO_2 based materials promoted the formation of hydration products. After 28 days of curing, the TNT-reinforced cement paste contained the maximum amount of hydration products ($Ca(OH)_2$). Furthermore, the cement paste containing TNT exhibited better photocatalytic effects than that containing n-TiO_2, but worse than that containing m-TiO_2.

Keywords: photocatalysis; titanium dioxide nanotube; anatase TiO_2; hydration products; cement paste

1. Introduction

TiO_2 and its nanostructured materials have been extensively investigated and applied in several fields, including photocatalysis [1,2], Li-ion batteries [3,4], and dye-sensitized solar cells [5,6] owing to their low cost, high chemical stability, nontoxicity, and super-hydrophobicity [1,7]. Typically, there are three polymorphs of TiO_2: anatase, rutile, and brookite. Anatase TiO_2 exhibits higher photocatalytic activity than the other phases, owing to its higher surface area, surface adsorption capacity, and lower charge carrier recombination rate [8]. Therefore, anatase TiO_2 is the most widely used photocatalyst capable of removing atmospheric pollutants, such as nitrogen oxides (NO_x), volatile organic oxides (VOCs), and nonvolatile organic residues, via redox reactions on the catalyst surface under natural sunlight (usually in the UV range) [9]. Hamidi et al. [10] suggested that photocatalytic efficiency is a complicated aspect influenced by several factors and concluded that the five factors that most affect the efficiency of photocatalytic processes are: (a) effective absorption of photon energy, (b) fast charge separation after absorption of photon energy to prevent electron–hole recombination, (c) separation of products from the surface of the photocatalyst, (d) long term photocatalyst stability, and (e) the redox potential of the electron–hole pair should be compatible with the redox potentials of the donor or acceptor species. These factors only take into account the TiO_2-based materials themselves. However, pure TiO_2 micro- and nanostructures are rarely directly combined into bulk materials for use; therefore, a convenient way to overcome this drawback is to attach the TiO_2-based material in/on different support materials [11].

Nano-reinforcement technology plays a critical role in a novel research direction in the field of construction and building materials [12]. Among the materials used in civil and architectural fields, cementitious materials, such as concrete, are considered to be one of the most popular materials used and consumed. In addition, the rich pore structure and chemical stability of cement hydrate products give cementitious materials a strong binding capacity and good processability [13]. These characteristics make cementitious materials suitable supports for loading nanomaterials [14]. On one hand, some literature reports suggest that nanomaterials have a reinforcement effect on cementitious materials, such as mechanical strength, durability, and ductility, and can accelerate the hydration reaction [15–17]. Chen et al. [18] found that after 28 days of curing, the compressive strength of the cement mortar increased from 40 MPa to 50 MPa with increasing the amount of nano-TiO_2 (up to 10% of cement weight). Li et al. [19] studied that the size effect of TiO_2 on the properties of cementitious materials and found that the nucleation effect and reinforcement of the mechanical properties of cement were more significant with smaller size TiO_2 (10 nm) than larger size TiO_2 (15 nm). Furthermore, previous studies [18,20] have shown that the incorporation of TiO_2 into cementitious materials can improve its mechanical strength owing to the ability of TiO_2 to accelerate the hydration process as well as the pore-refining effect.

On the other hand, the incorporation of nanomaterials into cementitious materials offers some novel features, such as thermal resistance [21], self-sensing [22], self-healing [23], and self-cleaning [24]. The incorporation of photocatalysts, such as TiO_2-based materials, into cementitious materials, is undoubtedly one of the most significant topics in the field of environmentally friendly building materials since the addition of photocatalysts contributes to the air-purifying, self-cleaning, self-sterilizing, and anti-fogging properties of the construction materials [25]. The advantage of this material is that, apart from the addition of TiO_2 particles to the building materials, only sunlight, oxygen, and water are required [26]. Seo et al. [27] researched that the NO removal rate for the photocatalytic cement-based materials and noted that the smaller size TiO_2 particles projected the higher NO absorption and removal rate in the dry condition. In addition, Zhang et al. [28] pointed out that particle size was important in nanocrystalline TiO_2-based catalysts mainly by influencing the kinetics of e^-/h^+ pair recombination. For methods of using photocatalytic materials in cementitious materials, besides adding the photocatalyst to the cementitious material, another way to provide a photocatalytic effect to the cementitious material is to attach the photocatalyst to the surface of the cementitious material as a thin film by coating. Feng et al. [14] investigated the photocatalytic performance of cement paste utilized smear method to prepare TiO_2 film on the cement paste surface, and the result showed that the photocatalytic cement paste could nearly completely discolor 10 mg/L Rhodamine B, methylene blue, and methylene orange solution in 50 min under UV light. In addition, Park et al. found that the incorporation of nano-TiO_2 into cement paste enhances the neutron shielding performance of the cement paste via Monte Carlo simulation at the thermal, slow, and intermediate neutron energy levels [29]. In summary, TiO_2 cannot only reinforce cementitious materials in terms of strength and hydration processes but can also add photocatalytic properties to cement-based materials. In 1998, Kasuga et al. [30] first reported the synthesis of TNT through a hydrothermal method by making use of anatase phase TiO_2 and revealed that TNT exhibits relatively larger specific surface areas than typical TiO_2 nanopowders. Therefore, TNT has been widely utilized in various applications, including photocatalysis [31,32], solar cells [33], and electrochromic devices [34]. According to our previous research [35], the incorporation of TNT into cementitious materials enhanced the compressive strength and flexural strength by 11.7% and 23.5%, respectively, after 28 days at low addition dosages (0.5% of cement weight) and accelerated the hydration reaction rate of the cementitious materials. This proved the possibility of using TNT as a nano-reinforcement agent for cement-based materials. However, there have been few studies on the photocatalytic properties of TNT in cementitious materials.

Thus, in this study, we aimed to investigate the photocatalytic performance of TNT-reinforced cement paste as well as cement paste blended with m-TiO_2 and n-TiO_2 particles for comparison. The TNT was synthesized through a hydrothermal method, and the surface area of TNT was checked by Brunauer–Emmett–Teller (BET) surface area analysis. The photocatalytic effects of the different TiO_2-based materials in the hardened cement were evaluated by determining the changes in methylene blue (MB) concentration by ultraviolet-visible (UV-vis) spectroscopy. Transmission electron microscopy (TEM) was used to inspect the morphology of the TiO_2 based materials. The effects of the TNT and TiO_2 particles on the hydration products were evaluated via X-ray diffraction (XRD) and thermogravimetric analysis (TG).

2. Experimental Procedures

2.1. Raw Materials

Ordinary Portland cement (OPC), provided by Sungshin Co. Ltd., Seoul, Korea, was used to produce the cement paste. The chemical composition, measured by X-ray fluorescence spectroscopy (XRF, ZSX Primusll, Rigaku, Tokyo, Japan), is shown in Table 1. The microsized anatase titanium (IV) dioxide (m-TiO_2) used in this work was purchased from FUJIFILM Wako Pure Chemical Corporation, Tokyo, Japan. Anatase titanium (IV) oxide nanopowder (n-TiO_2), purchased from Sigma Aldrich (Seoul, Korea), was used to synthesize TNT. The physical properties of the different TiO_2 powders are summarized in Table 2.

Table 1. Chemical composition (%) of ordinary Portland cement (OPC).

SiO_2	Al_2O_3	Fe_2O_3	CaO	MgO	K_2O	SO_3	TiO_2	LOI	Total
18.43	2.83	2.17	68.17	2.37	1.11	3.03	0.15	1.72	100

LOI: loss of ignition.

Table 2. Physical properties of anatase TiO_2 powders.

Name	Particle Size	Surface Area	Purity
m-TiO_2	0.15–0.25 μm	10.89 m^2/g	98.5%
n-TiO_2	<25 nm	45–55 m^2/g	99.7%

2.2. Hydrothermal Synthesis of TNT

The TNT was synthesized via the hydrothermal method developed by Kasuga et al. [30,36]. This method is a simple experimental route to obtain a tubular TNT structure [37]. Figure 1 illustrates the hydrothermal TNT synthetic process, detailed as follows: 1.0 g of TiO_2 anatase nanopowder was mixed with 50 mL of 10 M NaOH solution. Next, the mixture was sonicated for 30 min in a 20 s work cycle using an ultrasonic liquid processor (Q700 Sonicator, Qsonica, Newtown, CT, USA, 20 kHz, amplitude: 50%). Then, the homogeneous suspension was transferred to a Teflon autoclave to start the hydrothermal synthesis process at 120 °C for 48 h. Subsequently, the obtained product was naturally cooled to room temperature and washed successively with 0.1 M HCl solution and distilled water until the pH was close to neutral. Finally, the precipitate was filtered and dried at 60 °C for 24 h. The morphology and crystal structure of the TNT was monitored by TEM and XRD.

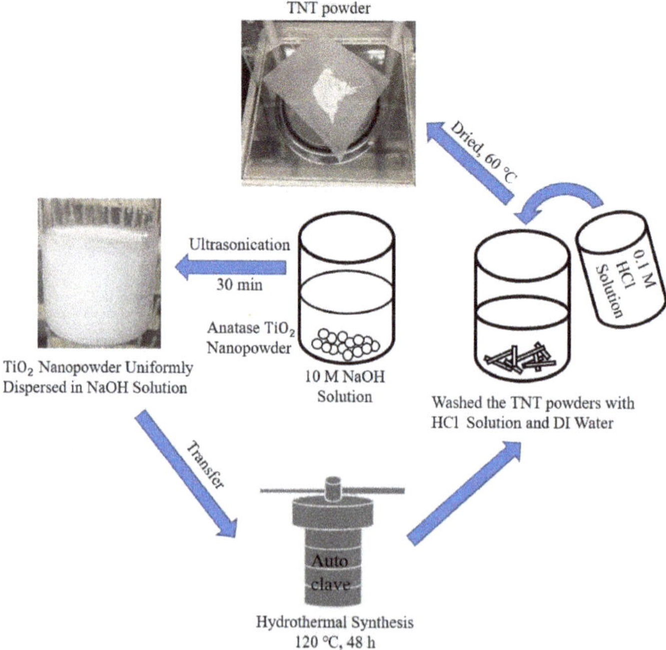

Figure 1. The titanium dioxide nanotube (TNT) hydrothermal synthesis process.

2.3. BET Analysis

The surface area of TNT was confirmed via N_2 adsorption on the 3Flex surface characterization analyzer at 77 K. The specific surface area was calculated by BET Equation (1).

$$\frac{1}{X[[P_0/P]-1]} = \frac{1}{X_m C} + \frac{C-1}{X_m C}\left(\frac{P}{P_0}\right) \quad (1)$$

where X is the weight of adsorbed N_2 at certain relative pressure (P/P_0), X_m is the volume of single-layer adsorption capacity of the gas at the standard temperature and pressure (273 K and 1 atm), C is constant.

2.4. Preparation of Cement Paste

Pure OPC paste and cement pastes containing photocatalyst powders were fabricated with a water-to-cement ratio (w/c) of 0.4. The mass content of TiO_2-based materials (m-TiO_2, n-TiO_2, and TNT) was 1 wt% of the mass of cement. Details of the mixed proportion of specimens are shown in Table 3, and the names of the three TiO_2-based materials were used to refer directly to the different specimens. To prepare the cement paste for testing, the different TiO_2-based materials were soaked in distilled water and ultrasonicated (Q700 Sonicator, Qsonica, Newtown, CT, USA, 20 kHz, amplitude: 50%) for 15 min to achieve a homogeneous solution before being added to the cement. The homogeneous solution was mixed with cement via a paste mixer (SPS-1, Malcom, Tokyo, Japan) for 12 min. The specimen was demolded after 24 h and cured at 25 °C and 65% relative humidity until being tested. The polylactic acid (PLA) molds used to obtain the specimens (50 mm × 50 mm × 5 mm) were designed using SolidWorks and fabricated with a fused-deposition modeling 3D printer (Guider II, Flashforge, Jinhua, China), as shown in Figure 2a. The reason for choosing the shape of the sample shown in Figure 2e was to obtain a relatively large surface that was easy to place and use in the photocatalytic test.

Table 3. Mix proportion of cement paste.

Samples	m-TiO_2 (g)	n-TiO_2 (g)	TNT (g)	W/C	Water (g)	Cement (g)
OPC (P)	-	-	-	0.4	16	40
m-TiO_2 (P)	0.4	-	-			
n-TiO_2 (P)	-	0.4	-			
TNT (P)	-	-	0.4			

(P) in the sample name denotes cement pastes.

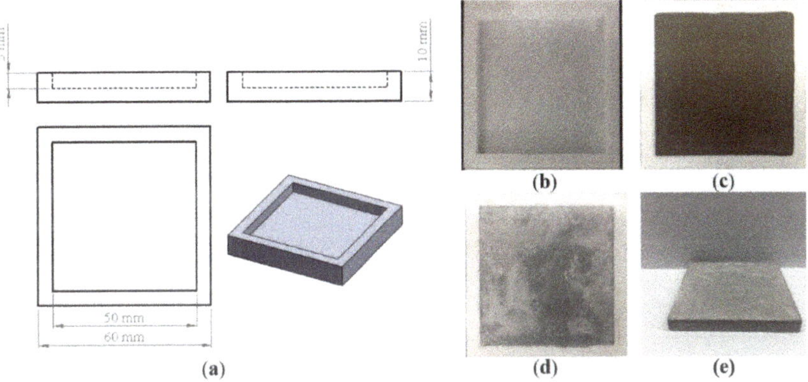

Figure 2. Schematic of the (**a**) three views of the cement mold, (**b**) real product of cement mold, (**c**) the fresh cement paste in the mold, (**d**) hardened cement paste in the mold, and (**e**) cement paste after 1 day of curing and demolding.

2.5. *X-ray Diffraction*

Phase characterization of the hardened cement pastes containing the nanomaterials was conducted using an XRD spectrometer (D2 PHASER, Bruker, Billerica, MA, USA) equipped with Cu Kα radiation (λ = 1.5406 Å). The measurement covered the angular range (2θ) from 7° to 70° with a step size of 0.01° at 1.5 sec per step. The phase was identified using the DIFFRAC.EVA software (Bruker, Billerica, MA, USA). The XRD samples were prepared by first grinding in a mortar and pestle and then sieved through a 400-mesh sieve.

2.6. *Thermogravimetric Analysis*

To determine the composition of the cement hydrates, TG and first derivative thermogravimetric analysis (DTG) curves were utilized [38]. The test was performed using a HITACHI STA7200 Simultaneous Thermogravimetric Analyzer (Tokyo, Japan) with a sensitivity of 0.2 µg. At the two test times (1 and 28 days), approximately 15 mg of the specimens were crushed and ground to obtain a particle size smaller than 75 µm. The test temperature ranged between 20 and 900 °C with a heating rate of 10 °C/min and an N_2 atmosphere of 200 mL/min.

2.7. *Transmission Electron Microscopy*

The morphologies of the m-TiO_2, n-TiO_2, and TNT were obtained by TEM (JEM2100F, JEOL, Tokyo, Japan). To prepare the samples for TEM, a suitable amount of powder was dissolved in ethanol by ultrasonic vibration, and then a few drops of the sonicated solution were placed on a carbon film Cu grid and dried for 48 h in a desiccator.

2.8. Photocatalytic Test by Methylene Blue Degradation

The photocatalytic activities of the prepared specimens (Table 3) were assessed by testing the decomposition of MB as an organic pollutant under UV light (PM-1600UVH, NDT Advance, Saitama, Japan) irradiation. Figure 3 shows a schematic diagram of the test equipment. The experimental conditions were as follows: after curing for 28 days, the hardened cement paste was immersed in 200 mL of MB solution in a 250 mL beaker. During the entire process, the solution was continuously stirred with a magnetic stirrer to keep the concentration of the solution consistent. Before irradiation, the MB solution containing the cement paste specimen was placed in the dark for 1 h to reach an adsorption–desorption equilibrium. Approximately 2 mL of MB solution was collected at 1 h intervals during the 5 h test. The change in MB concentration was monitored by UV-vis spectroscopy (Genesys 180, Thermo Scientific, Waltham, MA, USA) at the characteristic λ_{max} = 665 nm. The wavelength range used in the test was 200–800 nm. The normalized concentration of the photocatalysts was calculated using Equation (2):

$$\text{Normalized concentration} = c/c_0 \tag{2}$$

where c represents the concentration of the MB solution at different time intervals, and c_0 represents the initial concentration of MB solution.

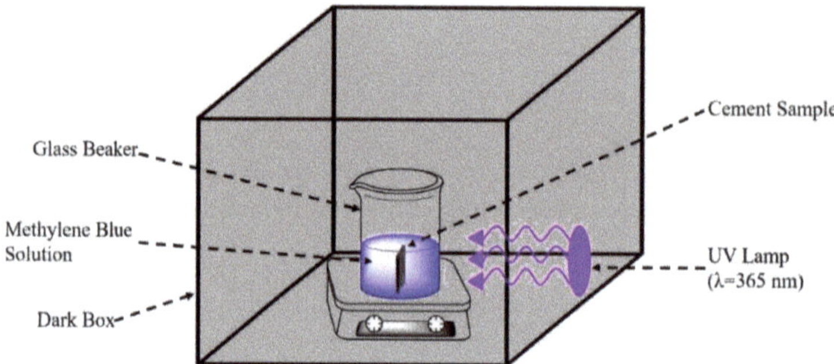

Figure 3. Schematic diagram of the photocatalytic test.

3. Results and Discussion

3.1. Characteristics of the Raw Materials

The crystallographic structure and relative crystallinity of the materials were identified via XRD. The XRD patterns of the three TiO_2 materials are shown in Figure 4. The peaks of m-TiO_2 and n-TiO_2 are situated at 25.3°, 37.8°, 48.0°, 53.8°, 55.0°, 62.6°, and 68.7°, which correspond to the (101), (004), (200), (105), (211), (204), and (116) crystallographic planes, respectively [35]. All the diffraction peaks were well established and perfectly matched to anatase TiO_2 [39]. The intensity of the n-TiO_2 diffraction peaks compares well to that of m-TiO_2; however, the diffraction peaks are broader owing to the decrease in particle size [40]. The diffraction pattern of TNT is similar to the results obtained by other researchers, with two clear peaks located at 24.5° and 48.3° [41–43]. Harsha et al. [44] reported the existence of peaks that could be identified as titanate, which suggests that Na^+ ions were replaced by H^+ ions through ion exchange during the washing process. Zavala et al. [41] demonstrated that the anatase peaks did not fit well owing to the low crystallinity of the TNT. This was attributed to the collapse of the nanotubes during the ion exchange process or the formation of nanosheets as a precursor to the nanotubes.

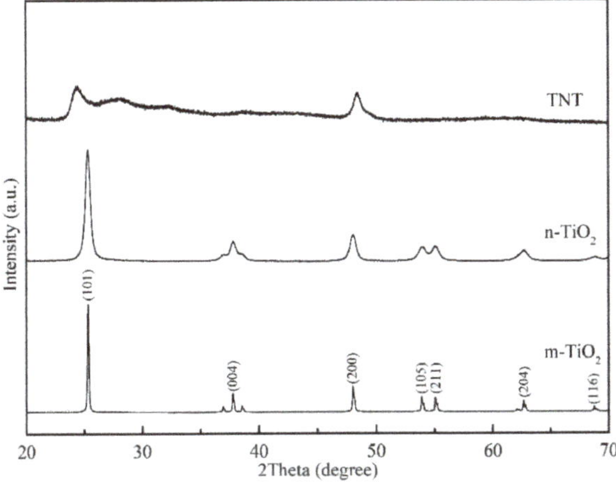

Figure 4. Comparison of the XRD patterns of m-TiO_2, n-TiO_2, and TNT.

Figure 5 shows the TEM micrographs of m-TiO_2, n-TiO_2, and TNT. The shape and the difference in size between the two TiO_2 particles are clearly shown in Figure 5a,b, respectively. In addition, clusters or aggregates of particles composed of smaller particles are observed [45]. Figure 5c,d illustrate the structure of the TNT synthesized through the hydrothermal method, and the nanotubular structure of the TNT is apparent, and the nanotubes are hollow and open-ended [46]. This is a typical characteristic that could be used to identify TNT [47]. The size of the TNT is summarized in Table 4; the synthesized TNT exhibited an average length of 105 nm. The N_2 adsorption isotherm of hydrothermally synthesized TNT is shown in Figure 6, and the curve was the hysteresis loop with a sharp inflection in N_2 adsorbed volume at P/P_0 was about 0.45, which suggested that the TNT was presented in tubular structures. The surface area of TNT obtained by BET was 196 m^2/g. Compared with the physical properties of nano-TiO_2 (45–55 m^2/g), as shown in Table 2, TNT had an approximately four-fold larger surface area.

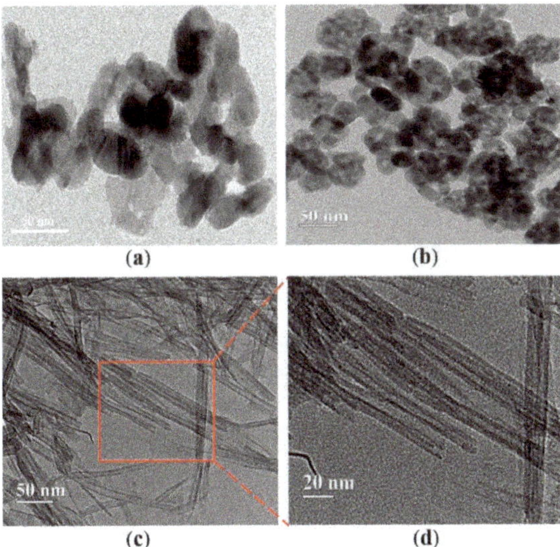

Figure 5. Morphology of (**a**) m-TiO_2, (**b**) n-TiO_2, and (**c**) TNT. (**d**) Higher magnification image of TNT.

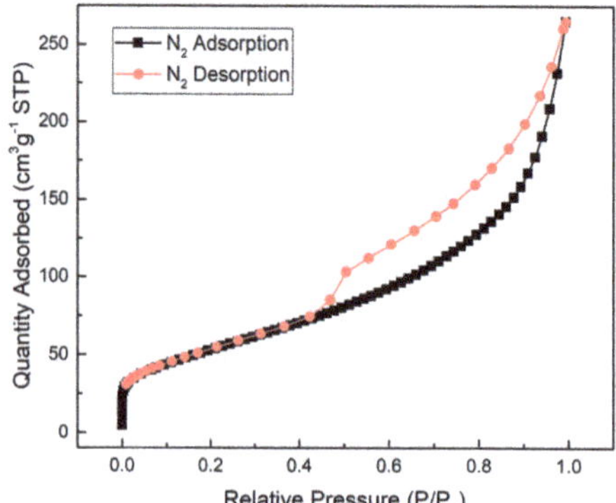

Figure 6. N_2 adsorption isotherm of hydrothermally synthesized TNT.

Table 4. The size of TNT synthesized by the hydrothermal method.

Product	Outside Diameter	Inside Diameter	Average Length	Max Length	Min Length	Surface Area
TNT	8 ± 3.4 nm	3 ± 2.5 nm	105 nm	200 nm	50 nm	196 m^2/g

3.2. Hydration Products

To further characterize the influence of the different TiO_2 powders and TNT on the hydration of the cement paste, hydration product analysis was performed on the cement pastes utilizing XRD and TG. Figure 7 shows the XRD patterns of the OPC and cement pastes containing the different TiO_2-based materials after 1 day and 28 days. The three main peaks of $Ca(OH)_2$ appear at 2θ = 18.0°, 28.6°, and 34.1°. After 1 day of hydration, n-TiO_2 and TNT accelerated the cement hydration because of their high surface areas providing nucleation sites for the hydration products, as previously reported [18,20]; this is in contrast to m-TiO_2, which could be due to the differences in the sizes of the nanoparticles. After 28 days of hydration, the diffraction peaks attributed to $Ca(OH)_2$ in all the XRD patterns of the cement pastes increased in intensity due to further hydration. In addition, the increase in the intensity of the $Ca(OH)_2$ peaks was more evident in the hardened cement paste containing the TiO_2 materials compared to OPC, especially for the hardened cement paste containing the TNT. This was possibly caused by the pronounced hollow tubular structure of the TNT compared to the TiO_2 particles, which provides more locations and nucleation sites for the formation of hydration products, further allowing more hydration products to be generated [35]. This was also reflected in the TG results of the hardened cement pastes (Figure 8). When the hydrated cement pastes are exposed to high temperatures, mass loss occurs at specific temperature boundaries due to the loss of free water, dehydration, dehydrogenation, and decarbonization of the hydration products [48]. In summary, the thermal decomposition of hydration products was classified into main stages as follows: the weight loss around 100 °C corresponds to some of the cement hydration products, such as C–S–H, ettringite and monosulfate, etc., lose part of chemically bound water; the weight loss at 400–450 °C corresponds to the decomposition of $Ca(OH)_2$; the weight loss above 600 °C corresponds to the decomposition of $CaCO_3$ into CaO and CO_2 [49]. It is evident from the TG results that the amount of $Ca(OH)_2$ produced increased in the cement paste containing TNT, whether at 1 or 28 days, compared to the others. This result implies that TNT has great potential to improve the mechanical properties of cement pastes by accelerating the hydration of cement clinker phases in early and long-term aging.

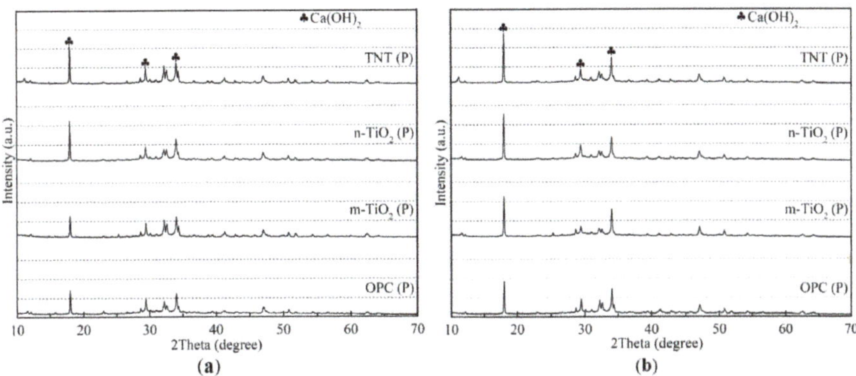

Figure 7. XRD patterns of the cement pastes after (**a**) 1 day and (**b**) 28 days of curing.

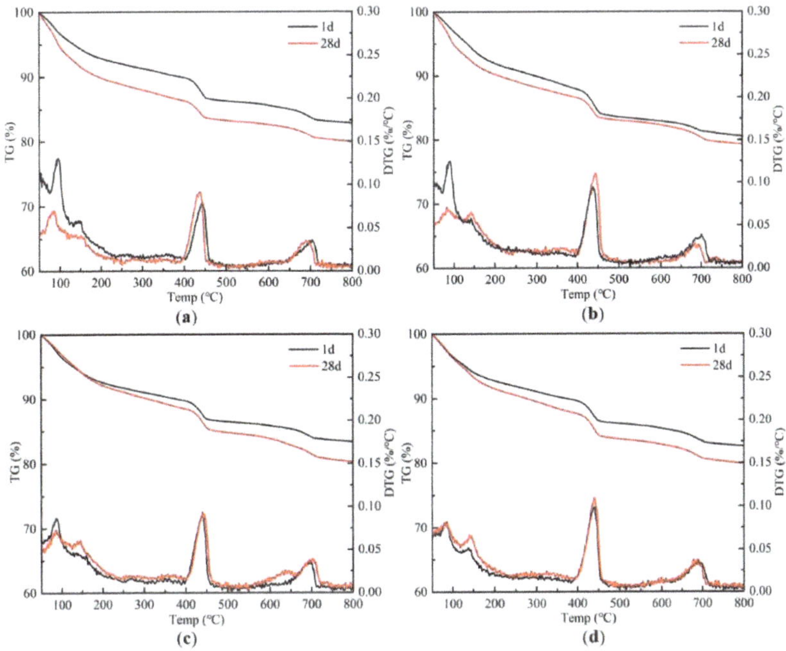

Figure 8. TG results of (**a**) OPC (P), (**b**) m-TiO_2 (P), (**c**) n-TiO_2 (P), and (**d**) TNT (P).

3.3. Photocatalytic Performance

According to other research, the concentration of MB solution decreased when cement paste was placed in MB solution; this may be attributed to the self-decomposition of MB and absorption by the cement paste [12,50,51]. Therefore, the change in concentration of MB solution with OPC and without OPC in dark conditions was measured, as shown in Figure 9. In our study, within 1 h, a decrease of MB concentration was found in OPC due to the porous structure of the OPC surface, whereas no significant self-decomposition of MB was found in pure MB solution.

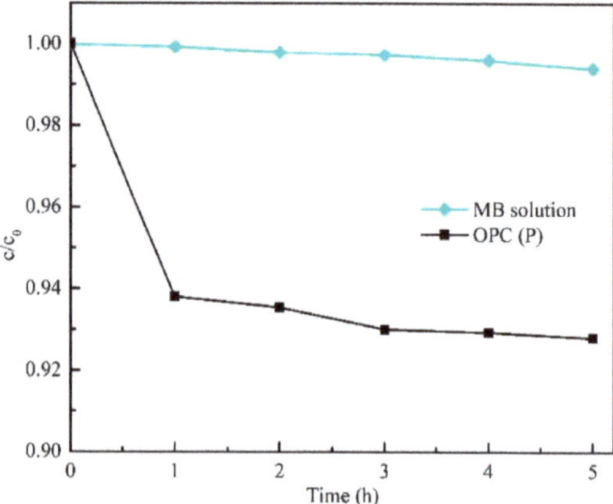

Figure 9. Concentration change of MB solution with and without OPC (P) under dark conditions.

Figure 10a shows the photocatalytic properties of the three pure TiO_2 photocatalysts (m-TiO_2, n-TiO_2, and TNT) in the degradation of MB. The data of decomposition of MB solution with pure photocatalysts is shown in Table 5. The concentration of the solution was almost unchanged in the dark condition (light off). The results indicate that m-TiO_2 exhibits the best photocatalytic performance over 5 h. Figure 10b shows the dye photodegradation behavior as it discolors from dark blue to nearly transparent, especially in the presence of TNT and m-TiO_2. The concentration of MB solution with TNT and m-TiO_2 rapidly decreased after the first hour of UV irradiation, and then the solution concentration further decreased during the following 4 h.

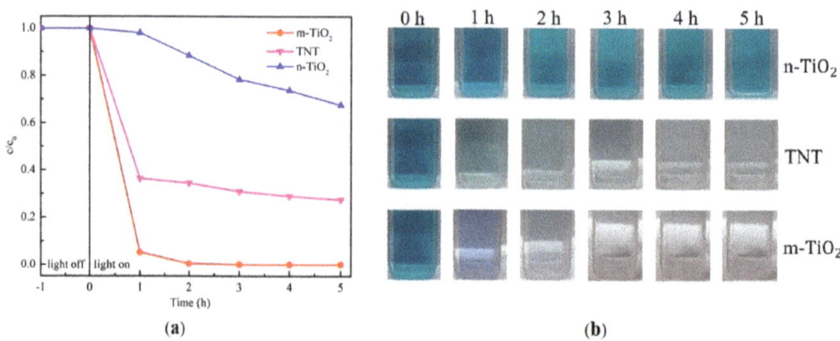

Figure 10. (**a**) Normalized concentration of methylene blue (MB) solution with pure photocatalysts and (**b**) the color change of the MB solution.

Table 5. The decomposition of MB solution with pure photocatalysts.

	Concentration of MB Solution (mg/mL)					
	0 h	**1 h**	**2 h**	**3 h**	**4 h**	**5 h**
m-TiO_2	1.51	0.02	0.00	0.00	0.00	0.00
n-TiO_2	1.51	1.32	1.18	1.13	1.05	0.97
TNT	1.51	0.51	0.5	0.45	0.43	0.40

In contrast, when the TiO_2 photocatalyst was added to the cement (Figure 11), there was a significant decrease in the photocatalytic effect due to the lack of direct contact between the photocatalyst and the organic pollutants. The decomposition of the MB solution with cement paste samples is shown in Table 6. Before UV irradiation, the rapid decrease in the concentration of MB solution under dark conditions was due to the absorption of MB on the surface of porous OPC, as already observed in Figure 9. The normalized concentration of TNT (P) was lower than n-TiO_2 (P); however, it was higher than that of m-TiO_2 (P). This result was similar to that observed in the experiments conducted using only the photocatalysts (Figure 10). This is possibly owing to the hollow tubular structure of TNT, which provides a shorter and broader diffusion path for the dye contaminants from the solution to enter the reactive area of the photocatalyst compared to n-TiO_2 [52]. m-TiO_2 showed the best results. This is similar to the results reported by Folli et al. [53,54], which they attributed to two factors: first, m-TiO_2 is more easily dispersed in the cement paste than the nanosized materials (TNT or n-TiO_2), which are more significantly agglomerated and difficult to disperse by physical stirring when mixed with cement. Second, large molecules, such as MB, cannot easily penetrate the cement, and thus the well-dispersed m-TiO_2 provides more available surface areas to adsorb and react with the large organic pollutants.

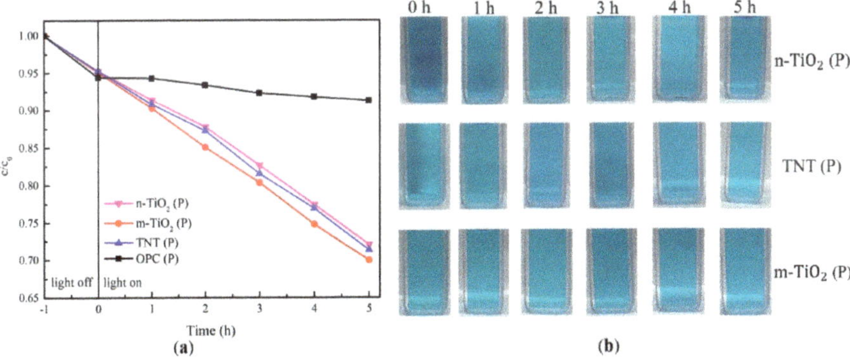

Figure 11. (**a**) Normalized concentration of MB solution with cement paste containing photocatalysts and (**b**) the color change of the MB solution.

Table 6. The decomposition of MB solution with cement paste samples.

	Concentration of MB Solution (mg/mL)					
	0 h	1 h	2 h	3 h	4 h	5 h
OPC (P)	1.41	1.41	1.39	1.38	1.37	1.36
m-TiO_2 (P)	1.42	1.32	1.24	1.15	1.09	1.03
n-TiO_2 (P)	1.42	1.39	1.33	1.25	1.17	1.08
TNT (P)	1.42	1.37	1.32	1.23	1.16	1.09

Recently, researchers have combined photocatalytic properties with cement-based materials by coating method. Feng et al. [14,55] fabricated TiO_2 film on the cement paste surface utilizing the smear and spray method and found that the cement paste with the TiO_2 was capable of degrading almost all of the dye solution. Shen et al. [56] prepared the concrete with an ultra-smooth surface covered with nano-TiO_2 and tested the photocatalytic effect, and the result showed that the degradation rate of MB solution increased with increasing UV irradiation. Although some other studies using the coating method showed a better photocatalytic effect than our results, TNT fully has potential as a reinforcing material since it not only provides a photocatalytic effect on the cement material but also improves the mechanical strength of cement paste owing to its large aspect ratio. In this study, to quantitatively evaluate the effect of the TiO_2 particles and nanotubes on the photocatalytic performance of cement paste, a specimen with a flat surface was utilized. To further improve the ability of the paste to degrade organic dyes, the contact surface area of the pastes could be increased. In this case, the use of TNT, which can improve the tensile strength of the cement paste [35], would be beneficial.

4. Conclusions

This study aimed to investigate the photocatalytic performance of TNT-reinforced cement paste and cement pastes modified with m-TiO_2 and n-TiO_2 for comparison. The TNT used in our work was prepared by the hydrothermal method using anatase TiO_2 as the raw material. From the TEM results, it was evident that a hollow, open-ended, tubular structure was successfully synthesized. The XRD and TG results indicated that the TNT provided the most significant contribution to the hydration process of the cement, which can be beneficial to the mechanical development of cement pastes. After the same curing time, the TNT-reinforced cement paste produced the largest amount of hydration products. Regarding the photocatalytic performance, due to the wrapping of the hardening cement around the photocatalyst, the photocatalytic effects of the TNT-reinforced cement paste and the other two modified cement pastes were more limited than that of the photocatalysts alone. The cement paste containing m-TiO_2 still showed the best capacity for photocatalytic degradation of MB solution, with about 30% of the MB solution was degraded. The TNT-reinforced cement paste followed closely behind. In summary, TNT-reinforced cement paste was able to produce more hydration products during the hydration process. Simultaneously, the tubular structure of TNT provided a better photocatalytic effect than n-TiO_2, which is commonly used in cement photocatalytic studies. This offers valuable support for further research on the application of TNT in environmentally friendly cementitious materials.

Author Contributions: Conceptualization, E.Z.N. and S.B.; formal analysis, J.L. and H.J.; resources, M.L. and J.H.K.; writing—origianl draft preparation, J.L. and H.J.; writing—review and editing, M.L., E.Z.N. and S.B.; supervision, J.H.K., S.J.K., K.M.L. and S.B. All authors have read and agreed to the published version of the manuscript.

Funding: This work was supported by the Korea Institute of Energy Technology Evaluation and Planning (KETEP) and the Ministry of Trade, Industry and Engergy (MOTIE), Republic of Korea. (NO. 20181110200070).

Conflicts of Interest: The authors declare no conflict of interest.

References

1. Nakata, K.; Fujishima, A. TiO_2 photocatalysis: Design and applications. *J. Photochem. Photobiol. C* **2012**, *13*, 169–189. [CrossRef]
2. Schneider, J.; Matsuoka, M.; Takeuchi, M.; Zhang, J.; Horiuchi, Y.; Anpo, M.; Bahnemann, D.W. Understanding TiO_2 photocatalysis: Mechanisms and materials. *Chem. Rev.* **2014**, *114*, 9919–9986. [CrossRef]
3. Tang, Y.; Zhang, Y.; Deng, J.; Wei, J.; Le Tam, H.; Chandran, B.K.; Dong, Z.; Chen, Z.; Chen, X. Mechanical force-driven growth of elongated bending TiO_2-based nanotubular materials for ultrafast rechargeable lithium ion batteries. *Adv. Mater.* **2014**, *26*, 6111–6118. [CrossRef] [PubMed]
4. Subramanian, V.; Karki, A.; Gnanasekar, K.I.; Eddy, F.P.; Rambabu, B. Nanocrystalline TiO_2 (anatase) for Li-ion batteries. *J. Power Sources* **2006**, *159*, 186–192. [CrossRef]
5. Roy, P.; Kim, D.; Lee, K.; Spiecker, E.; Schmuki, P. TiO_2 nanotubes and their application in dye-sensitized solar cells. *Nanoscale* **2010**, *2*, 45–59. [CrossRef] [PubMed]
6. Tan, B.; Wu, Y. Dye-Sensitized Solar Cells Based on Anatase TiO_2 Nanoparticle/Nanowire Composites. *J. Phys. Chem. B* **2006**, *110*, 15932–15938. [CrossRef] [PubMed]
7. Hashimoto, K.; Irie, H.; Fujishima, A. TiO_2 Photocatalysis: A Historical Overview and Future Prospects. *Jpn. J. Appl. Phys.* **2005**, *44*, 8269–8285. [CrossRef]
8. Zhang, J.; Zhou, P.; Liu, J.; Yu, J. New understanding of the difference of photocatalytic activity among anatase, rutile and brookite TiO_2. *Phys. Chem. Chem. Phys.* **2014**, *16*, 20382–20386. [CrossRef]
9. Macphee, D.E.; Folli, A. Photocatalytic concretes—The interface between photocatalysis and cement chemistry. *Cem. Concr. Res.* **2016**, *85*, 48–54. [CrossRef]

10. Hamidi, F.; Aslani, F. TiO_2-based Photocatalytic Cementitious Composites: Materials, Properties, Influential Parameters, and Assessment Techniques. *Nanomaterials* **2019**, *9*, 1444. [CrossRef]
11. Yang, J.; Wang, G.; Wang, D.; Liu, C.; Zhang, Z. A self-cleaning coating material of TiO_2 porous microspheres/cement composite with high-efficient photocatalytic depollution performance. *Mater. Lett.* **2017**, *200*, 1–5. [CrossRef]
12. Yuenyongsuwan, J.; Sinthupinyo, S.; O'Rear, E.A.; Pongprayoon, T. Hydration accelerator and photocatalyst of nanotitanium dioxide synthesized via surfactant-assisted method in cement mortar. *Cem. Concr. Compos.* **2019**, *96*, 182–193. [CrossRef]
13. Chen, J.; Poon, C.-S. Photocatalytic Cementitious Materials: Influence of the Microstructure of Cement Paste on Photocatalytic Pollution Degradation. *Environ. Sci. Technol.* **2009**, *43*, 8948–8952. [CrossRef] [PubMed]
14. Feng, S.; Liu, F.; Fu, X.; Peng, X.; Zhu, J.; Zeng, Q.; Song, J. Photocatalytic performances and durability of TiO_2/cement composites prepared by a smear method for organic wastewater degradation. *Ceram. Int.* **2019**, *45*, 23061–23069. [CrossRef]
15. Silvestre, J.; Silvestre, N.; de Brito, J. Review on concrete nanotechnology. *Eur. J. Environ. Civ. Eng* **2015**, *20*, 455–485. [CrossRef]
16. Cao, M.; Ming, X.; He, K.; Li, L.; Shen, S. Effect of Macro-, Micro- and Nano-Calcium Carbonate on Properties of Cementitious Composites—Review. *Materials* **2019**, *12*, 781. [CrossRef]
17. Mendoza Reales, O.A.; Dias Toledo Filho, R. A review on the chemical, mechanical and microstructural characterization of carbon nanotubes-cement based composites. *Constr. Build. Mater.* **2017**, *154*, 697–710. [CrossRef]
18. Chen, J.; Kou, S.-C.; Poon, C.-S. Hydration and properties of nano-TiO_2 blended cement composites. *Cem. Concr. Compos.* **2012**, *34*, 642–649. [CrossRef]
19. Li, Z.; Wang, J.; Li, Y.; Yu, X.; Han, B. Investigating size effect of anatase phase nano TiO_2 on the property of cement-based composites. *Mater. Res. Express* **2018**, *5*, 1591. [CrossRef]
20. Zhang, R.; Cheng, X.; Hou, P.; Ye, Z. Influences of nano-TiO_2 on the properties of cement-based materials: Hydration and drying shrinkage. *Constr. Build. Mater.* **2015**, *81*, 35–41. [CrossRef]
21. Sikora, P.; Abd Elrahman, M.; Stephan, D. The Influence of Nanomaterials on the Thermal Resistance of Cement-Based Composites—A Review. *Nanomaterials* **2018**, *8*, 465. [CrossRef] [PubMed]
22. Yoo, D.-Y.; You, I.; Zi, G.; Lee, S.-J. Effects of carbon nanomaterial type and amount on self-sensing capacity of cement paste. *Measurement* **2019**, *134*, 750–761. [CrossRef]
23. Siad, H.; Lachemi, M.; Sahmaran, M.; Mesbah, H.A.; Hossain, K.A. Advanced engineered cementitious composites with combined self-sensing and self-healing functionalities. *Constr. Build. Mater.* **2018**, *176*, 313–322. [CrossRef]
24. Ruot, B.; Plassais, A.; Olive, F.; Guillot, L.; Bonafous, L. TiO_2-containing cement pastes and mortars: Measurements of the photocatalytic efficiency using a rhodamine B-based colourimetric test. *Sol. Energy* **2009**, *83*, 1794–1801. [CrossRef]
25. Jimenez-Relinque, E.; Llorente, I.; Castellote, M. TiO_2 cement-based materials: Understanding optical properties and electronic band structure of complex matrices. *Catal. Today* **2017**, *287*, 203–209. [CrossRef]
26. Folli, A.; Pade, C.; Hansen, T.B.; De Marco, T.; Macphee, D.E. TiO_2 photocatalysis in cementitious systems: Insights into self-cleaning and depollution chemistry. *Cem. Concr. Res.* **2012**, *42*, 539–548. [CrossRef]
27. Seo, D.; Yun, T.S. NOx removal rate of photocatalytic cementitious materials with TiO_2 in wet condition. *Build. Environ.* **2017**, *112*, 233–240. [CrossRef]
28. Zhang, Z.; Wang, C.-C.; Zakaria, R.; Ying, J.Y. Role of Particle Size in Nanocrystalline TiO_2-Based Photocatalysts. *J. Phys. Chem. B* **1998**, *102*, 10871–10878. [CrossRef]
29. Park, J.; Suh, H.; Woo, S.M.; Jeong, K.; Seok, S.; Bae, S. Assessment of neutron shielding performance of nano-TiO_2-incorporated cement paste by Monte Carlo simulation. *Prog. Nucl. Energy* **2019**, *117*, 103043. [CrossRef]
30. Kasuga, T.; Hiramatsu, M.; Hoson, A.; Sekino, T.; Niihara, K. Formation of Titanium Oxide Nanotube. *Langmuir* **1998**, *14*, 3160–3163. [CrossRef]
31. Smith, Y.R.; Kar, A.; Subramanian, V. Investigation of Physicochemical Parameters That Influence Photocatalytic Degradation of Methyl Orange over TiO_2 Nanotubes. *Ind. Eng. Chem. Res.* **2009**, *48*, 10268–10276. [CrossRef]

32. Zhou, X.; Liu, N.; Schmuki, P. Photocatalysis with TiO_2 Nanotubes: "Colorful" Reactivity and Designing Site-Specific Photocatalytic Centers into TiO2 Nanotubes. *ACS Catal.* **2017**, *7*, 3210–3235. [CrossRef]
33. Ananthakumar, S.; Ramkumar, J.; Babu, S.M. Semiconductor nanoparticles sensitized TiO_2 nanotubes for high efficiency solar cell devices. *Renew. Sustain Energy Rev.* **2016**, *57*, 1307–1321. [CrossRef]
34. Lv, H.; Wang, Y.; Pan, L.; Zhang, L.; Zhang, H.; Shang, L.; Qu, H.; Li, N.; Zhao, J.; Li, Y. Patterned polyaniline encapsulated in titania nanotubes for electrochromism. *Phys. Chem. Chem. Phys.* **2018**, *20*, 5818–5826. [CrossRef] [PubMed]
35. Jee, H.; Park, J.; Zalnezhad, E.; Jeong, K.; Woo, S.M.; Seok, S.; Bae, S. Characterization of Titanium Nanotube Reinforced Cementitious Composites: Mechanical Properties, Microstructure, and Hydration. *Materials* **2019**, *12*, 1617. [CrossRef] [PubMed]
36. Kasuga, T.; Hiramatsu, M.; Hoson, A.; Sekino, T.; Niihara, K. Titania Nanotubes Prepared by Chemical Processing. *Adv. Mater.* **1999**, *11*, 1307–1311. [CrossRef]
37. Liu, N.; Chen, X.; Zhang, J.; Schwank, J.W. A review on TiO_2-based nanotubes synthesized via hydrothermal method: Formation mechanism, structure modification, and photocatalytic applications. *Catal. Today* **2014**, *225*, 34–51. [CrossRef]
38. Scrivener, K.; Snellings, R.; Lothenbach, B. *A Practical Guide to Microstructural Analysis of Cementitious Materials*; CRC Press: Boca Raton, FL, USA, 2018.
39. Wei, X.; Zhu, G.; Fang, J.; Chen, J. Synthesis, Characterization, and Photocatalysis of Well-Dispersible Phase-Pure Anatase TiO_2 Nanoparticles. *Int. J. Photoenergy* **2013**, *2013*, 726872. [CrossRef]
40. Li, D.; Song, H.; Meng, X.; Shen, T.; Sun, J.; Han, W.; Wang, X. Effects of Particle Size on the Structure and Photocatalytic Performance by Alkali-Treated TiO_2. *Nanomaterials* **2020**, *10*, 546. [CrossRef]
41. Lopez Zavala, M.A.; Lozano Morales, S.A.; Avila-Santos, M. Synthesis of stable TiO_2 nanotubes: Effect of hydrothermal treatment, acid washing and annealing temperature. *Heliyon* **2017**, *3*, e00456. [CrossRef]
42. Wang, D.; Zhou, F.; Liu, Y.; Liu, W. Synthesis and characterization of anatase TiO_2 nanotubes with uniform diameter from titanium powder. *Mater. Lett.* **2008**, *62*, 1819–1822. [CrossRef]
43. Jiang, F.; Zheng, S.; An, L.; Chen, H. Effect of calcination temperature on the adsorption and photocatalytic activity of hydrothermally synthesized TiO_2 nanotubes. *Appl. Surf. Sci.* **2012**, *258*, 7188–7194. [CrossRef]
44. Harsha, N.; Ranya, K.R.; Babitha, K.B.; Shukla, S.; Biju, S.; Reddy, M.L.; Warrier, K.G. Hydrothermal processing of hydrogen titanate/anatase-titania nanotubes and their application as strong dye-adsorbents. *J. Nanosci. Nanotechnol.* **2011**, *11*, 1175–1187. [CrossRef] [PubMed]
45. Zhou, C.-H.; Xu, S.; Yang, Y.; Yang, B.-C.; Hu, H.; Quan, Z.-C.; Sebo, B.; Chen, B.-l.; Tai, Q.-D.; Sun, Z.-H.; et al. Titanium dioxide sols synthesized by hydrothermal methods using tetrabutyl titanate as starting material and the application in dye sensitized solar cells. *Electrochim. Acta* **2011**, *56*, 4308–4314. [CrossRef]
46. Chen, J.; Wang, H.; Wei, X.; Zhu, L. Characterization, properties and catalytic application of TiO_2 nanotubes prepared by ultrasonic-assisted sol-hydrothermal method. *Mater. Res. Bull.* **2012**, *47*, 3747–3752. [CrossRef]
47. Wang, Y.Q.; Hu, G.Q.; Duan, X.F.; Sun, H.L.; Xue, Q.K. Microstructure and formation mechanism of titanium dioxide nanotubes. *Chem. Phys. Lett.* **2002**, *365*, 427–431. [CrossRef]
48. Snehal, K.; Das, B.B.; Akanksha, M. Early age, hydration, mechanical and microstructure properties of nano-silica blended cementitious composites. *Constr. Build. Mater.* **2020**, *233*, 117212. [CrossRef]
49. Suh, H.; Jee, H.; Kim, J.; Kitagaki, R.; Ohki, S.; Woo, S.; Jeong, K.; Bae, S. Influences of rehydration conditions on the mechanical and atomic structural recovery characteristics of Portland cement paste exposed to elevated temperatures. *Constr. Build. Mater.* **2020**, *235*, 117453. [CrossRef]
50. Lu, Z.; Wang, Q.; Yin, R.; Chen, B.; Li, Z. A novel TiO_2/foam cement composite with enhanced photodegradation of methyl blue. *Constr. Build. Mater.* **2016**, *129*, 159–162. [CrossRef]
51. Cerro-Prada, E.; Garcia-Salgado, S.; Quijano, M.A.; Varela, F. Controlled Synthesis and Microstructural Properties of Sol-Gel TiO_2 Nanoparticles for Photocatalytic Cement Composites. *Nanomaterials* **2018**, *9*, 26. [CrossRef]
52. Macak, J.M.; Zlamal, M.; Krysa, J.; Schmuki, P. Self-organized TiO_2 nanotube layers as highly efficient photocatalysts. *Small* **2007**, *3*, 300–304. [CrossRef] [PubMed]
53. Folli, A.; Jakobsen, U.H.; Guerrini, G.L.; Macphee, D.E. Rhodamine B Discolouration on TiO_2 in the Cement Environment: A Look at Fundamental Aspects of the Self-cleaning Effect in Concretes. *J. Adv. Oxid. Technol.* **2009**, *12*, 126–133. [CrossRef]

54. Folli, A.; Pochard, I.; Nonat, A.; Jakobsen, U.H.; Shepherd, A.M.; Macphee, D.E. Engineering Photocatalytic Cements: Understanding TiO_2 Surface Chemistry to Control and Modulate Photocatalytic Performances. *J. Am. Ceram. Soc.* **2010**, *93*, 3360–3369. [CrossRef]
55. Feng, S.; Song, J.; Liu, F.; Fu, X.; Guo, H.; Zhu, J.; Zeng, Q.; Peng, X.; Wang, X.; Ouyang, Y.; et al. Photocatalytic properties, mechanical strength and durability of TiO_2/cement composites prepared by a spraying method for removal of organic pollutants. *Chemosphere* **2020**, *254*, 126813. [CrossRef] [PubMed]
56. Shen, W.; Zhang, C.; Li, Q.; Zhang, W.; Cao, L.; Ye, J. Preparation of titanium dioxide nano particle modified photocatalytic self-cleaning concrete. *J. Clean. Prod.* **2015**, *87*, 762–765. [CrossRef]

 materials

Article

CO_2 Uptake and Physicochemical Properties of Carbonation-Cured Ternary Blend Portland Cement–Metakaolin–Limestone Pastes

Rizwan Hameed [1], Joonho Seo [1], Solmoi Park [2], Issam T. Amr [3] and H.K. Lee [1,*]

1 Department of Civil and Environmental Engineering, Korea Advanced Institute of Science and Technology, 291 Daehak-ro, Yuseong-gu, Daejeon 34141, Korea; rhameed@kaist.ac.kr (R.H.); junhoo11@kaist.ac.kr (J.S.)
2 Department of Civil Engineering, Pukyong National University, 45 Yongso-ro, Nam-gu, Busan 48513, Korea; solmoi.park@pknu.ac.kr
3 Carbon Management Division, Research & Development Center, Saudi Aramco, Dhahran 31311, Saudi Arabia; issam.amr@aramco.com
* Correspondence: haengki@kaist.ac.kr; Tel.: +82-42-350-3623

Received: 14 September 2020; Accepted: 15 October 2020; Published: 19 October 2020

Abstract: The feasibility of carbonation curing of ternary blend Portland cement–metakaolin–limestone was investigated. Portland cement was substituted by the combination of metakaolin and limestone at levels of 15%, 30%, and 45% by the mass. The ternary blends were cured with four different combinations of ambient and carbonation curing. The mechanical property, CO_2 uptake, and mineralogical variations of the ternary blend pastes were investigated by means of compressive strength test, thermogravimetric analysis, and X-ray diffractometry. In addition, volume of permeable voids and sorptivity of the ternary blends were also presented to provide a fundamental idea of the pore characteristics of the blends. The test results showed that the increasing amount of metakaolin and limestone enhanced the CO_2 uptake, reaching 20.7% for the sample with a 45% cement replacement level at 27 d of carbonation. Meanwhile, the compressive strength of the samples was reduced up to 65% upon excessive incorporation of metakaolin and limestone. The samples with a replacement level of 15% exhibited a comparable strength and volume of permeable voids to those of the sample without substitution, proving that the ternary blend Portland cement–metakaolin–limestone can be a viable option toward the development of eco-friendly binders.

Keywords: Portland cement; limestone; metakaolin; carbonation curing; CO_2 uptake; LC^3

---

1. Introduction

Portland cement (PC) has a global production of more than 4,000 Mt per year, which is 25 times higher than what it was in 1950 [1,2]. The huge production of PC accompanies a significant CO_2 emission, which accounts for nearly 5–8% of entire CO_2 emissions [2,3]. Of these emissions, 50–60% are attributed to the calcination of limestone during the manufacturing process of PC, while the other share is from the burning of fossil fuels [1]. Due to the CO_2 emissions, from the cement industry and other sources, the concentration of CO_2 in the Earth's atmosphere has been increased from 280 ppm, in the preindustrial era, to 414.5 ppm in 2020 [4,5]. Different solutions have been adopted to reduce the carbon footprint of the construction industry such as partial replacement of PC with waste materials [6–8], alternative clinkers [9], and the use of alkali-activated binders [10–13]. Recently, the carbonation curing of PC-based materials has become a focus of attention as a potential means of reducing the atmospheric CO_2 concentration [14,15].

Carbonation curing is the introduction of elevated CO_2 concentrations into the fresh or premature state concrete [3,15,16]. Although the idea of carbonation curing was first proposed in the 1970s [17,18], it was reluctantly adopted afterwards. However, in recent years amid increasing concerns towards global warming, the interest in the carbonation curing of PC-based materials has been reignited [3,14,15,19–21]. In carbonation curing, CO_2 is exposed to both anhydrates and hydrates. Portlandite is carbonated more significantly as compared to other hydrates till 28 d of carbonation, while the carbonation of C-S-H continues even after 28 d of carbonation. Anhydrous materials carbonate barely in comparison to hydrates [22]. Carbonation curing exhibits certain advantages such as rapid gain in mechanical strength at an early age and enhanced durability [3,15,16]. It is also considered as a potential approach to sequester CO_2 in PC-based materials [15]. In contrast to conventional steam curing, which requires elevated temperatures of 50–70 °C with higher humidity levels, carbonation curing is less energy-intensive [14].

Supplementary cementitious materials (SCMs) are used to reduce the clinker factor of PC due to their lower energy inputs than PC [6,8]. Conventional SCMs have limited amounts to replace the clinkers as global PC alternatives. Slag is available at around 10% of PC production and this proportion is not expected to be increased in the future. Fly ash is produced at around 30% of PC production and is anticipated to be reduced due to the growing environmental concerns related to the coal power generation [6,7,23,24]. On the other hand, clays have a wide availability globally [24]. Metakaolin is formed by dehydroxylation of clays, rich in kaolinite, when calcined at 700–850 °C [24–26]. Metakaolin is highly pozzolanic and forms aluminum-containing phases in PC-based systems [24–28]. In addition, carbo-aluminates may also be formed in the presence of freely available carbonates [26]. Limestone is also available worldwide which accelerates the hydration of PC by providing nucleation sites due to additional surface area [29,30]. In addition, the incorporated limestone forms monocarboaluminate and hemicarboaluminate, which helps in the stabilization of ettringite [29].

PC-based systems containing SCMs more than a threshold level, i.e., around 30% of PC, showed declined mechanical performances at an early age [28]. To enhance the early mechanical properties of systems with higher amounts of substitution, an economic solution can be the addition of limestone in the PC-based systems since the limestone provides additional nucleation sites and promotes early hydration [28,31]. Limestone reacts with alumina-containing phases and produce carbonate-AFm phases, yet this reaction pathway is blocked if available alumina is limited in the system [31]. As metakaolin provides a high amount of alumina, combination of limestone and metakaolin can provide better properties even at higher levels of PC substitution [28]. Ternary binder, referred to as limestone calcined clay cements (LC^3) is gaining attention from the last decade [24,32–36]. LC^3 can replace PC up to 45% while maintaining comparable performances [28]. Due to the utilization of abundantly available materials in the LC^3 system, it has the potential to replace PC at a global scale. Pilot production of LC^3 has been tested in some countries [37–39].

Carbonation curing of PC-based materials has been extensively studied during the last decade [3,14,16,19,21]. Studies on the effect of mineral admixtures on carbonation curing suggest that mineral admixtures can enhance CO_2 uptake [20,40–42], particularly a study conducted by Zhang et al. [20] showed that higher levels (50%) of substitution by fly ash can further improve the CO_2 uptake capacity. To the authors' knowledge, however, the carbonation curing of the LC^3 systems has never been tested. This approach can give two-fold benefits, i.e., high substitution of PC and sequestration of CO_2. With this background, the present research work was aimed at the investigation on the carbonation curing of ternary blends of PC–metakaolin–limestone. Samples of pure PC systems and ternary blends were prepared at the substitution levels of 15%, 30%, and 45% by mass. These samples were tested under four different curing regimes of ambient and carbonation curing. The CO_2 uptake and physicochemical properties of the blends under various combinations of ambient and carbonation curing were evaluated by compressive strength, X-ray diffraction (XRD), thermogravimetry analysis (TGA), volume of permeable voids, and sorptivity tests.

2. Experimental Program

2.1. Materials and Sample Preparation

Portland cement (PC), conforming to the ASTM C150, was obtained from Sungshin Cement Co., Ltd., South Korea. Metakaolin, branded as MKC100, was supplied by Nycon Materials Co., Ltd., South Korea, and commercially available limestone powder was procured from Duksan Reagents Company, Co., Ltd., South Korea. The chemical composition determined by the qualitative X-ray fluorescence (XRF) analysis is presented in Table 1. Note that the loss-on-ignition was not tabulated here since Table 1 shows qualitative XRF results, yet it is anticipated that the limestone might exhibit 40–50 wt % of loss-on-ignition due to the presence of carbonates. Basic properties of PC used in this study are presented in Table 2.

Table 1. Chemical composition of the raw materials used in this study.

wt %	Portland Cement	Metakaolin	Limestone
CaO	62.50	0.92	99.20
SiO_2	21.00	50.10	0.08
Al_2O_3	5.90	38.40	0.01
Fe_2O_3	3.20	5.69	0.03
MgO	0.11	0.11	0.28
R_2O	0.80	0.62	0.01
SO_3	2.10	0.05	0.01
TiO_2	0.38	3.45	-
P_2O_3	0.14	0.09	0.01
Mn_2O_5	0.10	0.01	-
SrO	0.15	0.06	0.23

Table 2. Properties of Portland cement provided by manufacturer.

Portland Cement	
Fineness	3450 cm^2/g
Initial setting time	225 min
Final setting time	345 min
Density	3.14 g/cm^2
Standard compressive strength development	
3 day	15.6 MPa
7 day	25.2 MPa
28 day	51.2 MPa

Mix proportion of the samples are shown in Table 3. Four different mixtures were prepared: one with PC only, while the other three mixtures had different replacement levels of PC by the combination of metakaolin and limestone. In these mixtures, the weight ratio of metakaolin and limestone powder was kept as 2:1 based on previous studies [28,31,32,35]. PC was replaced with the combinations of metakaolin and limestone at varying weight percentages of 15%, 30%, and 45%. The sample ID was designated based on the replacement levels of PC, for instance, ML15 indicates a mixture replacing PC with 10 wt % of metakaolin and 5 wt % of limestone. Paste samples with a constant water-to-binder ratio of 0.5 were fabricated. Dry materials were mixed for 3 min. After dry mixing, water was added and further mixed for 5 min in order to make uniform and homogenized pastes. Fresh slurry was poured into 50-mm cubes, 40 × 40 × 160 mm prisms, and Φ100 × 50 mm cylinders for compressive strength tests, carbonation degree measurement, and durability tests including volume of permeable voids and sorptivity, respectively. All samples were sealed with plastic wraps immediately after casting to avoid the evaporation of water.

Table 3. Mixture proportion of the samples expressed as mass ratio.

Sample ID	Portland Cement	Metakaolin	Limestone	Water/Powder [1] Ratio
OPC	1	0.00	0.00	0.5
ML15	0.85	0.10	0.05	0.5
ML30	0.70	0.20	0.10	0.5
ML45	0.55	0.30	0.15	0.5

[1] Powder denotes the summation of Portland cement, metakaolin, and limestone.

2.2. Curing Conditions and Test Methods

The curing regimes used in this study are summarized in Figure 1. All samples were commonly allowed 24 h of initial curing at 20 °C. After initial curing, the samples underwent four different curing regimes. W-series samples were cured at ambient conditions for 28 d. L-, M-, and H-series samples were carbonation-cured for 6 h, 13 d, and 27 d, respectively. After carbonation curing, L- and M-series samples were cured at ambient conditions until 28 d. Ambient conditions here describe the sealed curing of the samples at 23 ± 2 °C. For carbonation curing, CO_2 concentration, temperature, and relative humidity were 10%, 20 °C, and 60%, respectively. Samples for chemical analyses were crushed and sieved by a 3 mm sieve before carbonation in order to get a uniform carbonation regardless of location. It should be noted here that complete nomenclature of the samples includes curing condition, for instance, ML30-M indicates a set of samples with a mixture of ML30 (70% PC, 20% metakaolin, and 10% limestone) with the curing regime following M-series.

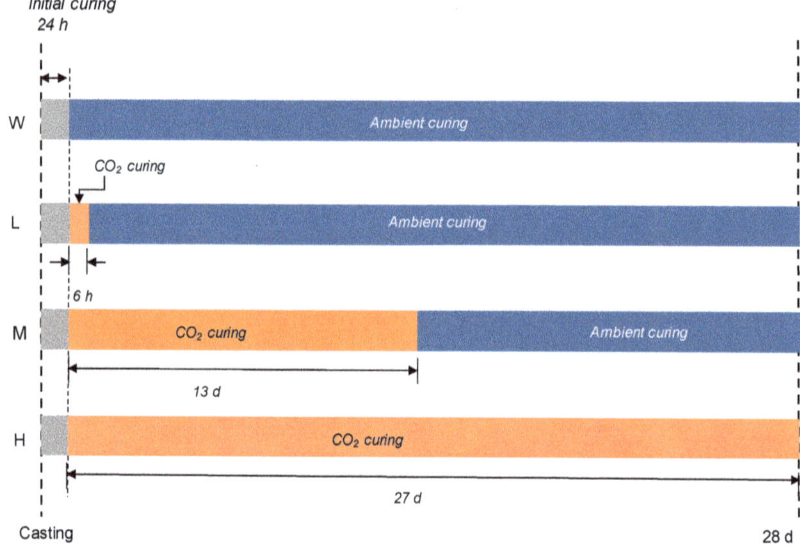

Figure 1. Curing regimes used in this study.

The compressive strength of samples at 28 d of curing was determined in accordance with ASTM C109 [43] by using a 250 kN universal testing machine at a loading rate of 0.5 MPa/s. The representative strength value was averaged from three replicas. The carbonation degree of the samples was measured by spraying a 1% phenolphthalein indicator onto the cross-section of the carbonated prisms [19,44]. The carbonation degree for the L-, M-, and H-series samples was determined immediately after carbonation curing, i.e., after 6 h, 13 d, and 27 d of carbonation for L-, M-, and H-series samples, respectively. In addition, carbonation degree of the samples at 3 and 7 d of carbonation was additionally provided to observe the progressive carbonation degree. The carbonation degree was determined by

Equation (1) [19,44]. The pH value of pore solution in the samples was measured using a suspension made with powdered sample and deionized water. A 2 g measure of a sample was immersed in 10 mL of deionized water and stirred at 200 rpm for 10 min before pH measurement.

$$\text{Carbonation degree } (\%) = \frac{\text{Carbonated area}}{\text{Total cross} - \text{sectional area}} \times 100 \tag{1}$$

Characterization of mineral phases was carried out by means of XRD analysis at 28 d of curing. The XRD was performed using an Empyrean instrument under a CuKα radiation with current and voltage of 30 mA and 40 kV, respectively. The XRD patterns were collected in a 2θ° range of 5–65 2θ° with a step size of 0.026 2θ° and a step time of 1.58s. Thermal evaluation of hydrates present in the samples was done by TGA at 28 d of curing. The weight variation of the samples was monitored in the temperature range of 25–1000 °C with a fixed heating rate of 10 °C/min. N_2 gas was constantly injected during the measurement so as to avoid oxidization of the samples. CO_2 uptake was measured from TGA curves. Percentage mass loss was quantified by the tangential method to calculate the mass loss associated with the decarbonation of $CaCO_3$ [36,45].

Sorptivity test for all samples was performed at 28 d of curing, in accordance with the ASTM C1585 [46]. The initial mass of the samples was determined after sealing the side surfaces. Then, these samples were immersed in 3 mm deep water. Mass of the samples in surface dry condition was frequently measured at time intervals specified in ASTM C1585. The absorption (I) was measured following Equation (2) [46].

$$I = \frac{m_t}{a \times d} \tag{2}$$

where m_t, a, and d denote variation of sample mass (g), sample area exposed to water (mm^2), and density of water (g/mm^3), respectively. Initial and final sorptivity coefficients were determined as the slope of the best fit line between I and the square root of time (s$^{0.5}$), from 1 min to 6 h, and 1 d to 7 d, respectively. Volume of permeable voids of the samples were tested at 28 d of curing, in accordance with the ASTM C642 [47].

3. Results and Discussion

3.1. Compressive Strength

The compressive strength of the samples after 28 d of curing is shown in Figure 2. The compressive strength of the W-series samples decreased as PC replacement level increased. Comparable compressive strength to the OPC-W sample was observed in the ML15-W sample. This agrees with the outcomes reported in a previous research [31]. While with the higher PC replacement level, a declined mechanical behavior was observed in comparison with the OPC-W sample. The ML30-W and ML45-W samples exhibited compressive strengths of 47.8 and 40.1 MPa, respectively. Previous studies of PC-metakaolin-limestone blends presented identical results to what reported in the present study [28,31]. The OPC samples exhibited comparable compressive strength for all curing regimes, i.e., ambient and carbonation curing regimes. Chen et al. [48] described that carbonation-cured PC-systems show improvement in compressive strengths at early ages while the positive effect weakens with longer age. Other than this, the optimal pre-curing duration before start of carbonation curing, also depends with the carbonation duration; it decreases with the increase in carbonation curing duration [48]. Blended samples showed more prominent behavior with an increase in carbonation duration. L-series samples showed comparable compressive strengths with their W-series counterparts. The duration of carbonation affected the mechanical strength. Among blended samples, compressive strength of the ML15 samples was observed to be comparable for the L- and M-series samples, but showed a notable reduction in strength of the H-series sample. The ML30 and ML45 samples showed declined compressive strengths upon an increment in the carbonation durations. The reduction in the compressive strength of the blended samples upon carbonation can be attributed to their increased overall porosity (explained in Section 3.5) and reduced amount of portlandite upon higher replacement of PC [42]. Zhang et al. [20] described that pozzolanic reaction is hindered by the early carbonation curing. Carbonation reaction reduces the alkalinity by consuming portlandite which is essential for pozzolanic reaction. This effect is more prominent with an increase in the carbonation time and higher substitution levels. Due to this hindrance in pozzolanic reaction, most portion of the SCM acts as a filler material in the paste which might be the reason for lower compressive strengths of blended pastes for longer carbonation durations.

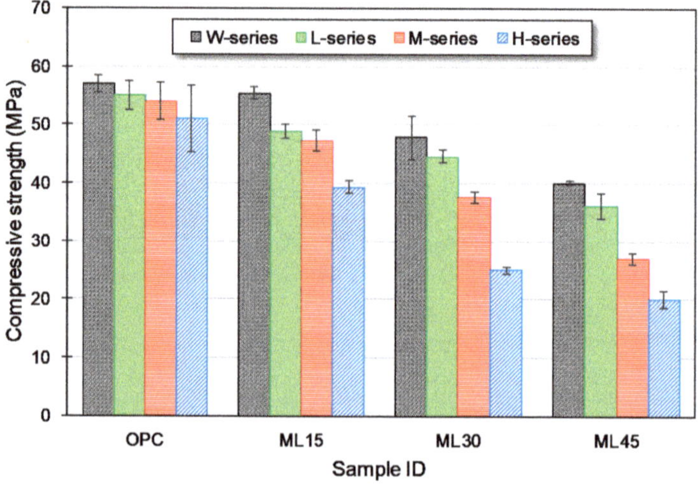

Figure 2. Compressive strength of samples at 28 d of curing.

3.2. *Carbonation Degree and pH Variation*

Carbonation degree of the samples is shown in Figure 3. After 6 h of carbonation curing, all samples showed almost similar extent of carbonation degree, i.e., 4–7%. With an increase in the carbonation duration, samples exhibited different aspect of carbonation degree. The OPC samples showed a remarkable increase in the carbonation degree from 6 h to 2 d of carbonation. The carbonation degree of the OPC samples kept increasing at a steady rate until 27 d of carbonation. At 27 d of carbonation curing, 80% of the cross-sectional area of the OPC samples was carbonated. The carbonation degrees of the OPC samples were similar to those reported in previous works [19,44]. Blended samples with limestone and metakaolin showed much higher rates of carbonation degree than OPC samples at all ages. For the ML15 samples, the carbonation degree reached a value of 71% and 89% at 2 d and 6 d of carbonation and almost completely carbonated at 13 d of carbonation. With the increase in the replacement level of PC, the extent of carbonation also surged. The carbonation degree of ML45 samples was 92% even at 2 d of carbonation curing. At 6 d of carbonation curing, both ML30 and ML45 samples were fully carbonated.

The pH value of the samples at 28 d of curing is shown in Figure 4. The pH value of the samples cured with the W-series regime was similar regardless of the substitution level. Only a slight reduction in the pH value of W-series samples was observed with an increase of substitution level. This trend became clear in the L-series samples due to the dilution effect associated with the substitution of PC with metakaolin and limestone. The M-series samples showed a notable reduction in the pH values, among which the OPC sample maintained pH value approximately at 10. The blended M-series samples were found to be almost fully carbonated as the pH value of them reached 8. This was reflected in the carbonation degree of the samples at 13 d of carbonation (see Figure 3). For H-series samples, all but OPC sample exhibited similar pH level, meaning the entire carbonation of the samples.

Testing age (carbonation time)	OPC	ML15	ML30	ML45
1.25 d (6 h)				
Carbonation degree (%)	4	5	7	5
3 d (2 d)	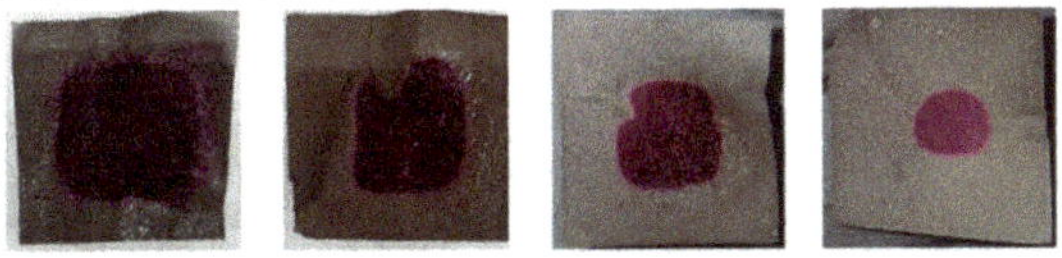			
Carbonation degree (%)	52	71	78	92
7 d (6 d)	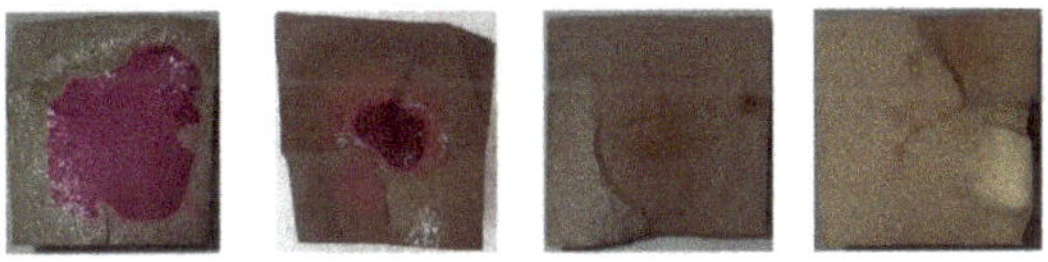			
Carbonation degree (%)	54	89	100	100
14 d (13 d)	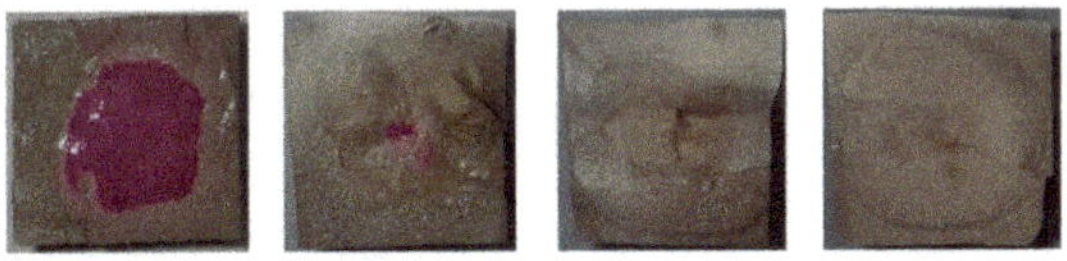			
Carbonation degree (%)	67	98	100	100
28 d (27 d)				
Carbonation degree (%)	80	100	100	100

Figure 3. Carbonation degree of the samples.

3.3. *Phase Identification by X-Ray Diffractometry*

The XRD patterns of W-series samples are presented in Figure 5a. The OPC-W sample showed peaks related to the presence of portlandite, calcite, C-S-H, and ettringite. Among the blended samples, the ML15-W sample showed portlandite peaks with the highest intensity which were still lesser than that of OPC-W peaks. The intensity of portlandite peaks got reduced for higher substitution levels of PC. Previous studies show that the reaction of metakaolin and limestone in blended systems consumes portlandite and overall reduction in the amount of PC in these blends also contributes in reduction of portlandite formation [28,31]. Peaks associated with two AFm phases—i.e., hemicarboaluminte and monocarboaluminate—were also observed at 10.7 2θ° and 11.6 2θ°, respectively, agreeing with available literatures [28,31,33]. The intensity of these phases was increased with higher replacement levels of PC in the blended samples [31]. Peaks related to ettringite were also present in the blended samples. Small peaks associated with the unreacted belite were also observed in the W-series samples.

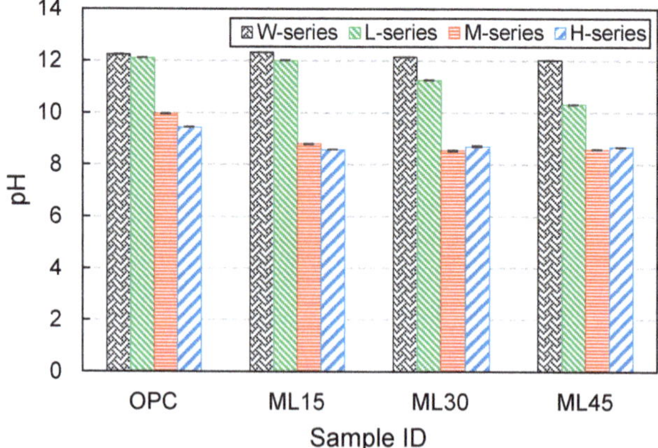

Figure 4. pH value of samples at 28 d of curing.

The XRD patterns of L-, M-, and H-series samples are shown in Figure 5b–d, respectively. All samples showed strong peaks related to calcite. Intensity of portlandite peaks was observed to be decreased for the OPC-L sample in comparison to the OPC-W sample, which further decreased in the OPC-M and OPC-H samples. Reduction in portlandite peak intensity verifies the conversion of portlandite to calcite due to the carbonation curing [42]. For blended samples, portlandite peaks were observed to be reduced in intensity with increase in the carbonation duration. This reduction was also proportional with the increase in substitution levels; samples with high substitution levels—i.e., ML30 and ML45 samples—mainly showed peaks related to calcium carbonate polymorphs. Peaks related to AFm and AFt phases also vanished with the carbonation curing. Carbonation-induced decomposition of these phases is evident from previous research works [48]. M- and H-seires samples also showed peaks related to brownmillerite, whose inetnsity also reduced with increased replacement levels of PC.

3.4. CO_2 *Uptake by Thermogravimetric Analysis*

TGA curves of the samples are shown in Figure 6. The W-series samples showed weight loss humps at around 100 °C and a shoulder around 140 °C due to the dehydration of chemically attached water from C-S-H, ettringite, and AFm phases [16,44,49,50]. It is reported that the weight loss hump at around 140–160 °C is mainly associated with the presence of monocarboaluminates and hemicarboalumiates [28]. Weight loss hump observed from 420 °C to 500 °C showed the dehydroxylation of portlandite [16,51]. The OPC-W sample showed the highest amount of portlandite. In the blended samples, a reduction of portalndite was observed; with higher substitution levels, higher consumption of portlandite was observed, which can also be seen in XRD results. Weight loss humps in the temperature range of 550–800 °C can be attributed to the decarbonation of $CaCO_3$ [52,53].

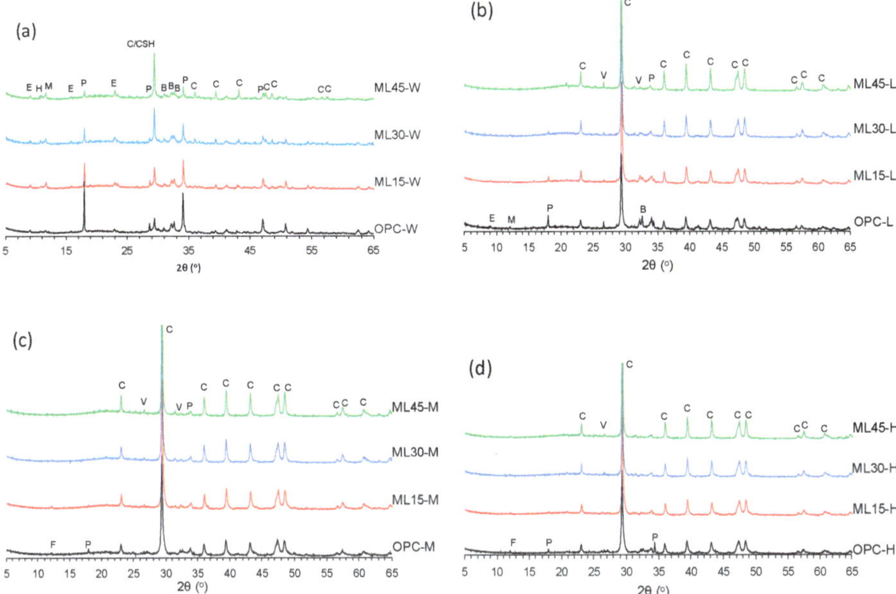

Figure 5. XRD patterns of (**a**) W-, (**b**) L-, (**c**) M-, and (**d**) H-series samples. The annotations are as follows: B—belite, C—calcite, CSH—calcium silicate hydrate, E—ettringite, F—brownmillerite, H—hemicarboaluminate, L—larnite, M—monocarboaluminate, P—portlandite, and V—vaterite.

The L-, M-, and H-series samples exhibited a reduced weight loss related to dehydration of water from C-S-H, ettringite, and AFm phases (Figure 6b–d); more reduction was observed at increased carbonation curing durations. The weight loss hump in the temperature range of 420–500 °C, associated with the presence of portlandite, disappeared for all the samples except for the OPC-L and ML15-L samples. For the M- and H-series samples, no weight loss humps were observed for portlandite, reflecting the complete consumption of portlandite by carbonation. All L-, M-, and H-series samples showed decarbonation of calcite with strong weight loss humps. CO_2 uptake for the carbonated samples is presented in Table 4. For the blends with limestone powder, percentage mass loss originated from the limestone was eliminated to genuinely identify the quantity of carbonation products. The OPC samples showed a similar carbonation uptake to those reported in the literature [19]. It was observed that the CO_2 uptake capacities of the samples have strong relation with carbonation-curing duration and PC replacement levels: highest CO_2 uptake capacity was observed for ML45-H samples. Tu et al. [54] explained that the increase in carbonation capacity of systems with limestone are due to two physical effects: dilution and nucleation. In dilution effect, the cement particles are more sparsely spread and, as a result, CO_2 access to particles become easier, while limestone powder in a system provide more nucleation sites on which carbonation products can precipitate. It is also reported that limestone shows higher affinity for carbonation products ($CaCO_3$) due to higher molecular recognition and improves the CO_2 uptake [54].

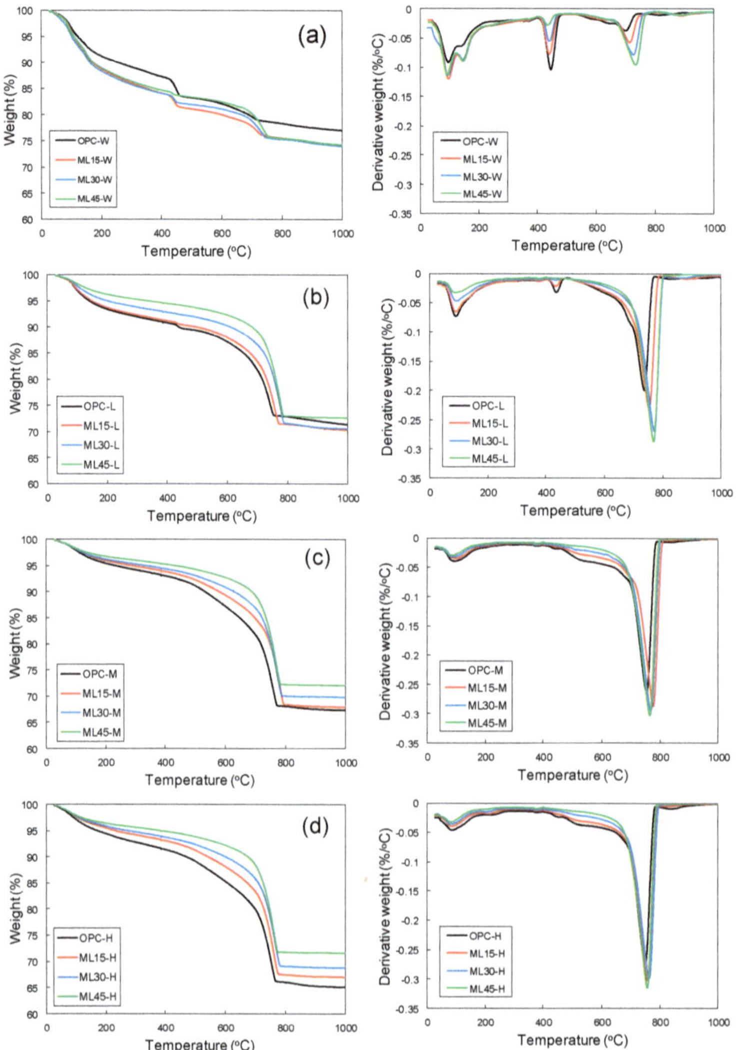

Figure 6. Thermogravimetry analysis curves of (**a**) W-, (**b**) L-, (**c**) M-, and (**d**) H-series samples.

Table 4. CO_2 uptake capacity of the carbonated samples.

Sample ID	CO_2 Uptake (g/100g of Powder [1])		
	L-Series	M-Series	H-Series
OPC	11.3	13.7	13.8
ML15	16.4	19.1	20.1
ML30	16.6	19.3	20.1
ML45	17.4	19.4	20.7

[1] Powder denotes the summation of Portland cement, metakaolin, and limestone.

3.5. Volume of Permeable Voids and Sorptivity

The volume of permeable voids and sorptivity coefficients of the samples under various curing regimes are shown in Table 5. The OPC samples showed a slight reduction in the volume of permeable voids with the increase in carbonation curing duration. The ML15-W samples exhibited almost similar volume of permeable voids as that of OPC-W samples, while ML30-W and ML45-W samples showed higher values of volume of permeable voids. Previous research works also described that the total porosity of blended systems was higher than that of pure PC systems and only blends with up to 15% replacement of PC by combined substitution by limestone and metakaolin exhibited similar porosity to that of PC systems [28,31]. Among carbonated blended systems, ML15 samples showed a slight reduction in the volume of permeable voids, while the ML30 and ML45 samples presented increase in the volume of permeable voids with the increase of carbonation curing duration. Initial and secondary sorptivity coefficients also explained the reduction in the volume of permeable voids for the OPC samples, while slight reduction can be observed for ML15 samples. In contrast, increase in sorptivity coeficients was observed for ML30 and ML45 samples with increase in carboantion curing durations. Qin et al. [42] reported an increase in the total porosity of blended systems upon carbonation curing. In general, carbonation of portlandite reduces the porosity of PC-based systems, but in systems where portlandite quantity is low due to pozzolanic reaction and reduced clinker content, carbonation of C-S-H takes place which coarsens the porosity [53]. The longer duration of carbonation exhibited more carbonation of the C-S-H phase which reflects the porosity results.

Table 5. Volume of permeable voids and sorptivity coefficients of the samples under various curing regimes.

Sample ID	Volume of Permeable Voids (%)	Initial Sorptivity Coefficient ($\times10^{-3}$ mm/s$^{1/2}$)	Secondary Sorptivity Coefficient ($\times10^{-3}$ mm/s$^{1/2}$)
OPC-W	32.6	9.5	0.12
ML15-W	33.7	8.8	0.11
ML30-W	34.8	6.5	0.13
ML45-W	35.2	6.1	0.14
OPC-L	30.4	8.3	0.10
ML15-L	32.1	7.9	0.11
ML30-L	35.9	8.5	0.13
ML45-L	36.3	9.4	0.15
OPC-M	27.4	7.3	0.10
ML15-M	31.0	7.8	0.13
ML30-M	34.6	11.1	0.17
ML45-M	38.6	12.3	0.18
OPC-H	24.9	6.8	0.11
ML15-H	30.4	7.7	0.13
ML30-H	35.1	10.6	0.17
ML45-H	38.6	11.8	0.18

4. Conclusions

The present study investigated the effect of carbonation curing on the PC-metakaolin-limestone ternary blends. Ternary blends replacing the PC with the combinations of metakaolin and limestone by mass levels of 15%, 30%, and 45% were exposed to four different combinations of ambient and carbonation curing regimes. Performances of these blends were evaluated by means of compressive strength, carbonation degree, XRD, TGA, volume of permeable voids, and sorptivity tests. Key findings obtained from the study are summarized below:

(1) The compressive strength of the blended samples exhibited a reduction in the strength as compared with that of the OPC sample. The loss of the compressive strength was increased as the substitution level increased from 15% to 45%. In addition, an increase in the duration of carbonation from 6 h to 27 d resulted in the significant loss of strength levels. The blends with a high substitution and longer exposure duration to CO_2 experienced significant changes in mechanical strength.

(2) Blended samples showed higher rates of carbonation than the OPC samples at all carbonation curing ages. At 27 d of carbonation curing, OPC sample showed carbonation degree of 80%, while the ML30 and ML45 samples exhibited complete carbonation even at 6 d of carbonation. Carbonation degree was governed both by carbonation duration and by cement replacement level.
(3) The XRD and TGA analyses showed the consumption of portlandite upon carbonation, which was proportional with the carbonation-curing duration. Upon carbonation, the main phases observed were $CaCO_3$ polymorphs.
(4) The replacement of the PC by metakaolin and limestone vastly improved the CO_2 uptake capacity, showing environmental benefits. The increase in the CO_2 uptake of the ML45 samples with respect to the OPC samples was 54%, 42%, and 50% for L-, M-, and H-series, respectively.
(5) An increase in the volume of permeable voids was observed upon exposure to CO_2 for the blended samples due to reduced portlandite amount which promoted carbonation of C-S-H which ultimately coarsens the porosity. The ML45-H sample showed volume of permeable voids of 38.6% which is 18.4% higher than that of the OPC-W sample.

Author Contributions: Conceptualization, R.H., J.S., S.P., and I.T.A.; Methodology, R.H.; Formal analysis, R.H., J.S., and S.P.; Investigation, R.H.; Writing—original draft preparation, R.H.; Writing—review and editing, J.S., S.P., I.T.A., and H.K.L.; Supervision, H.K.L.; Funding acquisition, H.K.L. All authors have read and agreed to the published version of the manuscript.

Funding: This study was supported by the Saudi Aramco-KAIST CO_2 Management Center to whom the authors are grateful.

Conflicts of Interest: No conflict of interest exists in the preparation and submission of this manuscript. All authors declare that they have no conflict of interest.

References

1. Shanks, W.; Dunant, C.F.; Drewniok, M.P.; Lupton, R.C.; Serrenho, A.; Allwood, J.M. How much cement can we do without? Lessons from cement material flows in the UK. *Resour. Conserv. Recycl.* **2019**, *141*, 441–454. [CrossRef]
2. Kajaste, R.; Hurme, M. Cement industry greenhouse gas emissions—Management options and abatement cost. *J. Clean. Prod.* **2016**, *112*, 4041–4052. [CrossRef]
3. Park, S.M.; Seo, J.H.; Lee, H.K. Thermal evolution of hydrates in carbonation-cured Portland cement. *Mater. Struct.* **2018**, *51*, 7. [CrossRef]
4. Kaliyavaradhan, S.K.; Ling, T.-C. Potential of CO_2 sequestration through construction and demolition (C & D) waste—An overview. *J. CO_2 Util.* **2017**, *20*, 234–242.
5. Earth's CO2 Homepage. Available online: https://www.co2.earth/ (accessed on 20 April 2020).
6. Lothenbach, B.; Scrivener, K.; Hooton, R.D. Supplementary cementitious materials. *Cem. Concr. Res.* **2011**, *41*, 1244–1256. [CrossRef]
7. Juenger, M.C.G.; Siddique, R. Recent advances in understanding the role of supplementary cementitious materials in concrete. *Cem. Concr. Res.* **2015**, *78*, 71–80. [CrossRef]
8. Juenger, M.C.G.; Snellings, R.; Bernal, S.A. Supplementary cementitious materials: New sources, characterization, and performance insights. *Cem. Concr. Res.* **2019**, *122*, 257–273. [CrossRef]
9. Gartner, E.; Sui, T. Alternative cement clinkers. *Cem. Concr. Res.* **2018**, *114*, 27–39. [CrossRef]
10. Juenger, M.C.G.; Winnefeld, F.; Provis, J.L.; Ideker, J.H. Advances in alternative cementitious binders. *Cem. Concr. Res.* **2011**, *41*, 1232–1243. [CrossRef]
11. Provis, J.L.; Bernal, S.A. Geopolymers and Related Alkali-Activated Materials. *Annu. Rev. Mater. Res.* **2014**, *44*, 299–327. [CrossRef]
12. Provis, J.L. Alkali-activated materials. *Cem. Concr. Res.* **2018**, *114*, 40–48. [CrossRef]
13. Provis, J.L.; Van Deventer, J.S.J. *Alkali Activated Materials: State-of-the-Art Reports, RILEM TC 224-AAM*; Springer/RILEM: Dordrecht, The Netherlands, 2014.
14. Zhang, D.; Ghouleh, Z.; Shao, Y. Review on carbonation curing of cement-based materials. *J. CO_2 Util.* **2017**, *21*, 119–131. [CrossRef]

15. Jang, J.G.; Kim, G.M.; Kim, H.J.; Lee, H.K. Review on recent advances in CO_2 utilization and sequestration technologies in cement-based materials. *Constr. Build. Mater.* **2016**, *127*, 762–773. [CrossRef]
16. Seo, J.H.; Park, S.M.; Lee, H.K. Evolution of the binder gel in carbonation-cured Portland cement in an acidic medium. *Cem. Concr. Res.* **2018**, *109*, 81–89. [CrossRef]
17. Berger, R.L.; Young, J.F.; Leung, K. Acceleration of hydration of calcium silicates by carbon dioxide treatment. *Nat. Phys. Sci.* **1972**, *240*, 16–18. [CrossRef]
18. Young, J.F.; Berger, R.L.; Breese, J. Accelerated Curing of Compacted Calcium Silicate Mortars on Exposure to CO_2. *J. Am. Ceram. Soc.* **1974**, 394–397. [CrossRef]
19. Jang, J.G.; Lee, H.K. Microstructural densification and CO_2 uptake promoted by the carbonation curing of belite-rich Portland cement. *Cem. Concr. Res.* **2016**, *82*, 50–57. [CrossRef]
20. Zhang, D.; Asce, S.M.; Cai, X.; Shao, Y. Carbonation Curing of Precast Fly Ash Concrete. *J. Mater. Civ. Eng.* **2016**, *28*, 1–9. [CrossRef]
21. Zhang, D.; Shao, Y. Early age carbonation curing for precast reinforced concretes. *Constr. Build. Mater.* **2016**, *113*, 134–143. [CrossRef]
22. Boumaaza, M.; Huet, B.; Turcry, P.; Aït-Mokhtar, A. The CO_2-binding capacity of synthetic anhydrous and hydrates: Validation of a test method based on the instantaneous reaction rate. *Cem. Concr. Res.* **2020**, *135*, 106113. [CrossRef]
23. Snellings, R. Assessing, Understanding and Unlocking Supplementary Cementitious Materials. *RILEM Tech. Lett.* **2016**, *1*, 50. [CrossRef]
24. Scrivener, K.; Martirena, F.; Bishnoi, S.; Maity, S. Calcined clay limestone cements (LC^3). *Cem. Concr. Res.* **2018**, *114*, 49–56. [CrossRef]
25. Ramezanianpour, A.A.; Bahrami Jovein, H. Influence of metakaolin as supplementary cementing material on strength and durability of concretes. *Constr. Build. Mater.* **2012**, *30*, 470–479. [CrossRef]
26. Sabir, B.; Wild, S.; Bai, J. Metakaolin and calcined clays as pozzolans for concrete: A review. *Cem. Concr. Compos.* **2001**, *23*, 441–454. [CrossRef]
27. Fernandez, R.; Martirena, F.; Scrivener, K.L. The origin of the pozzolanic activity of calcined clay minerals: A comparison between kaolinite, illite and montmorillonite. *Cem. Concr. Res.* **2011**, *41*, 113–122. [CrossRef]
28. Antoni, M.; Rossen, J.; Martirena, F.; Scrivener, K. Cement substitution by a combination of metakaolin and limestone. *Cem. Concr. Res.* **2012**, *42*, 1579–1589. [CrossRef]
29. Lothenbach, B.; Le Saout, G.; Gallucci, E.; Scrivener, K. Influence of limestone on the hydration of Portland cements. *Cem. Concr. Res.* **2008**, *38*, 848–860. [CrossRef]
30. Mohamed, A.R.; Elsalamawy, M.; Ragab, M. Modeling the influence of limestone addition on cement hydration. *Alexandria Eng. J.* **2015**, *54*, 1–5. [CrossRef]
31. Nguyen, Q.D.; Khan, M.S.H.; Castel, A. Engineering Properties of Limestone Calcined Clay Concrete. *J. Adv. Concr. Technol.* **2018**, *16*, 343–357. [CrossRef]
32. Ferreiro, S.; Herfort, D.; Damtoft, J.S. Effect of raw clay type, fineness, water-to-cement ratio and fly ash addition on workability and strength performance of calcined clay – Limestone Portland cements. *Cem. Concr. Res.* **2017**, *101*, 1–12. [CrossRef]
33. Dhandapani, Y.; Santhanam, M. Assessment of pore structure evolution in the limestone calcined clay cementitious system and its implications for performance. *Cem. Concr. Compos.* **2017**, *84*, 36–47. [CrossRef]
34. Dhandapani, Y.; Sakthivel, T.; Santhanam, M.; Gettu, R.; Pillai, R.G. Mechanical properties and durability performance of concretes with Limestone Calcined Clay Cement (LC^3). *Cem. Concr. Res.* **2018**, *107*, 136–151. [CrossRef]
35. Nguyen, Q.D.; Castel, A. Reinforcement corrosion in limestone flash calcined clay cement-based concrete. *Cem. Concr. Res.* **2020**, *132*, 106051. [CrossRef]
36. Shi, Z.; Lothenbach, B.; Geiker, M.R.; Kaufmann, J.; Leemann, A.; Ferreiro, S.; Skibsted, J. Experimental studies and thermodynamic modeling of the carbonation of Portland cement, metakaolin and limestone mortars. *Cem. Concr. Res.* **2016**, *88*, 60–72. [CrossRef]
37. Vizcaíno-Andrés, L.M.; Sánchez-Berriel, S.; Damas-Carrera, S.; Pérez-Hernández, A.; Scrivener, K.L.; Martirena-Hernández, J.F. Industrial trial to produce a low clinker, low carbon cement. *Mater. Construcción* **2015**, *65*, e045.
38. Bishnoi, S.; Maity, S.; Mallik, A.; Joseph, S.; Krishnan, S. Pilot scale manufacture of limestone calcined clay cement: The Indian experience. *Indian Concr. J.* **2014**, *88*, 22–28.

39. Emmanuel, A.C.; Haldar, P.; Maity, S.; Bishnoi, S. Second pilot production of limestone calcined clay cement in India: The experience. *Indian Concr. J.* **2016**, *90*, 57–63.
40. Huang, H.; Guo, R.; Wang, T.; Hu, X.; Garcia, S.; Fang, M.; Luo, Z.; Maroto-Valer, M.M. Carbonation curing for wollastonite-Portland cementitious materials: CO_2 sequestration potential and feasibility assessment. *J. Clean. Prod.* **2019**, *211*, 830–841. [CrossRef]
41. Sharma, D.; Goyal, S. Accelerated carbonation curing of cement mortars containing cement kiln dust: An effective way of CO_2 sequestration and carbon footprint reduction. *J. Clean. Prod.* **2018**, *192*, 844–854. [CrossRef]
42. Qin, L.; Gao, X.; Chen, T. Influence of mineral admixtures on carbonation curing of cement paste. *Constr. Build. Mater.* **2019**, *212*, 653–662. [CrossRef]
43. American Society for Testing and Materials. *ASTM C109/109M-20a: Standard Test Method for Compressive strength of Hydraulic Cement Mortars (Using 2-in. or [50-mm] Cube Specimens)*; ASTM International: West Conshohocken, PA, USA, 2020.
44. Siddique, S.; Naqi, A.; Jang, J.G. Influence of water to cement ratio on CO2 uptake capacity of belite-rich cement upon exposure to carbonation curing. *Cem. Concr. Compos.* **2020**, *111*, 103616. [CrossRef]
45. Scrivener, K.; Snellings, R.; Lothenbach, B. *A Practical Guide to Microstructural Analysis of Cementitious Materials*, 4th ed.; CRC Press: Boca Raton, FL, USA, 2015.
46. American Society for Testing and Materials. *ASTM C1585-13, Standard Test Method for Measurement of Rate of Absorption of Water by Hydraulic-Cement Concretes*; ASTM International: West Conshohocken, PA, USA, 2013.
47. American Society for Testing and Materials. *ASTM C642-13, Standard Test Method for Density, Absorption, and Voids in Hardened Concrete*; ASTM International: West Conshohocken, PA, USA, 2013.
48. Chen, T.; Gao, X. Effect of carbonation curing regime on strength and microstructure of Portland cement paste. *J. CO2 Util.* **2019**, *34*, 74–86. [CrossRef]
49. Mehta, P.K.; Monteiro, P.J.M. *Concrete: Microstructure, Properties, and Materials*, 4th ed.; McGraw-Hill Education: New York, NY, USA, 2014.
50. Ben Haha, M.; Le Saout, G.; Winnefeld, F.; Lothenbach, B. Influence of activator type on hydration kinetics, hydrate assemblage and microstructural development of alkali activated blast-furnace slags. *Cem. Concr. Res.* **2011**, *41*, 301–310. [CrossRef]
51. Silva, D.A.; Roman, H.R.; Gleize, P.J.P. Evidences of chemical interaction between EVA and hydrating Portland cement. *Cem. Concr. Res.* **2002**, *32*, 1383–1390. [CrossRef]
52. Jeong, Y.; Park, H.; Jun, Y.; Jeong, J.-H.; Oh, J.E. Microstructural verification of the strength performance of ternary blended cement systems with high volumes of fly ash and GGBFS. *Constr. Build. Mater.* **2015**, *95*, 96–107. [CrossRef]
53. Borges, P.H.R.; Costa, J.O.; Milestone, N.B.; Lynsdale, C.J.; Streatfield, R.E. Carbonation of CH and C-S-H in composite cement pastes containing high amounts of BFS. *Cem. Concr. Res.* **2010**, *40*, 284–292. [CrossRef]
54. Tu, Z.; Guo, M.Z.; Poon, C.S.; Shi, C. Effects of limestone powder on $CaCO_3$ precipitation in CO_2 cured cement pastes. *Cem. Concr. Compos.* **2016**, *72*, 9–16. [CrossRef]

Article

Elucidation of the Hydration Reaction of UHPC Using the PONKCS Method

Hyunuk Kang [1], Nankyoung Lee [1] and Juhyuk Moon [1,2,*]

[1] Department of Civil and Environmental Engineering, Seoul National University, 1 Gwanak-ro, Gwanak-gu, Seoul 08826, Korea; kanghu93@snu.ac.kr (H.K.); nankyoung@snu.ac.kr (N.L.)

[2] Institute of Construction and Environmental Engineering, Seoul National University, 1 Gwanak-ro, Gwanak-gu, Seoul 08826, Korea

* Correspondence: juhyukmoon@snu.ac.kr; Tel.: +82-2880-1524

Received: 25 September 2020; Accepted: 16 October 2020; Published: 19 October 2020

Abstract: This study explored the hydration reaction of ultra-high-performance concrete (UHPC) by using X-ray diffraction (XRD), nuclear magnetic resonance (NMR), and thermogravimetric analysis (TGA) as analysis methods. The partial- or no-known crystal structure (PONKCS) method was adopted to quantify the two main amorphous phases of silica fume and C-S-H; such quantification is critical for understanding the hydration reaction of UHPC. The measured compressive strength was explained well by the degree of hydration found by the PONKCS method, particularly the amount of amorphous C-S-H. During heat treatment, the pozzolanic reaction was more intensified by efficiently consuming silica fume. After heat treatment, weak but continuous hydration was observed, in which the cement hydration reaction was dominant. Furthermore, the study discussed some limitations of using the PONKCS method for studying the complicated hydration assemblage of UHPC based on the results of TGA and NMR. Generally, the PONKCS method underestimated the content of silica fume in the early age of heat treatment. Furthermore, the structural evolution of C-S-H, confirmed by NMR, should be considered for more accurate quantification of C-S-H formed in UHPC. Nevertheless, PONKCS-based XRD could be useful for understanding and optimizing the material properties of UHPC undergoing heat treatment.

Keywords: hydration reaction; nuclear magnetic resonance; thermogravimetry; ultra-high-performance concrete; X-ray diffraction; calcium silicate hydrate

---

1. Introduction

Cement is a mixture of the crystalline phases of substances such as alite, belite, and aluminate. The mixture undergoes different hydration reactions that result in the various material properties of concrete. To elucidate the chemical reactions and understand the evolution of the material properties, characterization methods, such as X-ray diffraction (XRD) and thermogravimetric analysis (TGA), are used to quantify the mineral phases during the hydration reaction [1–5]. In TGA, quantitative analysis is possible only for specific phases, such as calcium hydroxide and calcium carbonate [6–8]. In contrast, XRD can provide quantitative information for most of the crystalline phases; however, analysis of the amorphous phases, which is critical for investigating the time-dependent phase assemblage of cement-based materials, is still problematic [9]. For example, if amorphous or less-crystalline materials, such as silica fume, ground granulated blast-furnace slag, or fly ash, are mixed with cement for specific purposes, accurate separation of the amorphous reaction product from the raw material is difficult [10,11].

Generally, the Rietveld refinement method (using an internal or external standard method) can be applied to quantify an amorphous phase using XRD [12,13]. Additionally, if the partial or no known crystal structure (PONKCS) method, which can simulate the intensity of the virtually generated amorphous pattern, is used, quantitative analysis of the amorphous phases is possible [14–16]. The internal or external standard method can resolve one unknown (i.e., amorphous content)—that is, the total amount of existing amorphous phases. In contrast, the PONKCS method can be theoretically applied to investigate the phase quantification of two or more different amorphous phases. This is a plausible approach because each amorphous phase has a distinct hump range, which can be individually fitted before the reaction occurs.

Ultra-high-performance concrete (UHPC) is a structural material that is implemented in construction [17–20]. Its durability is especially outstanding due to its excellent compressive strength and dense microstructure [21,22]. Typical high-strength concrete-like UHPC contains amorphous silica fume as a raw material, and its main hydration product is C-S-H, the main hydration product of general cement hydration. Since both the raw material and resulting main hydration product are in non-crystalline phases, the XRD-based characterization of the hydration mechanism of UHPC (e.g., the degree of consumption of the silica fume or production of C-S-H) is difficult [23]. Additionally, the lower degree of hydration of UHPC also makes accurate quantitative phase identification more difficult [13,24,25].

This study aimed to elucidate the hydration reactions of UHPC using the PONKCS method due to UHPC's amorphous phase in the raw mixture (i.e., silica fume) and in hydrated form (i.e., C-S-H). By fitting both phases using the PONKCS method, accurate hydration assemblage of the complex system of UHPC was obtained. For example, it included the degree of consumption of silica fume and the production of C-S-H. In addition to the quantitative XRD (QXRD) analysis, TGA and nuclear magnetic resonance (NMR) analysis were also performed to verify the time-dependent hydration characteristics of UHPC. Finally, the study investigated the strength development of UHPC via interpretation by mineralogical characterizations.

2. Materials and Experimental Details

2.1. Sample Preparation

OPC (Ordinary Portland cement) (Union Cement Co., Ltd., Chungcheong-do, Korea), silica fume (Grade 940U, Elkem, Olso, Norway), silica powder (S-SIL 10, SAC, Ulsan, Korea), silica sand (Saeron Co., Ltd., Gangwon-do, Korea), and polycarboxylate ether-based superplasticizer (Flowmix 3000U, Dongnam, Gyeonggi-do, Korea) were used based on the authors' previous UHPC studies [26–29]. The mix design used in this study is shown in Table 1. The particle size distributions of the OPC and silica fume (Figure 1) were obtained using a microparticle size analyzer (Malvern Instruments., LTD., Malvern, UK). To measure compressive strength, quartz powder and silica sand were included, as shown in Table 1. Meanwhile, quartz powder and silica sand were excluded in samples for XRD, TGA, and NMR because the components are chemically inert.

Table 1. Mix proportions of specimens by weight ratio.

Mixture	OPC	Silica Fume	Quartz Powder	Silica Sand	Water	Superplasticizer
UHPC	1000	250	350	1100	225	40

Solid-state superplasticizer was used.

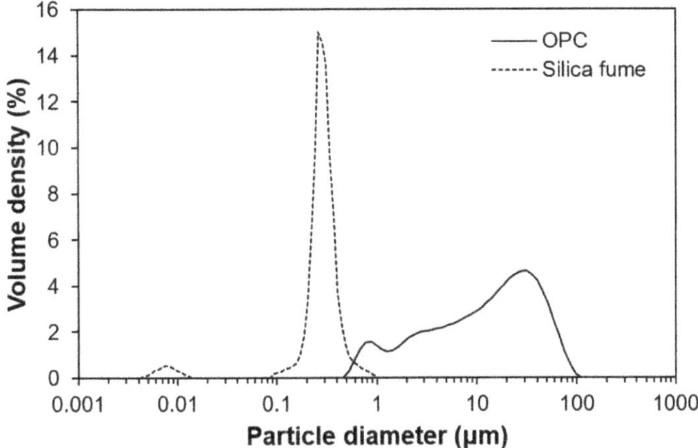

Figure 1. Particle size distributions of ordinary Portland cement and silica fume.

The paste and UHPC samples were prepared by dry-mixing for 1 min, followed by mixing with the polycarboxylate ether-based superplasticizer and water for 5 min. Then, they were placed into the $50 \times 50 \times 50$ mm^3 cubic mold and cured at room temperature for 1 day, and then cured at 90 °C for 2 days. Subsequently, they were cured at 20 °C until subsequent testing.

2.2. Experimental Details

To measure compressive strength, three identical cubes were tested to determine the average compressive strength of the specimens cured for 1, 2, 4, and 28 days. For XRD, TGA, and NMR measurements, the paste was soaked first in isopropyl alcohol and then ethyl ether to stop the hydration (by removing free water) at the target date of investigation (i.e., 1, 2, 4, and 28 days). The remaining ethyl ether was removed by heating the specimen at 40 °C.

An X-ray diffractometer (D2 Phaser, Bruker Co. Ltd., Land Baden-Württemberg, Germany) equipped with Cu-Kα radiation (λ = 1.5418 Å) was used to measure the powder's XRD pattern in the range of 2θ between 5° and 60°. The acquired diffraction patterns of the paste specimens were analyzed using the TOPAS software version 7.0 (Bruker Co. Ltd., Land Baden-Württemberg, Germany). The XRD results for the raw OPC and silica fume are presented in Figure 2. The main constituents of the OPC were alite (51.6%), belite (0.5%), dolomite (18.1%), and limestone (29.7%). A hump was observed for the silica fume between 14° and 30° [30].

TGA was performed in an N_2 environment using a DSC/TG system (SDT Q600, TA Instrument Ltd., Newcastle, DE, USA) at a heating rate of 10 K/min up to 1050 °C. The mineral phases of calcium hydroxide and calcium carbonate were calculated based on their decomposition temperatures of 400–500 °C and 600–800 °C, respectively [12]. The ^{29}Si MAS NMR experiments were performed using Advance III HD (Bruker, Karlsruhe, Germany) at 119.182 MHz. The NMR spectra were obtained using a 5-mm HX CPMAS probe and a 5-mm zirconia rotor with a rotation speed of 10.0 MHz, a pulse width of 2.2 μs, and a relaxation delay of 22 s. The ^{29}Si chemical shifts referenced external samples of −135.5 ppm of tetramethyl silane (TMS) and −135.5 ppm of tetrakis silane to 0 ppm of aqueous $AlCl_3$, respectively.

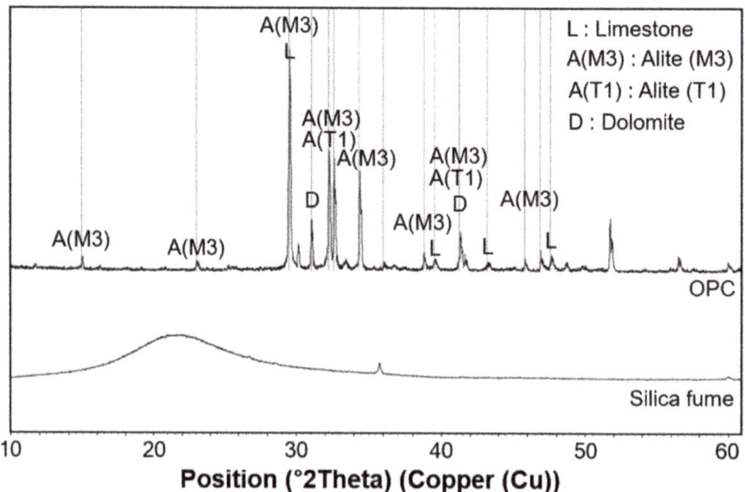

Figure 2. Measured X-ray diffraction patterns of raw materials.

3. Experimental Results

3.1. Compressive Strength Results

The average and standard deviation of the compressive strengths are shown in Figure 3. The average compressive strengths cured for 1, 2, 4, and 28 days were 50.7, 129.3, 136.9, and 117.2 MPa, respectively. The strength significantly increased during the 2 days of the heat treatment period, and then it slightly decreased from 4 to 28 days.

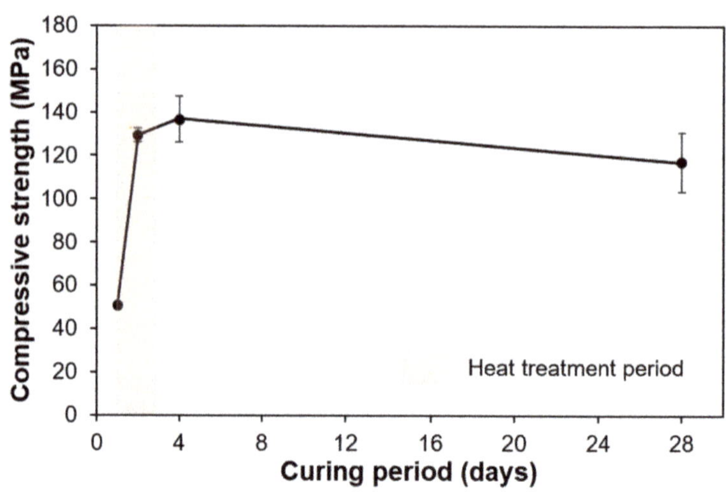

Figure 3. Compressive strength of the specimens.

3.2. TGA Results

The results of the TGA are shown in Figure 4. Regardless of curing time, calcium hydroxide (decomposition range between 400 and 500 °C) was hardly detected; this finding was similar to the authors' previous hydration studies on UHPC [13,31,32]. The main weight loss occurred between 600 and 800 °C, which indicated the decomposition of calcium carbonate [33]. In this study, normalization to anhydrous was performed based on the following equation

$$W_{corrected} = \frac{W_{Rietveld}}{1 - H_2O_{Bound,TGA}} \quad (1)$$

where $W_{corrected}$, $W_{Rietveld}$, and $H_2O_{Bound,TGA}$ indicate the normalized result, the weight percentage from XRD analysis, and the chemically bound water (CBW), respectively.

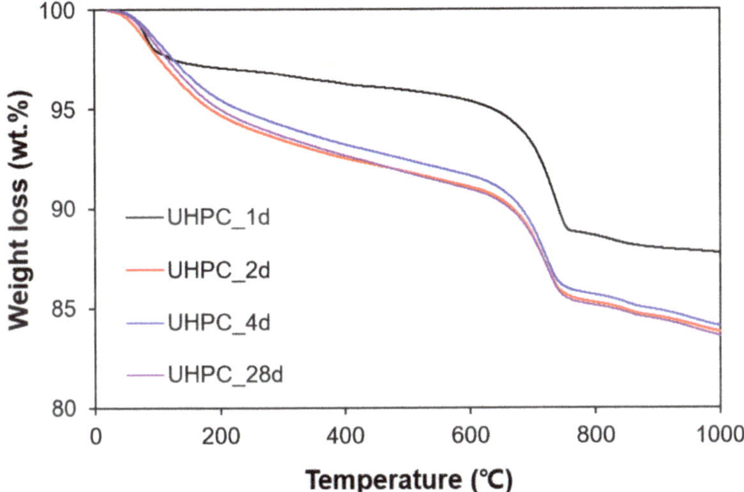

Figure 4. Measured TGA of all specimens.

The CBW was calculated from the weight loss value at 600 °C. The CBW for 1, 2, 4, and 28 days of the samples was 4.6%, 8.9%, 8.3%, and 9.0%, respectively. Generally, the amount of CBW increased as the curing time increased; the CBW represented the degree of hydration. However, during the heat treatment period, the CBW was slightly reduced (from 8.9% to 8.3%), possibly because partial CBW in pre-formed, and formulating C-S-H can be lost at 90 °C [16,34].

3.3. XRD Results

To use the PONKCS method for QXRD, a preliminary experiment was conducted on a mixture of silica fume and quartz powder. A sample mixed with a weight ratio of quartz powder to silica fume of 1:1 was prepared, and the database was generated as follows [35]

$$(ZMV)_{Amorphous} = \frac{W_{Amorphous}}{W_{Quartz}} \times \frac{S_{Quartz}}{S_{Amorphous}} \times (ZMV)_{Quartz} \quad (2)$$

where W_x, S_x, ZM, and V indicate the weight percentage of x, scale factor of x, cell mass, and unit cell volume, respectively. Based on this method, the scale factor, virtual Miller constant, and corresponding intensity of silica fume were probabilistically calculated to generate the hypothetical space group for the diffraction pattern of silica fume, and a similar method was used to generate the C-S-H pattern. The XRD pattern obtained from the OPC cured for 7 years was used after excluding peak contributions from the known crystalline phases [36]. Then, the fitted patterns of the silica fume and C-S-H were used for the quantitative phase analysis of the UHPC to evaluate the weight variation of the two amorphous phases. The virtual patterns of the silica fume and C-S-H generated with the PONKCS method are presented in Figures 5 and 6, respectively.

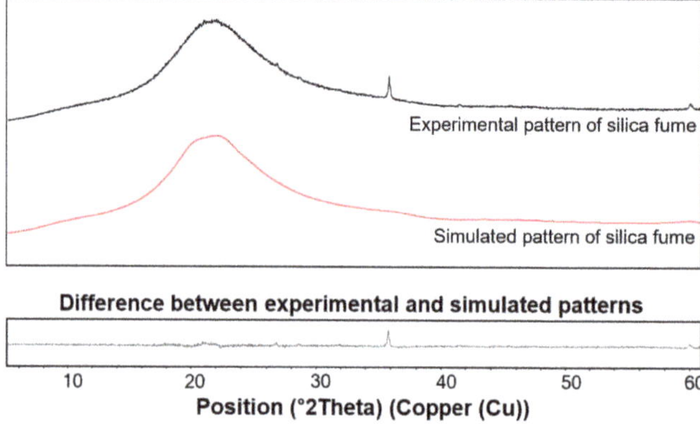

Figure 5. Experimental and simulated diffraction patterns of silica fume.

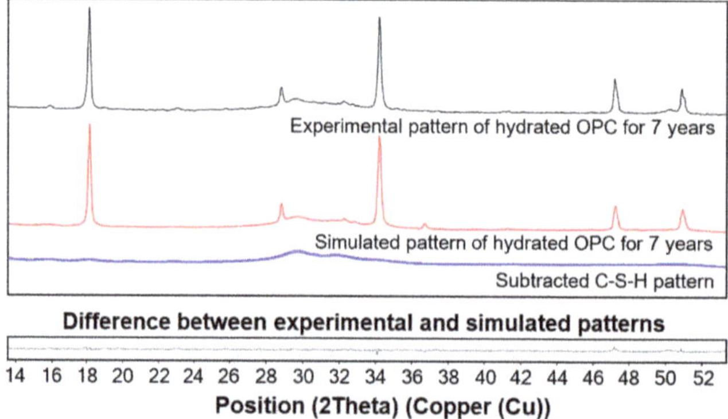

Figure 6. Experimental and simulated diffraction patterns of C-S-H. The experimental pattern of hydrated OPC for 7 years was reproduced from a previous study [23].

For the PONKCS method and subsequent refinement processes, the background was excluded by using a polynomial function of an order less than 5. Previous studies indicated that inaccurate fitting may occur when a low angle range is included in the PONKCS method or Rietveld analysis [36]. In the present study, the refinement was performed from 8° 2θ degrees, considering the presence of ettringite. Then, to reduce scattering at a low angle, a slit was adopted to enhance the fitting accuracy. Neither the internal standard nor the external standard method was used in this study, as the pattern corresponding to the background and all humps would have been assigned to the weight of an amorphous phase, thereby significantly overestimating the amount of actual amorphous content in the samples. Furthermore, separating the phase content of the two main amorphous phases (i.e., silica fume and C-S-H) would not have been possible in UHPC hydration with either method.

The measured XRD patterns (UHPC_1 day, UHPC_2 days, UHPC_4 days, and UHPC_28 days) and the corresponding refined patterns are shown in Figure 7. The results of the QXRD are shown in Figure 8. Note that the free water content in the Figure 8 was calculated by the subtraction of the scaled solid amount (using the CBW value at each date) from the initial mix proportion. The main hydration products were C-S-H and ettringite, while raw materials of silica fume, alite, limestone, and dolomite were still identified in all ages. However, minor phases in OPC, such as belite and aluminate, were not identified after hydration began. In the samples cured for 2 days and 4 days, ettringite was found at around 9°, but it disappeared at 4 days, indicating that it was completely decomposed through curing at 90 °C for 2 days [37]. The reduction in the intensity of the peaks corresponding to alite indicated the continuous progress of hydration at all ages. After normalization to anhydrous content, the silica fume was measured as 16.9%, 15.1%, 12.3%, and 12.1%, while C-S-H was about 35.3%, 46.2%, 51.8%, and 52.8% in samples cured for 1, 2, 4, and 28 days, respectively. Thus, the quantitative separation of the two amorphous phases was possible based on the adopted PONKCS method.

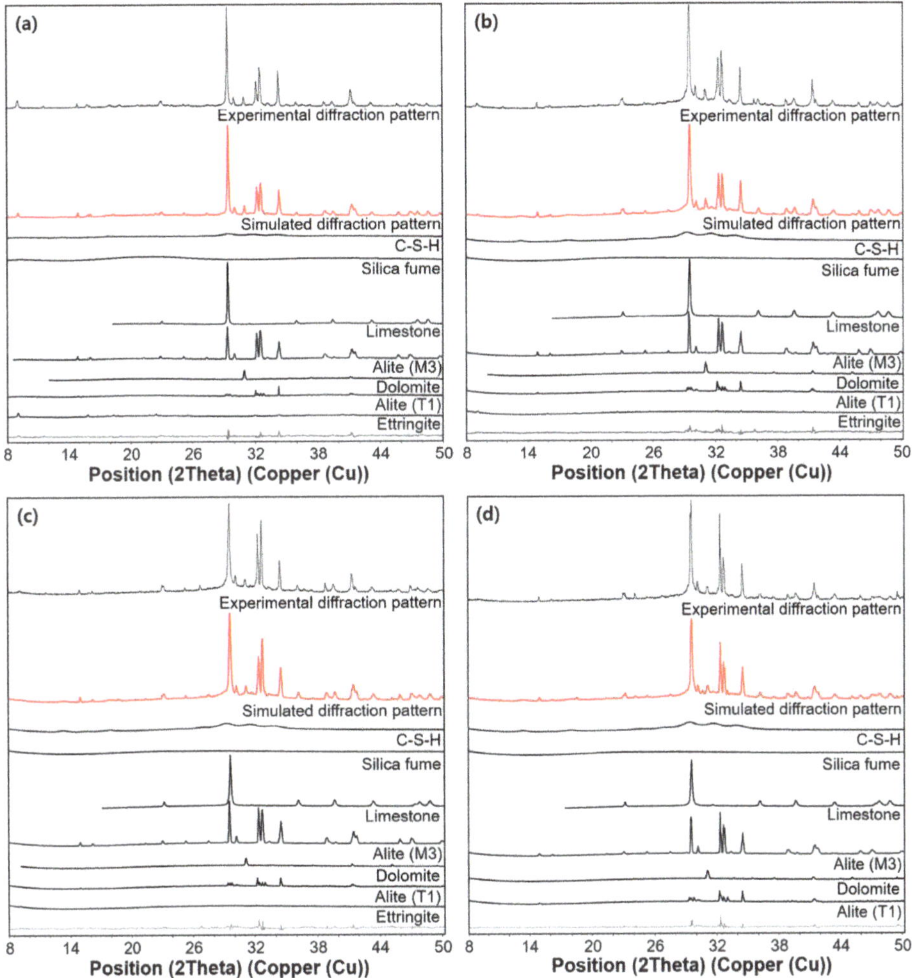

Figure 7. Rietveld refinement results of all samples: (**a**) UHPC_1 day, (**b**) UHPC_2 days, (**c**) UHPC_4 days, (**d**) UHPC_28 days.

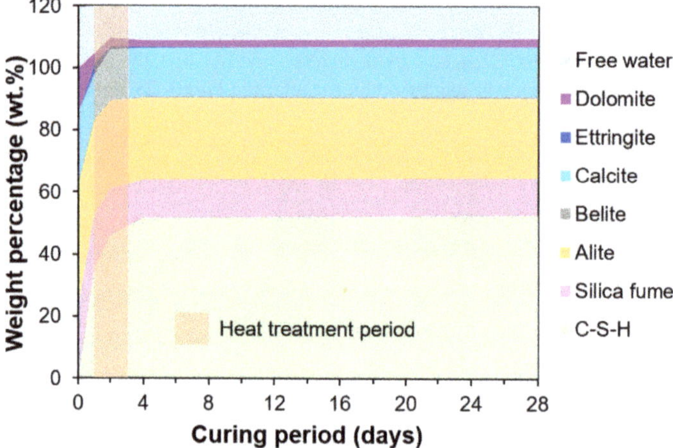

Figure 8. QXRD results after normalization to anhydrous content.

3.4. NMR Results

Figure 9 shows the deconvolution data from the NMR measurements. Similar to the results obtained with typical cement-based materials, a set of Q^0 peaks (−71 and −74 ppm), a Q^1 peak (−79 ppm), a Q^2(1Al) peak (−81 ppm), a Q^2 peak (−85 ppm), and a Q^4 peak (−110 ppm) were identified. The peaks were deconvoluted using Gaussian and Lorentzian functions, and the results are presented in Table 2 [25,38,39]. The Q^0 and Q^4 peak represented alite and silica fume, respectively [39]. The intensity of both peaks generally decreased, while the intensity of the peaks corresponding to the C-S-H tended to increase as hydration progressed.

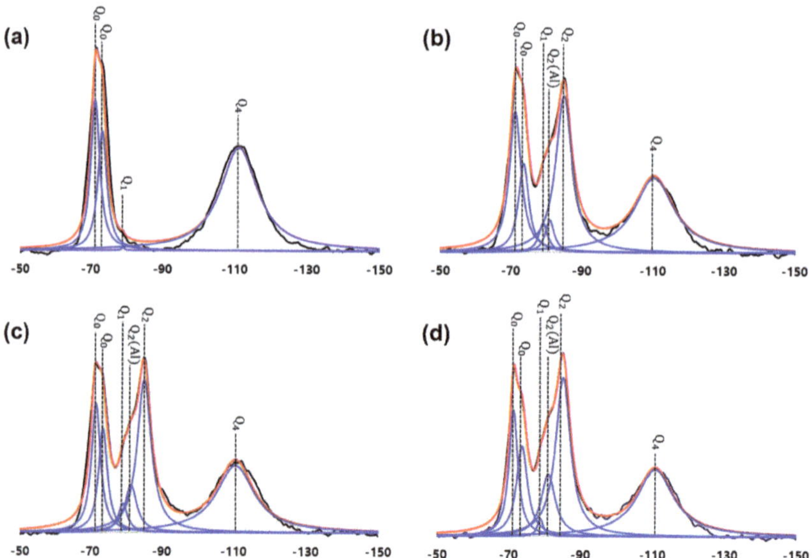

Figure 9. ^{29}Si magic angle spinning NMR (^{29}Si MAS NMR) spectra and de-convoluted patterns: (**a**) UHPC_1 day, (**b**) UHPC_2 days, (**c**) UHPC_4 days, and (**d**) UHPC_28 days.

Table 2. Deconvolution results for ^{29}Si MAS NMR spectra, wt.%.

Specimen	Q^0 −71 ppm	Q^0 −74 ppm	Q^1 −79 ppm	Q^2(1Al) −81 ppm	Q^2 −85 ppm	Q^4 −110 ppm	Al/Si	MCL
UHPC_1day	40.00	31.70	0.90	0	0	27.40	0	2.00
UHPC_2days	26.93	16.97	5.66	6.18	29.95	14.31	0.074	15.86
UHPC_4days	24.38	19.56	5.47	9.04	28.87	12.68	0.10	17.51
UHPC_28days	24.19	17.25	3.33	11.90	30.47	12.88	0.13	31.02

The Q^1 peak represented the end of the silicate chain, while the Q^2(1Al) and Q^2 peaks represented the connected silicate tetrahedral chains of C-S-H. The length of the formulated silicate chain (i.e., the mean chain length (MCL)) can be estimated as follows [40]

$$MCL_{nc} = \frac{2\left[Q^1 + Q^2 + \frac{3}{2}Q^2(1\text{Al})\right]}{Q^1} \tag{3}$$

$$\frac{\text{Al}}{\text{Si}_{nc}} = \frac{\frac{1}{2}Q^2(1\text{Al})}{Q^1 + Q^2 + Q^2(1\text{Al})} \tag{4}$$

The estimated MCL increased as the hydration reaction progressed (Table 2). At 1 day, C-S-H with a small MCL was formed during the early stages of hydration. During the heat treatment, partial decomposition of C-S-H may have occurred (as determined by the reduced CBW in Figure 4); this could be associated with the water loss in the newly formed C-S-H at the perimeter of pre-existing C-S-H. Thus, the MCL and the fractal dimension of the C-S-H were significantly increased [13].

After the heat treatment period, continuous hydration of OPC could have still occurred, resulting in the higher stability of the formed C-S-H structure. Therefore, the crystallinity of C-S-H was relatively high at later ages. The estimated Al/Si ratio had a similar tendency to that of the MCL, as determined by Equation (4) [25]. However, more rigorous investigation must be performed to refine the equation for estimating the nanostructure of C-(A)-S-H, especially in restrained environments or for UHPC [13,40].

4. Discussion

4.1. C-S-H Formation Confirmed by XRD and TGA

This section discusses the C-S-H formation of UHPC upon curing based on the experimental results of XRD and TGA. The amount of C-S-H increased as hydration progressed, and the increase was the largest during the heat treatment process. TGA can be used to estimate the amount of C-S-H by measuring the weight loss of water in the C-S-H. However, the precise range of temperatures at which C-S-H decomposes (i.e., the evaporation of physically or chemically bound water) has not been fully clarified. For example, Taylor et al. reported the range as 115–125 °C, while Odelson et al. used the range of 200–400 °C to estimate the amount of water loss associated with C-S-H [41,42].

Figure 10 compares the variations of the measured amount of C-S-H from the PONKCS method and weight loss related to the C-S-H in the two different temperature ranges. The XRD yielded up to a C-S-H weight of 52% in the UHPC (excluding the weight contributions of quartz powder and silica sand) at 28 days. As expected, the net amount of formation of C-S-H was the largest during the heat treatment period. However, the potential error of the XRD-based quantification of the C-S-H was that the used C-S-H model was subtracted from the 7-year-old hydrated OPC. Thus, it was not simulated using the C-S-H pattern in UHPC, which might have been different considering the complex microstructure of UHPC due to heat treatment as well as the pozzolanic reaction [13].

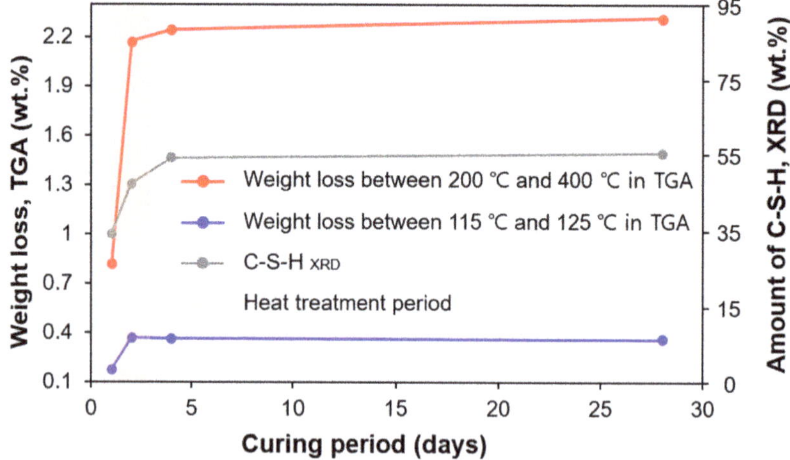

Figure 10. Comparison of C-S-H formation measured from XRD and TGA.

According to the TGA results, the calculated weight loss in both temperature ranges quickly increased during the heat treatment period, and then stabilized. Interestingly, the absolute weight loss, as well as the degree of increase in 200–400 °C, was calculated to be much larger than that in the range from 115 to 125 °C. Weakly bound water in C-S-H may be easily lost during the heat treatment, so the weight loss measured after the heat treatment can be small. Even so, the trend measured by XRD analysis (i.e., the weight percentage of C-S-H) was similar to that found by TGA (i.e., weight loss of water in C-S-H).

4.2. Relationship between Compressive Strength and C-S-H Formation

The relationship between compressive strength and C-S-H formation in the samples cured for 1, 2, 4, and 28 days is shown in Figure 11. Except for the UHPC_28 days sample, the relationship between the development of strength and the amount of C-S-H was directly proportional. In the early stages of heat treatment, the produced amount of C-S-H was the greatest due to the acceleration of the hydration reaction caused by heat treatment. Thus, the compressive strength was also greatly enhanced [13,31]. However, the relationship between compressive strength and C-S-H in the samples of UHPC_4 days and UHPC_28 days was not clear. The main cause of this phenomenon was due to the slight strength reduction at 28 days along with a slightly increased production of C-S-H. As was previously explained, after heat treatment in UHPC, the available space for subsequent hydration is limited. Therefore, the later age of hydration in dense microstructures may cause micro-cracking in the matrix, thereby leading to a slight decrease in strength at later ages [31].

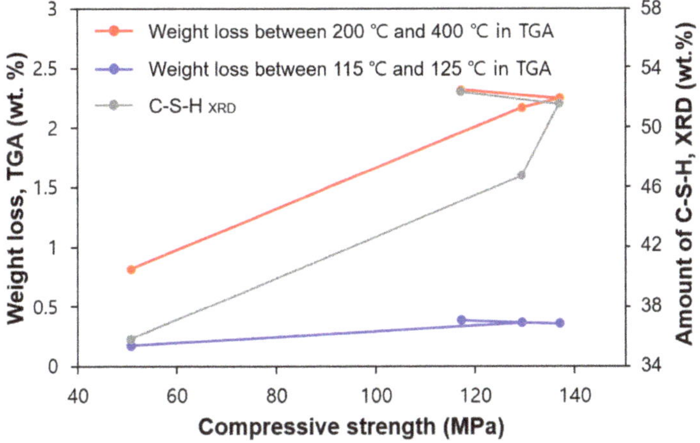

Figure 11. Relationship between strength development and the estimated amount of C-S-H.

4.3. Relationship between Consumption of Silica Fume and Formation of C-S-H

Figure 12 shows the relationship between the formation of C-S-H and the consumption of silica fume. First, the consumption of silica fume was the greatest during the heat treatment period, which was associated with the largest formation of C-S-H in that period (at 2 and 4 days). The consumption of silica fume can be explained by the pozzolanic reaction, which can be efficiently accelerated by the consumption of portlandite during heat treatment. This was also consistent with a recent hydration study of UHPC [13]. As a result of the enhanced pozzolanic reaction, the detected portlandite after the heat treatment was found to be less than 0.5 wt.%.

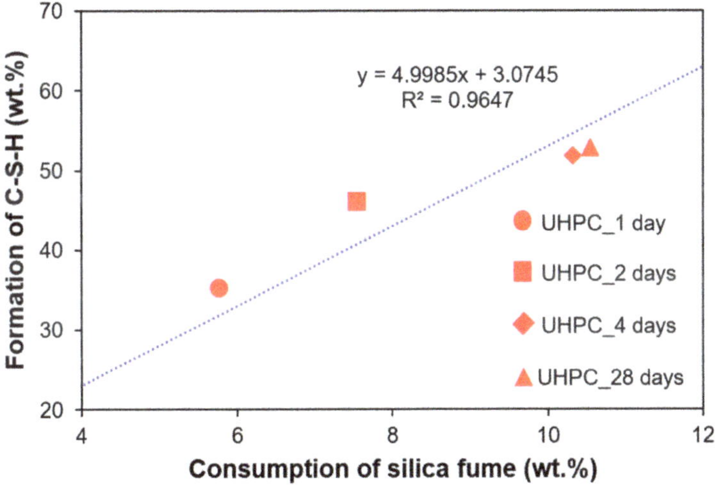

Figure 12. Relationship between consumption of silica fume and formation of C-S-H.

Heat treatment is the conventional curing method for UHPC to expedite the production in the precast form [24]. However, the degree of hydration in UHPC is very low, due to the limited amount of water and space availability [13,31]. In this study, silica fume was still identified at 28 days (Figure 9), while the portlandite was almost completely consumed during the heat treatment. Therefore, the pozzolanic reaction during heat treatment was mostly terminated due to the lack of portlandite.

Further hydration after heat treatment was mostly based on the OPC hydration, which is not very efficient in terms of filling space via hydration reaction. While the pozzolanic reaction can occur in the surrounding hydration products, cement hydration at that stage requires the diffusion of water through pre-hydrated products. This subsequent hydration may induce micro-cracking in the matrix, thus slightly reducing the compressive strength.

As seen in Figure 12, the relationship between the formation of C-S-H and the consumption of silica fume was generally linear. Greater C-S-H formation at 1 and 2 days (points above the trend line) indicated the relatively higher OPC hydration (i.e., less consumption of silica fume but more formation of C-S-H). The points at 4 and 28 days lie below the trend line, showing greater consumption of silica fume compared to the produced amount of C-S-H. This analysis also confirmed that the pozzolanic reaction was intensified during the heat treatment period. However, it also may have suggested that the crystallinity between C-S-H due to OPC hydration and pozzolanic reaction can be different. This potential difference could impact the accuracy of fitting C-S-H with the PONKCS method, which only relied on the C-S-H pattern subtracted from OPC hydration at 7 years [36].

4.4. Comparison of the Reactivity of Silica Fume and Clinker Materials from XRD and NMR

As discussed, the potential difference in the crystallinity of the C-S-H formed by the pozzolanic reaction (mainly during heat treatment) and that formed by OPC hydration may affect the quantitative content of C-S-H. Furthermore, as confirmed by the NMR experiment, the nanostructure of C-S-(A)-H also evolved as the length of the MCL and Al/Si changed. The impact of the MCL and Al/Si on the crystallinity of C-S-H should be considered for more accurate quantification of the content of C-S-H [13]. Instead, the relative reactivity of silica fume and clinker (i.e., alite) can be seen in the NMR and QXRD results. Figure 13 shows the variation in the ratios of Q^0/Q^4, as found by NMR (Q^0 and Q^4 indicate alite and silica fume, respectively) and alite/silica fume, as found by QXRD. Except for the sample of UHPC_2 days, the similar tendency of an abrupt increase at early ages and a steady decrease at later ages was confirmed.

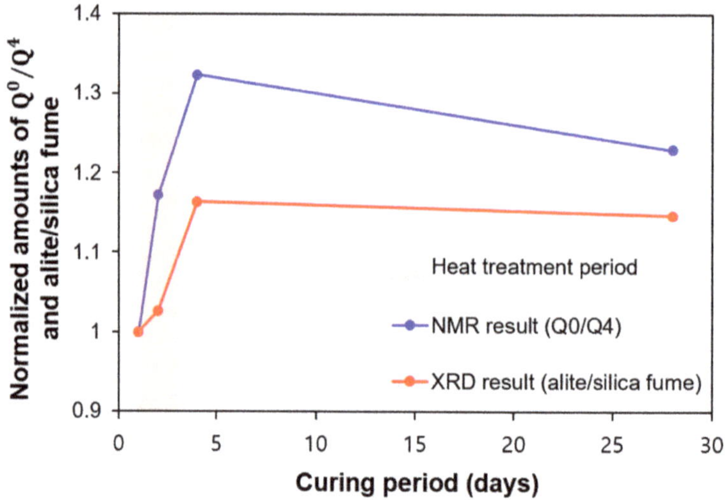

Figure 13. Normalized amounts of Q^0/Q^4 and alite/silica fume as a function of curing time.

The abrupt increase in both Q^0/Q^4 and alite/silica fume directly indicated greater consumption of silica fume at early ages. Again, this can be related to the efficiently accelerated pozzolanic reaction due to heat treatment. Additionally, the steady decrease in those values at later ages can be interpreted as a more active OPC hydration (instead of the pozzolanic reaction). Although the general trend is similar, the NMR showed a higher value compared to XRD at 2 days due to the difference caused by the PONKCS method, which was adopted to quantify the silica fume in a mixture.

Compared to the measurement accuracy of NMR in identifying silica fume and alite, QXRD certainly had a limitation in accurately quantifying the relative amount of silica fume intrinsically, represented by a small hump in the XRD pattern. Although this study was able to capture the silica fume using the PONKCS method, the accuracy of quantifying the silica fume may have been lower, especially during the accelerated pozzolanic reaction period. During the heat treatment, the consumption of well-dispersed silica fume particles occurred first. Later, an additional reaction proceeded for the bulk type of silica fume, which had a stronger intensity in the XRD experiment. Therefore, during the early stages of heat treatment, the PONKCS method may underestimate the amount of silica fume.

5. Conclusions

This study elucidated the complicated hydration reaction of UHPC using various experiments. Using the PONKCS method, the amorphous phases of silica fume and C-S-H were successfully separated and quantified. To confirm the reliability of the QXRD results, TGA and NMR were also used. The conclusions of this study are as follows:

1. The compressive strength of UHPC cured for 1, 2, 4, and 28 days was 50.7, 129.3, 136.9, and 117.2 MPa, respectively. The 2 days of heat treatment significantly enhanced the pozzolanic reaction, effectively increasing the compressive strength. However, it showed a slight decrease at later ages, which was explained by the micro-cracking that may have been caused by the steady formation of C-S-H on the pre-formed dense microstructure of UHPC. This is plausible, since the formation was dominated by OPC hydration (rather than pozzolanic reaction) at later ages. which should involve the micro-expansion of existing hydration products to secure space for subsequent hydration;
2. In terms of the degree of hydration of UHPC, the QXRD and TGA showed similar results. At 28 days, both QXRD and NMR indicated that a significant amount of silica fume was left, which would produce a physical filling effect on the UHPC. The NMR also showed the evolution of the nanostructure of C-S-H, as the MCL of C-S-H increased over time. This structural change was not considered in the C-S-H model adopted in the PONKCS method. Therefore, the variation of crystallinity of C-S-H or the structural difference of C-S-H in regular OPC hydration and UHPC can produce an error when quantifying C-S-H using the PONKCS method;
3. This study was able to quantify the two main amorphous phases (i.e., silica fume and C-S-H) in UHPC. During the heat treatment of UHPC, the consumption of silica fume was dominant, as shown by the more active pozzolanic reaction. The evolution of the quantitative content of C-S-H was well correlated with the development of strength. At a very early stage of heat treatment, QXRD underestimated the content of silica fume; the dispersed silica fume with a small contribution of XRD peak intensity was consumed first during the heat treatment.

Author Contributions: Conceptualization, H.K. and J.M.; methodology, H.K. and J.M.; software, H.K. and N.L.; validation, H.K. and J.M.; formal analysis, H.K.; investigation, H.K.; resources, H.K.; data curation, H.K. and N.L.; writing—original draft preparation, H.K.; writing—review and editing, N.L. and J.M.; visualization, H.K.; supervision, J.M. All authors have read and agreed to the published version of the manuscript.

Funding: This work was supported by the Research Resettlement Fund for the new faculty of Seoul National University.

Acknowledgments: The Institute of Engineering Research at Seoul National University provided research facilities for this work.

Conflicts of Interest: The authors declare no conflict of interest.

References

1. Jeon, D.; Jun, Y.; Jeong, Y.; Oh, J.E. Microstructural and strength improvements through the use of Na_2CO_3 in a cementless Ca (OH) 2-activated class F fly ash system. *Cem. Concr. Res.* **2015**, *67*, 215–225. [CrossRef]
2. Jeong, Y.; Hargis, C.W.; Chun, S.-C.; Moon, J. The effect of water and gypsum content on strätlingite formation in calcium sulfoaluminate-belite cement pastes. *Constr. Build. Mater.* **2018**, *166*, 712–722. [CrossRef]
3. Mehta, P.K.; Monteiro, P.J. *Concrete Microstructure, Properties and Materials*; McGraw-Hill: New York, NY, USA, 2017; pp. 23–32.
4. Schöler, A.; Lothenbach, B.; Winnefeld, F.; Zajac, M. Hydration of quaternary Portland cement blends containing blast-furnace slag, siliceous fly ash and limestone powder. *Cem. Concr. Compos.* **2015**, *55*, 374–382. [CrossRef]
5. Scrivener, K.; Füllmann, T.; Gallucci, E.; Walenta, G.; Bermejo, E. Quantitative study of Portland cement hydration by X-ray diffraction/Rietveld analysis and independent methods. *Cem. Concr. Res.* **2004**, *34*, 1541–1547. [CrossRef]
6. Alahrache, S.; Winnefeld, F.; Champenois, J.-B.; Hesselbarth, F.; Lothenbach, B. Chemical activation of hybrid binders based on siliceous fly ash and Portland cement. *Cem. Concr. Compos.* **2016**, *66*, 10–23. [CrossRef]
7. Carmona-Quiroga, P.; Blanco-Varela, M. Ettringite decomposition in the presence of barium carbonate. *Cem. Concr. Res.* **2013**, *52*, 140–148. [CrossRef]
8. Haha, M.B.; Le Saout, G.; Winnefeld, F.; Lothenbach, B. Influence of activator type on hydration kinetics, hydrate assemblage and microstructural development of alkali-activated blast-furnace slags. *Cem. Concr. Res.* **2011**, *41*, 301–310. [CrossRef]
9. Rietveld, H. A profile refinement method for nuclear and magnetic structures. *J. Appl. Crystallogr.* **1969**, *2*, 65–71. [CrossRef]
10. Jansen, D.; Stabler, C.; Goetz-Neunhoeffer, F.; Dittrich, S.; Neubauer, J. Does Ordinary Portland Cement Contain Amorphous Phase? A Quantitative Study Using an External Standard Method. *Powder Diffr. J.* **2011**, *26*, 31–38. [CrossRef]
11. Naber, C.; Stegmeyer, S.; Jansen, D.; Goetz-Neunhoeffer, F.; Neubauer, J. The PONKCS method applied for time-resolved XRD quantification of supplementary cementitious material reactivity in hydrating mixtures with ordinary Portland cement. *Constr. Build. Mater.* **2019**, *214*, 449–457. [CrossRef]
12. Jeong, Y.; Hargis, C.W.; Kang, H.; Chun, S.-C.; Moon, J. The effect of elevated curing temperatures on high ye'elimite calcium sulfoaluminate cement mortars. *Materials* **2019**, *12*, 1072. [CrossRef] [PubMed]
13. Lee, N.; Jeong, Y.; Kang, H.; Moon, J. Heat-induced acceleration of pozzolanic reaction under restrained conditions and consequent structural modification. *Materials* **2020**, *13*, 2950. [CrossRef] [PubMed]
14. Bruce, P.G.; Le Bail, A.; Madsen, I.; Cranswick, L.M.; Cockcroft, J.K.; Norby, P.; Zuev, A.; Fitch, A.; Rodriguez-Carvajal, J.; Giacovazzo, C. Powder Diffraction: Theory and Practice. *R. Soc. Chem.* **2015**, 541–548.
15. Madsen, I.; Scarlett, N.; Kleeberg, R.; Knorr, K. Quantification of Phases with Partial or No Known Crystal Structures. *Int. Tables Crystallogr.* **2019**, 356–360.
16. Kang, H.; Kang, S.-H.; Jeong, Y.; Moon, J. Quantitative analysis of hydration reaction of GGBFS using X-ray diffraction method. *J. Korea Concr. Inst.* **2020**, *32*, 241–250. [CrossRef]
17. Zhou, M.; Lu, W.; Song, J.; Lee, G.C. Application of ultra-high performance concrete in bridge engineering. *Constr. Build. Mater.* **2018**, *186*, 1256–1267. [CrossRef]
18. Klemens, T. Flexible concrete offers new solutions. *Concr. Constr.* **2004**, *49*, 72.
19. Tu'ma, N.H.; Aziz, M.R. Flexural performance of composite ultra-high performance concrete-encased steel hollow beams. *Civ. Eng. J.* **2019**, *5*, 1289–1304. [CrossRef]
20. Yang, R.; Yu, R.; Shui, Z.; Gao, X.; Xiao, X.; Fan, D.; Chen, Z.; Cai, J.; Li, X.; He, Y. Feasibility analysis of treating recycled rock dust as an environmentally friendly alternative material in ultra-high performance concrete (UHPC). *J. Clean. Prod.* **2020**, *258*, 120673. [CrossRef]
21. Ghasemi, S.; Zohrevand, P.; Mirmiran, A.; Xiao, Y.; Mackie, K. A super lightweight UHPC–HSS deck panel for movable bridges. *Eng. Struct.* **2016**, *113*, 186–193. [CrossRef]
22. Luo, J.; Shao, X.; Fan, W.; Cao, J.; Deng, S. Flexural cracking behavior and crack width predictions of composite (steel+ UHPC) lightweight deck system. *Eng. Struct.* **2019**, *194*, 120–137. [CrossRef]

23. Snellings, R.; Salze, A.; Scrivener, K. Use of X-ray diffraction to quantify amorphous supplementary cementitious materials in anhydrous and hydrated blended cements. *Cem. Concr. Res.* **2014**, *64*, 89–98. [CrossRef]
24. Huang, W.; Kazemi-Kamyab, H.; Sun, W.; Scrivener, K. Effect of cement substitution by limestone on the hydration and microstructural development of ultra-high performance concrete (UHPC). *Cem. Concr. Compos.* **2017**, *77*, 86–101. [CrossRef]
25. Lee, N.K.; Koh, K.; Kim, M.O.; Ryu, G. Uncovering the role of micro silica in hydration of ultra-high performance concrete (UHPC). *Cem. Concr. Res.* **2018**, *104*, 68–79. [CrossRef]
26. ISO, I. *Cement-Test Methods-Determination of Strength*; ISO: Geneva, Switzerland, 2009.
27. Kang, S.-H.; Hong, S.-G.; Moon, J. Performance comparison between densified and undensified silica fume in ultra-High performance fiber-reinforced concrete. *Materials* **2020**, *13*, 3901. [CrossRef] [PubMed]
28. Kang, S.-H.; Jeong, Y.; Tan, K.H.; Moon, J. High-volume use of limestone in ultra-high performance fiber-reinforced concrete for reducing cement content and autogenous shrinkage. *Constr. Build. Mater.* **2019**, *213*, 292–305. [CrossRef]
29. Kang, S.-H.; Hong, S.-G.; Moon, J. The use of rice husk ash as reactive filler in ultra-high performance concrete. *Cem. Concr. Res.* **2019**, *115*, 389–400. [CrossRef]
30. Mills, K. The Estimation of Slag Properties. Short Course Presented as Part of Southern African Pyrometallurgy. 2011, pp. 35–36. Available online: https://www.pyro.co.za/KenMills/KenMills.pdf (accessed on 19 October 2020).
31. Kang, S.-H.; Lee, J.-H.; Hong, S.-G.; Moon, J. Microstructural investigation of heat-treated ultra-high performance concrete for optimum production. *Materials* **2017**, *10*, 1106. [CrossRef]
32. Lee, N.; Koh, K.; Park, S.; Ryu, G. Microstructural investigation of calcium aluminate cement-based ultra-high performance concrete (UHPC) exposed to high temperatures. *Cem. Concr. Res.* **2017**, *102*, 109–118. [CrossRef]
33. Jeong, Y.; Park, H.; Jun, Y.; Jeong, J.-H.; Oh, J.E. Microstructural verification of the strength performance of ternary blended cement systems with high volumes of fly ash and GGBFS. *Constr. Build. Mater.* **2015**, *95*, 96–107. [CrossRef]
34. Abdulkareem, O.M.; Fraj, A.B.; Bouasker, M.; Khelidj, A. Effect of chemical and thermal activation on the microstructural and mechanical properties of more sustainable UHPC. *Constr. Build. Mater.* **2018**, *169*, 567–577. [CrossRef]
35. Ferg, E.; Simpson, B. Using PXRD and PONKCS to determine the kinetics of crystallisation of highly concentrated NH 4 NO 3 emulsions. *J. Chem. Crystallogr.* **2013**, *43*, 197–206. [CrossRef]
36. Li, X.; Snellings, R.; Scrivener, K.L. Quantification of amorphous siliceous fly ash in hydrated blended cement pastes by X-ray powder diffraction. *J. Appl. Crystallogr.* **2019**, *52*, 1358–1370. [CrossRef]
37. Zhou, Q.; Glasser, F.P. Thermal stability and decomposition mechanisms of ettringite at <120 C. *Cem. Concr. Res.* **2001**, *31*, 1333–1339.
38. Metz, K.; Lam, M.; Webb, A. Reference deconvolution: A simple and effective method for resolution enhancement in nuclear magnetic resonance spectroscopy. *Concepts Magn. Reson.* **2000**, *12*, 21–42. [CrossRef]
39. Pena, P.; Mercury, J.R.; De Aza, A.; Turrillas, X.; Sobrados, I.; Sanz, J. Solid-state 27Al and 29Si NMR characterization of hydrates formed in calcium aluminate–silica fume mixtures. *J. Solid State Chem.* **2008**, *181*, 1744–1752. [CrossRef]
40. Andersen, M.D.; Jakobsen, H.J.; Skibsted, J. Characterization of white Portland cement hydration and the CSH structure in the presence of sodium aluminate by 27Al and 29Si MAS NMR spectroscopy. *Cem. Concr. Res.* **2004**, *34*, 857–868. [CrossRef]
41. Odelson, J.B.; Kerr, E.A.; Vichit-Vadakan, W. Young's modulus of cement paste at elevated temperatures. *Cem. Concr. Res.* **2007**, *37*, 258–263. [CrossRef]
42. Taylor, H.F. *Cement Chemistry*; Thomas Telford London: London, UK, 1997; Volume 2.

Article

Effect of Clinker Binder and Aggregates on Autogenous Healing in Post-Crack Flexural Behavior of Concrete Members

Kwang-Myong Lee [1], Young-Cheol Choi [2], Byoungsun Park [3], Jinkyo F. Choo [4,*] and Sung-Won Yoo [2,*]

[1] Department of Civil and Environmental System Engineering, Sungkyunkwan University, 2066 Seobu-ro, Jangan-gu, Suwon-si 16419, Gyeonggi-do, Korea; leekm79@skku.edu
[2] Department of Civil and Environmental Engineering, Gachon University, 1342 Seongnamdae-ro, Sujeong-gu, Seongnam-si 13120, Gyeonggi-do, Korea; zerofe@gachon.ac.kr
[3] Construction Technology Research Center, Korea Conformity Laboratories, 199 Gasan digital 1-ro, Geumcheon-gu, Seoul 08503, Korea; pbs0927@kcl.re.kr
[4] Department of Energy Engineering, Konkuk University, 120 Neungdong-ro, Gwangjin-gu, Seoul 05029, Korea
* Correspondence: jfchoo@konkuk.ac.kr (J.F.C.); imysw@gachon.ac.kr (S.-W.Y.); Tel.: +82-31-750-5723 (S.-W.Y.)

Received: 16 September 2020; Accepted: 9 October 2020; Published: 12 October 2020

Abstract: Crack healing has been studied extensively to protect reinforced concrete structures from the ingress of harmful ions. Research examining the regain in the mechanical properties of self-healing composites has focused mostly on the computation of the healing ratio based on the measurement of the tensile and compressive strengths but with poor regard for the flexural performance. However, the regain in the flexural performance should also be investigated for design purposes. The present study performs flexural testing on reinforced concrete members using crushed clinker binder and aggregates as well as crystalline admixtures as healing agents. Healing ratios of 100% for crack widths smaller than 200 μm and 85% to 90% for crack widths of 250 μm were observed according to the admixing of clinker binder and aggregates. Water flow test showed that the members replacing binder by 100% of clinker achieved the best crack healing performance. The crack healing property of concrete improved to some extent the rebar yield load, the members' ultimate load and energy absorption capacity and ductility index. The crack distribution density from the observed crack patterns confirmed the crack healing effect provided by clinker powder. The fine grain size of clinker made it possible to replace fine aggregates and longer healing time increased the crack healing effect.

Keywords: crack healing; clinker binder and aggregate; flexural performance; water flow test; curing

---

1. Introduction

Reinforced concrete can deteriorate due to a variety of reasons among which the corrosion of reinforcing steel is a leading cause. Reinforced steel corrodes due to the ingress of harmful ions like chlorides and sulfates. Once steel starts to corrode, the resulting rust occupies a larger volume, which creates tensile stresses in concrete and eventually leads to cracking, delamination and spalling of concrete. In turn, such cracking of concrete accelerates the penetration of harmful ions that worsens the degradation of the reinforced concrete structure [1]. Implementing repair of concrete cracking in due time is thus fundamental in extending the lifespan of the structure. However, apart from being costly and labor-intensive, manual repair is often inaccessible in offshore and underground concrete structures, which are likely to experience aggravated degradation and a shortened lifespan.

The restless search for durability and resilience of concrete structures has led some researchers to focus on self-healing concrete composites with the built-in ability of repairing narrow cracks without human or external intervention [2–5]. Two major types of self-healing concretes have emerged: the autogenous type [4], which is achieved by autogenous healing materials such as mineral admixtures like ground-granulated blast-furnace slag (GGBFS), silica fume or fly ash, fibers and nanofillers [6–12]; and, the autonomous type [3], which is realized by unconventional engineered additions such as shape memory alloys, capsules, polymers or bacteria to seal the cracks [13–24]. Autogenous healing is an old and well-known phenomenon that originates naturally from the cementitious material like the hydration of clinker minerals or the carbonation of calcium hydroxide while autonomous healing requires a trigger to activate the process. Both types have been proven to restore the mechanical properties and durability of the concrete structure to some extent but are believed to be only capable of repairing cracks within a few hundreds of micrometers, meaning structural damage cannot be repaired. To date, the autonomous healing method has shown better performance in healing cracks than most of the autogenous healing methods which have been seen to heal cracks with widths narrower than 150 μm [5].

Most studies assessing the self-healing performance of concrete have focused on the recovery of durability through the evaluation of the crack filling ratio and the reduction of penetration [25,26]. The methods usually adopted are chloride permeability test, water permeability test, isothermal calorimetry, crack closing test, etc. [12,25]. However, as pointed out by Guo and Chidiac [27], self-healing of concrete is made up of two concurrent concepts: self-sealing of cracks, which requires plugging of openings for durability, and self-healing of cracks, which refers to recovery of mechanical properties for strength. Therefore, some studies also conducted tensile and compressive strength tests, flexural and ultrasonic pulse velocity tests, etc. to assess the recovery of mechanical properties. When it comes to flexure, studies on self-healing concrete focus mainly on crack repair rather than mechanical performance. It is also noteworthy that the regain in mechanical properties may become meaningless if the repaired crack is weaker than the concrete matrix, subsequent cracks may develop at the same location when the healing agent is exhausted [2,27].

Considering the lifespan of the concrete structure and to be free of the exhaustion of the healing agent, this study pays attention to the autogenous healing. A recent study showed that the autogenous healing could be enhanced using GGBFS and crystalline admixture rather than using fly ash [28]. The results of this work confirmed that concretes incorporating supplementary cementitious materials like GGBFS can develop superior self-healing properties owing to the significant amount of unhydrated particles present in its microstructures as well as improved mechanical and permeation properties when stressed by mechanical loads [29]. On the other hand, another study investigated the proper distribution of cement particle sizes providing a suitable amount of $Ca(OH)_2$ and unhydrated cement to improve the self-healing ability of concrete [30]. Moreover, Berger [31] and Allahverdi [32] reported that the use of cement clinker aggregates significantly improves the concrete properties like the compressive strength, chloride penetration depth and water absorption. Besides, the phases present in the cement clinker binder are known to yield a strong solid with low porosity and offer protection to chloride ingress.

Accordingly, this study examines the autogenous healing of concrete containing clinker binder and aggregates as well as crystalline admixtures as healing agents. Crack healing effect is investigated through crack closing and water flow tests. In addition, even if the flexural behavior depends to a large extent on the steel reinforcement, bending tests are conducted on full-scale test members to investigate any eventual recovery in the flexural mechanical performance brought by the autogenous healing. Fo the tests, three different concrete mixes and four different loading and curing conditions were chosen as test variables. Cracking was induced by pre-loading the specimens at 28 days. The concrete mixes differ by the maximum particle size of the clinker powder replacing the binder and the aggregates. The curing conditions are air-dry curing for 28 days and additional water curing for 90 days. The loading conditions consider pre-loading and loading at 28 days, pre-loading at 28 days and loading at 28 + 90 days.

2. Experimental Methods

2.1. Material Test of Mortar

The first stage of the investigation started with material test on the mortar mixes to be used later in the fabrication of the concrete specimens. Table 1 presents the chemical composition of the raw materials. The grain size analysis of ground granulated blast-furnace slag (GGBFS) revealed grain sizes between 0.011 and 58.953 μm with an average of 10 μm (Figure 1). The grain distribution and chemical composition of GGBFS and clinker appeared to present no significant difference. Table 2 arranges the mix proportions of the 3 series of mortar considered in this study. Plain series is the control series with mortar made of ordinary Portland cement (OPC) and sand. Series 2.5 and 0.85 correspond to mortars with clinker binder and aggregates in which clinker was crushed to have particle sizes of 2.5 mm and 0.85 mm for replacing sand and cement, respectively. A water-to-binder ratio of 0.4 was used in all the mixes.

Table 1. Chemical composition of raw materials (in weight ratio).

Material	CaO	SiO_2	Al_2O_3	Fe_2O_3	MgO	K_2O	Na_2O	SO_3
GGBFS	42.51	29.13	15.82	0.67	4.43	0.52	0.28	3.59
Clinker	64.34	22.87	4.96	2.94	1.28	0.82	0.23	0.44

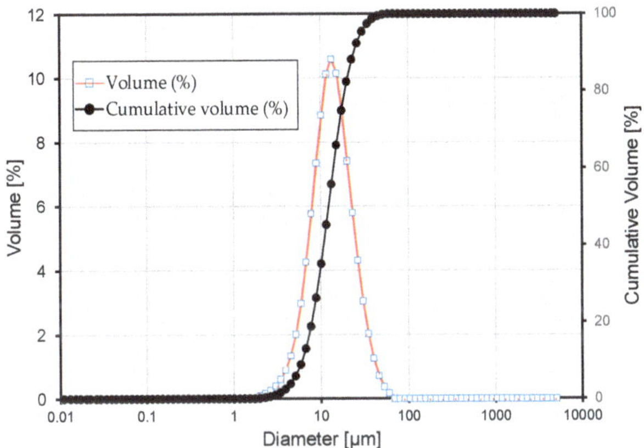

Figure 1. Grain size analysis results of GGBFS.

Table 2. Mix proportions of mortar (in kg/m^3, W/B = 0.4).

Series	Binder					Aggregate	
	OPC	GGBFS	Na_2SO_4	Anhydrite	Clinker Binder	Clinker Sand	Sand
Plain	690	-	-	-	-	-	1380
2.5	428	173	10	10	69	69	1311
0.85	428	173	10	10	138	-	1311

The self-healing performance of pre-cracked mortar specimens was assessed by constant head water flow test (Figure 2) on sets of three ϕ100 × 50 mm cylinders fabricated for each considered mix proportions. The fabricated specimens were stored for 1 day in a constant temperature and humidity chamber at temperature of 20 ± 1 °C and relative humidity of 100% prior to water curing for 27 days at 20 ± 1 °C. Regular cracking was induced by splitting test on the specimens notched in both ends. The crack width was adjusted to 0.25 mm and 0.30 mm by disposing silicon sheets on both ends of the specimens. The cylinders were installed in an acryl mold to fix them during the water flow test. The experimental setup follows the method proposed by Choi et al. [33], which is based on the work of Lepech and Li [34].

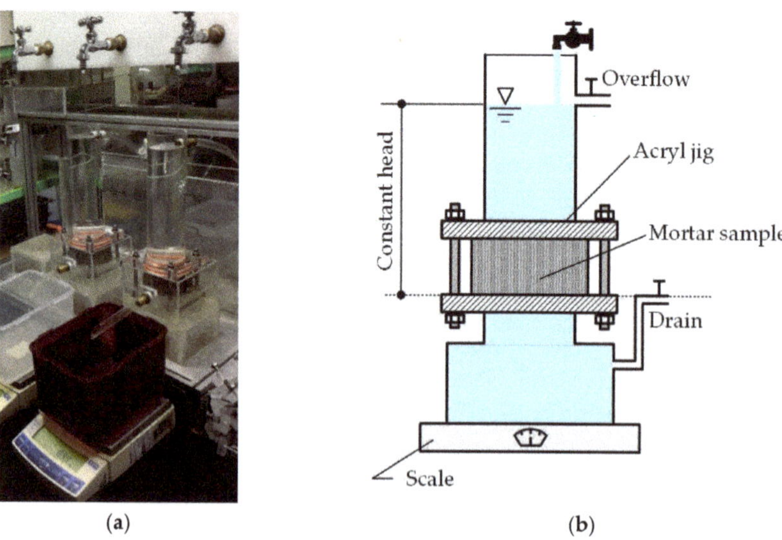

Figure 2. Water flow test: (**a**) photograph of actual test on mortar; (**b**) schematic illustration.

2.2. *Material Test of Concrete*

Table 3 summarizes the mix proportions of concrete using the 3 different types of mortar of this study. As mentioned above, two different sizes of clinker powder were considered to replace cement and sand. Among the ingredients, 0.85 stands for clinker binder (particle size < 0.85 mm) and 2.5 for clinker aggregate (grain size < 2.5 mm). SP represents superplasticizer of which the proportion corresponds to 1.5 weight percent.

Table 3. Mix proportions of concrete (in kg/m^3).

Series	Water	OPC	GGBFS	Na_2SO_4	Anhydrite	Clinker		Aggreg.	SP
						0.85	2.5		
Plain	275	495	172	10	-	-	-	1375	9
2.5	276	428	173	10	10	69	69	1311	9
0.85	276	428	173	10	10	138	-	1311	9

2.3. *Test for Fexural Behavior*

2.3.1. Test Variables

The three different concrete mixes listed in Table 3 and four different loading and curing conditions were chosen as test variables. The concrete mixes differ by the maximum particle size of the clinker powder replacing the binder and the aggregates: the mix with particle size of 0.85 mm for the clinker replacing aggregates; the mix with 50% of 0.85-mm clinker and 50% of 2.5-mm clinker replacing binder and aggregates; and, the plain series (control). The four loading and curing conditions are:

- V_28D series for the members at 28 days loaded until the ultimate state;
- PLRL_28D series for the members pre-loaded up to 50% of the ultimate load at 28 days followed by loading until the ultimate state;
- RL_28 + 90D series for the members pre-loaded up to 50% of the ultimate load at 28 days, followed by 90 days of water curing prior to loading until the ultimate state;
- V_28 + 90D series for the members loaded up to the ultimate load after 28 days and after 28 + 90 days.

The combination of these test variables gives a total of 12 test members. The members were fabricated with the mixes presented in Table 3 and were subjected to air-dry curing until 28 days. Table 4 arranges the designation and features of the test members. Note that this additional period of 90 days was decided following the work of Alyousif [35], which stated that engineered cementitious composite beams exhibit strength recovery after 90 days of extended moist curing regardless of their size.

Table 4. Designation and specifications of test members.

Designation	Clinker Particle Size	Curing Condition	Loading and Pre-Damage Status
V_28D-2.5	2.5 mm	28 days of air-drying	Loading up to ultimate state at 28 days
PLRL_28D-2.5		28 days of air-drying	Pre-loading up to 50% of ultimate load at 28 days + loading up to ultimate state at 28 days
RL_28 + 90D-2.5		28 days of air-drying followed by partial water curing for 90 days	Pre-loading up to 50% of ultimate load at 28 days + loading up to ultimate load 90 days after
V_28 + 90D-2.5		Air-drying until 28 + 90 days	Loading up to ultimate state after 28 + 90 days
V_28D-0.85	0.85 mm	28 days of air-drying	Loading up to ultimate state at 28 days
PLRL_28D-0.85		28 days of air-drying	Pre-loading up to 50% of ultimate load at 28 days + loading up to ultimate state at 28 days
RL_28 + 90D-0.85		28 days of air-drying followed by partial water curing for 90 days	Pre-loading up to 50% of ultimate load at 28 days + loading up to ultimate load 90 days after
V_28 + 90D-0.85		Air-drying until 28 + 90 days	Loading up to ultimate state after 28 + 90 days
P-V_28D	OPC	28 days of air-drying	Loading up to ultimate state at 28 days
P-PLRL_28D		28 days of air-drying	Pre-loading up to 50% of ultimate load at 28 days + loading up to ultimate state at 28 days
P-RL_28 + 90D		28 days of air-drying followed by partial water curing for 90 days	Pre-loading up to 50% of ultimate load at 28 days + loading up to ultimate load 90 days after
P-V_28 + 90D		Air-drying until 28 + 90 days	Loading up to ultimate state after 28 + 90 days

2.3.2. Fabrication of Test Members and Test Setup

Figure 3 depicts the shape of the flexural test members for examining the stiffness recovery. The reinforced concrete members present rectangular cross-section with width of 200 mm and height of 300 mm and were designed to have pure bending section of 600 mm. The steel reinforcement uses SD400 bars with a 19 mm-diameter as tensile reinforcement, a 13 mm-diameter as compressive reinforcement and a 13 mm-diameter as shear reinforcement. The materials were first dry-mixed prior to the introduction of water and the admixtures. Wet curing was conducted for two days in a laboratory by covering the members with a curing tent and supplying continuously appropriate humidity. Thereafter, dry-air curing was performed until 28 days.

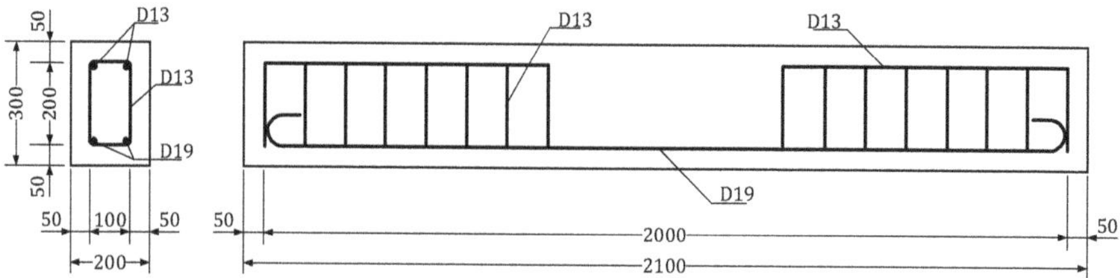

Figure 3. Dimensions and shape of flexural test members.

Strain gages were installed on the tensile and shear reinforcing bars before placing the concrete. Four strain gages were disposed on the tensile reinforcement with two sensors at mid-span and two sensors at 1/3 positions, that is at 600 mm in both sides from mid-span. Four strain gages were attached to the shear reinforcement with one sensor in each fourth and fifth shear rebar on the left and right-hand sides of the critical section at mid-span. The layout of the reinforcement strain gages is shown in Figure 4.

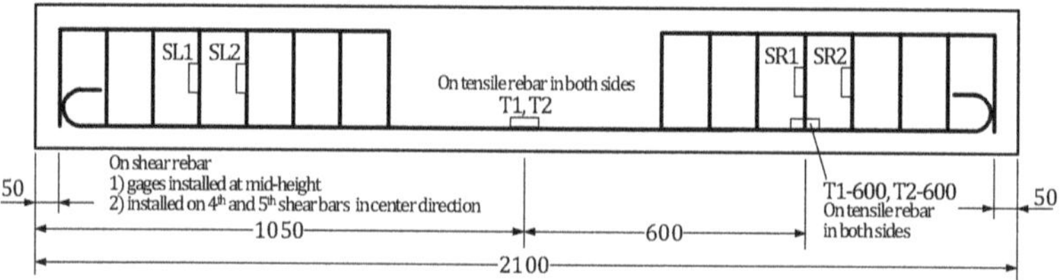

Figure 4. Layout of reinforcement strain gages in test members.

Figure 5 shows the setup for the 4-point bending test of the members as prescribed by ASTM C1609. The test members were simply supported by disposing them on supports located 150 mm from the ends of the members to achieve a supported length of 1800 mm. An LVDT was installed at the loading point for measuring the deflection to achieve displacement control for loading applied at speed of 0.5 mm/min. Figure 6 presents a photograph and a schematic illustration of the 4-point bending test of the flexural test members.

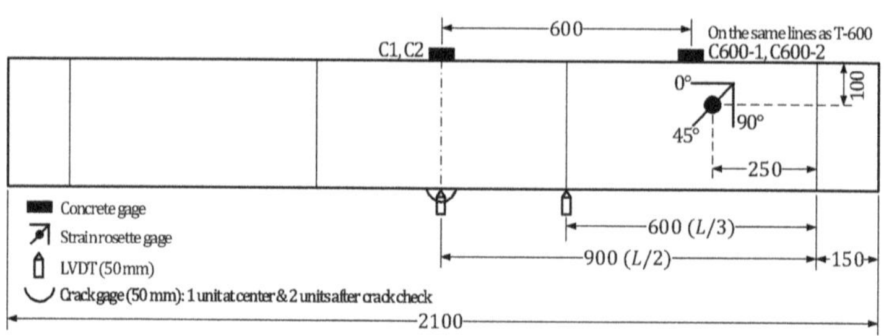

Figure 5. Setup for 4-point loading test.

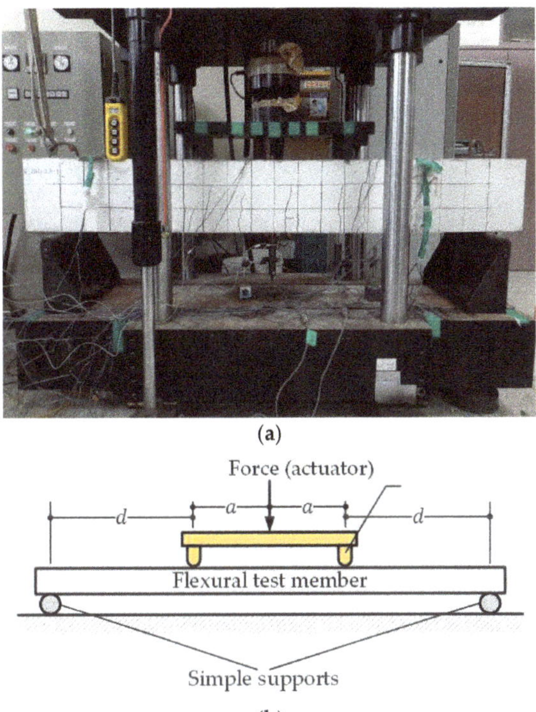

Figure 6. Four-point bending test of flexural members: (**a**) photograph of actual bending test; (**b**) schematic illustration.

As explained in Table 4, pre-loading was applied up to 50% of the ultimate load followed by loading until the ultimate state at 28 days for members PLRL_28D. Members RL_28 + 90D were pre-loaded up to 50% of the ultimate load at 28 days and 150 mm of their lower part was then subjected to partial water curing for 90 days to examine the crack healing performance of pre-damaged members (Figure 7) before loading until the ultimate state. Members V_28 + 90D experienced dry-air curing until 118 days (28 + 90 days) before being loaded up to the ultimate load to examine the stiffness recovery according to the maximum size of the clinker powder replacing the binder and the aggregates.

Figure 7. Partial water curing of pre-damaged test members.

3. Test Results

3.1. Material Test Results

The material test results of mortar and concrete are presented. Table 5 shows the compressive strength and slump flow measured for each of the considered mortars listed in Table 2. In Table 5, the values in parentheses represent the standard deviation, σ.

Table 5. Compressive strength and slump flow of considered mortars.

Series	Compressive Strength (σ)		Slump Flow (σ) (mm)
	7 Days (MPa)	28 Days (MPa)	
Plain	38.32 (1.12)	47.46 (1.48)	203 (8.2)
2.5	37.27 (0.89)	48.92 (1.59)	207 (8.0)
0.85	37.89 (1.03)	49.33 (1.56)	208 (8.1)

The water flow test results are plotted in Figure 8. The reduction ratio in Figure 8 indicates the water flow at 7 or 28 days as compared to that at day zero. This reduction ratio appears to be smaller as the crack width is large and the age is young. On the whole, the reduction ratio of the 0.85-mm series is higher than that of the 2.5-mm series samples, which shows that the replacement of binder by clinker powder brings a greater crack healing effect.

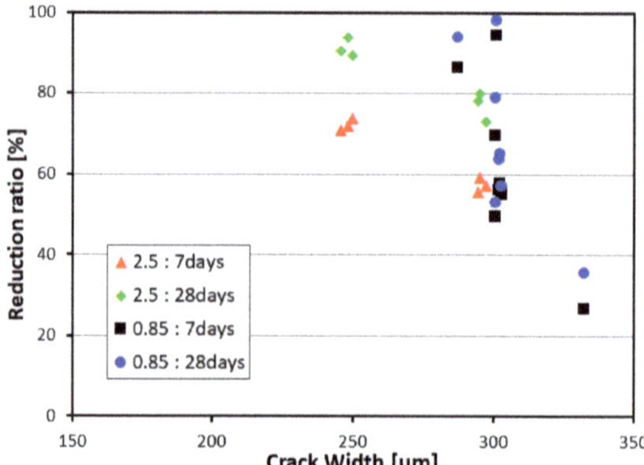

Figure 8. Water flow test results.

The basic physical properties of the concrete mixes are presented in Tables 6 and 7. Recalling that some test members experienced pre-damage at 28 days and were reloaded 90 days later, measurement of the physical properties was conducted at 28 days and 28 + 90 days. In Table 5, the compressive strength, f_{ck}, was measured on prismatic specimens at 28 days and 90 days later in compliance with KS L ISO 679. Similar levels of compressive strength were developed in all the mixes. The elastic modulus, E_c, was measured as the slope of the linear 10–40% region in the stress–strain diagram using triaxial concrete strain rosette gages attached to the ϕ100 × 200-mm mold. Results similar to those of the compressive strength were obtained. The 0.85-mix series with a relatively meaningful binder replacement ratio by 0.85 mm of clinker exhibited the best physical properties. The slump and air content measured on fresh concrete are also indicated in Table 6. Table 7 arranges the flexural strength, f_b, measured using the third-point loading method in compliance with KS F 2048. In Table 7, the values in parentheses represent the standard deviation, σ. Similarly to the results of Table 6, similar flexural strength was measured in all the mixes. Here also, the 0.85-mix series with a relatively meaningful binder replacement ratio by 0.85 mm of clinker exhibited the best physical properties. In Tables 6 and 7, the values in parentheses represent the standard deviation, σ.

Table 6. Physical properties of concrete mixes.

Series	f_{ck} (σ)		E_c (σ)		Slump (σ) (mm)	Air (%)
	28 Days (MPa)	28 + 90 Days (MPa)	28 Days (MPa)	28 + 90 Days (MPa)		
Plain	57.55 (2.01)	59.65 (1.98)	26,713 (1,219)	29,453 (1,410)	42.5 (2.10)	7.5
2.5	60.79 (2.08)	67.58 (2.11)	21,670 (1,002)	25,538 (987)	44.2 (2.06)	4.6
0.85	62.88 (2.14)	63.63 (2.18)	28,190 (1,145)	29,087 (1,328)	45.2 (1.97)	8.2

Table 7. Flexural test results of concrete mixes.

Series	Width of Failure Section (mm)		Height of Failure Section (mm)		Load (σ) (kN)		f_b (σ) (MPa)	
	28 Days	28 + 90 Days	28 Days	28 + 90 Days	28 Days	28 + 90 Days	28 Days	28 + 90 Days
Plain	99.8	100.0	101.0	98.8	16.0 (0.92)	16.6 (1.01)	7.1 (0.31)	7.6 (0.18)
2.5	101.7	100.0	100.0	100.0	13.8 (0.73)	18.2 (1.12)	6.1 (0.27)	8.4 (0.31)
0.85	98.0	100.0	100.5	100.0	17.6 (0.77)	19.3 (1.09)	8.0 (0.29)	8.7 (0.28)

3.2. Flexural Test Results

3.2.1. Crack, Yield, Ultimate Loads and Failure Pattern

All the test members exhibited negligible difference in the crack, yield and ultimate loads apart from some slight variability. Moreover, all the test members failed through flexure. Table 8 summarizes the measured crack, yield and ultimate loads as well as the failure pattern of the 12 test members listed in Table 4.

Table 8. Crack, yield and ultimate loads and failure patterns of test members.

Test Members	Crack Load (kN)	Yield (kN, mm)		Ultimate (kN, mm)		Ultimate/Yield		Failure Pattern
		Load	Displ.	Load	Displ.	Load	Displ.	
V_28D-2.5	5.63	242.18	7.93	267.90	18.10	1.11	2.28	Flexure
PLRL_28D-2.5	4.16	246.20	6.52	283.33	21.32	1.15	3.27	Flexure
RL_28 + 90D-2.5	11.54	238.97	8.39	288.00	38.58	1.21	4.60	Flexure
V_28 + 90D-2.5	9.96	245.02	7.48	281.01	26.02	1.15	3.48	Flexure
V_28D-0.85	2.78	234.94	7.93	277.88	20.02	1.18	2.52	Flexure
PLRL_28D-0.85	4.71	251.36	5.51	290.98	21.92	1.16	3.98	Flexure
RL_28 + 90D-0.85	11.79	233.91	5.81	291.90	41.94	1.25	7.22	Flexure
V_28 + 90D-0.85	9.42	245.59	7.28	289.55	37.02	1.18	5.09	Flexure
P-V_28D	6.17	240.44	8.52	268.29	20.23	1.12	2.37	Flexure
P-PLRL_28D	5.37	236.81	5.47	279.62	18.66	1.18	3.41	Flexure
P-RL_28 + 90D	11.32	237.08	6.94	282.60	41.78	1.19	6.02	Flexure
P-V_28 + 90D	5.92	239.54	8.26	278.75	33.72	1.16	4.08	Flexure

3.2.2. Load-Deflection Relations

Figures 9 and 10 plot the load-deflection relations measured after 28 days of curing. Figure 11 shows the load-deflection relations after pre-loading. Figure 12 plots the load-deflection relations after pre-loading and additional 90 days of curing. All the test members show linear load-deflection relations until early cracking followed by a nonlinear increase in the deflection after cracking and finally the increase in the load until the ultimate load. The members that experienced additional 90 days of curing after 28 days appear to undergo ultimate deflection nearly larger than 100% of that of the members loaded after 28 days of curing due to the plastic deformation effect induced by their age and the yield of steel reinforcement. This result confirms the increasing trend in the flexural resistance observed by Alyousif [35].

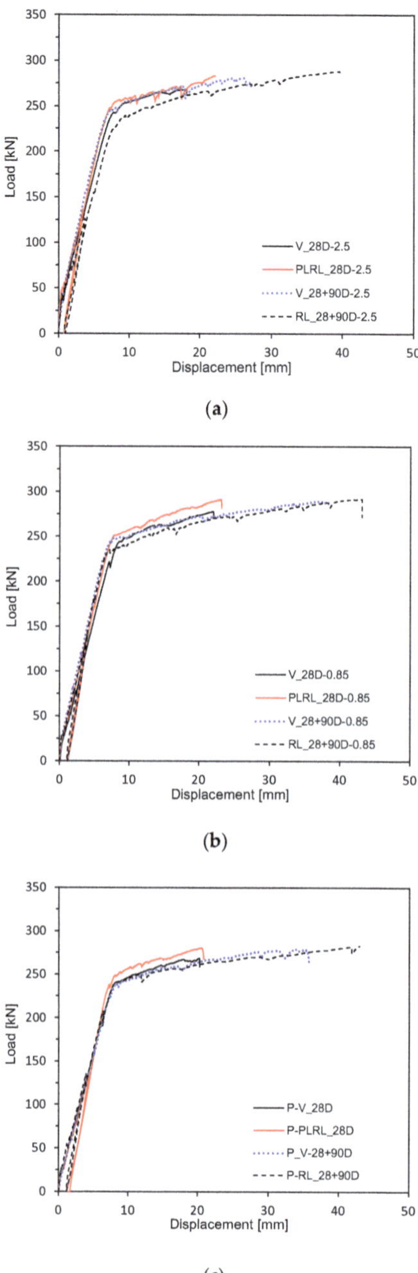

Figure 9. Load-deflection measurement by mixes: (**a**) 2.5 series; (**b**) 0.85 series; (**c**) Plain series.

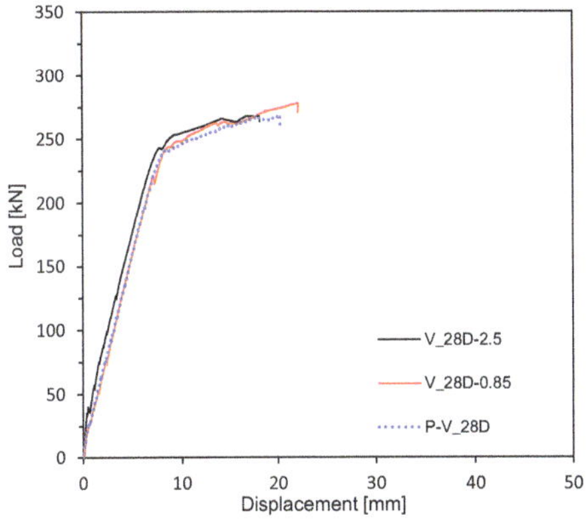
Load [kN]
Displacement [mm]
V_28D-2.5
V_28D-0.85
P-V_28D

(**a**)

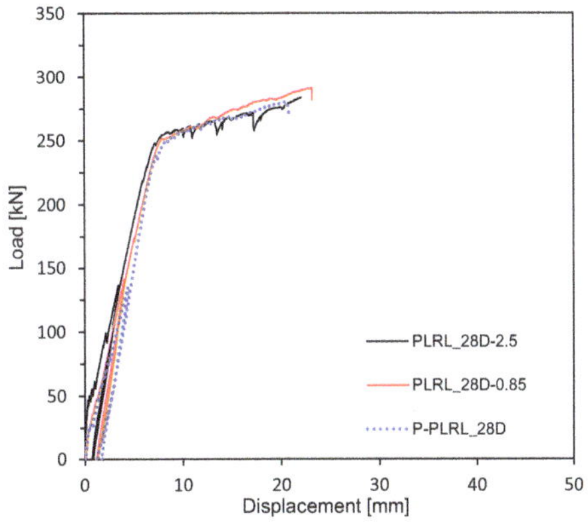
Load [kN]
Displacement [mm]
PLRL_28D-2.5
PLRL_28D-0.85
P-PLRL_28D

(**b**)

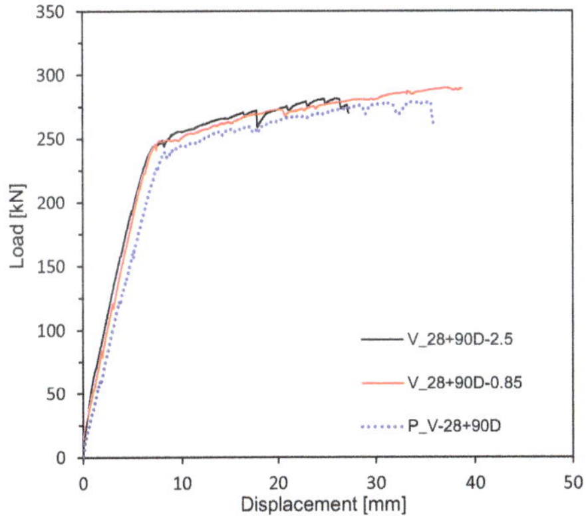
Load [kN]
Displacement [mm]
V_28+90D-2.5
V_28+90D-0.85
P_V-28+90D

(**c**)

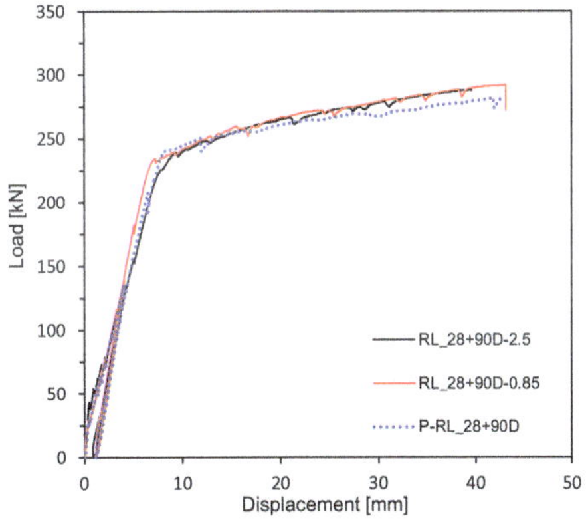
Load [kN]
Displacement [mm]
RL_28+90D-2.5
RL_28+90D-0.85
P-RL_28+90D

(**d**)

Figure 10. Load-deflection measurement by loading conditions: (**a**) V_28D series; (**b**) PLRL_28D series; (**c**) RL_28D series; (**d**) RL_28 + 90D series.

Load [kN]
Displacement [mm]
PLRL_28D-2.5
: initial loading
PLRL_28D-2.5
: re-loading

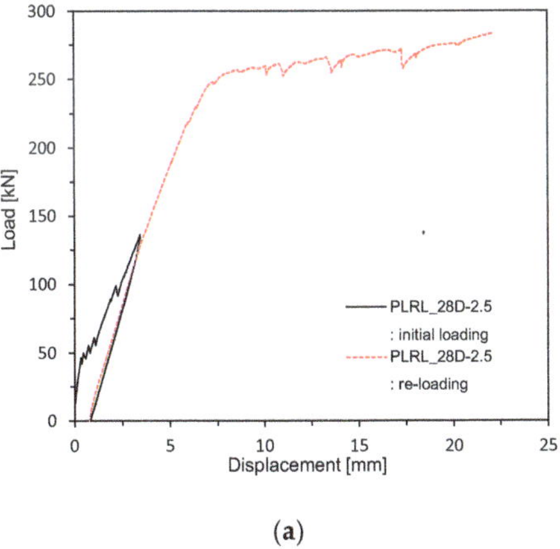

(**a**)

Figure 11. *Cont.*

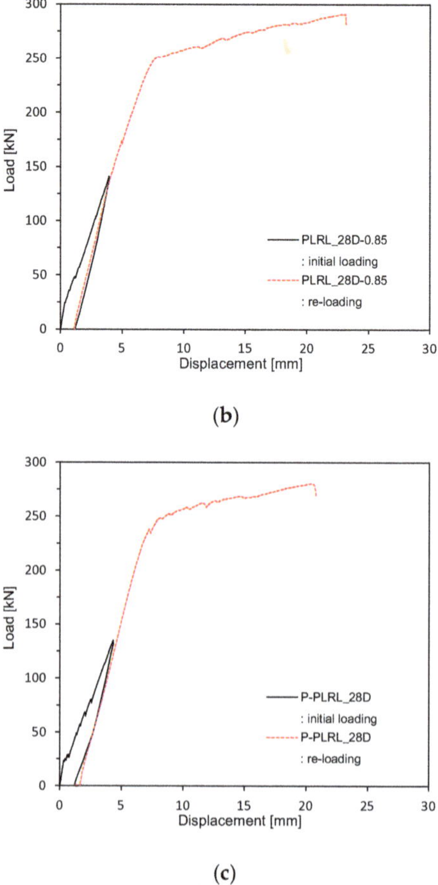

(b)

(c)

Figure 11. Load-deflection measurement after 28 days (up to ultimate load after pre-loading): (**a**) 2.5 series; (**b**) 0.85 series; (**c**) Plain series.

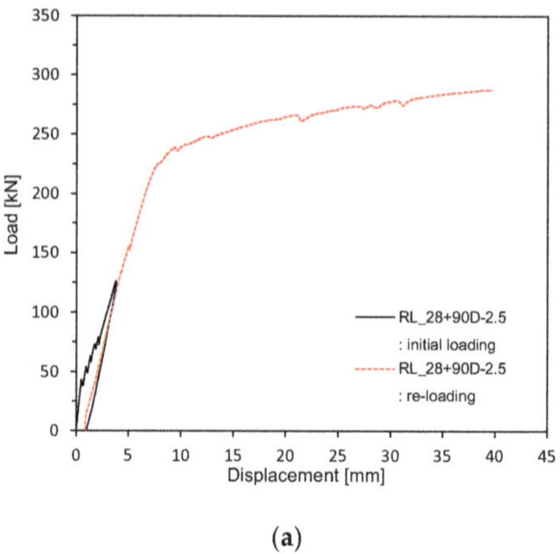

(a)

Figure 12. *Cont.*

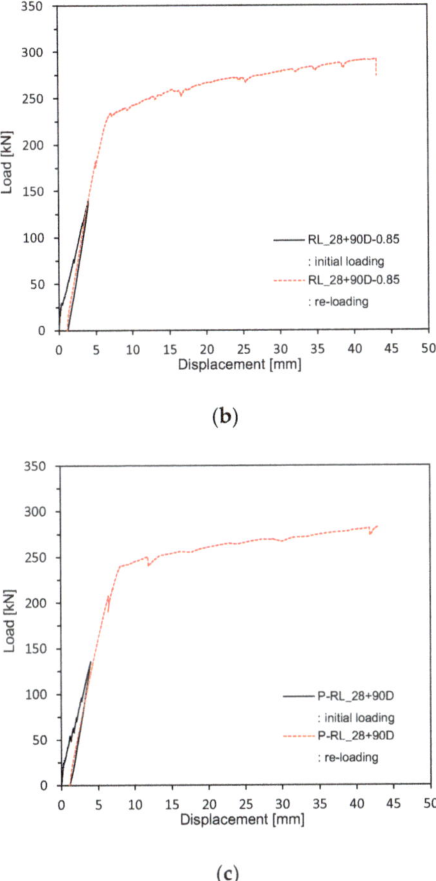

Figure 12. Load-deflection measurement after 28 + 90 days (up to ultimate load after pre-loading and additional 90 days of curing after pre-loading): (**a**) 2.5 series; (**b**) 0.85 series; (**c**) Plain series.

3.2.3. Load–Rebar Strain Relations

Figures 13 and 14 plot the load–rebar strain relations measured after 28 days of curing. Figure 15 shows the load–rebar strain relations after pre-loading. Figure 16 plots the load–rebar strain relations after pre-loading and an additional 90 days of curing. Here, the rebar strain practically did not develop before cracking and started to increase linearly after the initiation of cracks to increase starkly after yielding.

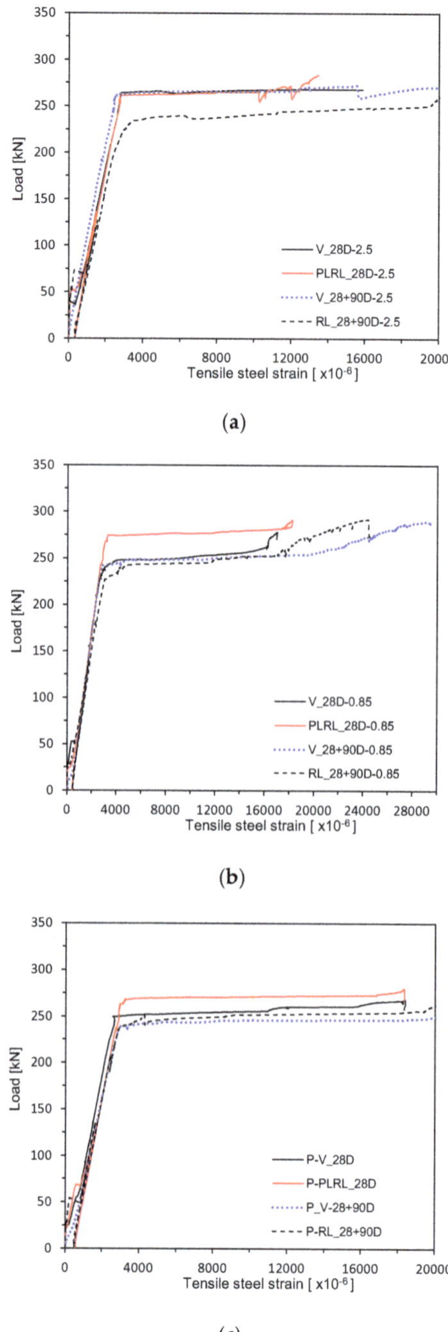

Figure 13. Load–steel reinforcement strain measurement by mixes: (**a**) 2.5 series; (**b**) 0.85 series; (**c**) Plain series.

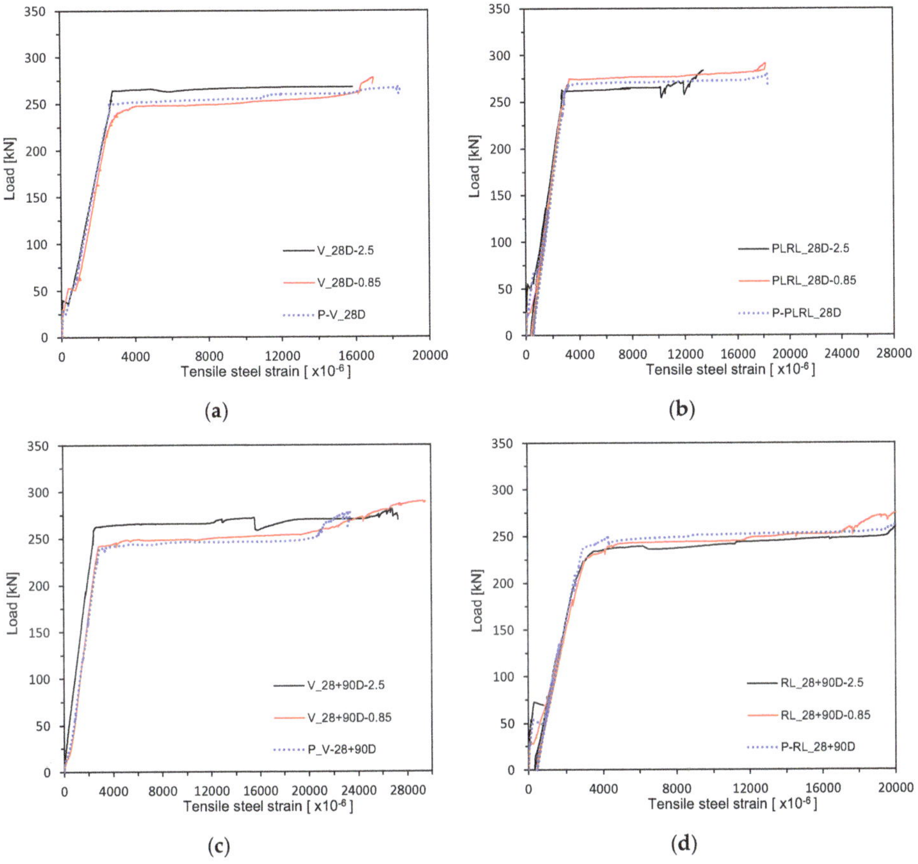

Figure 14. Load–steel reinforcement strain measurement by loading conditions: (**a**) V_28D series; (**b**) PLRL_28D series; (**c**) RL_28D series; (**d**) RL_28 + 90D series.

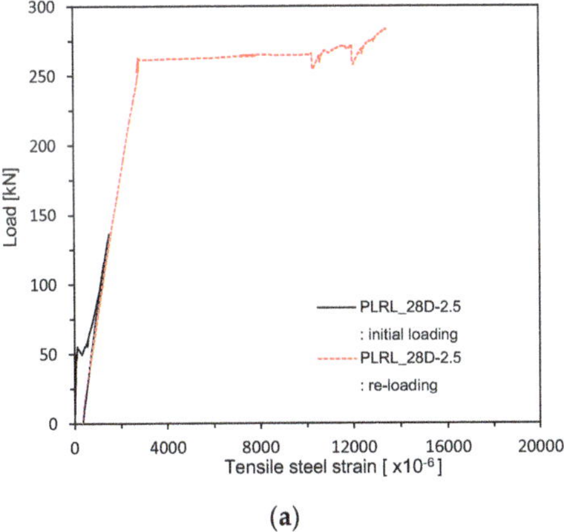

(**a**)

Figure 15. *Cont.*

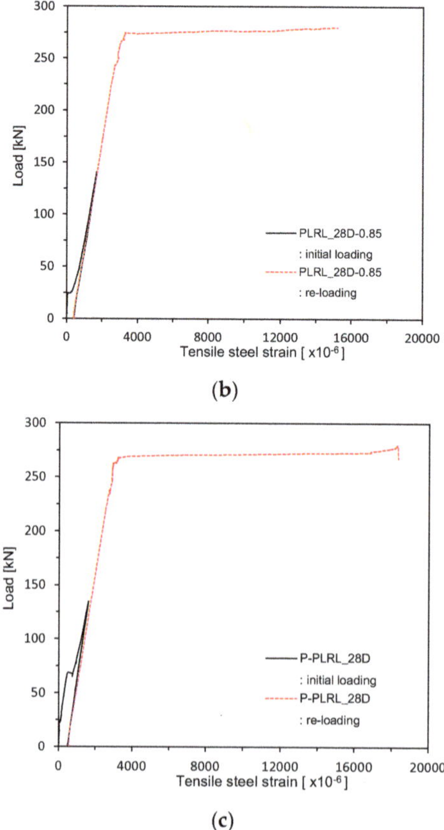

(b)

(c)

Figure 15. Load–steel reinforcement strain measurement after 28 days (up to ultimate load after pre-loading): (**a**) 2.5 series; (**b**) 0.85 series; (**c**) Plain series.

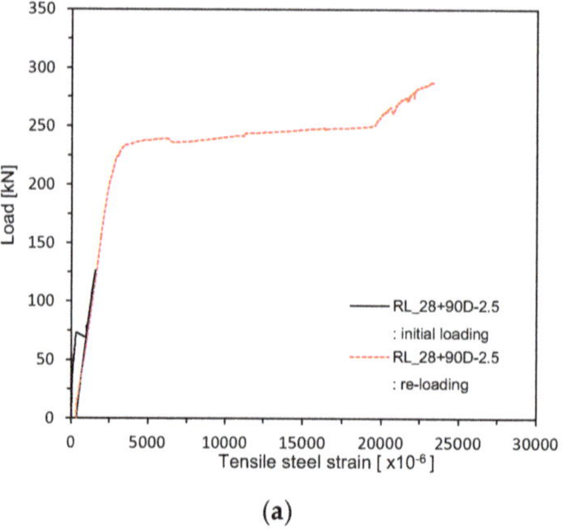

(a)

Figure 16. *Cont.*

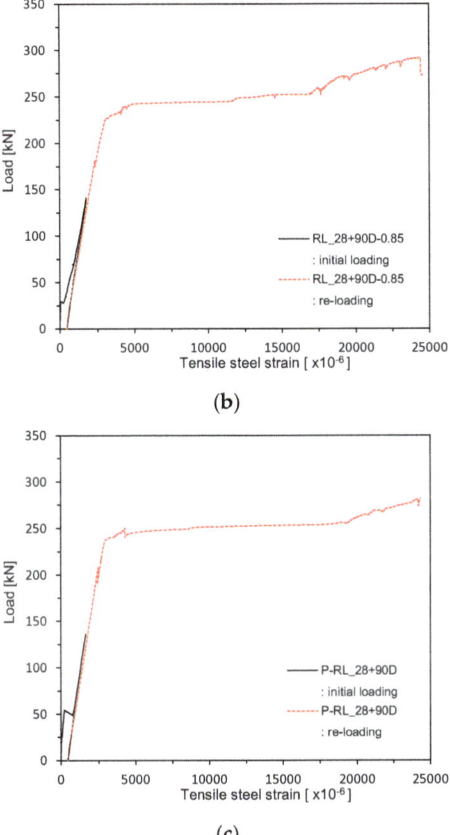

Figure 16. Load–steel reinforcement strain measurement after 28 + 90 days (up to ultimate load after pre-loading and additional 90 days of curing after pre-loading): (**a**) 2.5 series; (**b**) 0.85 series; (**c**) Plain series.

3.2.4. Load–Concrete Strain Relations

Figures 17 and 18 plot the load–concrete strain relations measured after 28 days of curing. Figure 19 shows the load–concrete strain relations after pre-loading. Figure 20 plots the load–concrete strain relations after pre-loading and additional 90 days of curing. The members loaded up to the ultimate at 28 days developed ultimate concrete strain of about 0.0030 but those which experienced additional 90 days of curing saw their ultimate concrete strain be around 0.0040 due to effects of aging and residual deformation.

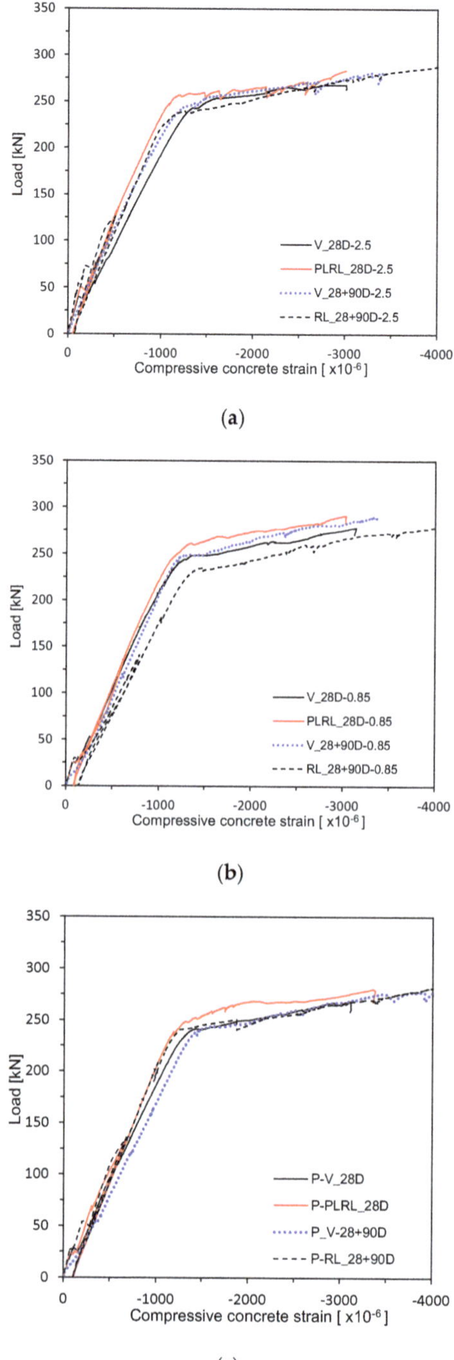

Figure 17. Load–concrete strain measurement by mixes: (**a**) 2.5 series; (**b**) 0.85 series; (**c**) Plain series.

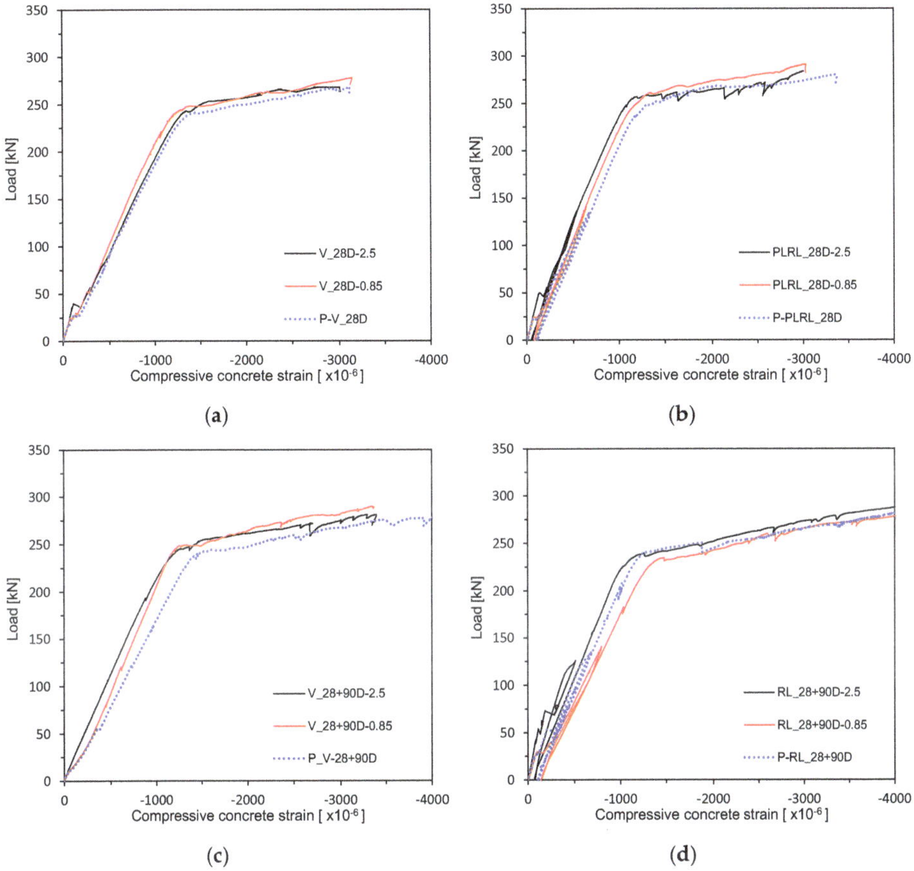

(a) (b)

(c) (d)

Figure 18. Load–concrete strain measurement by loading conditions: (**a**) V_28D series; (**b**) PLRL_28D series; (**c**) RL_28D series; (**d**) RL_28 + 90D series.

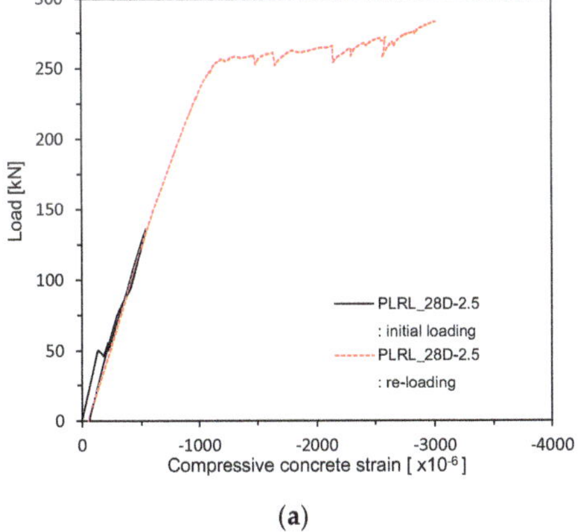

(**a**)

Figure 19. *Cont.*

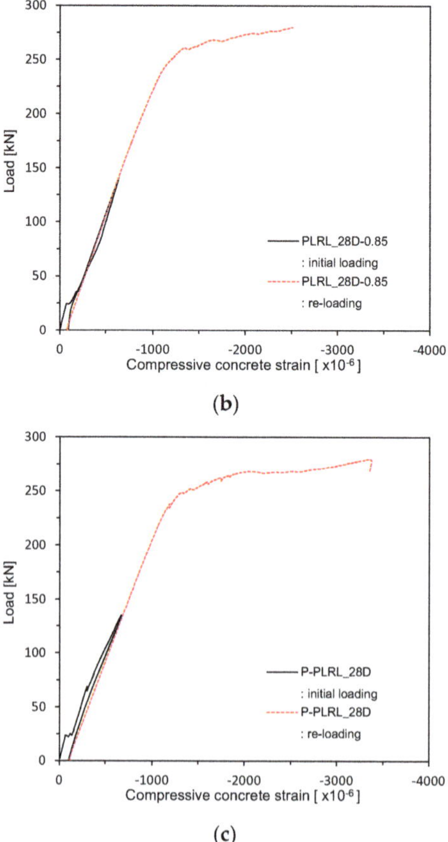

(**b**)

(**c**)

Figure 19. Load–concrete strain measurement after 28 days (up to ultimate load after pre-loading): (**a**) 2.5 series; (**b**) 0.85 series; (**c**) Plain series.

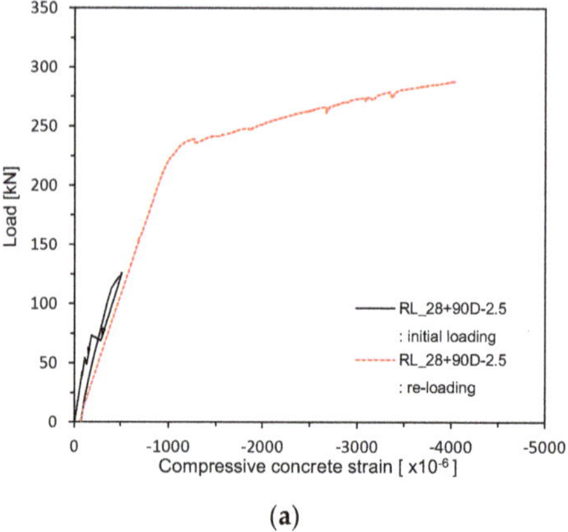

(**a**)

Figure 20. *Cont.*

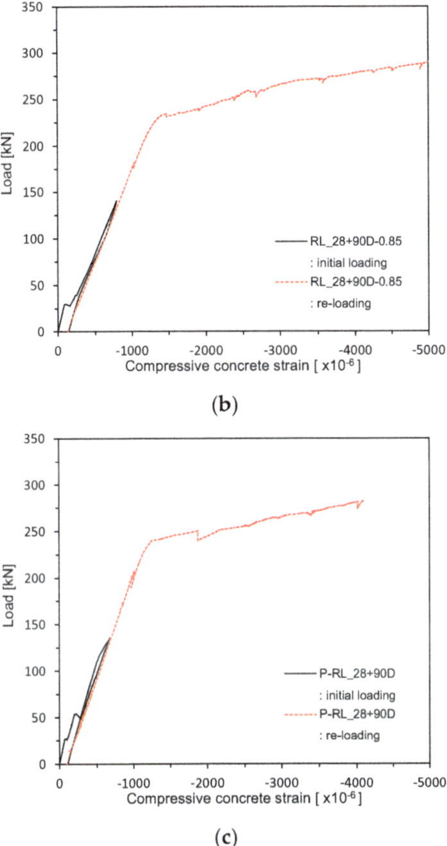

Figure 20. Load–concrete strain measurement after 28 + 90 days (up to ultimate load after pre-loading and additional 90 days of curing after pre-loading): (**a**) 2.5 series; (**b**) 0.85 series; (**c**) Plain series.

3.2.5. Crack Patterns

The number of vertical cracks observed in the test members ran between 10 and 15 with 11 cracks for member V_28D-2.5, 15 for member PLRL_28D-2.5, 14 for member RL_28 + 90D-2.5, 10 for member V_28 + 90D-2.5, 13 for member V_28D-0.85, 14 for member PLRL_28D-0.85, 12 for member RL_28 + 90D-0.85, 15 for member V_28 + 90D-0.85, 13 for member P-PLRL_28D, 14 for member P-RL_28 + 90D, and 13 for member P-V_28 + 90D. The crack pattern after the flexural test of the test members is shown in Figure 21. All the members experienced flexural cracking and flexural failure.

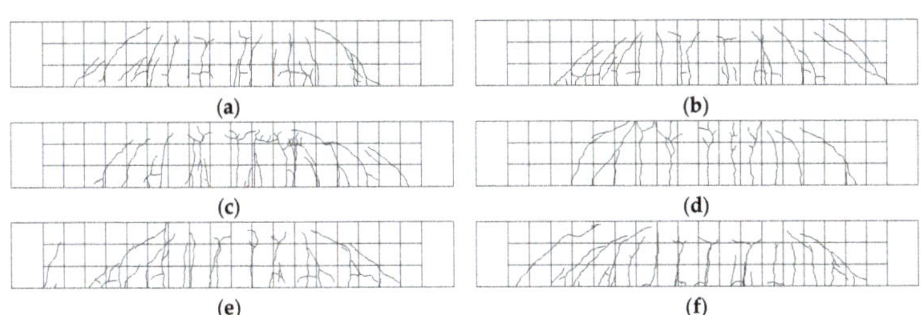

Figure 21. *Cont.*

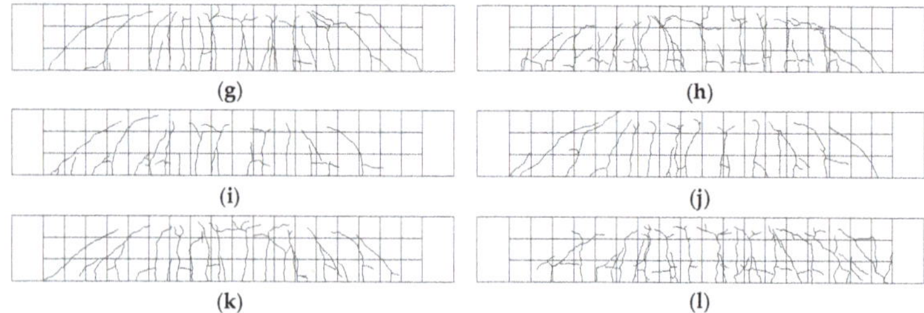

Figure 21. Crack patterns of test members: (**a**) V_28D-2.5; (**b**) PLRL_28D-2.5; (**c**) RL_28 + 90D-2.5; (**d**) V_28 + 90D-2.5; (**e**) V_28D-0.85; (**f**) PLRL_28D-0.85; (**g**) RL_28 + 90D-0.85; (**h**) V_28 + 90D-0.85; (**i**) P-V_28D; (**j**) P-PLRL_28D; (**k**) P-RL_28+90D; (**l**) P-V_28 + 90D.

4. Discussion of Results

4.1. Visual Observation of Crack Healing Effect

Figures 22–24 illustrate the cracks and healing of the test members after pre-damage at 28 days and after the additional 90 days of curing. The target crack width before loading of the members pre-loaded up to 50% of the ultimate load was 0.16 mm. In view of the figures, the maximum crack width after 90 days of additional curing ranged between 0.019 and 0.053 mm. Such a result can be attributed to various factors like the crack closure generated by the removal of the load and the autogenous healing.

The enlarged images of the cracks reveal that the 0.85-series members replacing binder by 100% of clinker achieved the best crack healing performance. The 2.5-series members replacing each binder and aggregates by 50% of clinker exhibited crack healing performance in-between that of the P-series members without clinker and that of the 0.85-series members.

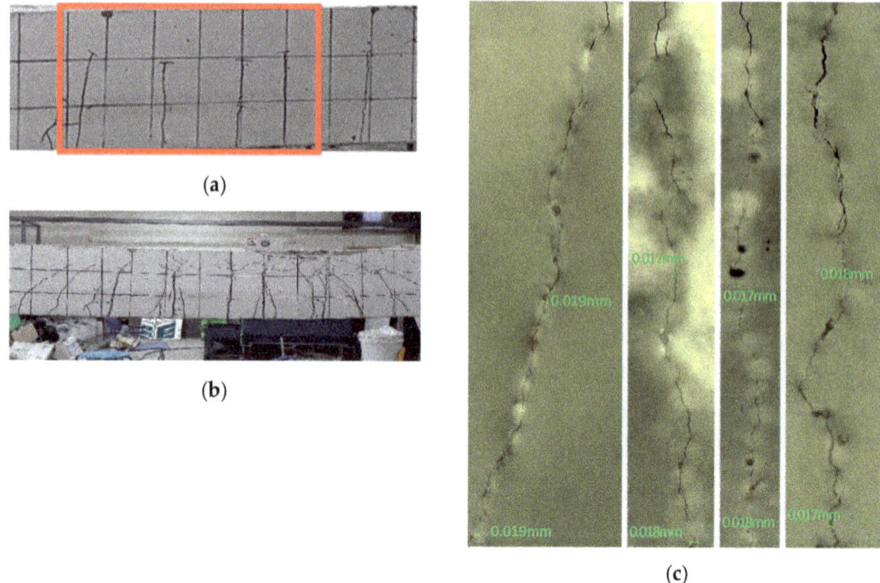

Figure 22. Crack healing of member RL_28 + 90D-2.5: (**a**) after pre-loading at 28 days + 90 days of additional curing; (**b**) after loading at 28 + 90 days; (**c**) enlargement of cracks after completion of test.

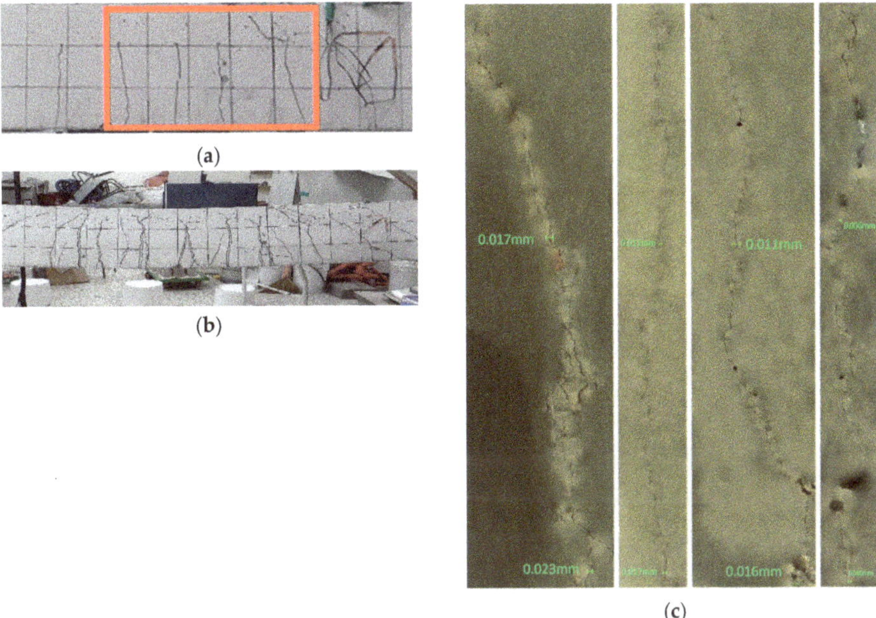

Figure 23. Crack healing of member RL_28 + 90D-0.85: (**a**) after pre-loading at 28 days + 90 days of additional curing; (**b**) after loading at 28 + 90 days; (**c**) enlargement of cracks after completion of test.

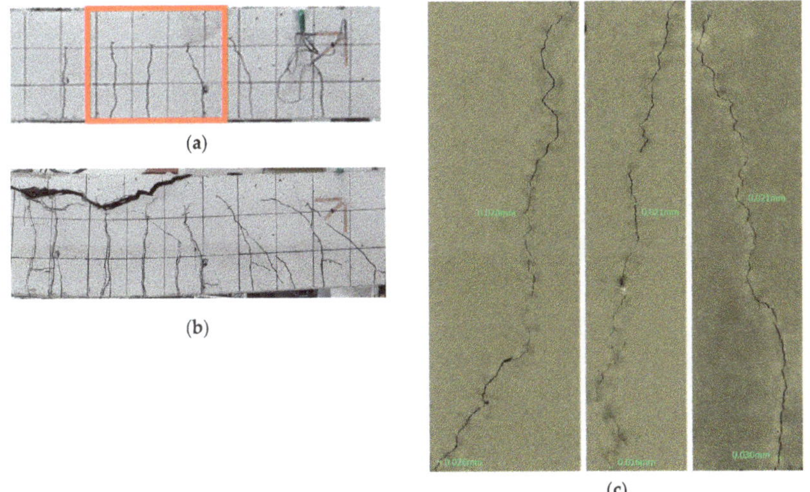

Figure 24. Crack healing of member P-RL_28 + 90D: (**a**) after pre-loading at 28 days + 90 days of additional curing; (**b**) after loading at 28 + 90 days; (**c**) enlargement of cracks after completion of test.

An attempt to objectively quantify the crack healing effect is achieved by measuring the crack distribution density from the observed crack patterns shown in Figure 21. Table 9 arranges the crack distribution density of the flexural specimens obtained by counting the number of cracks at the bottom of the members and dividing this number by the member's length (2000 mm). All the specimens experienced a similar number of vertical cracks. Moreover, the overall crack pattern observed in the beam members were similar to those of the experimental results reported by Alyousif [35].

Table 9. Crack distribution density at bottom of specimens (/mm).

Test Member	2.5-Series	0.85-Series	P-Series
V_28D	0.0055	0.0065	0.0050
V_28 + 90D	0.0050	0.0075	0.0065
PLRL_28D	0.0075	0.0070	0.0065
RL_28 + 90D	0.0070	0.0060	0.0070

A closer look reveals that, for the 0.85-series members, the RL-series members that suffered pre-damage exhibit relatively smaller crack density than the V-series members which did not experience pre-damage. Besides, the opposite occurs for the P-series and 2.5-series members. This situation can be credited to the fact that, compared to the 2.5-series members, the 0.85-series members exhibit a strong healing effect for cracks thinner than 200 μm, which reduces the number of cracks, but does not heal wider cracks, which experience crack opening under reloading. This phenomenon becomes more acute in older members. According to the studies by Sunayana and Barai [36] and Sturm et al. [37], the closer cracking spacing in the beams might be caused by the effect of high shrinkage in beams. In other words, the healing effect improves with time. Consequently, the fine grain size of clinker makes it possible to replace fine aggregates and longer healing time increases the crack healing effect.

4.2. *Evaluation of Crack Healing Effect by Water Flow Test*

The crack healing effect was also examined by water flow test. Figure 25 shows the water flow test performed to evaluate quantitatively the crack healing effect achieved in the test members after pre-damage at 28 days and an 90 additional days of curing. The water flow test was performed using cylindrical tubes made of acryl with a diameter of 100 mm and height of 1000 mm. The tubes were installed over the cracked and healthy parts of the members that were waterproofed. Water was filled in the tubes up to the specified height and the difference in water level in the tube was measured after 4 h.

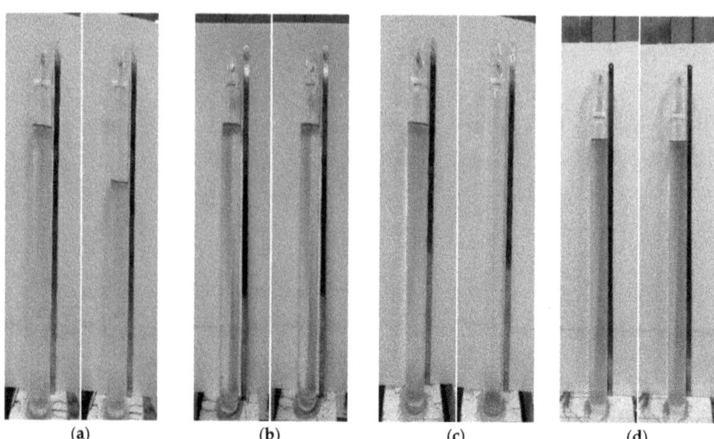

Figure 25. Results of water flow test: (**a**) 2.5 series; (**b**) 0.85 series; (**c**) Plain series; (**d**) healthy part.

The results of the water flow test are arranged in Table 10. There is no way to provide direct figures about the crack healing effect but the drop of the water level in the water flow test can give indirect insight on this effect. It appears that the 0.85-series members replacing binder by 100% of clinker achieved the best crack healing performance whereas the 2.5-series members replacing each binder and aggregates by 50% of clinker exhibited crack healing performance in-between that of the P-series members without clinker and that of the 0.85-series members. This verifies the previous observation done in view of the enlarged images of the cracks.

Table 10. Water flow test results.

Test Member	Initial Water Height (mm)	Final Water Height (mm)	Difference of Water Height (mm)
Healthy part	800	798	2
RL_28 + 90D-2.5		642	158
RL_28 + 90D-0.85		798	2
P-RL_28 + 90D		0	800

The results arranged in Table 10 support the conclusion of Escoffres et al. [12], although it concerned high-performance fiber reinforced concrete, that the addition of crystalline admixture slowed down the self-healing process measured by water permeability but provided greater mechanical recovery under tension to the material. In other words, this means that the improvement of durability by self-healing will remain in presence of crystalline admixture until a significant variation of the applied load.

4.3. Evaluation of Crack Healing Effect on Mechanical Performance

4.3.1. Yield and Ultimate Loads

Figure 26 compares the yield load of the steel reinforcement and the ultimate load of the members after pre-damage at 28 days. All the test members exhibit similar tendency. Compared to member V_28D which experienced 28 days of air-dry curing and loaded up to the ultimate state at 28 days, the test members showed a maximum difference of 7% for the yield load and approximately 8% for the ultimate load. Considering that the difference in the load was 4% to 5% between 28 days and 28 + 90 days as well as the corresponding difference in the strength, the P-series members without crack healing agent seem to undergo an actual difference of 2% to 3% in the yield and ultimate load with the same tendency as those observed in the other members. This last result may indicate that the crack healing effect has practically no effect in improving the yield and ultimate loads of the reinforced concrete member.

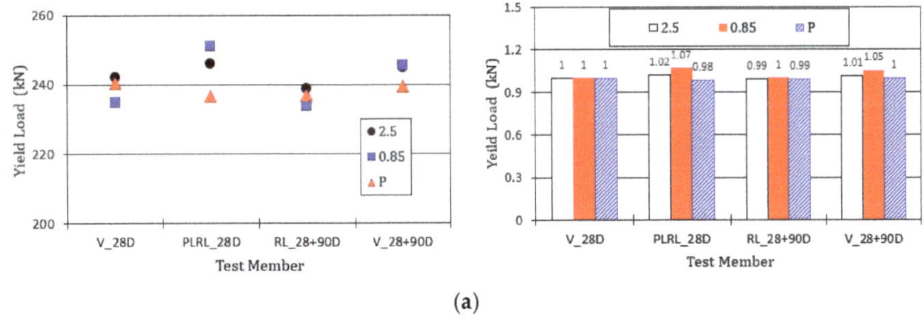

(a)

Figure 26. *Cont.*

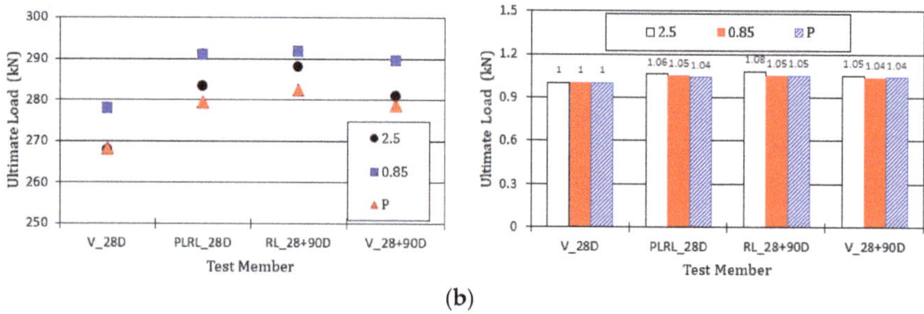

(**b**)

Figure 26. Comparison of reinforcement yield load and members' ultimate load: (**a**) yield load of steel reinforcement; (**b**) ultimate load of test members.

4.3.2. Energy Absorption Capacity and Ductility Index

Figure 27 compares the energy absorption capacity and ductility index of the test members after pre-damage at 28 days. It appears that the energy absorption capacity increased with the age. Member RL_28 + 90D, which was pre-damaged and experienced additional curing for 90 days, exhibited the highest energy absorption capacity. Such an increase in the energy absorption capacity can be attributed to the increase in the concrete strength with the age and the continuous evolution of cementitious composites with time [34]. Moreover, the members that were pre-damaged and reloaded developed higher energy absorption capacity because of the relatively larger displacement sustained during reloading due to the rebar yield and plastic deformation induced by pre-damage compared to that of the not pre-damaged members. The ductility index of all the test members shows a tendency resembling that of the energy absorption capacity. Compared to member V_28D, the difference in the energy absorption capacity reached a maximum of 249% and that for the ductility index reached approximately 287%. The crack healing had some effect on the increase in the energy absorption capacity and the ductility index but this increasing effect appears to be minimal when considering the increase in the energy absorption capacity brought by the age and the larger displacement sustained during reloading caused by the plastic deformation induced by pre-damage.

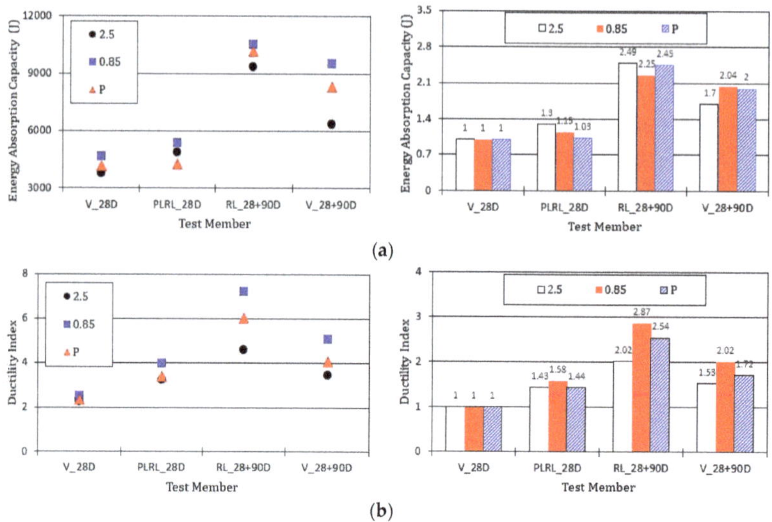

(**b**)

Figure 27. Comparison of energy absorption capacity and ductility index: (**a**) energy absorption capacity; (**b**) ductility index.

5. Conclusions

The present study examined the effect of self-healing on the flexural behavior of reinforced concrete members. The considered healing agents were clinker binder and aggregates as well as crystalline admixtures. A total of 12 test members were fabricated with respect to 3 different concrete mixes and 4 different loading and curing conditions. The following conclusions can be drawn from the experimental results.

1. The utilization of mineral admixtures was seen to improve the self-healing performance of mortar. Especially, healing ratios of 100% for crack widths smaller than 200 μm and 85% to 90% for crack widths of 250 μm were observed according to the admixing of clinker binder and aggregates. All the considered mixes exhibited similar results in terms of the compressive strength, elastic modulus and flexural strength. The series with a relatively meaningful binder replacement ratio by 0.85-mm clinker exhibited the best physical properties.
2. A negligible difference was observed in the crack, yield and ultimate loads. Failure occurred through flexure with a similar number of vertical cracks. All the test members showed linear load-deflection relationship until the initiation of cracks followed by a nonlinear increase in the deflection beyond the crack initiation and the load continued to increase until the ultimate load. For the test members cured additionally for 90 days after 28 days, the ultimate deflection reached more than 100% of that of the members loaded after 28 days of curing due to the plastic deformation caused by the yield of the rebar and the aging.
3. Rebar strain practically did not occur prior to the initiation of cracks and grew linearly after cracking to experience a large increase after yielding. The ultimate concrete strain in the members loaded up to the ultimate load at 28 days reached 0.0030 but that in the members that were additionally cured for 90 days after 28 days reached 0.0040 due to the aging and the effect of residual deformation.
4. The maximum crack width observed after 90 days of additional curing ranged between 0.019 and 0.053 mm. These values can be attributed to various factors like crack closure following the removal of loading as well as crack healing. The enlarged images of the cracks revealed that the 0.85-series members replacing binder by 100% of clinker achieved the best crack healing performance. The 2.5-series members replacing each binder and aggregates by 50% of clinker exhibited crack healing performance in-between that of the P-series members without clinker and that of the 0.85-series members. This observation was confirmed by the results of the water flow test performed on the test members.
5. In terms of mechanical performance, the crack healing property of concrete increased, to some extent, the rebar yield load, the members' ultimate load and energy absorption capacity and ductility index. However, this effect appeared to be minimal when considering the increase in the energy absorption capacity brought by the age and the larger displacement sustained during reloading caused by the plastic deformation induced by pre-damage.
6. An attempt to objectively quantify the crack healing effect by measuring the crack distribution density from the observed crack patterns was performed and this confirmed the crack healing effect provided by clinker powder.
7. Finally, the fine grain size of clinker made it possible to replace fine aggregates and the longer healing time increased the crack healing effect.

Author Contributions: Conceptualization, K.-M.L. and S.-W.Y.; methodology, Y.-C.C. and S.-W.Y.; experiment, B.P. and S.-W.Y.; validation, B.P. and Y.-C.C.; investigation, K.-M.L. and J.F.C.; data curation, B.P. and S.-W.Y.; writing—original draft preparation, J.F.C. and S.-W.Y.; writing—review and editing, Y.-C.C. and J.F.C.; visualization, B.P.; supervision, S.-W.Y.; project administration, K.M.L. and S.-W.Y.; funding acquisition, K.-M.L. All authors have read and agreed to the published version of the manuscript.

Funding: This research was supported by a grant (20SCIP-C158977-01) from Construction Technology Research Program funded by Ministry of Land, Infrastructure and Transport of Korean government. And this research was also supported by the Gachon University research fund of 2020 (GCU-2020-03030001).

Conflicts of Interest: The authors declare no conflict of interest.

References

1. James, A.; Bazarchi, E.; Chiniforush, A.A.; Aghdam, P.P.; Hosseini, M.R.; Akbarnezhad, A.; Martek, I.; Ghodoosi, F. Rebar corrosion detection, protection, and rehabilitation of reinforced concrete structures in coastal environments: A review. *Constr. Build. Mater.* **2019**, *224*, 1026–1039. [CrossRef]
2. Van Tittelboom, K.; De Belie, N. Self-healing of cementitious materials—A review. *Materials* **2013**, *6*, 2182–2217. [CrossRef]
3. Wang, W.; Zhang, T.; Wang, X.; He, Z. Research status of self-healing concrete. *Earth Environ. Sci.* **2019**, *218*, 012037. [CrossRef]
4. Rajczakowska, M.; Habermehl-Cwirzen, K.; Hedlund, H.; Cwirzen, A. Autogenous self-healing: A better solution for concrete. *J. Mater. Civ. Eng.* **2019**, *31*, 03119001. [CrossRef]
5. Zhang, W.; Zheng, Q.; Ashour, A.; Han, B. Self-healing cement concrete for resilient infrastructures: A review. *Compos. Part B* **2020**, *189*, 107892. [CrossRef]
6. Camara, L.A.; Wons, M.; Esteves, I.C.A.; Medeiros-Junior, R.A. Monitoring the self-healing of concrete from the ultrasonic pulse velocity. *J. Compos. Sci.* **2019**, *3*, 16. [CrossRef]
7. Wang, X.; Fang, C.; Li, D.; Han, N.; Xing, F. A self-healing cementitious composite with mineral admixtures and built-in carbonate. *Cem. Concr. Comp.* **2018**, *92*, 216–229. [CrossRef]
8. Park, B.; Choi, Y.C. Investigating a new method to assess the self-healing performance of hardened cement pastes containing supplementary cementitious materials and crystalline admixtures. *J. Mater. Res. Technol.* **2019**, *8*, 6058–6073. [CrossRef]
9. Lee, M.W. Prospects and future directions of self-healing fiber-reinforced composite materials. *Polymers* **2020**, *12*, 379. [CrossRef]
10. Kim, S.; Yoo, D.Y.; Kim, M.J.; Banthia, N. Self-healing capability of ultra-high-performance fiber-reinforced concrete after exposure to cryogenic temperature. *Cem. Concr. Comp.* **2019**, *104*, 103335. [CrossRef]
11. Wang, J.; Ding, S.; Han, B.; Ni, Y.; Ou, J. Self-healing properties of reactive powder concrete with nanofillers. *Smart Mater. Struct.* **2018**, *27*, 115033. [CrossRef]
12. Escoffres, P.; Desmettre, C.; Charron, J.P. Effect of a crystalline admixture on the self-healing capability of high-performance fiber reinforced concretes in service conditions. *Constr. Build. Mater.* **2018**, *173*, 763–774. [CrossRef]
13. Kua, H.W.; Gupta, S.; Aday, A.N.; Srubar, W.V., III. Biochar-immobilized bacteria and superabsorbent polymers enable self-healing of fiber-reinforced concrete after multiple damage cycles. *Cem. Concr. Compos.* **2019**, *100*, 35–52. [CrossRef]
14. Xu, J.; Yao, W. Multiscale mechanical quantification of self-healing concrete incorporating non-ureolytic bacteria-based healing agent. *Cem. Concr. Res.* **2014**, *64*, 1–10. [CrossRef]
15. Zhang, Z.; Zhang, Q.; Li, V.C. Multiple-scale investigations on self-healing induced mechanical property recovery of ECC. *Cem. Concr. Compos.* **2019**, *103*, 293–302. [CrossRef]
16. Deng, H.; Liao, G. Assessment of influence of self-healing behavior on water permeability and mechanical performance of ECC incorporating superabsorbent polymer (SAP) particles. *Constr. Build. Mater.* **2018**, *170*, 455–465. [CrossRef]
17. Xue, C.; Li, W.; Li, J.; Tam, V.W.Y.; Ye, G. A review study on encapsulation-based self-healing for cementitious material. *Struct. Concr.* **2019**, *20*, 198–212. [CrossRef]
18. Gupta, S.; Pang, S.D.; Kua, H.W. Autonomous healing in concrete by bio-based healing agents—A review. *Constr. Build. Mater.* **2017**, *146*, 419–428. [CrossRef]
19. Park, B.; Choi, Y.C. Self-healing capability of cementitious materials with crystalline admixtures and super absorbent polymers (SAPs). *Constr. Build. Mater.* **2018**, *189*, 1054–1066. [CrossRef]
20. Hong, G.; Song, C.; Park, J.; Choi, S. Hysteretic behavior of rapid self-sealing of cracks in cementitious materials incorporating superabsorbent polymers. *Constr. Build. Mater.* **2019**, *195*, 187–197. [CrossRef]

21. Lv, L.; Guo, P.; Liu, G.; Han, N.; Xing, F. Light induced self-healing in concrete using novel cementitious capsules containing UV curable adhesive. *Cem. Concr. Compos.* **2020**, *105*, 103445. [CrossRef]
22. Qureshi, T.S.; Kanellopoulos, A.; Al-Tabbaa, A. Encapsulation of expansive powder minerals within a concentric glass capsule system for self-healing concrete. *Constr. Build. Mater.* **2016**, *121*, 629–643. [CrossRef]
23. Gilabert, F.A.; Van Tittelboom, K.; Van Stappen, J.; Cnudde, V.; De Belie, N.; Van Paepegem, W. Integral procedure to assess crack filling and mechanical contribution of polymer-based healing agent in encapsulation-based self-healing concrete. *Cem. Concr. Compos.* **2017**, *77*, 68–80. [CrossRef]
24. Snoeck, D.; Pel, L.; De Belie, N. Autogenous healing in cementitious materials with superabsorbent polymers quantified by means of NMR. *Sci. Rep.* **2020**, *10*, 642. [CrossRef] [PubMed]
25. Park, B.; Choi, Y.C. Quantitative evaluation of crack self-healing in cement-based materials by absorption test. *Constr. Build. Mater.* **2018**, *184*, 1–10. [CrossRef]
26. Azarsa, P.; Gupta, R.; Biparva, A. Assessment of self-healing and durability parameters of concretes incorporating crystalline admixtures and Portland Limestone Cement. *Cem. Concr. Compos.* **2019**, *99*, 17–31. [CrossRef]
27. Guo, S.; Chidiac, S. Self-healing concrete: A critical review. In Proceedings of the 2019 CSCE Annual Conference, Laval, QC, Canada, 12–15 June 2019.
28. Choi, Y.C.; Park, B. Enhanced autogenous healing of ground granulated blast furnace slag blended cements and mortars. *J. Mater. Res. Technol.* **2019**, *8*, 3443–3452. [CrossRef]
29. Takagi, E.M.; Lima, M.G.; Helene, P.; Meideros-Junior, R.A. Self-healing of self-compacting concretes made with blast furnace slag cements activated by crystalline admixture. *Int. J. Mater. Prod. Tec.* **2018**, *56*, 169–186. [CrossRef]
30. Yuan, L.; Chen, S.; Wang, S.; Huang, Y.; Yang, Q.; Liu, S.; Wang, J.; Du, P.; Cheng, X.; Zhou, Z. Research on the improvement of concrete self-healing based on the regulation of cement particle size distribution (PSD). *Materials* **2019**, *12*, 2818. [CrossRef]
31. Berger, R.L. Properties of concrete with cement clinker aggregate. *Cem. Concr. Res.* **1974**, *4*, 99–112. [CrossRef]
32. Allahverdi, J.S.A. Using PC clinker as aggregate-enhancing concrete properties by improving ITZ microstructure. *Mag. Concr. Res.* **2020**, *72*, 173–181. [CrossRef]
33. Choi, S.W.; Bae, W.H.; Lee, K.M.; Shin, K.J. Correlation between crack width and water flow of cracked mortar specimens measured by constant water head permeability test. *J. Korea Concr. Inst.* **2017**, *29*, 267–273. [CrossRef]
34. Lepech, M.D.; Li, V.C. Water permeability of engineered cementitious composites. *Cem. Concr. Compos.* **2009**, *31*, 744–753. [CrossRef]
35. Alyousif, A. Self-Healing Capability of Engineered Cementitious Composites Incorporating Different Types of Pozzolanic Materials. Ph.D. Thesis, Ryerson University, Toronto, ON, Canada, 2016.
36. Sunayana, S.; Barai, S.V. Flexural performance and tension-stiffening evaluation of reinforced concrete beam incorporating recycled aggregate and fly ash. *Constr. Build. Mater.* **2018**, *174*, 210–223. [CrossRef]
37. Sturm, A.B.; Visintin, P.; Oehlers, D.J. Time dependent serviceability behavior of reinforced concrete beam: Partial interaction tension stiffening mechanics. *Struct. Concr.* **2018**, *19*, 508–523. [CrossRef]

 materials

Article

Strength, Drying Shrinkage, and Carbonation Characteristic of Amorphous Metallic Fiber-Reinforced Mortar with Artificial Lightweight Aggregate

Se-Jin Choi *, Ji-Hwan Kim, Sung-Ho Bae and Tae-Gue Oh

Department of Architectural Engineering, Wonkwang University, 460 Iksan-daero, Iksan 54538, Korea; 3869kjh@naver.com (J.-H.K.); caos1344@naver.com (S.-H.B.); alkjd3@naver.com (T.-G.O.)
* Correspondence: csj2378@wku.ac.kr; Tel.: +82-63-850-6789

Received: 2 September 2020; Accepted: 1 October 2020; Published: 7 October 2020

Abstract: This paper investigates the strength, drying shrinkage, and carbonation characteristic of amorphous metallic fiber-reinforced mortar with natural and artificial lightweight aggregates. The use of artificial lightweight aggregates has the advantage of reducing the unit weight of the mortar or concrete, but there is a concern that mechanical properties of concrete such as compressive strength and tensile strength may deteriorate due to the porous properties of lightweight aggregates. In order to improve the mechanical properties of lightweight aggregate mortar, we added 0, 10, 20, and 30 kg/m^3 of amorphous metallic fibers to the samples with lightweight aggregate; the same amount of fiber was applied to the samples with natural aggregate for comparison. According to this investigation, the flow of mortar decreased as the amount of amorphous metallic fiber increased, regardless of the aggregate type. The compressive strength of lightweight aggregate mortar with 10 kg/m^3 amorphous metallic fiber was similar to that of the LAF0 sample without amorphous metallic fiber after 14 days. In addition, the flexural strength of the samples increased as the amount of amorphous metallic fiber increased. The highest 28-d flexural strength was obtained as approximately 9.28 MPa in the LAF3 sample, which contained 30 kg/m^3 amorphous metallic fiber. The drying shrinkage of the samples with amorphous metallic fiber was smaller than that of the sample without amorphous metallic fiber.

Keywords: compressive strength; flexural strength; drying shrinkage; amorphous metallic fiber; carbonation; mortar

1. Introduction

In general, concrete exerts strong performance against compression, but its tensile strength is very weak—about 8% to 12% of compressive strength—and there is increasing interest in steel-fiber-reinforced concrete to improve such properties as low flexural strength and impact strength [1]. Recently, studies of concrete with thin-shaped amorphous metallic fibers have been actively conducted [2–4]. Amorphous metallic fiber has the advantage of reducing CO_2 and energy consumption in the manufacturing process because this process is simpler than that using normal steel fibers, and there is no subsequent process after the molten iron [4]. In particular, the corrosion or destruction of iron occurs through crystal grain boundaries appearing in the crystalline metal, and the amorphous metal is known to have excellent corrosion resistance and excellent tensile strength [5].

Meanwhile, in order to reduce the weight of ultra-high-rise and large-scale concrete structures, interest in artificial lightweight aggregates having a lighter weight than general aggregates is increasing [6]. The use of artificial lightweight aggregates has the advantage of reducing the unit weight of the cement composite, but there is a concern that mechanical properties such as compressive strength and tensile strength may deteriorate due to the porous properties of lightweight aggregates. Existing studies of cement composites reinforced with amorphous metal fibers have mostly focused on the reduction of the plastic shrinkage of cement composites [7,8] or the improvement of the toughness and impact resistance of high-strength concrete [9–13].

Choi et al. [7] investigated technology for reducing the shrinkage of concrete reinforced with amorphous metallic fibers. According to this study, amorphous metallic fiber is a material developed to improve the shortcomings of steel fibers that has excellent mechanical properties and corrosion resistance. In addition, it can effectively reduce the amount of shrinkage of the mortar and concrete by properly mixing.

Won et al. [8] evaluated the bonding properties between thin amorphous micro-steel fibers and cement composite materials. The bond strength test results with mortar showed that the maximum pull-out load of amorphous micro-steel fiber was larger than that of hooked-type steel fiber.

Lee et al. [10] studied the impact resistance of amorphous-steel-fiber-reinforced cement composites, finding that the composite had a large inhibitory effect on backside destruction by the high-speed impact. It was found that the impact resistance of the cement composites using 30-mm amorphous steel fibers was excellent.

Kim et al. [12] examined the effect that the amorphous metallic fibers have on the attainability of the mixing conditions, the static mechanic properties, and the impact resistance of concrete to those of hooked-end steel fibers. The test results showed that the concrete reinforced with amorphous metallic fibers was more effective at resisting cracking than the concrete reinforced with hooked-end steel fibers.

There are few studies on amorphous-metallic-fiber-reinforced mortar using artificial lightweight aggregates, and it is expected to be helpful in increasing the use of lightweight aggregate concrete if the engineering characteristics of mortar or concrete with artificial lightweight aggregates can be improved using amorphous metallic fibers. In this study, the flow, unit weight, compressive strength, flexural strength, split tensile strength, drying shrinkage, and carbonation characteristic of amorphous-metallic-fiber-reinforced mortar using natural aggregate and artificial lightweight aggregates were investigated in order to improve the mechanical properties of mortar with artificial lightweight aggregates.

2. Materials and Methods

2.1. Materials

The cementitious materials used in this study were ASTM type I OPC manufactured by the Asia Cement Co. (Seoul, Korea); the blast furnace slag powder was obtained from Daehan Slag Co., Ltd., in Gwangyang, Korea. Natural fine aggregate with a density of 2.60 and a fineness modulus of 2.89 was used. As an artificial lightweight aggregate, the lightweight fine aggregate of KOEN in Jinju, Korea—manufactured by calcining coal ash and dredged soil at about 1100 to 1200 °C—was used.

Figure 1 shows the shape and scanning electron microscope (SEM) image of the artificial lightweight aggregate, which contains numerous voids.

(a) (b)

Figure 1. Artificial lightweight aggregate sample: (**a**) Shape, (**b**) SEM image.

In general, the surface of the artificial lightweight aggregate first reaches a high temperature in the calcining process, and the surface part is first liquefied and shows a denser shell structure than the inner part [14]. The artificial lightweight aggregate used in this study also showed a similar trend.

Figure 2 shows the shape and SEM image of the amorphous metallic fiber used in this study.

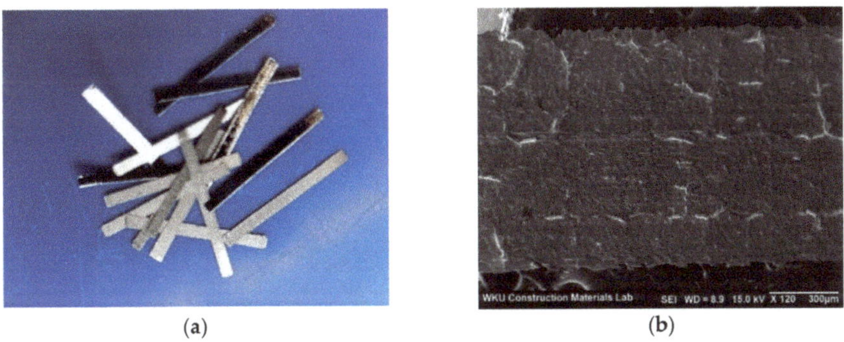

(a) (b)

Figure 2. Amorphous metallic fiber: (**a**) Shape, (**b**) SEM image.

Amorphous metallic fiber with a specific gravity of 7.2 g/cm^3, a tensile strength of 1400 N/mm^2, and a length of 15 mm manufactured by SG Group in France was used.

Tables 1 and 2 show the chemical composition of the cement and BFS (blast furnace slag powder) and the physical properties of the aggregates used.

Table 1. Chemical composition of cement and blast furnace slag powder (BFS).

Components	SiO_2	Al_2O_3	Fe_2O_3	CaO	MgO	K_2O
Cement	17.43	6.50	3.57	64.40	2.55	1.17
BFS	30.61	13.98	0.32	40.47	6.43	0.60

Table 2. Physical properties of aggregates.

Type of Sand	Fineness Modulus	Surface Dry Density (g/cm^2)	Oven Dry Density (g/cm^2)	Water Absorption Ratio (%)	Unit Weight (kg/L)
Natural sand (NS)	2.89	2.60	-	1.00	1427
Lightweight sand (LS)	4.61	1.77	1.63	8.71	1010

Table 3 shows the sieve passing ratio of natural and artificial lightweight aggregates used in this study.

Table 3. Sieve passing ratio of aggregates.

Type of Sand	Sieve Passing Ratio (%)						
	10 mm	**5 mm**	**2.5 mm**	**1.2 mm**	**0.6 mm**	**0.3 mm**	**0.15 mm**
Standard	100 100	100 95	100 80	85 50	60 25	30 10	10 2
NS	100	100	93.75	73.50	52	20	6
LS	100	99.75	38.50	1.25	0.25	0.20	0.20

2.2. Mixing Proportions and Specimen Preparation

In this study, we prepared amorphous-metallic-fiber-reinforced mortar using either natural aggregate or artificial lightweight aggregate. The amount of amorphous metallic fiber used was 0, 10, 20, and 30 kg/m^3. The water-to-binder ratio was fixed at 0.5. In all mixtures, blast furnace slag powder was used to replace 40% of the weight of cement. The mixing proportions of the mortar mixtures are summarized in Table 4.

Table 4. Mix proportions of mortar.

Mix.	W/(C+BFS)	Water (kg/m^3)	Cement (kg/m^3)	BFS (kg/m^3)	Natural Sand (kg/m^3)	Lightweight Sand (kg/m^3)	Fiber (kg/m^3)
NAF0	0.5	170	204	136	735	-	0
NAF1	0.5	170	204	136	735	-	10
NAF2	0.5	170	204	136	735	-	20
NAF3	0.5	170	204	136	735	-	30
LAF0	0.5	170	204	136	-	502	0
LAF1	0.5	170	204	136	-	502	10
LAF2	0.5	170	204	136	-	502	20
LAF3	0.5	170	204	136	-	502	30

The components of the samples were blended in a mechanical mixer, and 50-mm cube molds were prepared for the compressive strength test and unit weight test. Cylindrical molds (⌀50 mm × 100 mm) were used for the split-tensile strength test. Bar-type molds (40 mm × 40 mm × 160 mm) were used to measure the flexural strength, drying shrinkage, and accelerated-carbonation test. After 24 h, the strength specimens were removed from their molds and cured at 20 °C in a water tank.

The flow and compressive strength tests of each mix were conducted in accordance with KS L 5105 [15]. The flexural strength and split-tensile strength tests of each mix were conducted in accordance with KS F 2408 [16] and KS F 2423 [17]. The strength test values were the average values of three samples. The unit weights of the samples of each mix were measured in accordance with KS F 2462 [18]. The drying shrinkage test was conducted according to KS F 2424 [19] using a mechanical strain gauge.

In the carbonation test, the carbonation depth of the specimen was measured using a phenolphthalein solution after the accelerated-carbonation process until the required age in the accelerated-carbonation chamber having a CO_2 concentration of 5% in accordance with KS F 2584 [20]. Figure 3 shows the carbonation test samples and accelerated-carbonation chamber by CK Corporation of South Korea, Seoul, South Korea, which was used for the carbonation test of the samples.

Figure 3. Accelerated-carbonation chamber and samples.

3. Results and Discussion

3.1. Mortar Flow

Figure 4 shows the flow value of amorphous-metallic-fiber-reinforced mortar using natural aggregate and artificial lightweight aggregate.

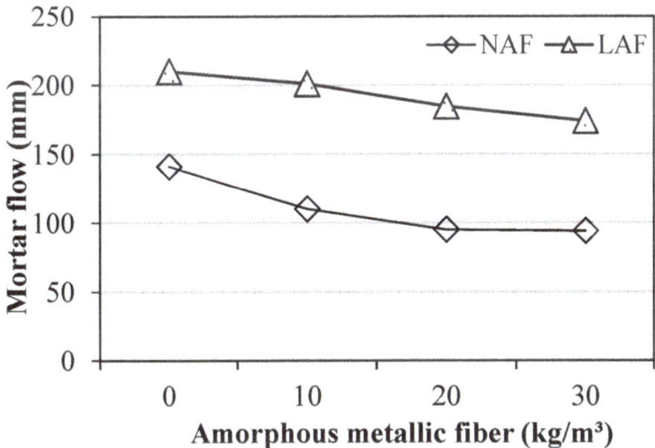

Figure 4. Variation of mortar flow (NAF; Natural aggregate mortar with fiber, LAF; Lightweight aggregate mortar with fiber).

As shown in Figure 4, in both the natural aggregate mixture and artificial lightweight aggregate mixture, the flow of the sample without amorphous metallic fibers was highest. When using natural aggregate, the flow of the samples with amorphous metallic fibers was about 21% to 33% lower than the flow of the NAF0 sample without amorphous metallic fibers. The flow of the sample using artificial lightweight aggregate was larger than that of the sample using natural aggregate, regardless of fiber content. This seems to be due to the round shape of the artificial lightweight aggregate and some of the water absorbed by the artificial lightweight aggregate during the pre-wetting process. In mixtures using artificial lightweight aggregates, the flow also decreased as the amount of amorphous metallic fibers increased. However, the flow reduction of amorphous-metallic-fiber-reinforced mortar using artificial lightweight aggregate was less than that when using natural aggregate.

3.2. Unit Weight

Figure 5 shows the variation in the unit weight of amorphous-metallic-fiber-reinforced mortar using natural aggregate and artificial lightweight aggregate.

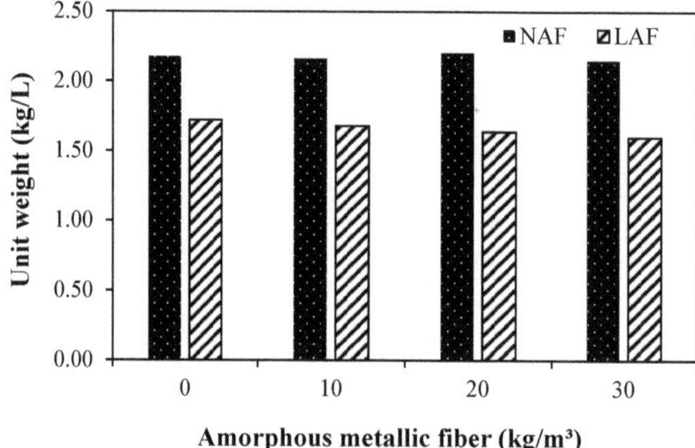

Figure 5. Unit weight.

When using natural aggregate, it shows a similar unit weight regardless of the amount of amorphous metallic fiber used. The unit weight of the NAF0 sample without amorphous metallic fiber was about 2.17 kg/L.

The unit weight of the fiber-reinforced natural aggregate mortar sample using 30 kg/m^3 of amorphous metallic fiber was slightly less than that of the NAF0 sample without fiber. In addition, the unit weight of the mortar samples using artificial lightweight aggregate was 20% or less than that of the mortar samples using natural aggregate. The unit weight of the artificial lightweight aggregate mortar sample without amorphous metallic fibers was about 1.72 kg/L. The unit weight of the sample using amorphous metallic fiber was about 1.60 to 1.68 kg/L, which was slightly less than that of the LAF0 sample without fiber. This seems to be due to the adhesion characteristics and pores in the interface between the amorphous metallic fiber and the mortar matrix.

3.3. Compressive Strength

Figure 6 shows the variation in the compressive strength of mortar using natural aggregates according to the amount of amorphous metallic fibers.

As can be seen in the Figure 6a, the 7-day compressive strength of the NAF0 sample without amorphous metallic fibers was about 31.7 MPa, which showed the highest compressive strength of all samples. The compressive strength of both the NAF1 sample and the NAF2 sample, in which 10 kg/m^3 and 20 kg/m^3 of amorphous metallic fibers were respectively mixed, was about 29.3 MPa—about 7.5% lower than that of the NAF0 sample. After 14 days, the compressive strength of the NAF0 sample without amorphous metallic fiber was about 37.1 MPa, showing a higher compressive strength than other mixtures, and the samples using amorphous metallic fiber showed a slightly lower compressive strength than the NAF0 sample. After 28 days, the compressive strength of the NAF1 sample with 10 kg/m^3 amorphous metallic fibers was about 43 MPa, similar to that of the NAF0 sample; the NAF2 sample and NAF3 sample with 20 kg/m^3 or more amorphous metallic fibers were about 7% to 23% lower than the NAF0 sample without amorphous metallic fibers.

In the case of artificial lightweight aggregate, the compressive strength of the LAF0 sample without amorphous metallic fiber was highest after 7 days, similar to the sample using natural aggregate (Figure 6b). After 14 days, the compressive strengths of the LAF0 sample without amorphous metallic fibers and the LAF1 sample with 10 kg/m^3 of amorphous metallic fibers were similar, at about 31 MPa. After 28 days, the compressive strength of the LAF0 sample was the highest, at about 35.8 MPa, and the compressive strength of the sample with amorphous metallic fibers decreased as the amount of amorphous metallic fibers increased. When using amorphous metallic fibers in both natural aggregate mortar and artificial lightweight aggregate mortar, the compressive strength of the samples tended to be somewhat reduced or similar to that of the samples without fiber, which is similar to the results of existing literature [1,21] in that the compressive strength enhancement effect of steel-fiber-reinforced concrete is not large.

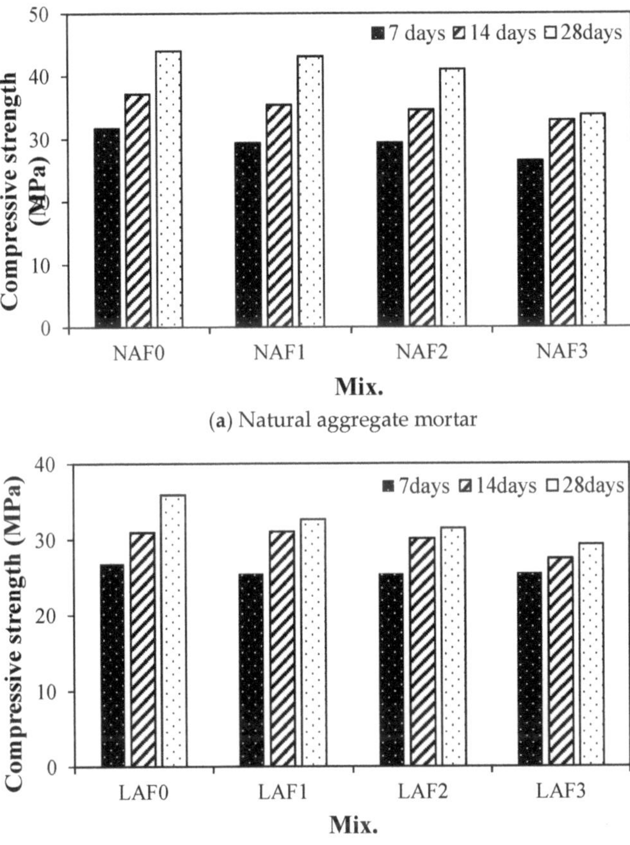

Figure 6. Compressive strength, (**a**) Natural aggregate mortar, (**b**) Artificial lightweight aggregate mortar.

3.4. Flexural Strength

Figure 7 shows the variation in flexural strength of amorphous-metallic-fiber-reinforced mortar using natural and artificial lightweight aggregates according to the amount of amorphous metallic fibers after 28 days.

In the case of the mixtures using natural aggregate, the flexural strength of the samples increased as the amount of amorphous metallic fiber increased. The flexural strength of the NAF3 sample using 30 kg/m^3 of amorphous metallic fiber was about 9.79 MPa, which was about 38.8% higher than that of the NAF0 sample without amorphous metallic fiber. When artificial lightweight aggregate was used, the LAF0 sample without amorphous metallic fiber showed the lowest flexural strength at 4.96 MPa, and the flexural strength of the mixtures increased as the amount of amorphous metallic fiber increased. The flexural strength of the LAF3 sample using 30 kg/m^3 of amorphous metallic fiber was about 9.28 MPa, which was about 87% higher than that of the LAF0 sample. In addition, the effect of enhancing the flexural strength of the sample due to amorphous metallic fiber reinforcement was higher in the mixtures using artificial lightweight aggregate than in the mixtures using natural aggregate.

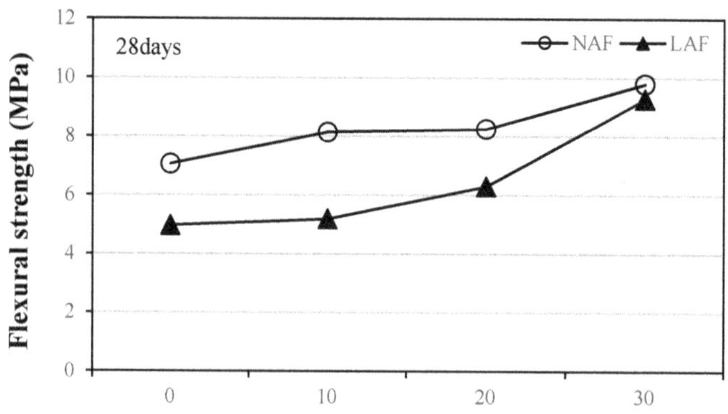

Figure 7. Flexural strength.

Figure 8 shows the variation of the ratio of flexural strength (Fb) and compressive strength (Fc) of the samples by the amount of amorphous metallic fiber.

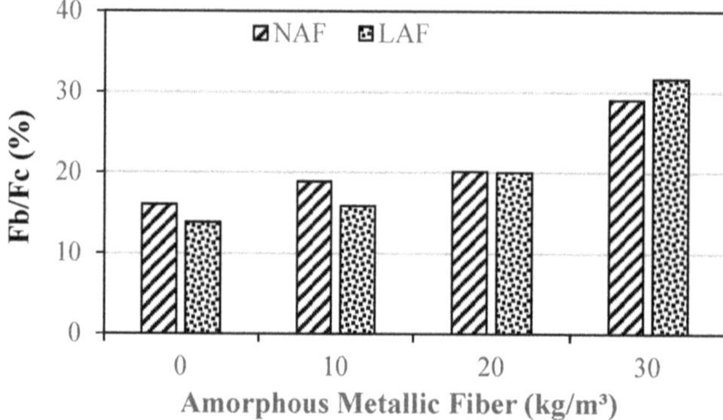

Figure 8. The ratio of flexural strength and compressive strength.

In the case of amorphous-metallic-fiber-reinforced mortar using natural aggregate, Fb/Fc was about 18.9% to 29.0%. This value is about 2.9% to 13.0% higher than that of the NAF0 sample without amorphous metallic fibers. In the case of the mixtures using artificial lightweight aggregate, the sample without amorphous metallic fiber showed an Fb/Fc value of about 13.8%, and the Fb/Fc of amorphous-metallic-fiber-reinforced mortar with artificial lightweight aggregate was about 15.9% to 31.7%, which was about 2.1% to 17.9% higher than that of the sample without amorphous metallic fiber. As can be seen in the Figure 8, when the amount of amorphous metallic fiber used was more than 20 kg/m^3, the Fb/Fc of the artificial lightweight aggregate mortar was similar to or higher than that of the natural aggregate mortar.

3.5. Tensile Strength

Figure 9 shows the variation in the split-tensile strength of the mortar sample using natural and artificial lightweight aggregates according to the amount of amorphous metallic fibers after 28 days.

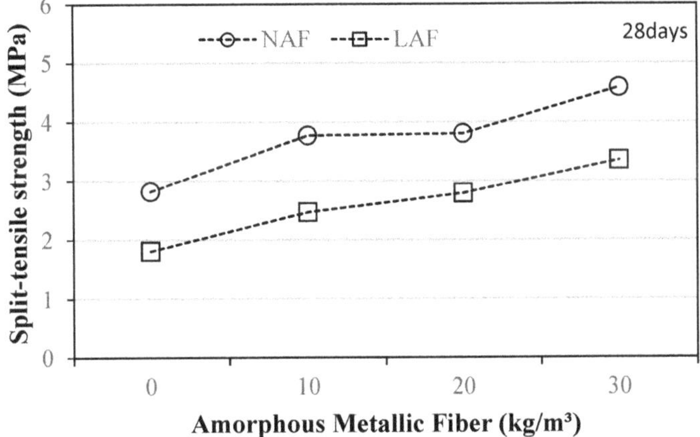

Figure 9. Split-tensile strength.

The split-tensile strength of the amorphous-metallic-fiber-reinforced samples increased as the amount of amorphous metallic fibers increased in both natural aggregate mortar and artificial lightweight aggregate mortar. When amorphous metallic fibers were not used, the tensile strength of the natural aggregate mortar was about 2.8 MPa, which was about 55.5% higher than that of the artificial lightweight aggregate mortar. The amorphous-metallic-fiber-reinforced mortar sample using natural aggregate showed a tensile strength of about 3.7 to 4.5 MPa, which was about 32.1% to 60.7% higher than that of the NAF0 sample without amorphous metallic fiber. In addition, the tensile strength of the amorphous-metallic-fiber-reinforced mortar sample using artificial lightweight aggregate was about 2.4 to 3.3 MPa, which was about 33.3% to 83.3% higher than that of the LAF0 sample without amorphous metallic fiber. Similar to the case of flexural strength, the enhancement of the tensile strength of the samples due to the reinforcement of amorphous metallic fibers was greater in the artificial lightweight aggregate mortar than in the natural aggregate mortar.

Figure 10 shows the variation in the ratio of the tensile strength (Ft) and compressive strength (Fc) of the samples by the amount of amorphous metallic fiber.

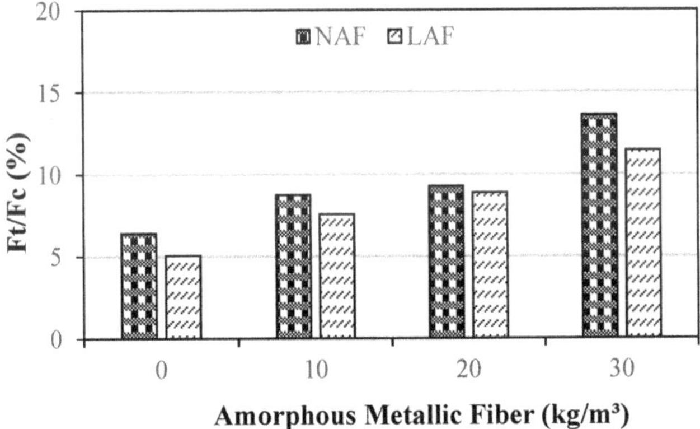

Figure 10. The ratio of tensile strength and compressive strength.

When amorphous metallic fibers were not used, the Ft/Fc values of natural aggregate mortar and artificial lightweight aggregate mortar were 6.4% and 5.0%, respectively. The Ft/Fc values of the amorphous-metallic-fiber-reinforced mortar samples using natural aggregate and artificial lightweight aggregate were 8.7% to 13.5% and 7.5% to 11.4%, respectively. Regardless of aggregate type, the Ft/Fc of the samples increased as the amount of amorphous metallic fiber increased.

3.6. Drying Shrinkage

Figure 11a shows the variation of drying shrinkage of amorphous-metallic-fiber-reinforced mortar using natural aggregate.

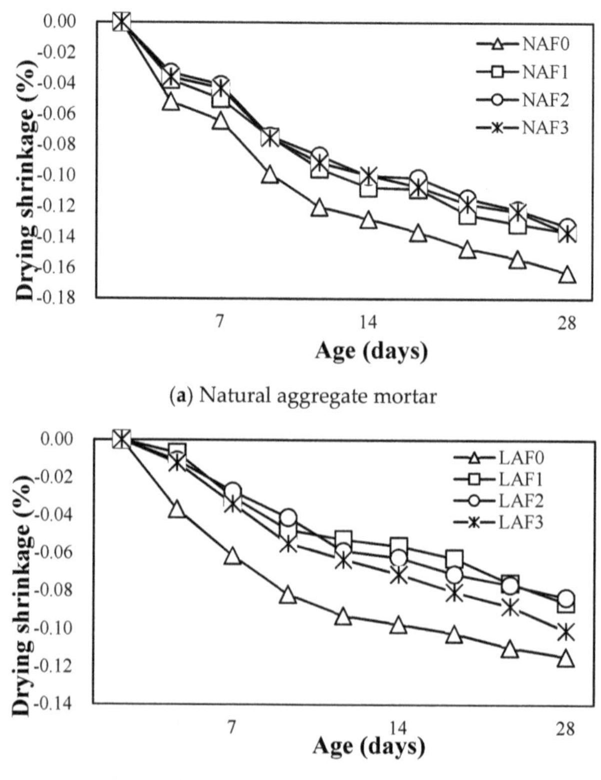

(**a**) Natural aggregate mortar

(**b**) Artificial lightweight aggregate mortar

Figure 11. Drying shrinkage, (**a**) Natural aggregate mortar, (**b**) Artificial lightweight aggregate mortar.

As can be seen in the Figure 11a, the drying shrinkage of the NAF0 sample without amorphous metallic fiber showed the highest value of about 0.163% after 28 days. In the case of amorphous-metallic-fiber-reinforced mortar, regardless of the amount of amorphous fiber used, the drying shrinkage of the samples was about 0.13%—about 20% lower than that of the NAF0 sample.

Figure 11b shows the variation of drying shrinkage of amorphous-metallic-fiber-reinforced mortar using artificial lightweight aggregate. Even in the case of artificial lightweight aggregate mortar, the LAF0 sample without amorphous metallic fibers showed the highest drying shrinkage, and LAF1 and LAF2 samples using 10 and 20 kg/m^3 of amorphous metallic fibers, respectively, had a drying shrinkage of 0.08% after 28 days—about 27.2% lower than that of the LAF0 sample. Therefore, it is suggested that the drying shrinkage of artificial lightweight aggregate mortar can be effectively reduced by appropriately using amorphous metallic fibers.

3.7. Carbonation Depth

The accelerated-carbonation depth of the samples using natural aggregate and artificial lightweight aggregate according to the amount of amorphous metallic fibers is shown in Figure 12.

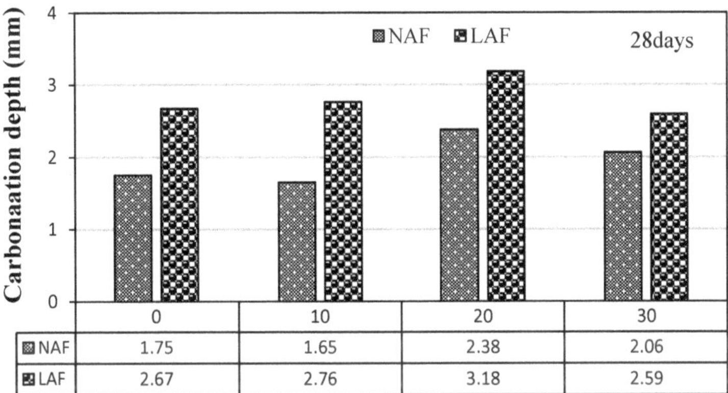

	0	10	20	30
NAF	1.75	1.65	2.38	2.06
LAF	2.67	2.76	3.18	2.59

Figure 12. Accelerated-carbonation depth.

In the case of natural aggregate mortar, the carbonation depth of the NAF0 sample without amorphous metallic fiber and the NAF1 sample using 10 kg/m^3 of amorphous metallic fiber was similar. The carbonation depth of the NAF2 sample and NAF3 sample using more than 20 kg/m^3 of amorphous metallic fiber was somewhat higher than that of the NAF0 sample. Even in the case of artificial lightweight aggregate mortar, the carbonation depth of the sample using amorphous metallic fibers was relatively higher than that of the sample without fiber. When the same amount of amorphous metallic fiber was used, the carbonation depth of artificial lightweight aggregate mortar was higher than that of natural aggregate mortar, which was presumed to be due to the porous characteristics of artificial lightweight aggregate.

4. Conclusions

The conclusions of this study can be summarized as follows:

(1) In both the natural aggregate mixture and the artificial lightweight aggregate mixture, the flow of the samples without amorphous metallic fibers was highest. In addition, the flow of the sample using artificial lightweight aggregate was larger than that of the sample using natural aggregate, regardless of fiber content.

(2) The unit weight of the mortar samples using artificial lightweight aggregate was 20% or less than that of the mortar samples using natural aggregate. After 14 days, the compressive strengths of the LAF0 sample without amorphous metallic fibers and the LAF1 sample with 10 kg/m^3 of amorphous metallic fibers were similar, at about 31 MPa.

(3) The flexural strength of the amorphous-metallic-fiber-reinforced lightweight aggregate mortar sample using 30 kg/m^3 of amorphous metallic fiber was about 9.28 MPa, which was about 87% higher than that of the LAF0 sample without fibers. In the case of amorphous-metallic-fiber-reinforced mortar using natural aggregates, Fb/Fc was about 18.9% to 29.0%—about 2.9% to 13.0% higher than that of the NAF0 sample without amorphous metallic fibers.

(4) The effect of enhancing the flexural strength and tensile strength of the sample due to amorphous metallic fiber reinforcement was higher in the mixtures using artificial lightweight aggregate than in those using natural aggregate.

(5) The LAF0 sample without amorphous metallic fibers showed the highest drying shrinkage, and the LAF1 and LAF2 samples using 10 and 20 kg/m^3 of amorphous metallic fibers, respectively, had a drying shrinkage of 0.08% after 28 days, which was about 27.7% lower than that of the LAF0 sample. Therefore, it is suggested that the drying shrinkage of artificial lightweight aggregate mortar can be effectively reduced by appropriately using amorphous metallic fibers.
(6) When the same amount of amorphous metallic fiber was used, the carbonation depth of the artificial lightweight aggregate mortar was somewhat higher than that of the natural aggregate mortar, which was presumed to be due to the porous characteristics of artificial lightweight aggregate.

However, further research is needed to establish the strength development mechanism and respective relationships between the strength properties of cement mortar containing various sizes and volumes of amorphous metallic fibers and water–binder ratio, binder content, durability, etc.

Author Contributions: S.-J.C. conducted all of the experimental studies, and analyzed the test data. J.-H.K. and S.-H.B. conducted some experiments and wrote the manuscript. T.-G.O. advised the experimental work and revised the manuscript. All authors have read and agreed to the published version of the manuscript.

Funding: This research was supported by the Basic Science Research Program through the National Research Foundation of Korea (NRF) funded by the Ministry of Education (NRF-2019R1I1A3A01049510). This work was also supported by a National Research Foundation of Korea (NRF) grant funded by the Korean government (MSIT) (No.2020R1A4A3079595).

Acknowledgments: The authors gratefully acknowledge the National Research Foundation of Korea and Ministry of Education for the financial support of this work. The authors would like to thank Editage for English language editing.

Conflicts of Interest: The authors declare no conflict of interest.

References

1. Song, P.S.; Hwang, S. Mechanical properties of high-strength steel fiber-reinforced concrete. *Constr. Build Mater.* **2004**, *18*, 669–673. [CrossRef]
2. Choi, S.J.; Hong, B.T.; Lee, S.J.; Won, J.P. Shrinkage and corrosion resistance of amorphous metallic-fiber-reinforced cement composites. *Compos. Struct.* **2014**, *107*, 537–543. [CrossRef]
3. Choi, S.J.; Sung, J.H.; Sung, J.M.; Kim, G.D. Estimation for the field application of amorphous metallic fiber reinforced concrete. *Mag. KCI* **2014**, *26*, 359–360.
4. Yang, J.M.; Yoon, S.H.; Choi, S.J.; Kim, G.D. Development and application of pig iron based amorphous fiber for concrete reinforcement. *Mag. KCI* **2013**, *25*, 38–41.
5. Nayar, S.K.; Gettu, R. Synergy in toughness by incorporating amorphous metal and steel fibers. *ACI Mater. J.* **2015**, *112*, 821–827. [CrossRef]
6. Yoon, J.Y.; Kim, H.; Kim, Y.J.; Sim, S.H. Prediction model for mechanical properties of lightweight aggregate concrete using artificial neural network. *Materials* **2019**, *12*, 2678. [CrossRef] [PubMed]
7. Choi, S.J.; Won, J.P. Technology for reducing of shrinkage of amorphous metallic fiber-reinforced concrete. *Mag. KCI* **2018**, *30*, 15–21.
8. Won, J.P.; Hong, B.T.; Lee, S.J.; Choi, S.J. Bonding properties of amorphous micro-steel fibre-reinforced cementitious composites. *Comp. Struct.* **2013**, *102*, 101–109. [CrossRef]
9. Kim, R.W.; Kim, C.K.; Hwang, U.J.; Lee, S.J.; Choi, S.J.; Won, J.P. Flexural performance and plastic shrinkage of amorphous metallic fiber reinforced cement composites. *Mag. KCI* **2013**, *25*, 143–144.
10. Lee, S.G.; Kim, G.Y.; Choi, G.C.; Kim, H.S.; Kim, J.H.; Kim, L.H. Evaluation of impact reistance of amorphous steel fiber-reinforced cement composites by casting direction. *Mag. KCI* **2015**, *27*, 441–442.
11. Lee, S.G.; Kim, G.Y.; Hwang, E.C.; Son, M.J.; Pyein, S.J.; Nam, J.S. Impact resistance performance of amorphous metallic fiber reinforced cement composite by fiber length. *Mag. Build Concr.* **2019**, *31*, 711–712.
12. Kim, H.; Kim, G.; Nam, J.; Kim, J.; Han, S.; Lee, S. Static mechanical properties and impact resistance of amorphous metallic fiber-reinforced concrete. *Comp. Struct.* **2015**, *134*, 831–844. [CrossRef]
13. Kim, Y.I.; Lee, Y.K.; Kim, M.S. Influence of steel fiber volume ratios on workability and strength characteristics of steel fiber reinforced high-strength concrete. *Korean Build. Constr. Mag. Build. Constr.* **2008**, *8*, 75–84. [CrossRef]

14. Kang, S.G. Effect of shell structure of artificial lightweight aggregates on the e mission rate of absorbed water. *J. Ceram. Soc.* **2008**, *11*, 750–754. [CrossRef]
15. KS L 5105. *Testing Method for Compressive Strength of Hydraulic Cement Mortars*; Korea Standards Association: Seoul, Korea, 2007.
16. KS F 2408. *Standard Test Method for Flexural Strength of Concrete*; Korea Standards Association: Seoul, Korea, 2016.
17. KS F 2423. *Standard Test Method for Tensile Splitting Strength of Concrete*; Korea Standards Association: Seoul, Korea, 2016.
18. KS F 2462. *Standard Test Method for Unit Weight of Structural Light Weight Concrete*; Korea Standards Association: Seoul, Korea, 2016.
19. KS F 2424. *Standard Test Method for Length Change of Mortar and Concrete*; Korea Standards Association: Seoul, Korea, 2015.
20. KS F 2584. *Standard Test Method for Accelerated Carbonation of Concrete*; Korea Standards Association: Seoul, Korea, 2015.
21. Gao, J.; Sun, W.; Morino, K. Mechanical properties of steel fiber-reinforced, high-strength, lightweight concrete. *Cem. Concr. Compos.* **1997**, *4*, 307–313. [CrossRef]

Article

Pore Structure Characteristics of Foam Composite with Active Carbon

Jungsoo Lee and Young Cheol Choi *

Department of Civil and Environmental Engineering, Gachon University, Seongnam 13120, Korea; dlwjdtn7509@naver.com

* Correspondence: zerofe@gachon.ac.kr; Tel.: +82-31-750-5721; Fax: +82-31-754-2772

Received: 18 August 2020; Accepted: 10 September 2020; Published: 11 September 2020

Abstract: Characterization of porous materials is essential for predicting and modeling their adsorption performance, strength, and durability. However, studies on the optimization of the pore structure to efficiently remove pollutants in the atmosphere by physical adsorption of construction materials have been insufficient. This study investigated the pore structure characteristics of foam composites. Porous foam composites were fabricated using foam composite with high porosity, open pores, and palm shell active carbon with micropores. The content was substituted 5%, 10%, 15%, and 20% by volume of cement. From the measured nitrogen adsorption isotherm, the pore structure of the foam composite was analyzed using the Brunauer–Emmett–Teller (BET) theory, Barrett–Joyner–Halenda (BJH) analysis, and Harkins-jura adsorption isotherms. From the analysis results, it was found that activated carbon increases the specific surface area and micropore volume of the foam composite. The specific surface area and micropore volume of the foam composite containing 15% activated carbon were 106.48 m^2/g and 29.80 cm^3/g, respectively, which were the highest values obtained in this study. A foam composite with a high micropore volume was found to be effective for the adsorption of air pollutants.

Keywords: foam composite; active carbon; micropore volume; specific surface area; pore structure

1. Introduction

Air and water pollution issues directly affect our lives and are becoming increasingly significant worldwide with every passing year. According to the National Air Pollutants Emission Service in Korea, in 2016, air pollutant emissions amounted to 1248 thousand tons of NOx, 359 thousand tons of SOx, 1024 thousand tons of volatile organic compounds (VOCs), and 795 thousand tons of CO [1]. In addition, the number of sewage treatment facilities are increasing owing to industrialization, thereby making it difficult to manage pollutant emissions and demanding countermeasures [2].

In recent years, numerous studies have been conducted on adsorbents capable of removing pollutants to improve these environmental problems [3–6]. Adsorbents for removing contaminants generally use a porous material, with various types of activated carbon being widely used. Activated carbon is a porous material, has a high specific surface area, and is an adsorbent with excellent physical and chemical stability. Activated carbon exhibits various adsorption properties depending on its micropore structure [7,8]. The most common methods for measuring or evaluating the pore structure of a porous material are mercury intrusion porosimetry (MIP) and nitrogen adsorption/desorption isotherms. The MIP measurement has the disadvantage that its accuracy can decrease if there are ink-bottle pores present and there is a possibility of pore destruction by high pressure [9]. However, it has the advantage of being able to measure a wide range of pore sizes with relatively ease [10]. Gas adsorption methods have also been used to analyze the pore structure of cement-based materials for decades [11,12]. Based on the amount of gas adsorbed, the interior surface area of the pores can be estimated using Langmuir's monolayer-based theory or the Brunauer–Emmett–Teller (BET) theory, which is based on multilayer adsorption. In addition, the pore size distribution can be obtained using Barrett–Joyner–Halenda (BJH) analysis based on capillary condensation [11].

According to the International Union of Pure and Applied Chemistry (IUPAC), pores of adsorbents such as activated carbon are classified into micropores of 2 nm or less, mesopores of 2–50 nm, and macropores of 50 nm or more. In general, mesopores act as a pathway for gas or liquid substances to move from macropores to micropores. Furthermore, macropores have been found to affect the diffusion rate [13]. The adsorption performance of activated carbon is greatly affected by the pore volume, size distribution, and specific surface area of the activated carbon. In addition, activated carbon with numerous micropores has a high adsorption capacity against low-molecular-weight pollutants [14–16].

Numerous studies on activated carbon have primarily focused on the analyses of the correlation between the surface area and the amount adsorbed, pore volume, and the amount of adsorbed pores formed on and inside the activated carbon [17–19]. Horgnies et al. [20] evaluated the reduction performance of NO_2 and CO_2 gas for hardened cement paste containing activated carbon. Based on the analysis results, it was confirmed that the NO_2 and CO_2 gas concentrations were reduced by reacting with activated carbon [20]. Recently, foam concrete with a high connection pore and specific surface area has been receiving a significant amount of attention for higher adsorption performance. Sun et al. [21] performed scanning electron microscopy (SEM) image analysis to analyze the pore structure of foam concrete based on the type of foaming agent. They found that the foam concrete with synthetic surfactant-based foam agents had a smaller pore size distribution and fewer connecting pores than that with plant and animal surfactant-based foaming agents. Kunhananda Nambiar et al. [22] investigated the correlation between the pore distribution and compressive strength of foamed concrete using optical microscope image analysis and found that the pore structure of the foam concrete greatly influenced its strength and density, and that the size of the pores and strength were inversely proportional.

Many researchers have focused on improving adsorption performance by controlling the pore size and distribution contained in activated carbon. However, there are few studies showing construction materials that efficiently remove pollutants in the atmosphere by physical adsorption using activated carbon. In this study, a foam composite containing activated carbon was fabricated to improve the adsorption performance of air pollutants by increasing the number of micropores. Activated carbon was substituted 5%, 10%, 15%, and 20% by volume of cement. The pore structures of the foam composite of each variable were analyzed via optical microscope image analysis and gas adsorption tests.

2. Experimental Program

2.1. Materials

In this study, porous foam composites were prepared to estimate the pore structure characteristics that have a significant effect on the pollutant adsorption capacity. Ordinary Portland cement (OPC) and palm-based activated carbon were used as raw materials. The densities of OPC and active carbon were 3.13 g/cm^3 and 1.90 g/cm^3, respectively. Table 1 shows the chemical oxide composition of OPC (obtained by X-ray fluorescence (XRF) analysis, (ZSX Primus II, Rigaku, Tokyo, Japan). The calculated Bogue phase compositions of the OPC were 53.6% C_3S, 18.4% C_2S, 7.0% C_3A, and 10.0% C_4AF by mass.

Table 1. Chemical oxide composition of Ordinary Portland cement (OPC) by XRF analysis.

	Chemical Compositions (wt.%)							
	CaO	SiO_2	Al_2O_3	Fe_2O_3	MgO	K_2O	Na_2O	SO_3
OPC	61.5	20.5	4.75	3.3	3.06	1.69	0.171	1.52

Figure 1 shows the SEM image of the palm-based activated carbon. As shown in Figure 1, the surface morphology of activated carbon has uneven cavities and fine pores. Activated carbon adsorbs pollutants as well as moisture through pores. In particular, activated carbon, which contains many micropores, has the advantage of removing gaseous pollutants.

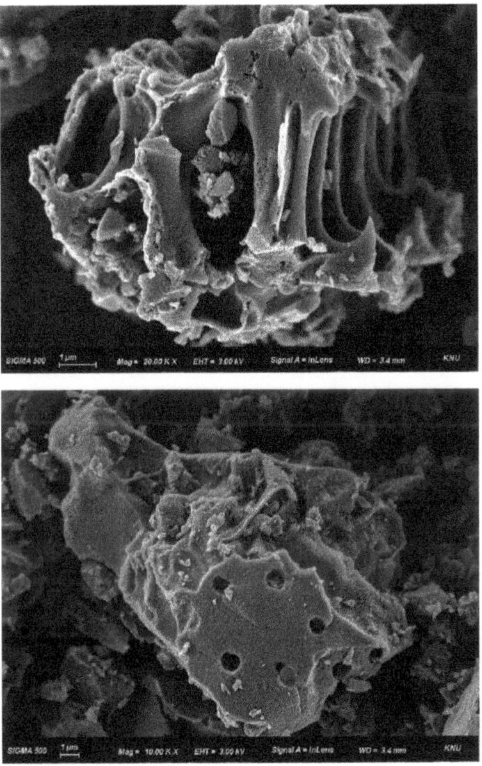

Figure 1. SEM image of palm-based activated carbon.

Figure 2 shows the particle size distribution of OPC and activated carbon obtained via laser diffraction analysis. The particle size distributions of OPC and activated carbon showed a monomodal distribution with peak values at 5.87 μm and 11.57 μm, respectively. The average sizes of OPC and activated carbon were 5.42 μm and 10.20 μm, respectively.

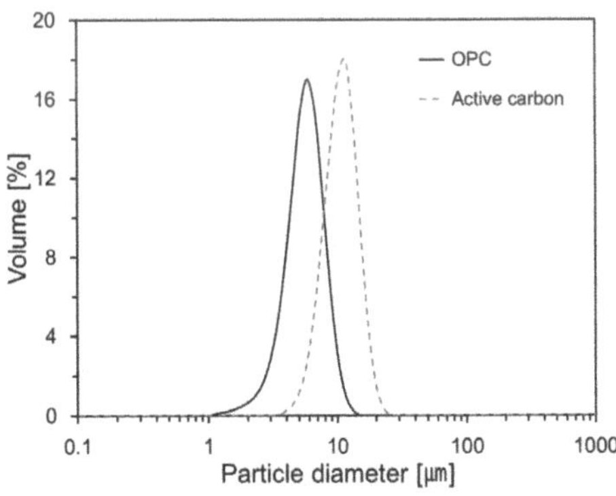

Figure 2. Particle size distributions of activated carbon and OPC.

2.2. Mix Proportions

The porous foam composites were fabricated using the pre-foaming method in accordance with the mix proportions as presented in Table 2. The main variable is the substitution ratio of activated carbon to cement. Foam composites were prepared by substituting 0, 5%, 10%, 15%, and 20% of the cement volume with activated carbon. The foaming agent for the production of foam composites was F-200, which was produced in South Korea, the main component of which is a peptide compound that is a natural polymer material. The density of the foaming agent was 1.19 g/cm^3. The foaming ratio, water-to-binder ratio, and binder content of foam composites were fixed at 69%, 0.3, and 400 kg/m^2, respectively. Foam composites of different mixtures, according to Table 2, were placed in a cylinder mold of Ø100 × 200 mm^3. The specimens were cured in a chamber at constant temperature and humidity chamber (temperature: 20 °C ± 1 °C, relative humidity: 60 ± 3%) for 24 h. Then, the mold was demolded and stored in a chamber at constant temperature and humidity under the same conditions. Subsequently, the specimens were de-molded and stored in a chamber, again under the same conditions.

Table 2. Mix proportions of foam composite.

	Target Density (kg/m^3)	Composition of Mixture (per m^3)				Slurry Density (kg/m^3)
		OPC (kg)	AC (kg)	Water (kg)	Foam (m^3)	
Plain	580.0	400.0	-	120	0.69	580.2
AC5	580.0	380.0	12.1	120	0.69	582.7
AC10	580.0	360.0	24.3	120	0.69	589.5
AC15	580.0	340.0	26.4	120	0.69	582.6
AC20	580.0	320.0	48.6	120	0.69	580.1

2.3. Test Methods

The particle size distributions of the raw materials were analyzed using Horbia's LA-950 equipment (HORIBA, Kyoto, Japan), and the surface morphology of activated carbon was analyzed by SEM (Signal 500 by Carl Zeiss, Oberkochen, Germany). A platinum coating was applied to the top surface of the specimen.

The macropore properties of the foam composite were analyzed using the ASTM C 457 test method. To analyze the macropore characteristics, test specimens with dimensions 50 × 50 × 20 mm^3 were prepared. After converting the optical microscope image of the foam composite into a gray-level image, the pores and matrices were prominent in black and white, simplifying the pore size measurement. The porosity was calculated by dividing the sum of the measured pore areas by the total area. To analyze the specific surface area and micropore characteristics of the foam composite, BET analysis was performed using the ASAP 2020 (Micromeritics, Norcross, GA, USA) equipment. The samples were degassed at a temperature of 573 K for 24 h under vacuum before analysis to remove moisture and adsorbed contaminants on the surface. After the degassing process, the adsorption isotherm was acquired by obtaining the relative pressure and the amount of adsorption at 77 K.

The measurement range of the adsorption isotherm was between 0 and 1, and the specific surface area was measured by applying the BET theory [23] to the measured adsorption isotherm. When measuring the specific surface area, N_2 gas was used as the adsorbent, and the cross-sectional area occupied by one molecule of N_2 was calculated to be 0.162 nm^2. The micropore volume and micropore surface area of the foam composite were obtained by the t-plot method [24,25] using the Harkins and Jura thickness equation of the following Equation (1). The Harkins and Jura thickness curve equation was derived by fitting the data for various metal oxides.

$$t = \sqrt{\frac{13.99}{0.034 - \log_{10}\frac{P}{P_0}}} \tag{1}$$

Here, t is the film thickness (Å) of nitrogen adsorbed on the sample surface, and P and P_0 are the absolute vapor pressure and saturation vapor pressure, respectively.

Figure 3 shows the results of the micropore volume and micropore surface area of the activated carbon used in this study. The micropore volume was calculated using the y-axis intercept value of a straight line obtained by linear regression of data in the range of 0.5–1 nm adsorption thickness. The micropore surface area was calculated using the slope of a straight line obtained by linear regression of the adsorption data of micropores with an adsorption thickness of 0.5 nm or less. The microporous volume and microporous surface area of activated carbon were 356.31 cm^3/g and 468.78m^2/g, respectively.

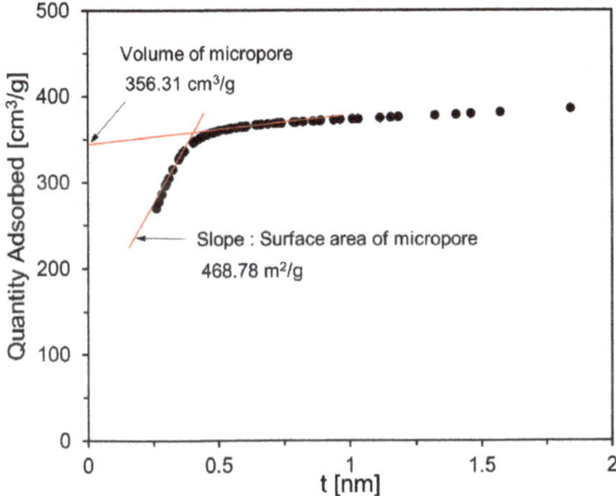

Figure 3. Micropore volume and micropore surface area of activated carbon.

The adsorption average pore diameter of the foam composite was calculated using the BJH method [11].

3. Results and Discussion

In this study, a new concept of a porous foam composite was developed to improve the performance of air pollutant adsorption. As illustrated in Figure 4, based on a foam composite with open pores and high porosity, activated carbon, having a large number of micropores for the adsorption of air pollutants, is dispersed throughout the pore surface of the foam composite to improve the adsorption performance. This concept of porous foam composite is expected to be used in various construction materials such as sidewalk blocks, soundproof panels, parking panels, and ceiling materials.

3.1. Pore Structures of Foam Composite

In order to analyze the macropores of the foam composite, the surface was polished with fine sandpaper, and optical microscope images of the foam composite cross section were taken. The optical microscope images were converted into gray-level images, as shown in Figure 5. The pore size was measured by distinguishing the pores and cement matrices in black and white—the pore and cement matrices were classified as black and white, respectively. The total pore content and average pore size were calculated by summing all the measured pores. The total pore content was calculated by dividing the sum of the areas of all measured pores by the total image area. In addition, the average pore size was calculated by assuming the shape of the pores to be a circle and calculating the total number of pores.

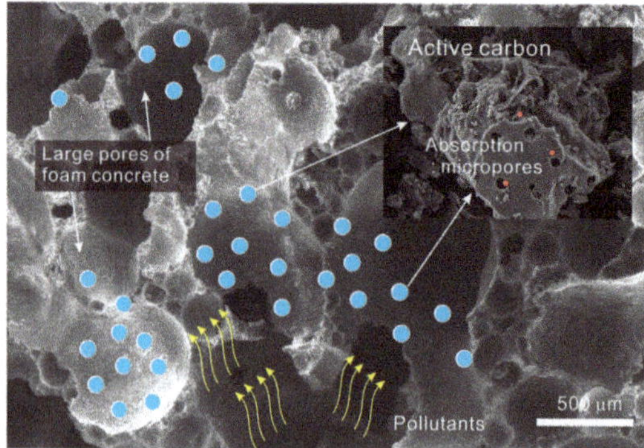

Figure 4. The concept of porous foam composite.

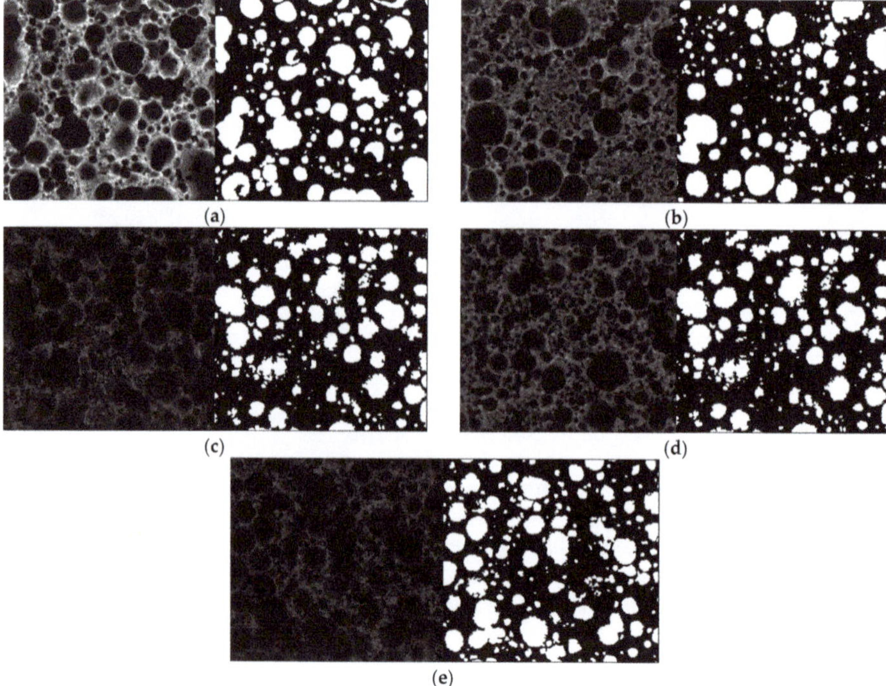

Figure 5. Optical microscope image analysis for porosity measurement of foam composite. (**a**) Plain; (**b**) AC5; (**c**) AC10; (**d**) AC15; (**e**) AC20.

Figure 6 shows the pore content according to the pore size of the foam composite for each variable. To calculate the pore content of the foam composite, 121 images were used for each variable. To calculate the pore content, 11 × 11 images were measured by dividing the cross section of each variable equally. Figure 5 shows the optical microscope image measured for each variable and the corresponding converted gray image sample. As shown in Figure 6, more than 95% of the pore size of the foam composite was less than 1 mm, and it exhibited a similar pore size distribution regardless of the amount of activated carbon.

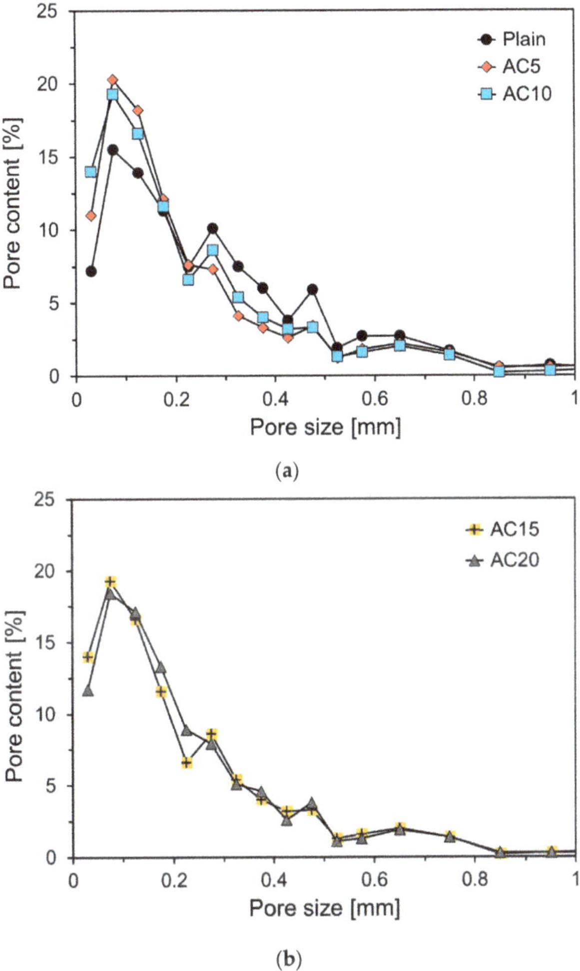

Figure 6. Pore contents of foam composite specimens. (**a**) Plain, AC5, AC10; (**b**) AC15 and AC20.

The total pore content and average pore size of Plain were 32.8% and 0.277 mm, respectively. The total pore content of AC5 mixed with 5% activated carbon was approximately 3.1% larger than that of Plain, but the average pore size was 0.028 mm smaller than that of Plain. It seems that the mixing of activated carbon prevented air bubbles from becoming larger and evenly dispersed in the foam composite. Furthermore, it seems that the activated carbon in the foam composite prevents air bubbles from becoming larger owing to the air bubbles being combined, allowing them to be evenly distributed. The total pore contents of AC10, AC15, and AC20 were 34.3%, 31.2%, and 26.1%, respectively, and the total pore content tended to decrease as the active carbon content increased. The average pore sizes of AC10, AC15, and AC20 were 0.224, 0.208, and 0.224 mm, respectively. Excluding AC20, the average pore size tended to decrease as the content of activated carbon increased. It seems that the incorporation of an appropriate amount of activated carbon smooths the distribution of air bubbles.

3.2. Specific Surface Area of Foam Composite

Figure 7 shows the nitrogen adsorption isotherm of the foam composite according to the relative pressure. In the IUPAC, graphs expressed according to adsorption characteristics are classified from type I to type V according to their shape [26]—the nitrogen adsorption isotherm in Figure 7 corresponds to a type II curve. This type of adsorption isotherm appears mainly in porous solids with micropores, and it can be seen that an inflection point appears when the pressure is low. The inflection point indicates the formation of monolayer adsorption. It was confirmed that the inflection point occurred in the range of relative pressure less than 0.1 in all the foam composites, as can be observed in Figure 7.

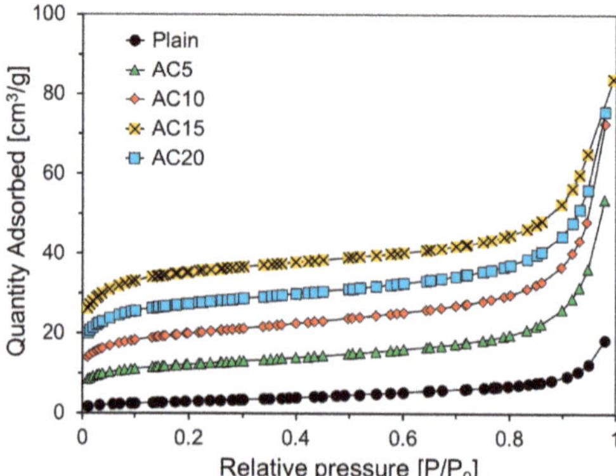

Figure 7. Nitrogen adsorption isotherm curves of foam composites.

For all the variables, there was an insignificant increase in adsorption, and a flat isotherm appeared in the range of 0.2 to 0.8 relative pressure. From these results, it was concluded that all the foam composites had fewer mesopores [27,28].

Figure 8 shows the results of the regression analysis for the BET model equation [23] from the nitrogen adsorption experiment results in Figure 7. The BET model is an extension of Langmuir theory, which is a theory for monolayer molecular adsorption and multi-layer adsorption.

$$\frac{1}{Q\left(\frac{P_0}{P}-1\right)} = \frac{C-1}{V_m} \times \frac{P_0}{P} + \frac{1}{V_m C} \tag{2}$$

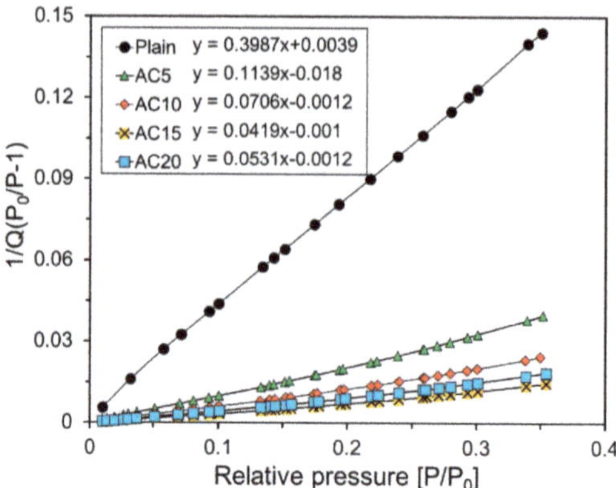

Figure 8. Specific surface areas of foam composites.

Here, P and P_0 represent the absolute pressure and saturation pressure, respectively, and Q represents the amount of gas adsorbed on the specimen. V_m and C are the amount of gas absorbed on the monolayer and the BET constant, respectively. The BET constant is a material-dependent dimensionless constant and usually has a value of 50 to 200.

V_m was calculated using the slope and y-axis intercept corresponding to each variable by regression analysis, as shown in Figure 8. The specific surface area of the foam composite was calculated using V_m regression analysis. The equation used to calculate the specific surface area is as follows [29]:

$$S_{BET}\left(\mathrm{m}^2/\mathrm{g}\right) = 4.355 \cdot V_m\left(\mathrm{cm}^3/\mathrm{g}\right) \tag{3}$$

The values of V_m of Plain, AC5, AC10, AC15, and AC20 calculated by regression analysis were 2.48, 8.92, 14.41, 24.45, and 19.27 cm^3/g, respectively. The specific surface areas of Plain, AC5, AC10, AC15, and AC20 calculated by Equation (3) from the results of V_m were 10.75, 38.85, 62.76, 106.48, and 83.92 m^2/g, respectively.

Figure 9 shows the measured specific surface area and average pore size of the foam composite according to the replacement level of activated carbon. The specific surface area and average pore size of Plain were 10.82 m^2/g and 10.64 nm, respectively. The value of the specific surface area of Plain is higher than that of general cement paste. According to Odler's research results [30], the specific surface area of cement paste at 1 year old with a water-cement ratio of 0.3 to 0.5 ranged from 4.4 to 9.5 m^2/g. The specific surface area of AC5 was 38.85 m^2/g, which was approximately 4 times larger than that of Plain, but the average pore size was 8.55 nm, which was 2.09 nm smaller than the average pore size of Plain.

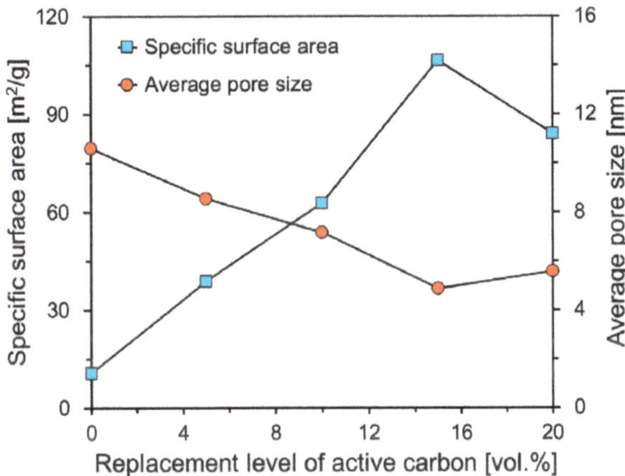

Figure 9. Specific surface areas and average pore sizes of foam composites according to replacement level of active carbon.

The specific surface areas of AC10, AC15, and AC20 were measured to be 62.75, 106.48, and 83.91 m^2/g, respectively. This was approximately 6 to 10 times larger than the specific surface area of Plain. This seems to be a result of the high specific surface area of the activated carbon itself, and similar results were also found in other studies [31,32]. The average pore sizes of AC10, AC15, and AC20 were 7.17, 4.87, and 5.58 nm, respectively. The replacement level of activated carbon was up to 15%, as the amount of activated carbon increased, the specific surface area increased, and the average pore size tended to decrease. In the case of AC20, the dispersion of activated carbon was not uniform compared to AC15 when the foam composite was blended, so the specific surface area decreased and the average pore size increased compared to AC15.

As shown in Figure 9, it was found that the specific surface area and average pore size of the foam composite containing activated carbon had an inverse relationship. It was confirmed that the foam composite had a relatively large specific surface area when the average pore size was small. This appears to have increased the specific surface area due to the large distribution of small pores on the surface of the foam composite. Meltem Asiltürk et al. [32] reported that the specific surface area increased as the activated carbon content increased, but the average pore size did not decrease significantly. Excluding AC20, as the amount of activated carbon increased, the specific surface area of the foam composite increased, but the average pore size of the foam composite tended to decrease.

3.3. Micropores of Foam Composite

The volume and surface area of the micropores of the foam composite were obtained from the measured adsorption isotherms by means of the t-plot method using the Harkins and Jura thickness equation [24,25]. Figure 10 shows the amount of nitrogen adsorbed on the foam composite according to the adsorption thickness.

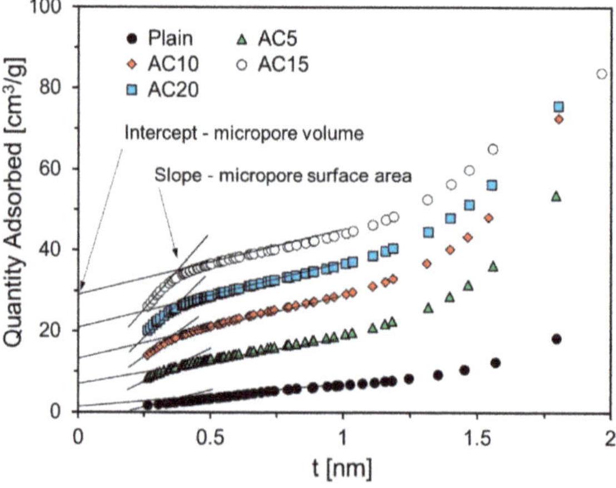

Figure 10. Relationship between quantity and film thickness adsorbed nitrogen of specimens.

All foam composites showed a narrow adsorption distribution at a thickness of 0.5 nm or less, and the amount of nitrogen adsorbed was large at a thickness of 1.5 nm or more, as shown in Figure 10. In addition, it was confirmed that the higher the replacement level of activated carbon, the greater the amount of nitrogen adsorption of the foam composite, which seems to be due to the micropore characteristics of the activated carbon itself.

It was confirmed that the foam composites containing 5%, 10%, 15%, and 20% of activated carbon—all except for Plain—had an inflection point in the range of 0.5 nm or less in thickness. It seems that AC10, AC15, and AC20 had a monolayer adsorption at a thickness of 0.5 nm or less.

Figure 11 shows the relationship between the micropore volumes and the surface areas of the foam composites according to the activated carbon content. In the case of Plain, the micropore volume and surface area were 1.06 cm^3/g and 7.04 m^2/g, respectively. The micropore volume and surface area of the foam composite containing 5% activated carbon were 7.40 cm^3/g and 18.67 m^2/g, respectively. Compared to Plain, the micropore volume was seven times and the micropore surface area was three times larger, and it seems that the incorporation of activated carbon developed the microporous structure of the foam composite.

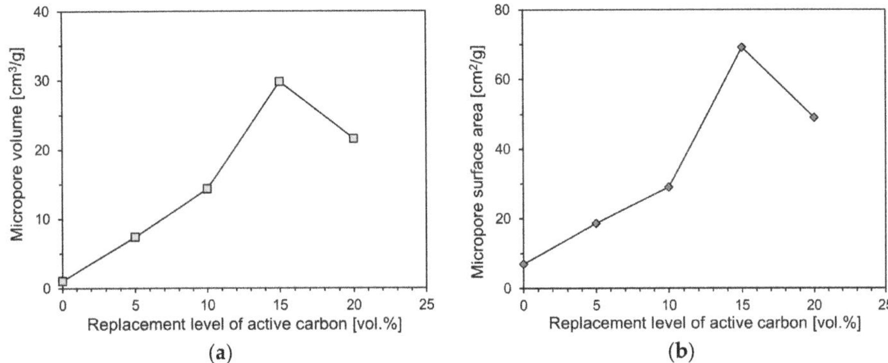

Figure 11. Micropore volume and surface area of specimens according to active carbon content. (**a**) Micropore volume; (**b**) Micropore surface area.

The micropore volumes of AC10, AC15, and AC20 were 14.28, 29.80, and 21.56 cm^3/g, respectively, which are about 14 to 29 times larger than that of Plain. However, the increase in the micropore surface area of the specimen containing activated carbon compared to Plain did not increase significantly compared to the volume of the micropores. The micropore surface areas of AC10, AC15, and AC20 were 29.01, 69.16, and 49.00 m^2/g, respectively, which increased by roughly 4 to 10 times compared to Plain. As shown in Figure 11, it was confirmed that the micropore volume and surface area of the foam composites, except for AC20, increased as the replacement level of activated carbon increased. It seems that the incorporation of an appropriate amount of activated carbon affects the development of the micropore structure.

3.4. Pore Size Distribution of Foam Composite

Figure 12 shows the dV/dlog(D) pore volume and cumulative pore volume according to the pore diameter of the foam composite. Foam composites mixed with activated carbon, AC5, AC10, AC15, and AC20 all show a narrow pore size distribution below 3 nm (see Figure 12a). The cumulative pore volume and average pore diameter of Plain were 0.028 cm^3/g and 9.99 nm, respectively. The cumulative pore volume of AC5 was 0.078 cm^3/g, which was 2.8 times larger than that of Plain. This is due to the micropore properties contained in the activated carbon itself (see Figure 3). The average pore diameter of AC5 was 11.10 nm, which is an increase of 1.12 nm over Plain. The cumulative pore volumes of AC10 and AC15 were 0.100 and 0.123 cm^3/g, respectively, and showed a tendency to increase as the replacement level of active carbon increased. In foam composites, a large amount of activated carbon is not well dispersed, and agglomeration occurs. Therefore, it seems that it had a negative effect on the micropore structure.

This trend was similar for the average pore diameter. The average pore diameters of AC10, AC15, and AC20 were 9.24, 5.23, and 6.79 nm, respectively, which decreased in size up to 15% replacement level of activated carbon, but increased by 20%. This result is similar to the micropore volume and surface area analysis results of the foam composite.

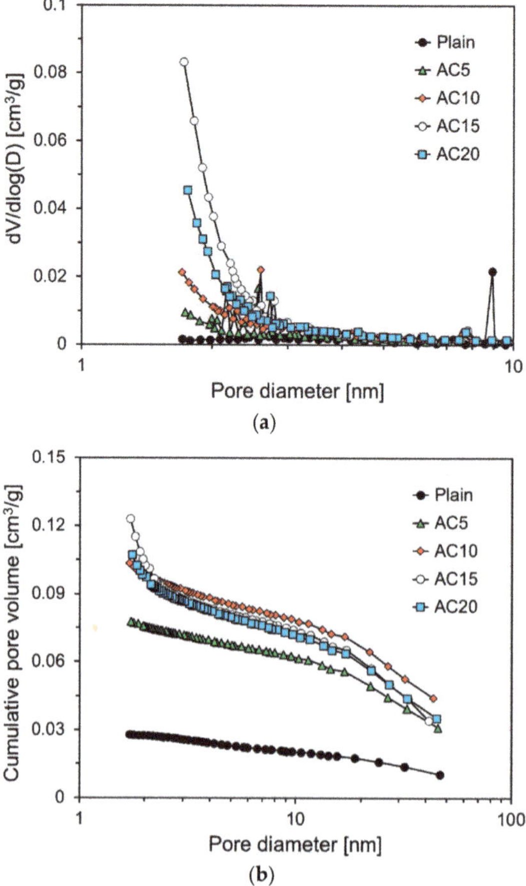

Figure 12. Pore size distributions and cumulative pore volume of specimens. (**a**) dV/log(D); (**b**) Cumulative pore volume.

4. Conclusions

In this study, a porous foam composite was fabricated using palm-based activated carbon with micropores in a foam composite with a large number of pores and open pores to improve the pollutant adsorption performance. The replacement level of activated carbon (5%, 10%, 15%, and 20%) to the volume of cement was used as the main variable, and the pore structures of the foam composites were analyzed. The research results obtained in this study can be summarized as follows.

Image analyses results show that the pore sizes of the foam composites were more than 95% and less than 1mm, thereby showing a similar pore size distribution regardless of the replacement level of activated carbon. Excluding AC20, the total pore content and average pore size tended to decrease as the replacement level of activated carbon increased. This appeared to be effective in dispersing air bubbles by mixing an appropriate amount of activated carbon.

As a result of the specific surface areas of the foam composites, the specific surface areas of the foam composites containing activated carbon increased by 4 to 10 times that of Plain. On the contrary, the average pore sizes of the foam composites containing activated carbon were lower than those of Plain. The higher the specific surface area of the foam composite, the smaller the average pore size. The specific surface area of the foam composite containing 15% activated carbon was the highest at

106.48 m^2/g, which appeared to have a reaction area capable of adsorbing contaminants larger than the other foam composites.

Excluding AC20, the micropore volume and surface area tended to increase as the replacement level of activated carbon increased. In addition, it was confirmed that the higher the micropore volume of the foam composite, the greater the specific surface area. The micropore volume and surface area of AC15 were the highest at 29.80 cm^3/g and 69.16 m^2/g, respectively. It is expected that AC15 will have a high adsorption performance for gaseous pollutants owing to its large pores.

All of the activated carbon-incorporated foam composites showed a narrow pore size distribution below 3 nm due to the micropore properties contained in the activated carbon itself. The cumulative pore volumes of the foam composite containing activated carbon were 2.8 to 4.4 times larger than that of Plain, but the average pore diameters were all lower than that of Plain.

Foam composite with AC has a high porosity and micropores, so it is considered to be effective in adsorbing gaseous pollutants such as NOx in the atmosphere. Therefore, the foam composite with AC developed in this research can be used for sidewalk blocks, soundproof panels, etc., which is thought to help improve the air quality.

Author Contributions: Conceptualization, Y.C.C.; methodology, J.L.; validation, J.L. and Y.C.C.; investigation, J.L. and Y.C.C.; resources, J.L.; writing—original draft preparation, J.L.; writing—review and editing, Y.C.C.; visualization, J.L. and Y.C.C.; supervision, Y.C.C.; project administration, Y.C.C.; funding acquisition, Y.C.C. All authors have read and agreed to the published version of the manuscript.

Funding: This work was supported by the Korea Institute of Energy Technology Evaluation and Planning (KETEP) and the Ministry of Trade, Industry & Energy (MOTIE) of the Republic of Korea (No. 20181110200070). This work was also supported by the Gachon University research fund of 2019 (GCU-2019-0345).

Conflicts of Interest: The authors declare no conflict of interest.

References

1. National Air Pollutants Emission Service, Emissions of Pollutants in 2016. Available online: https://airemiss.nier.go.kr/ (accessed on 6 May 2020).
2. Moon, D.C.; Kim, C.S.; Park, I.Y.; Kim, M.R.; H, S.S.; Lee, K.H.; Lee, C.G. Liquid Phase Adsorption of Activated Carbon Fibers. *J. Anal. Sci. Technol.* **2000**, *13*, 573–583.
3. Pak, S.H.; Jeon, M.J.; Jeon, Y.W. Study of sulfuric acid treatment of activated carbon used to enhance mixed VOC removal. *Int. Biodeterior. Biodegrad.* **2016**, *113*, 195–200. [CrossRef]
4. Kang, K.H.; Kam, S.K.; Lee, S.W.; Lee, M.G. Adsorption Characteristics of Activated Carbon Prepared From Waste Citrus Peels by NaOH Activation. *J. Environ. Sci.* **2007**, *16*, 1279–1285. [CrossRef]
5. Pap, S.; Radonic, J.; Trifunovic, S.; Adamovic, D.; Mihajlovic, I.; Miloradov, M.V.; Sekulic, M.T. Evaluation of the adsorption potential of eco-friendly activated carbon prepared from cherry kernels for the removal of Pb^{2+}, Cd^{2+} and Ni^{2+} from aqueous wastes. *J. Environ. Manag.* **2016**, *184*, 297–306. [CrossRef] [PubMed]
6. Son, H.K.; Sivakumar, S.; Rood, M.J.; Kim, B.J. Electrothermal adsorption and desorption of volatile organic compounds on activated carbon fiber cloth. *J. Hazard. Mater.* **2016**, *301*, 27–34. [CrossRef]
7. Yavuz, R.; Akyildiz, H.; Karatepe, N.; Çetinkaya, E. Influence of preparation conditions on porous structures of olive stone activated by H_3PO_4. *Fuel Process. Technol.* **2010**, *91*, 80–87. [CrossRef]
8. Hagemann, N.; Spokas, K.; Schmidt, H.P.; Kägi, R.; Böhler, M.; Bucheli, T. Activated carbon, biochar and charcoal: Linkages and synergies across pyrogenic carbon's ABCs Water. *J. Water.* **2018**, *10*, 1–19. [CrossRef]
9. Diamond, S. Mercury porosimetry: An inappropriate method for the measurement of pore size distributions in cement-based materials. *Cem. Concr. Res.* **2000**, *30*, 1517–1525. [CrossRef]
10. Léon, C.A.L. New perspectives in mercury porosimetry. *Adv. Colloid Interface Sci.* **1998**, *76*, 341–372. [CrossRef]
11. Barrett, E.P.; Joyner, L.G.; Halenda, P.P. The determination of pore volume and area distributions in porous substances. I. Computations from nitrogen isotherms. *J. Am. Chem. Soc.* **1951**, *73*, 373–380. [CrossRef]
12. Powers, T.C.; Brownyard, T.L. Studies of the physical properties of hardened Portland cement paste. *ACI J. Proc.* **1947**, *43*, 845–880. [CrossRef]

13. Choi, Y.J.; Lee, Y.S.; Im, J.S. Effect of pore structure of Activated Carbon Fiber on Mechanical Properties. *Appl. Chem. Eng.* **2018**, *29*, 318–324. [CrossRef]
14. Snoeyink, V.L. Adsorption of organic compounds. In *Water Quality and Treatment*; Pontius, F.W., Ed.; McGraw-Hill: New York, NY, USA, 1990; pp. 781–875.
15. Corcho-Corral, B.; Olivares-Marin, M.; Fernandez-Gonzalez, C.; Gomez-Serrano, V.; Marcias-Garcia, A. Preparation and textural characterization of activated carbon from vine shoots (Vitis vinifera) by H3PO4-chemical activation. *Appl. Surf. Sci.* **2006**, *252*, 5961–5966. [CrossRef]
16. Wu, F.C.; Tseng, R.L.; Juang, R.S. Preparation of highly microporous carbons from fir wood by KOH activation for adsorption of dyes and phenol from water. *Sep. Purif. Technol.* **2005**, *47*, 10–19. [CrossRef]
17. Lee, S.W.; Cheon, J.K.; Park, H.J.; Lee, M.G. Adsorption characteristics of binary vapors among acetone, MEK, benzene, and toluene. *Korean J. Chem. Eng.* **2008**, *25*, 1154–1159. [CrossRef]
18. Lim, J.K.; Lee, S.W.; Kam, S.K.; Lee, D.W.; Lee, M.G. Adsorption characteristics of toluene vapor in fixed-bed activated carbon column. *J. Environ. Sci.* **2005**, *14*, 61–69. [CrossRef]
19. Lee, S.W.; Kam, S.K.; Lee, M.G. Comparison of breakthrough characteristics for binary vapors composed of acetone and toluene based on adsorption intensity in activated carbon fixed-bed reactor. *J. Ind. Eng. Chem.* **2007**, *13*, 911–916. [CrossRef]
20. Horgnies, M.; Dubois-Brugger, I.; Krou, N.J.; Belin, T.; Mignard, S.; Batonneau-Gener, I. Reactivity of NO_2 and CO_2 with hardened cement paste containing Active carbon. *Eur. Phys. J. Special Top.* **2015**, *224*, 1985–1994. [CrossRef]
21. Sun, C.; Zhu, Y.; Guo, j.; Zhang, Y.; Sun, G. Effects of foaming agent type on the workability, drying shrinkage, frost resistance and pore distributionwnd of foamed concrete. *Constr. Build. Mater.* **2018**, *186*, 833–839. [CrossRef]
22. Nambiar, E.K.K.; Ramamurthy, K. Air-void characterization of foam concrete. *Cem. Concr. Compos.* **2007**, *37*, 221–230. [CrossRef]
23. Brunauer, S.; Emmett, P.H.; Teller, E. Adsorption of Gases in Multimolecular Layers. *J. Am. Chem. Soc.* **1938**, *60*, 309–319. [CrossRef]
24. Harkins, W.D.; Jura, G. Surfaces of Solids. XI. Determination of Decrease (π) of Free Surface Energy of a Solid by an Adsorbed Film. *J. Am. Chem. Soc.* **1944**, *66*, 1356–1362. [CrossRef]
25. Gregg, S.J.; Sing, K.S.W. *Adsorption, Surface Area and Porosity*, 2nd ed.; Academic Press: London, UK, 1982.
26. Lee, S.W.; Min, K.M.; Kim, S.D.; Kim, D.K. Adsorption Characteristics of Hydrogen Sulfide on Iron-activated Carbon Composite Prepared by Ferric Nitrate and Ferric Chloride. *J. Korea Soc. Waste Manag.* **2015**, *32*, 772–779. [CrossRef]
27. Nan, D.; Liu, J.; Ma, W. Electrospun phenolic resin-based carbon ultrafine fibers with abundant ultra-small micropores for CO_2 adsorption. *Chem. Eng. J.* **2015**, *276*, 44–50. [CrossRef]
28. Diez, N.; Alvarez, P.; Granda, M.; Blanco, C.; Santamaria, R.; Menendez, R. CO_2 adsorption capacity and kinetics in nitrogen-enriched activated carbon fibers prepared by different methods. *Chem. Eng. J.* **2015**, *281*, 704–712. [CrossRef]
29. Langmuir, I. The constitution and fundamental properties of solids and liquids Part I. Solids. *J. Am. Chem. Soc.* **1916**, *38*, 2263. [CrossRef]
30. Odler, I. The BET-specific surface area of hydrated Portland cement and related materials. *Cem. Concr. Res.* **2003**, *33*, 2049–2056. [CrossRef]
31. Paušová, Š.; Riva, M.; Baudys, M.; Krýsa, J.; Barbieriková, Z.; Brezová, V. Composite materials based on active carbon/TiO_2 for photocatalytic water purification. *Catal. Today* **2019**, *328*, 178–182. [CrossRef]
32. Asiltürk, M.; Şener, Ş. TiO_2-activated carbon photocatalysts: Preparation, characterization and photocatalytic activities. *Chem. Eng. J.* **2012**, *180*, 354–363. [CrossRef]

MDPI
St. Alban-Anlage 66
4052 Basel
Switzerland
Tel. +41 61 683 77 34
Fax +41 61 302 89 18
www.mdpi.com

Materials Editorial Office
E-mail: materials@mdpi.com
www.mdpi.com/journal/materials

www.ingramcontent.com/pod-product-compliance
Lightning Source LLC
LaVergne TN
LVHW071607170726
843515LV00008B/2340

* 9 7 8 3 0 3 6 5 6 8 6 1 4 *